BOTANICA MARINA HANDBOOKS

Volume 1

Marine Algae

A Survey of Research and Utilization

By

TORE LEVRING
University of Gothenburg

HEINZ A. HOPPE
Studiengesellschaft zur Erforschung von Meeresalgen e. V., Hamburg

OTTO J. SCHMID
Studiengesellschaft zur Erforschung von Meeresalgen e. V., Hamburg

1969

CRAM, DE GRUYTER & CO., HAMBURG

By Gebr. Rasch & Co., Bramsche

Preface

Numerous excellent reviews and more or less specialized books in the field of marine botany have been published in later years. The reason for writing this book was the opinion of the authors that there was need for a synopsis which might be useful both for scientists and for those interested in the practical use of marine plants. This idea was kindly taken up by the publishing house Cram, de Gruyter & Co., who also in different ways supported our work.

Most of the original manuscript was written in German. In view of the international importance of marine algae as material for scientific research and for industrial utilization it was found suitable to publish our book in a language with a distribution as wide as possible. The manuscript was therefore translated into English. For the final revision of this English manuscript the authors and the publishing house are very grateful to Dr. M. R. Droop, Millport (Scotland), who kindly undertook this work. The difference and obscurities in both the botanical and chemical literature dealing with marine algae have caused certain problems. There may therefore be cases where disputed species or chemical substances have been mentioned.

The autheors will be happy to receive any suggestion for future corrections or improvements in cas this attempt of ours turns out to be useful to the readers.

T. Levring, H. A. Hoppe, O. Schmid

Contents

The Vegetation in the Sea

By Tore Levring

Marine vegetation consists, for the most part, of algae, but phanerogams are also found. The genera *Zostera*, *Ruppia*, *Cymodocea*, and *Posedonia* should be mentioned, which frequently form green meadows on the sea bottom in shallow regions with a soft surface.

The algae can be divided biologically into two large groups:

1) Benthic algae, i.e. the sedentary forms.
2) Planktonic algae (phytoplankton) which float in the water and are in most cases unicellular organisms.

This classification is, of course, not a systematic one, but purely a biological, though on the whole benthos and phytoplankton belong to different systematic classes. There are, however, representatives of both groups in the Chlorophyceae and also in some other cases.

ENVIRONMENTAL FACTORS

The Substratum

Reinke (1889) expressed the relation between the nature of the substratum and the benthic vegetation in the following way: hard bottom is covered with vegetation, whereas soft bottom has no vegetation. That is to say, the chemical composition of the substratum is without significance, and it is therefore of no interest, whether it consists of granite, lime, basalt, wood or only of a glass jar. This phenomenon is, of course, due to the fact that algae do not possess any true roots by means of which they could draw nutrient salts from the substratum, as the terrestrial plants draw them from the soil. The algae take their mineral salts direct from their surrounding medium, i.e. from the sea water. The only factor of importance is that the surface of the substratum should be suited for the algae to germinate on and to allow them to cling to it.

Though algae in general need a firm substrate, the soft bottoms are not always without vegetation. Some of the marine phanerogams grow there, for with their rhizomes and true roots they can find a hold in the soft bed. *Zostera marina*, *Posedonia*, and *Cymodocea* are good examples. A few subtropic and tropic algae should also be mentioned, forms for the most part which take a hold in the sands (or on coral reefs) with their root-like excrescences and rhizoids; for instance, species of *Rhipocephalus*, *Halimeda*, *Caulerpa*, *Avrainvillea*, *Udotea*, *Penicillus*, *Turbinaria*.

Apart from these forms there are also numerous species, perhaps the majority, which grow on other sea plants as epiphytes or in some cases even as parasites. They are mostly several decimetres to a few centimetres or even microscopic in size. Some of the epiphytic forms appear to demand certain species of plants as substratum, whereas in other cases the species of the host plant seems to be of no significance.

The Medium

Sea water is an aqueous solution of various mineral salts (sodium chloride and small quantities of magnesium sulphate and others). There are also dissolved gases (carbon dioxide, oxygen) and organic substances in the solution. The salinity of ocean water amounts to approximately 36 parts per mille (Atlantic Ocean 35 to 36 per mille; between the trade winds 37 per mille; Mediterranean 37 to 38 per mille; Red Sea 40 per mille; Shark's Bay (Northwest Australia) approximately 60 per mille). In certain littoral regions and inland waters the salinity may be lower, as, for instance, in the Baltic region. Thus in the northernmost part of the Gulf of Bothnia it amounts to less than 3 per mille, but it increases gradually until, around Southwest Finland and the Stockholm skerries, it is 6 per mille, and along the line of Öland-Gotland-Westcoast of Estonia 7 per mille. A salt content of 8 per mille is only reached in the West of Bornholm and in the Öre-Sound, while in the Great Belt and the Little Belt it varies between 10 and 20 per mille (cf. Pettersson and Ekman, 1897). The water can also have a different salinity in different depths. Algal vegetation is in this geographical area only to be found between 0 and 30 (40) metres; there is no vegetation in greater depths in the Baltic, Kattegat and Skagerrak areas.

Ignoring for the moment the deeper layers of water, the Kattegat and the Skagerrak have a surface layer with a salinity of less than 30 per mille. This layer, which in the Kattegat is of greater depth, originates from a current flowing from the Baltic through the two Belts and the Sound and expanding in the Kattegat and the Skagerrak into a surface current, called the Baltic Current. This current carries varying quantities of water during the various season of the year so that towards the end of winter and in spring it has its greatest depths. In the southern Kattegat it is up to 20 to 30 metres deep, in the Skagerrak 10 to 20 metres. Beneath this surface current there is a layer of shelf water which has a salinity of 32 to 34 per mille. In the Baltic the outflow of surface water[1]) is opposed by an inflow of water at a greater depth and of a higher salt content. The inflow, however, is not as strong as the outflow. Interpenetration between these two layers of water takes place. In the western regions of the Baltic the higher salt content of the deeper current becomes apparent. Thus for depths of 25 to 30 metres a salinity of approximately 20 to 30 per mille is stated (cf. Reinke, 1889, p. 14; Ruppin l.c.; Lakowitz, 1929, p. 409). In the central regions of the Baltic the surface layer is said to be of a depth which does not allow the deeper layers with their higher salinity to have a direct influence on the algal vegetation. That the varying salinity in the different depths of the sea can be of significance for the distribution and occurrence of algal vegetation in the deeper regions has already been pointed out by Reinke (l.c. p. 14). There are several examples which show that a species normally growing in the surface zones of waters of a higher salinity will in areas of lower salinity be met with in greater depths were it finds the higher salinity required for its growth.

---

[1]) North Sea water

The salinity determines the osmotic pressure of the sea water. In normal water the osmotic pressure amounts to 25 Atm. (= 25 x 14.7 lbs. per squ. in.). Cells of the plants living in this kind of medium must have a somewhat higher osmotic pressure. The sensitiveness of various species to a lower or a higher salinity differs greatly and is, of course, of great importance for the spreading and distribution of algae. The phenomenon of this so-called osmotic resistance was investigated particularly by BIEBL, HOEFLER, HOFFMANN, and KYLIN. It is a striking fact that the Chlorophyceae are less susceptible than the Rhodophyceae. BIEBL, moreover, stresses the larger range of osmotic resistance in surface algae as compared to the algae living in the deeper regions.

It can be regarded as an established fact that the lower salinity in the Baltic Sea as well as in the Kattegat and in the Skagerrak is in several cases the cause of reduced forms. According to experimental investigations photosynthesis is in some cases reduced by a lower salinity, e.g. in *Laminaria, Fucus serratus*, and *F.vesiculosus;* but not in the case of *Enteromorpha* (cf. MONTFORT 1931; STEEMANN NIELSEN, 1944). According to HOFFMANN (1929) the rate of respiration is increased due to similar causes (in *Fucus serratus* and *Laminaria saccharina;* but only slightly or not at all in species of *Enteromorpha* and in *Fucus vesiculosus).* Some, such as *Enteromorpha spp.* seem thus to be independent of variations in salinity. However, in the majority of cases a diminished salinity will most probably also cause a reduction in the photosynthetic activity and an increase in the rate of respiration. This is in all likelihood the explanation of the origin of many reduced forms. On the Swedish west coast some species, which on the Atlantic coasts of Europe grow in low depths, prefer those greater depths in which the water has a higher salinity than the surface waters. There (in approximately 15 to 30 metres) conditions of illumination are, however, so unfavourable that these forms cannot develop in their normal way, e.g. *Laminaria hyperborea, Callophyllis laciniata.* With regard to their susceptibility to a varying salinity the algae can be subdivided into four groups:

Euryhaline species – of little susceptibility
Stenohaline species – live in a certain range of salinity
Hyperhaline species – require a high salinity
Hypohaline species – to be found in brackish water

While in some regions the salinity varies, the proportions of the more important elements in the water remain constant. The major ions are shown on Table 1.

In addition there is a large number of trace elements so that altogether 44 elements are shown to exist in sea water (Table 2). To these we must add hydrogen, oxygen, and the inert gases neon, helium, and argon. The total number of elements amounts, therefore, to 49.

Some of these elements which are of special importance for the growth of algae will be dealt with later in this chapter.

Most of the elements necessary for plant growth are always to be found in sufficient quantity in sea water. The content of phosphate and nitrate, however, shows great variation. Marine plants – phytoplankton and benthos – absorb nitrate and phosphate and after their decay these substances are again dissolved in the water. In surface waters, where the photosynthetic activity is to be found, the concentrations of nitrate and phosphate ions reveal great annual variations. Below the photosynthetic layer the concentration increases. The water in depths greater than 100 to 150 metres is always rich in

Table 1

Principal Ions of Sea Water (Cl=19.00‰)

The percentage of the various ions are also shown. From SVERDRUP, JOHNSON, and FLEMING

	‰	%		‰	%
Na^+	10,556	30,61	Cl^-	18,980	55,2
Mg^{++}	1,272	3,69	SO_4^{--}	2,649	7,68
Ca^{++}	0,400	1,16	Br	0,065	0,19
K^+	0,380	1,10	H_3BO_3	0,026	0,07
Sr^{++}	0,013	0,04	HCO_3^-	0,140	0,41
			F^-	0,001	0,00

Table 2

Elements Dissolved in Sea Water (Cl=19.00‰)

(Except dissolved gases). According to SVERDRUP, JOHNSON and FLEMING (1946), NODDACK and NODDACK (1939) and HARVEY (1955).

Element	mg per kg of sea water	Element	mg per kg of sea water
Chlorine	18980	Copper	0,001–0,01
Sodium	10561	Zinc	0,005
Magnesium	1272	Lead	0,004
Sulphur	884	Selenium	0,004
Calcium	400	Caesium	0,002
Potassium	380	Uranium	0,0016–0,00015
Bromine	65	Molybdenum	0,0005
Carbon	28	Gallium	0,0005
Strontium	13	Thorium	0,0005
Boron	4,6	Cerium	0,0004
Silicon	0,02 – 4,0	Silver	0,0003
Fluorine	1,4	Vanadium	0,0003
Nitrogen (comp.)	0,01 – 0,7	Lanthanum	0,0003
Aluminium	0,5	Yttrium	0,0003
Rubidium	0,2	Tin	0,0003
Lithium	0,1	Bismuth	0,0002
Phosphorus	0,001 – 0,1	Nickel	0,0001
Barium	0,05	Cobalt	0,0001
Iodine	0,05	Scandium	0,00004
Arsenic	0,01 – 0,02	Mercury	0,00003
Iron	0,002 – 0,02	Gold	0,000006
Manganese	0,001 – 0,01	Radium	$0,2–3x10^{-10}$

Small quantities of the following elements may also be found: Antimony, Cadmium, Chromium, Germanium, Thallium, and Titanium.

Table 3
Distribution of Physiological Elements
According to KALLE

structural elements	contents of 100 g organic substances (dry) N	contents of 1 m³ seawater 35‰ S A	proportion A/N	catalytic elements	contens of 100 g organic substances (dry) N	contens of 1 m³ seawater 35‰ S A	proportion A/N
Hydrogen	7 g	—	—	Copper	5 mg	10 mg	2
Sodium	3 g	10,75 kg	3600	Zinc	20 mg	5 mg	4
Potassium	1 g	390 g	390	Boron	2 mg	5 mg	2500
Magnesium	0,4 g	1,3 kg	3300	Vanadium	3 mg	0,3 mg	0,1
Calcium	0,5 g	416 g	830	Arsenic	0,1 mg	15 mg	150
Carbon	30 g	28 g	1	Manganese	2 mg	5 mg	2,5
Silicon a)	0,5 g	500 mg	1	Fluorine	1 mg	1,4 g	1400
Silicon b)	10 g	500 mg	0,05	Bromine	2,5 mg	66 g	26000
Nitrogen	5 g	300 mg	0,06	Iron a)	1 g	50 mg	0,05
Phosphorus	0,6 g	30 mg	0,05	Iron b)	40 mg	50 mg	1,3
Oxygen				Cobalt	0,05 mg	0,1 mg	2
as O_2 and CO_2	47 g	90 g	2	Aluminium	1 mg	120 mg	120
Sulphur	1 g	900 g	900	Titan	100 mg	—	—
Chlorine	4 g	19,3 kg	4800	Radium	4.10^{-12}	10^{-10} g	25

nutrient salts which can be carried into the surface layer by currents. In littoral regions phosphate and nitrate are carried into the sea in large quantities through rivers. The relation between the development of phytoplankton and the content of nitrate and phosphate has been thoroughly studied. As regards benthos, many questions are still unanswered concerning this relation. Oxygen is a necessity for most organisms. Near the surface of the sea water, where the algae thrive, the water is nearly always saturated with oxygen or occasionally even super-saturated. Normally the water contains 6 millilitres of oxygen per litre at a temperature of 20 °C. The content is largely dependent on the temperature and to some extent on the salinity.

Carbon dioxide is an essential agent in the photosynthetic process and substrate. Sea water contains carbon dioxide as bicarbonate and carbonate ions in a rather complicated system combined with hydrogen ions:

$$H_2O + CO_2 \text{ (dissolved)} \leftrightharpoons H_2CO_3 \leftrightharpoons HCO_3^- + H^+ \leftrightharpoons CO_3^{--} + H^+$$

As can be seen from this, there is a relation between the concentration of hydrogen ions and dissolved carbon dioxide. The latter is either absorbed by the atmosphere or is discharged into it. In this process of exchange the atmospheric carbon dioxide is in nearly all cases in equilibrium with the carbonic acid system in the surface layer of the water. As can be seen from Fig. 1 the carbonic acid system is also dependent on the salinity.

The concentration of hydrogen ions corresponds to a pH value of 8. 2 (as a rule to pH 8.1–8.3). With increasing depth the pH falls and reaches a value of 8.0 between 150 to 1,500 metres, because in these regions the high rate of respiration of various organisms heighten the concentration of CO_2. In greater depths the pH rises again to approximately 8 (Fig. 2).

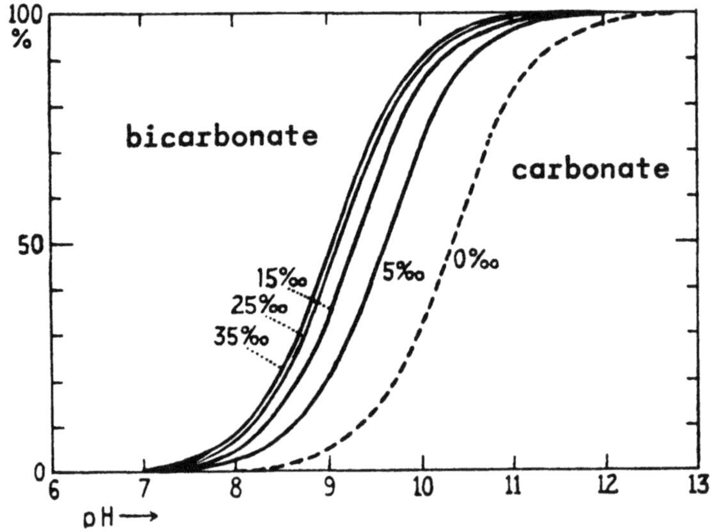

Figure 1: Balance between carbonate and bicarbonate at different pH and salinities (as °/₀₀). (From Buch)

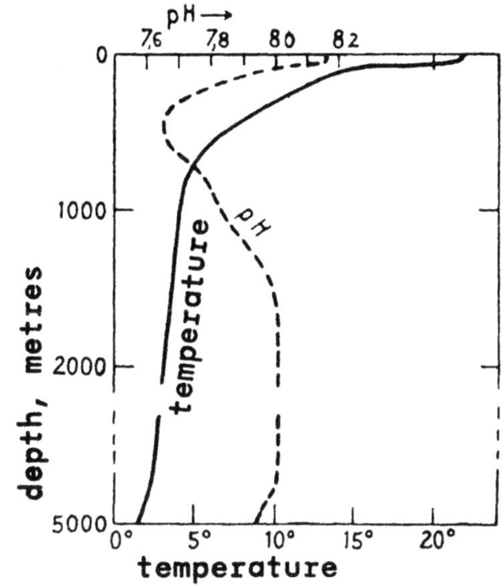

Figure 2: Correspondence between pH and depth on a station in the Atlantic (from WATTENBERG)

In littoral pools the pH can occasionally rise to 9.7. This is probably due to the great photosynthetic activity of certain species of algae *(Ulva)*. Such a great rise in the pH causes the perishing of many susceptible species of Rhodophyceae (ATKINS, 1923).

There are also dissolved organic substances to be found in sea water, according to KROGH (1935) in positively surprisingly large quantities in deep oceanic waters, roughly corresponding to 4 grammes of carbohydrate and 1.5 grammes of albumin per cubic metre, a figure which in the Atlantic alone amounts to 20,000 times of the annual wheat crops of the whole world. These substances, however, seem to be more or less unsuited for bacteria and play, in all probability, no major rôle as nutrients.

The colour of the water, especially in littoral regions, depends to some extent on the so-called "Gelbstoff" (KALLE, 1938).[1] This "yellow substance", which is probably a humus compound, has a great bearing on the illumination conditions in the water.

Synthetic sea water may be prepared according to various recipes. For ordinary purposes it is not necessary to prepare a mixture which contains all the elements known to be components of sea water. One should first prepare a basic solution containing the most essential elements, cf. Table 4 (according to LEVRING, 1946). All substances except $NaHCO_3$ should at first be dissolved in doubly distilled water and should be sterilized.

[1]) N. B. The German term "*Gelbstoff*" is commonly used in marine biological circles for the yellow mucus like compounds in sea water.

Table 4
The most Essential Elements of Synthetic Seawater

	30 ‰ water	35 ‰ water
Na Cl	23.60	27.53
K Cl	0.64	0.75
Mg Cl_2, $6H_2O$	4.53 (Mg Cl_2 : 2.12)	5.28 (2.47)
Mg SO_4, $7H_2O$	5.94 (Mg SO_4 : 2.92)	6.93 (3.41)
Ca Cl_2	0.98	1.14
Na NO_3	0.1	0.1
Na_2 HPO_4 $12H_2O$	0.02	0.02
NaH CO_3	0.192	0.192

The pH and the CO_2-content should then be adjusted by adding 5 millilitres of a solution of sodium bicarbonate (3.84 grammes of $NaHCO_3$ dissolved in 100 millilitres of water) to one litre of the prepared solution. 0.1 normal solution of NaOH would then be added until the pH of 8.2 is arrived. The trace elements of boron, copper, zinc, manganese, iron etc. could then be added if necessary or desirable[1]).

Table 5
Mean Surface Temperature of the Oceans

Northern latitude	Atlantic Ocean	Indian Ocean	Pacific Ocean
70°–60°	5,60		
60°–50°	8,66		5,74
50°–40°	13,16		9,99
40°–30°	20,40		18,62
30°–20°	24,16	26,14	23,38
20°–10°	25,81	27,23	26,42
10°– 0°	26,66	27,88	27,20
Southern latitude			
0°–10°	25,18	27,41	26,01
10°–20°	23,16	25,85	25,11
20°–30°	21,20	22,53	21,53
30°–40°	16,90	17,00	16,98
40°–50°	8,68	8,67	11,16
50°–60°	1,76	1,63	5,00
60°–70°	–1,30	–1,50	–1,30

---

1) For further details concerning artificial media see Provasoli, McLaughlin, and Droop (1957).

Temperature

The temperature of the water is of great importance for the development, the growth and the geographical distribution of marine algae. The annual variation in temperature differs greatly in different regions. In tropical waters these variations are always slight and amount to only a few degrees. In contrast, there are littoral regions, such as the Atlantic coast of North America from Cape Hatteras to New Foundland, or the coasts of the Kattegat and the Skagerrak, where the variations can amount to 18 to 20° C or more. In the Mediterranean the variations lie between 14 and 17° C. Generally the variations in temperature are greater in littoral regions than in the open sea. The variations in temperature are, naturally most marked at the surface; this is particularly true of sections of the sea more or less sealed-off, such as the Kattegat, the Skagerrak, and the Baltic Sea.

Water temperatures of various oceanic regions are presented in Table 5. Not only are annual mean temperatures important for algal growth and geographical distribution, but also the maximum and the minimum temperatures measured in these regions are decisive factors.

With respect to the phytoplankton the oceanic regions are divided into four large groups (Table 6).

Table 6
The Distribution of Phytoplankton

Temperature in degrees centigrade	Group	Corresponds to
< 10°	Oligothermic (cold region)	I Arctic
0–20°	Mesothermic α (cold temperate region)	II Boreal
14–24°	Mesothermic β (warm temperate region)	III Subtropical
> 24°	Polythermic (warm region)	IV Tropical

The same classification can be used in connection with benthic vegetation (cf. STEEMANN NIELSEN, 1944).

The algae are divided into two groups according to their susceptibility to variations in temperature (SETCHELL, 1920):

1) stenothermal species with a low resistance to changes in temperature (range of tolerance 10 degrees Centigrade)

2) eurythermal species with a high resistance to changes in temperature.

Water Movements

The following four factors mainly cause the movements of the sea water: the tides, the wind, changes of temperature, and the inflow of fresh water. These water movements – the vertical as well as the horizontal currents, wave motions, and intermixing between different water masses – are well known.

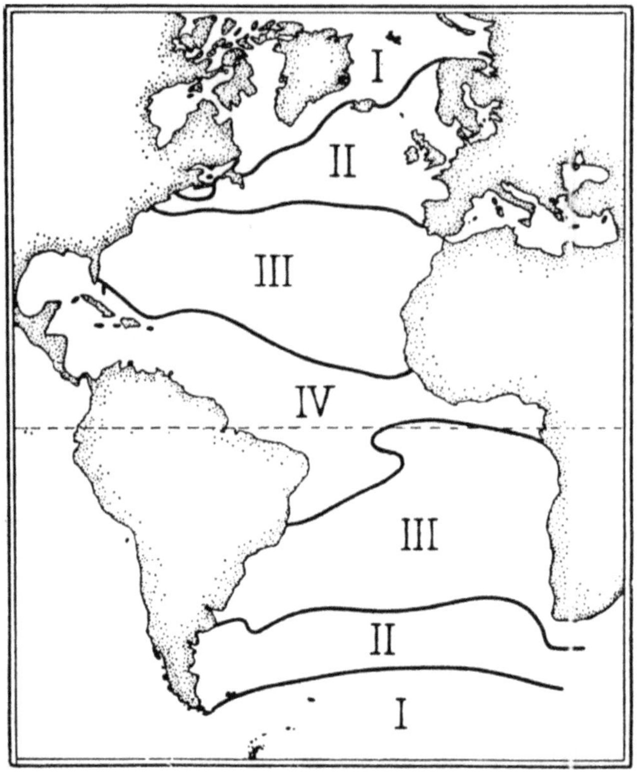

Figure 3: Temperature areas of the Atlantik surface water: I = oligothermic, II = mesothermic α, III = mesothermic β, IV = polythermic (from STEEMANN NIELSEN)

The tide in the open sea never amounts to more than 1 metre. The tidal wave is enlarged when it reaches areas of lower depth. In bays where special conditions become effective, enormous differences in the tide levels can be registered, e.g. in the central parts of the bay between Brittany and Normandy in France (St. Malo and Mont Saint Michel), approximately 12 metres, and in the Bay of Fundy on the inner coast of Nova Scotia in Canada 21 metres (Pl. II: 3). In open coastal areas the differences in the tide levels are approximately 1.5 to 3 metres.

There is hardly any tide in the Mediterranean because the Straits of Gibraltar are so narrow.

The North Sea area offers another example worth to be remarked. There the differences in the tide are relatively great, but as the depth in the North Sea varies, so the tidal waves move with different velocities and cause an interference phenomenon at the opening of the Kattegat and the Skagerrak which as a result are more or less tidefree. The tide difference on the southernmost point of the Norwegian coast is zero, whereas on the Swedish west coast it amounts to approximately 10 to 20 centimetres. There can, of course, occur differences in the water level caused by other factors.

This regular change in the water level plays, naturally, a great rôle for the vertical distribution of the vegetation in tidal zones.

As a consequence of the tide currents frequently occur, which in certain coastal areas, between islands or in narrow sounds, can have a relatively high velocity; a speed of 3 to 4 knots is not infrequently met with, in particular areas it can even exceed 10 knots.

Such currents – and of course other currents too – can often lead to turbulences in the water, and water movements of this kind have a balancing effect on water temperatures. The extreme contrast is represented by ponds and pools where the water is more or less stagnant and can become so warm by the power of the sun rays that stenothermal species, for instance, cannot thrive there.

A turbulence may also cause water that is lacking in nutrient salts to be mixed with water rich in them. In general, conditions of vegetation are favourable for benthos as well as for plankton in areas with such water movements.

Very strong currents may also cause the evolution of ecological forms of various benthic algae (Pl. I).

The complex effect of the waves can easily be observed if one studies the vegetation of two neighbouring locations, one of which is exposed and the other sheltered. The species occurring in the two localities are often very different, or the same species may show different development.

The Light

As for terrestrial plants light (radiant energy) – its intensity and spectral composition – is a factor of utmost importance for marine plants. A brief survey on light conditions in the sea and the investigations made into that field of study seems therefore to be appropiate at this point.

Whenever light penetrates into water, the radiant flux (the time rate of flow of radiant energy)[1]) will be reduced. This reduction (attenuance) is caused by absorptance (the ratio of the radiant flux lost from a beam by means of absorption, to the incident flux) and scatterance (the ratio of the radiant flux scattered from a beam, to the incident flux) by water molecules and by dissolved and suspended particles. This attenuance (C) is equal to the sum of absorptance (A) and scatterance (B).

$$C = A + B$$

In the case when a parallel beam of monochromatic light penetrates into water the extinction can be calculated mathematically according to

$$I_r = I_o \cdot e^{-cr}$$

where I_o is the original irradiance and I_r, when having passed a water layer of the thickness r. c represents the extinction coefficient, which is a characteristic of the water, and has different values for different wave-lenghts.

---

[1]) Concerning terminology cf. JERLOV (1964).

If the **transmittance** is defined as the percentage of radiant energy which remains as a balance, after the light has passed through a water layer of r meters (which is not equal to the depth) thickness, the transmittance can be expressed mathematically in the following manner

$$I_r = I_o \cdot T^r \quad \text{or} \quad T = \sqrt[r]{\frac{I_r}{I_o}}$$

where I_o stands for the incident irradiance, I_r is the irradiance remaining after the flux has passed through a water layer of r metres in thickness. T stands for the transmittance.

The determination of the spectral composition and of the value of the total irradiance in different depths of water are very complicated ventures. A standardized irradiance meter may be used in these operations (a selenium-photovoltaic cell combined with different filters). Omitting all technical and mathematical details, suffice it to say that we obtain as the result of such measurements curves which show the spectral distribution of the radiant energy. The value of the total irradiance at a certain depth can be arrived at by mathematical procedure, by the integration of the corresponding curves.

$$E_x = \int I_x(\lambda) \cdot d\lambda$$

E_x stands for the total irradiance at a depth of x metres, I_x irradiance at a certain wave-length λ.

Ocean water, which is particularly clear and translucent, has a maximum of transmittance for the blue light. In coastal areas, however, where the water always contains suspended and dissolved substances of varying quantity, the short wave-lengths are quickly absorbed, whereas the longer ones are less subject to this extinction (cf. LEVRING 1947). Of the dissolved substances what has been called the "Yellow substance" (Gelbstoff) discovered by KALLE, should be mentioned here. It will be found in a particularly high degree in coastal waters, and the green colour which the water often has in these regions is due to the occurence of this substance (cf. KALLE 1938). But it is also found in

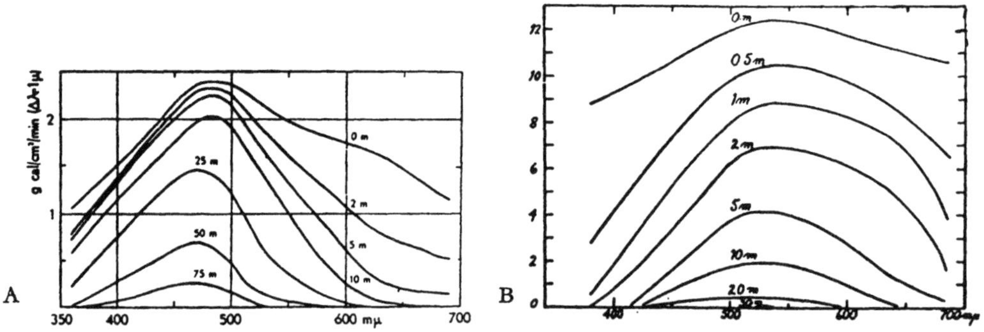

Figure 4: A. Spectral composition of radiant energy at different depths in clear oceanic water (East Mediterranean) (from JERLOV, 1951) – B. Spectral composition of radiant energy in coastal water (off west coast of Sweden) (from LEVRING, 1947)

ocean water and seems to be absent only in waters of the highest clarity (cf. JERLOV 1951, 1964).

When rays of light pass through the water, the direct and straight character of the flux becomes gradually indistinct due to the effect of scatterance so that with increasing depth the submarine illumination becomes increasingly diffuse.

Red light (i.e. radiant energy of long wave-lengths) is absorbed almost entirely by a water layer of a thickness of 5 metres. This absorption is mainly caused by water molecules. The blue light (the short wave-lengths irradiance), however, is mainly absorbed by suspended and dissolved (yellow substance) substances, while ultraviolet irradiance is particularly susceptible to absorption. In coastal areas (e.g. in north-west Europe) the water, therefore, has a maximum of transmittance for greenish light. Thus submarine illumination below a depth of 10 meters is of a greenish colour, while in particularly clear ocean water, on the other hand, it is of bluish tint. With increasing depth the irradiance will be more or less logarithmically reduced.

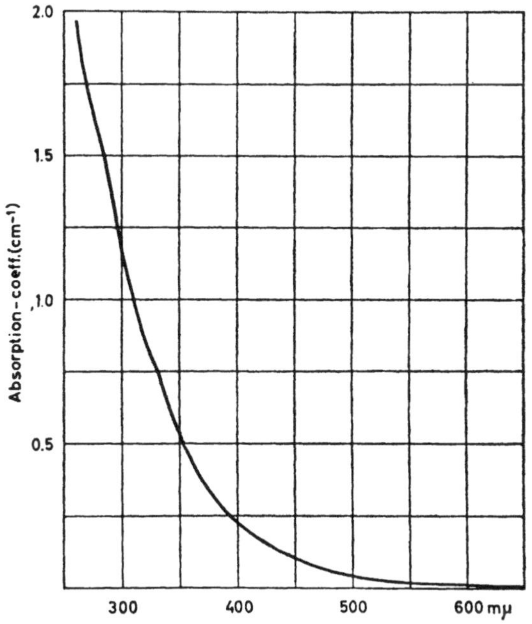

Figure 5: Absorption curve of yellow substance (from JERLOV)

It must be pointed out here that measurements, though carried out on submarine illumination with ordinary luxmeters, are of no practical value. Luxmeters are instruments designed for measurements of normal daylight and consequently have a great sensitivity to green light but only a slight sensitivity to red and blue light. If they are used for measurements of light of a different composition than normal daylight, the results thus obtained will be completely misleading.

A method for measuring the total irradiance in various depths of water, which is not too difficult to be carried out in practice and which offers satisfactory results for biological purposes at least, has been described by JOHNSON and KULLENBERG (1946). A correspon-ding theoretic calculation of the value, however, is extremely complicated.

THE BENTHIC VEGETATION

Growth and Periodicity

For the growth of algae their photosynthetic activity is, of course, of decisive importance, and that also makes the intensity as well as the spectral composition of the submarine light, which constitutes the energy source for the photosynthetic process, attain great significance.

The green algae *(Chlorophyceae)* are able to use red and blue light for photosynthesis, whereas green light is of little importance to them. The two photosynthetic maxima correspond very well to the absorption maxima of chlorophyll (Fig. 6, Fig. 7). Thus it could be maintained that the green leaves of the higher-order plants behave in a similar manner. Yet this is not so. A maximum photosynthesis is obtained in the presence of red light. The photosynthetic effect decreases when the wave-lengths become shorter, so that in the range of blue light it will reach its lowest value (cf. GABRIELSEN, 1940). This is due

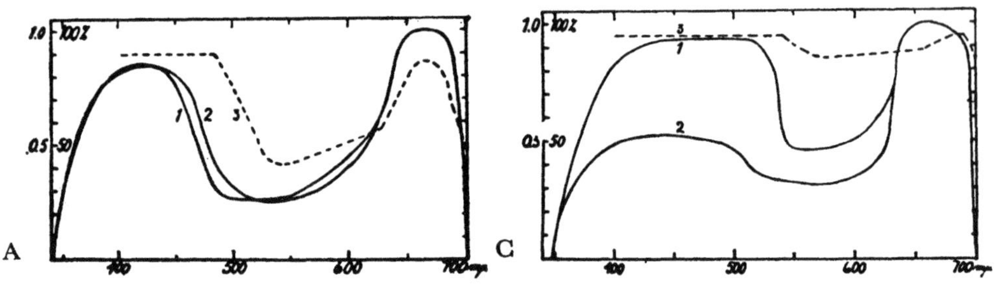

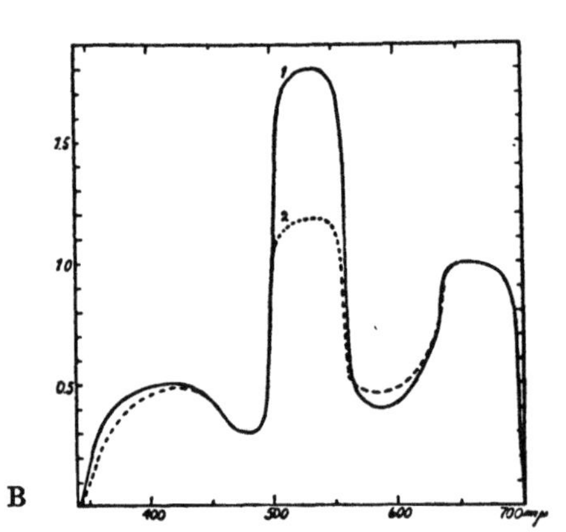

Figure 6: A–C. Spectral action curves (spectrum of equal energy); A green algae: *Ulva lactuca* (1), *Ulva linza* (2) and absorption curve for *U. lactuca* (3) (according to SEYBOLD and WEISSWEILER); B Red algae: *Ceramium pedicellatum*, from 20 M depth (1) and *C. rubrum*, from 0–0,5 M depth (2); C brown algae: *Fucus serratus* (1) and *F. vesiculosus* (2), absorption curve for *F. serratus* (3) (according to SEYBOLD and WEISSWEILER); (from LEVRING, 1947)

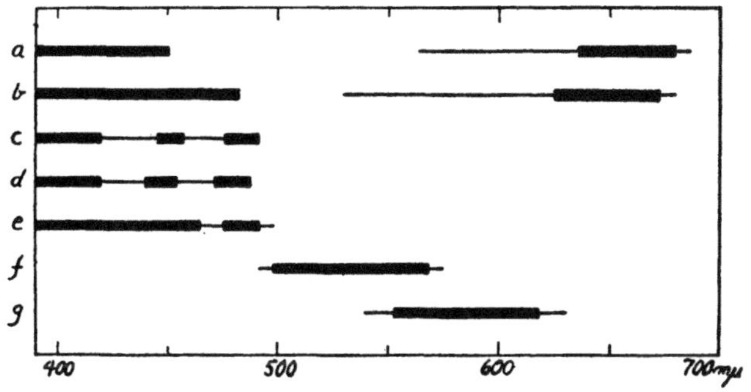

Figure 7: Shematic representation of the absorption of different pigments of the algae: Chlorophyll *a* (a) and *b* (b), carotene (c), xanthophyll (d), fucoxanthin (e), phycoerythrin (f) and phycocyanine (g) (from LEVRING)

to the unproductive absorption of the shorter wave-lengths by the foliage of anatomically complicated structure (thick cell walls, the presence of air in the mesophyll etc.; cf. SEYBOLD and WEISSWEILER, 1942–1943).

It is a striking fact that the red algae *(Rhodophyceae)* are able to use green as well as red and blue light as an energy source for photosynthesis (LEVRING, 1947). Certain species of red algae of deep water habitat have a marked maximum for green light, corresponding to the absorption of phycoerythrin, the typical pigment of red algae. The energy of the green light is absorbed by the phycoerythrin for photosynthesis. Orange light is absorbed by the blue phycocyanin of the Rhodophyceae and, especially, of the Cyanophyceae in an analogous manner (cf. FRENCH and YOUNG, 1952). Apart from this, these algae also absorb red and blue light by means of chlorophyll.

Like the Chlorophyceae, the Phaeophyceae also have two maxima; the maximum of the shorter wave-lengths has a wider range than that of the longer. Thus, greenish-blue light is absorbed for photosynthesis in addition to the red and blue. This is due to the additional absorption by the brown pigment fucoxanthin (cf. WASSNIK and KERSTEN, 1946). Algae with a high content of fucoxanthin have a particularly high short wave maximum (cf. SCHMIDT, 1937, MONTFORT, 1938, LEVRING, l.c.). The content of various pigments (chlorophyll, phycoerythrin, phycocyanin, and fucoxanthin) may strongly differ in a certain species of algae, depending on locality, depth, and also on seasonal changes. Hence it can be concluded that the maxima of a curve showing the values of spectral photosynthesis of certain species can be quite different, depending on individual circumstances.

The photosynthetic effect of mixed light equals the sum of photosynthetic processes of the various spectral ranges (cf. LEVRING, l.c.)[1]. The photosynthetic activity of algae, thriving in a certain depth, therefore depends not only on the total value of the radiant energy, but also on spectral composition of the latter. It can be proved relatively easily

[1]) According to EMERSON this is not quite true if the ranges are very narrow. The effect is then decreased.

by calculation that, when the depth increases, the photosynthetic activity of the green algae is inhibited more quickly than that of red or brown algae, provided that the water has its transmission maximum in the green light. This fact provides a good support for the theory of what is called complementary chromatic adaptation. This theory was put forward by ENGELMANN (1882–1884) and GAIDUKOV (1902–1906), and is based primarily on laboratory experiments with Cyanophyceae. It was established that light of different colours influences the colouring of Cyanophyceae to such an extent that gradually these algae assume the complementary colour of the light employed. These findings lead to the conclusion that most of the red algae are deep water species because they are able to use green light, i. e. their own complementary colour, for photosynthesis.

The vertical distribution of algae, on the other hand, was explained by BERTHOLD (1882) and OLTMANNS (1892) as being primarily due to the light intensity, not, however, to the colour of the light. Their theory is based on experiments and observations carried out in the marine environment. There are some forms which are adapted to strong light, and others that are adapted to dim light. If algae of a different colouring adapted to strong light and growing in shallow environments are transplanted to greater depths, their photosynthetic activity is reduced to such an extent that the usual differences between red, green, and brown algae are hardly perceptible. It is only by means of a very careful analysis that it can be proved that red algae are able to utilize the light in greater depths better than the other types. Determination of what is called the compensation depth, reveal that such forms can only thrive in shallow waters, i. e. in environments with strong light. Thus it can be said that these facts are corroborating evidence for the theory advanced by BERTHOLD and OLTMANNS. This, however, does not apply to the dim-light species which have their origin in deep water habitats. The presence of certain

A
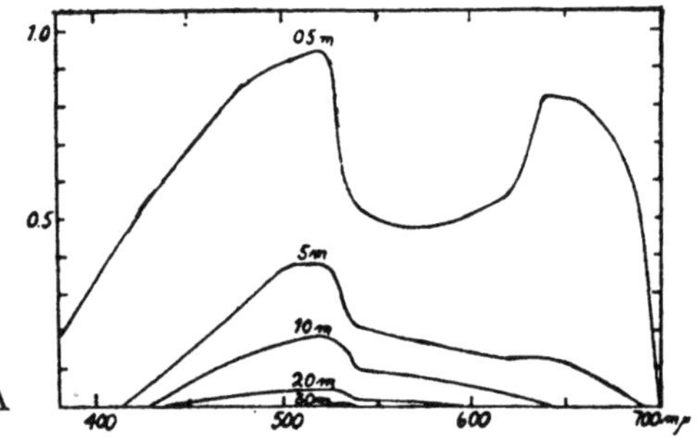

B
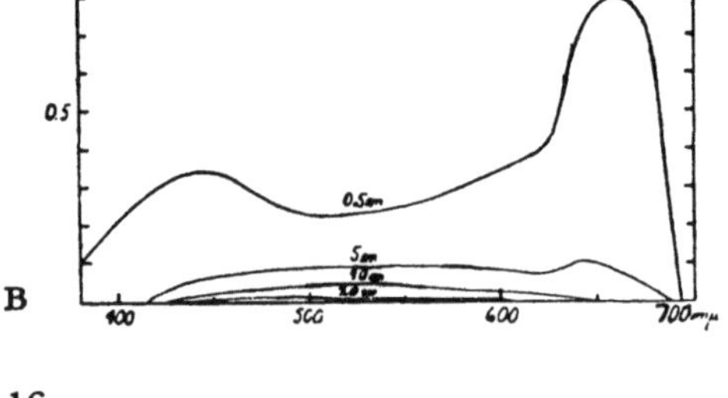

C
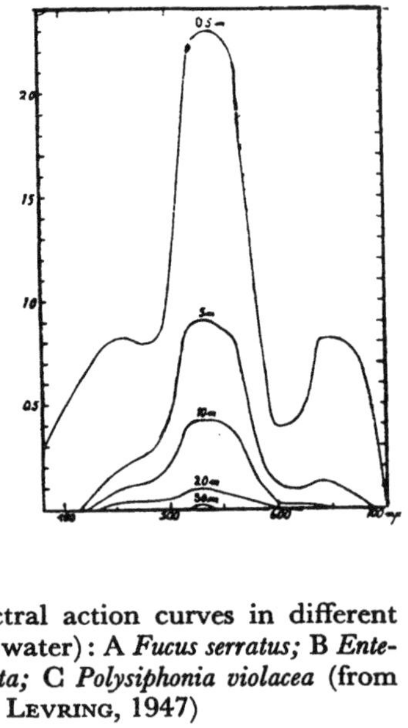

Figure 8: Spectral action curves in different depths (coastal water): A *Fucus serratus;* B *Enteromorpha clathrata;* C *Polysiphonia violacea* (from LEVRING, 1947)

pigments (of phycoerythrin in red algae and of fucoxanthin in brown algae) makes it possible for such forms to exist at all in regions below a depth of approximately 10 metres. In tropical and subtropical waters some Chlorophyceae are to be found in greater depths (cf. TAYLOR, 1961). Obviously this is due to the fact that the blue light is able to penetrate into the deeper regions, as the water is more translucent here than in other areas[1]). The energy of the blue light is thus absorbed by the chlorophyll. It should also be pointed out that deep-water algae, if exposed to the strong light prevailing in a depth of approximately half a metre, will rarely be able to utilize all available energy. In many cases this even will be an inhibitive factor. According to experiments made on the Swedish west coast (cf. LEVRING op. cit.) in a depth of half a metre, the maximal effect was already

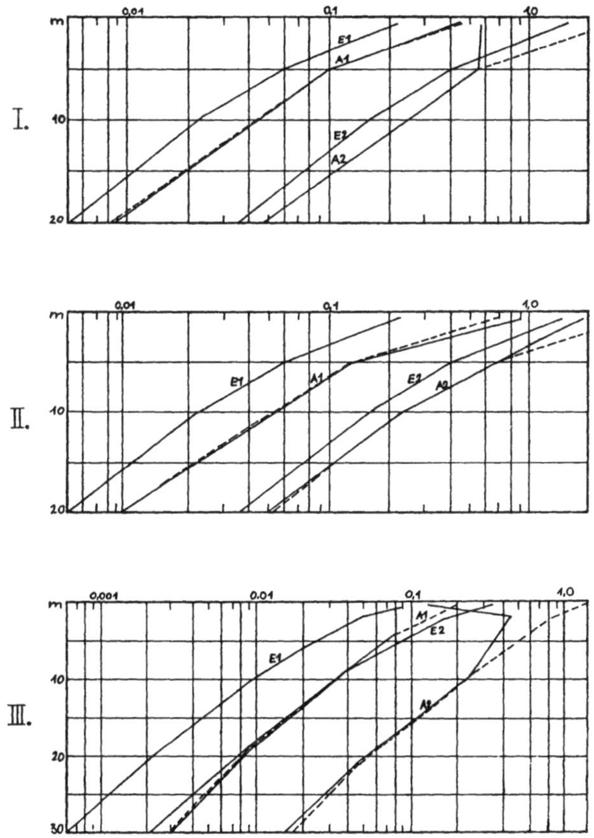

Figure 9: Total irradiance and photosynthesis in different depths (coastal water). E irradiance; A photosynthesis (---- theoretical curve). 1 and 2 are two different experiments. I. *Cladophora crystallina* (green alga, shallow water); II. *Dictyosiphon foeniculaceus* (brown alga, shallow water); III. *Delesseria sanguinea* (red alga, deep water) (from LEVRING).

[1]) During a stay in 1966 at Duke University marine laboratory, Beaufort, N. C. (U.S.A.) the author was able to show experimentally that the photosynthesis of i.e. *Ulva* was quite considerable even below 30 metres in the extremely clear water of the Gulf Stream and the Sargasso Sea off the shores of the Carolinas (LEVRING, 1968).

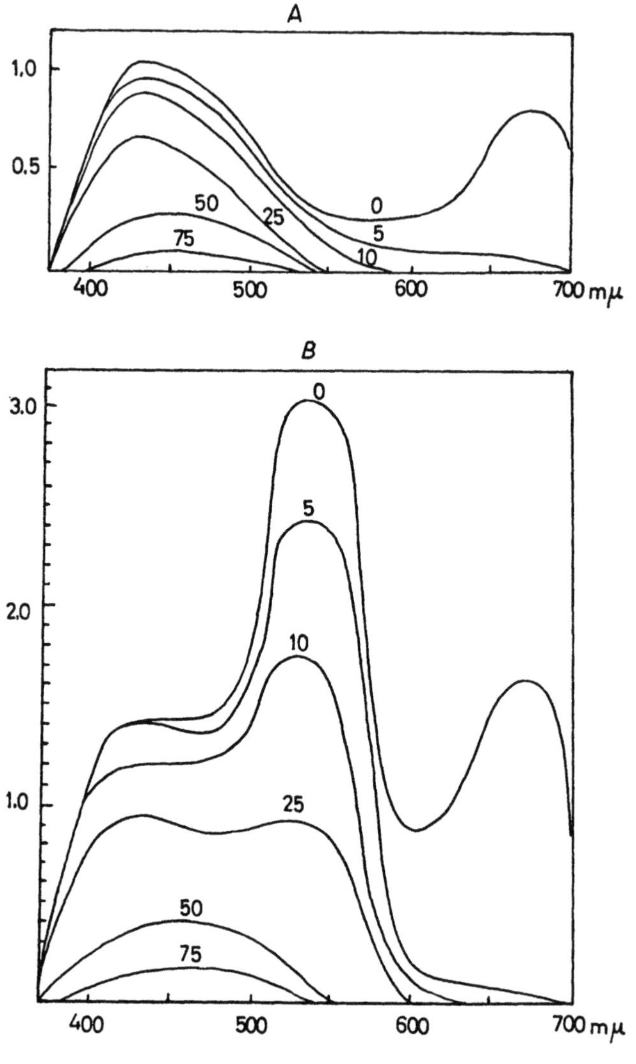

Figure 10: Action curves for *Ulva* (A) and *Polysiphonia* (B) in different depths computed for very clear oceanic water

reached with an illumination intensity (irradiance) corresponding to one of 0.1 to 0.3 gcal/ cm^2/min (gramme calories per square centimetre per minute) on the water surface. In this context it should be mentioned that the radiant energy on a cloudless summer day equals approximately 0.5 gcal/cm^2/min. It has already been pointed out that littoral algae are able to utilize higher intensities of light than algae living in deep waters. The latter soon reach the point of maximal photosynthetic activity. Even semi-dim light can provide the necessary energies for the photosynthesis in a depth of approximately 10 metres. It is a known fact that the majority of Chlorophyceae are littoral forms; the Phaeophyceae are mainly to be found in depths of about O to 15 metres, while the Rhodophyceae thrive in regions of all depths, although they predominantly occur in depths of 15 to 30 metres. Though there are distinct stronglight and dimlight forms of algae, it should be pointed out, in view of the spectral composition of the submarine light at various depths, that green algae have a greater aptitude for life in shallow water while brown algae preferrably occur at depths down to 15 metres and red algae in deep waters with greenish illumination. This, however, does not exclude the occurrence of red algae in shallower water.

The relation between submarine light and the vertical distribution of algal vegetation may be summarized in the following way: There are different types of algae adapted to different light intensities. The various pigments play an important rôle in the constitution of these types, those are pigments which permit the utilization of varied spectral composition. In other words, these types are not only adapted to different intensities, but also to different spectral composition of the radiation.

Thus light as the energy source of the photosynthetic process is an important factor in the growth of algae. Photosynthesis, however, is also dependent on temperature. Yet growth is not only determined by photosynthesis, but also by respiration, i.e. growth corresponds to the difference between photosynthesis and respiration. The rate of respiration is especially dependent on temperature. With increasing temperature the rate of respiration increases more quickly than photosynthetic activity. This explains the well-known phenomenon that algae have their great growth period in spring. At that time of the year the water in many areas is still cold and consequently the rate of respiration is still low. Light conditions, on the other hand, are so favourable that they permit an almost maximum degree of photosynthesis which, therefore, results in a large assimilation surplus.

The content of pigments in algae may also increase during the darker seasons (cf. LUBIMENKO – HUBINET), which results in increased photosynthetic activity. In experiments made with *Chlorella* it was ascertained by EMERSON (1929) that photosynthesis is proportional to chlorophyll concentration. Another factor favouring algal growth in spring and in late winter is the fact that during these seasons the sea is particularly rich in nitrate and phosphate.

As among terrestrial plants, there are both therophytes and perennials in marine vegetation. The therophytes mostly develop during a few months (generally in spring) and then wither away. Typical North-European examples are *Ulothrix* and *Urospora* sp., *Bangia fuscopurpurea*, *Scytosiphon*, *Chordaria*, *Dumontia incrassata*, and *Nemalion multifidum*.

It is worth remarking that species found on the shores of the Skagerrak and the Kattegat as typical spring forms, develop in colder regions (e.g. in the Arctic Ocean and Baltic Sea) as late as early summer or even midsummer. KYLIN (1907) distinguishes three different types of periodicity:

1) Forms that are vegetative and fertile all the year round *(Hildenbrandia)*.
2) Forms with vegetative development during the whole year, but with the period of fertility limited to the summer for some *(Fucus vesiculosus, Polysiphonia nigrescens)*, but to winter and spring for the majority (e.g. *Fucus serratus)*.
3) Forms showing vegetative growth and fertility only during a limited period. Often the vegetative organs for assimilation are shed after the termination of the period of growth *(Desmarestia aculeata, Polysiphonia elongata)*.

As mentioned earlier, growth takes place during the late winter and during spring. This applies, of course, only to regions with distinct differences in the seasons. A pronounced periodicity in vegetation is to be found in these regions. In regions with slight variations in temperature conditions are different, as, for example, in Greenland and near the Faroe Islands, where the algae grow all through the summer in the low temperatures, or in the subtropics and the tropics where the water always is warm.

Padina pavonia is a well-known species living in warm waters, and it can be found in the Mediterranean all the year round. It is also found on the shores of the English Channel, but there its period of development is limited to the summer months. Similarly, such cold water (boreal) species as *Ulothrix flacca* can be found all the year round in the northern oceans, whereas in the Mediterranean it occurs only during the winter and spring months.

On the whole conditions with only slight annual variations in temperature result in a less pronounced periodicity.

Finally, mention should be made of two interesting investigations into photoperiodism in algae published some years ago. The Norwegian plant physiologist HYGEN (1948) found out that *Ulothrix flacca* on long days produces swarmers which copulate and give rise to stationary zygotes. On short days these swarmers also copulate, but grow immediately into new filaments. The zygotes of the swarmers produced on long days do not germinate before the days become shorter again. These experiments carried out in cultures are in accordance with the natural life-cycle. FÖYN (1956) has carried out experiments on cultures of *Ulva* taken from various parts of Europe. In these experiments it became obvious that what was formerly known as *Ulva lactuca*, are really two species which in their habits are very similar to each another.

1) *Ulva lactuca* is anisogamous and requires long days for its normal development.

2) *Ulva thuretii* from southern Europe is isogamous and can develop in short days.

Thus photoperiodism, which for the higher terrestrial plants is of great importance, also plays a rôle for algae and certainly has to be taken into consideration.

Life Forms

It is well-known of terrestrial plants that many species being systematically entirely different may assume the same forms, if they thrive under equal ecological conditions. Many scientists have developed morphologic-ecological systems in an attempt to categorize the plants as biological types (e.g. the system of RAUNKIAER).

Similar attempts have been made in the field of marine algae. The first system of this kind was suggested by OLTMANNS (1905), and further developed by FUNK (1927), and by GISLÉN (1930). Such a system of basic morphological forms was also established by NIENBURG (1930), cf. Table 7.

The various groups of basic forms comprise species belonging to the most different systematic groups. The idea is that with such systems the vegetation of an area can be characterized without consideration of the various systematic groupings.

Several objections may be raised against such systems: the most important characteristics of a species do not always become evident as morphological criteria. Furthermore

Table 7

Life Forms
(According to NIENBURG)

I. Seaweeds – Large and coarse forms, exclusively belonging to the brown algae.
  a) Leather-type seaweeds – with a flat laminar thallus; most Laminariaceae belong to this group.
  b) Ligulate seaweeds – with ribbon-like, usually bifurcated thallus, comprising the Fucaceae.

II. Fine algae – comprising all other algae, with the exclusion of those with incrusted lime.
  a) Crust algae – with loose or closed fronds adhering to base.
  b) Filamentous algae – consisting of thin, usually uniseriate unbranched threads of at most a finger's length.
  c) Laminar algae – consisting of irregularly shaped leaf-like fronds, which in early development stages may be of tubular growth.
  d) Thread algae – branched or unbranched threads of the thickness of thick string, several decimetres up to 1.5 metres long.
  e) Shrublike algae – widely branched, with some branches either in one row, or leaf-like structure, 1–50 centimetres long.
  f) Cushion algae – solid or hollow cushions, diameter 1–10 centimetres.

III. Calcarious algae (Corallinaceae) – incrusted with lime.
  a) Calcarious crust algae – 1–10 centimetres in diameter, growing on other substrata.
  b) Calcarious bush-like algae – 1–10 centimetres in diameter.
  c) Calcarious nodule algae – 1–10 centimetres in diameter, of irregular shape, freely rolling on the ocean bottom, like pebbles.

IV. Higher Plants
  a) Sea grass and weeds
  b) Queller (Salicornia herbacea)

the morphological criteria on which these systems are based are not in any direct relation to the various ecological conditions.

In greater accordance with the concept of RAUNKIAER is the system established by KNIGHT and PARKE (1931), which was further developed by FELDMANN (1937), cf. Table 8. The various groupings here are based on the length of life cycle and on the manner in which the marine algae survive unfavourable seasons.

Now it must be said that the unfavourable and the favourable season of an area is by no means the same for different species of algae, as it normally is for terrestrial plants. Among the algae there are many species which grow during the cold season of the year and rest in summer, whereas others follow an opposite course of life cycle. As was suggested by STEEMANN NIELSEN (1944), the schedule set up by FELDMANN could, therefore, be developed further by dividing the groups IB (a-b) (Table 8) according to whether the unfavourable period was in the cold or in the warm season.

Algae, however, cannot be fully characterized ecologically by these systems unless the nature of the substratum, light, water etc. is also taken into consideration. In this connection, a very useful terminology has already been introduced by SETCHELL (1924, 1926).

I) With regard to the substratum:
  1) Thriving on rocks: epilithic
  2) Thriving within rocks: endolithic
  3) Thriving on sandy soils: psammophilous

4) Thriving on mud: pelophilic
5) Growing upon or in other plants: epiphytic and/or endophytic
6) Living upon or within an animal: epizooic and/or endozooic.

II) With regard to temperature:
1) Species not depending on temperature: eurythermic
2) Species confined to a certain range of temperature: stenothermic.

The latter category is subdivided into mesothermic and megathermic species according to the susceptibility to temperature.

Similarly, euryphotic and stenophotic species can be distinguished by their sensitivity to illumination. Stenophotic species may be further subdivided according to the light intensity required by the various forms into the following three groups: sciophilic, mesophotic, and heliophylic.

Euryhaline and stenohaline species may be distinguished according to their dependence on salinity. Hyperhalinic species thrive in very salt water, whereas hypohalinic species thrive in brackish water.

Cymathophilic algae require the strong movement of waves; pheophilic algae require the movement caused by strong currents of water, while galenophilic algae require the quiet waters of very sheltered places.

Geographical Distribution and Occurrence of the Marine Vegetation

Although during the last century several scientists have published descriptions of the vegetation and flora of various parts of the world, our present knowledge of the various floral regions is still too incomplete and fragmentary to enable us to give a universal survey. Of the factors mentioned earlier – light, salinity and temperature – temperature must play a decisive part in the geographical distribution of different algae. Apart from regions like the Kattegat, the Baltic Sea, the Black Sea etc., where there is a strong inflow of fresh water from rivers, the salinity throughout the oceans is of the same order and can, therefore, play no major part in the distribution of algae. In special cases, as those just mentioned, salinity may, naturally, assume great significance.

The composition of the algal flora of one region cannot be explained by climatic or hydrographic factors alone. Historical factors have also to be taken into account. In order to explain the occurrence of certain species in different regions which are separated by large sea areas, as, for instance, on the various shores of the oceans, it is necessary to bear in mind also former connections of coastal areas which have vanished y today (cf. SVEDELIUS, 1924). The distribution of different species along a more or less continuous coast line, as, for example, on the west coast of Europe, is in the first place undoubtedly due to temperature conditions. The work by BOERGESEN and JONSSON (1905) on the distribution of algae in northern Europe and in the Arctic Ocean may be regarded as a classical authority. These two scientists divide the algal species into five groups according to their distribution: The arctic, subarctic, boreal-arctic, cold boreal, and warm boreal group. This is a subdivision which has generally been accepted and used by other algologists, and it has been completed and enlarged (cf. KYLIN, 1907; van GOOR, 1923; ROSENVINGE, 1935 etc.). NIENBURG (1930) calls the fivc groups: Arctic, Arctic-French, North-European, West-European-Baltic,

Table 8

Life Forms
(According to FELDMANN)

I. ANNUAL ALGAE

			Examples
A. Thriving all the year round. One to several generations. Spores and zygotes germinating directly . . . Ephemerophyceae			*Cladophora*, *Enteromorpha*, Certain *Polysiphonia*
B. Thriving only during a certain season. a) Tiding over the rest of the year as microscopic vegetative form . . . Eclipsiophyceae	Microscopic form is a prothallium		*Phyllaria*, *Sporochnus*, *Nereia*
	Microscopic form is a plethysmothallus		*Asperococcus* and many other *Phaeophyta*
b) Tiding over the unfavourable season as resting bodies . . . Hypnophyceae	The resting bodies consisting of:	spores:	*Spongomorpha lanosa*
		zygotes:	*Vaucheria*
		hormogonia:	*Rivularia bullata*
		akinetes:	*Ulothrix pseudoflacca*
		protonema:	*Porphyra*
		germs:	*Dudresnaya*, *Halymenia*

II: PERENNIAL ALGAE

			Examples
A. Perennation of the entire thallus a) Thallus erect . . . Phanerophyceae			*Codium*, *Halimeda*, *Fucus*, *Vidalia*, *Phyllophora*
b) Thallus a crust . . . Chamaeophyceae			*Palmophyllum crassum*, *Hildenbrandia*, *Peyssonelia*, *Melobesiaceae*
B. Perennation only of a part of the thallus a) Only part of the erect thallus persistent . . . Hemiphanerophyceae			*Cladophora pellucida*, *Cystoseira*, *Sargassum*, *Laminaria hyperborea*, *Polysiphonia elongata*, *Rodriguezella*, *Sphaerococcus*
b) Only the basal portion persistent . . . Hemicryptophyceae	The basal portion consisting of:	a disc	*Cladostephus*, *Rissoella*, *Alsidium corallinum*, *Gymnogongrus griffithsiae*
		creeping filaments	*Udotea*, *Acetabularia* *Bryopsis muscosa*, *Gymnogongrus nicaeensis*

Meridional-Norwegian, and adds a more southerly group, the Meridional. In the present study the terminology of BOERGESEN and JONSSON has been used. The arctic group comprises species which occur only in the Polar Sea. In the subarctic group there are those whose occurrence in the Polar Sea is quite common, in the North Atlantic less frequent, but which can be found down to the Faroe Islands or England or the west of France. The boreal-arctic species occur in the Polar Sea and in the boreal regions of the Atlantic, at least down to North Africa. The cold boreal species spread from the north of Norway, the south of Iceland and the Faroe Islands, to England and the west of France, but some occasionally have been found in the Polar Sea (in the White Sea and the Murmansk Sea). The warm boreal species can be found from the north of England, the west of Norway, and some from the south of Iceland, the Faroe Islands, and the north of Norway at least down to North Africa and the Mediterranean. Finally, the meridional species (according to NIENBURG) occur in the south of the middle coasts of England, and from the south of Ireland (in isolated cases even in the North Sea) down to the Mediterranean.

The boundaries of the various groups are not clearly marked. The areas are more or less overlapping, so that they cannot be regarded as different, adjacent floral areas.

Several scientists have studied the same problems in different regions. Unfortunately there exists a certain confusion regarding nomenclature (cf. Table 9). It is, of course, highly desirable that a universal system of nomenclature should be used. Such a system (cf. Table 10) was suggested by CHAPMAN (1946), and it could well be used until further details have been established.

Several authors deal with the subject of endemism in various regions. The interesting problem of vicariating species is mentioned, among others, by SVEDELIUS (1924).

It is a well-known fact that for plankton algae the absolute temperature tolerance (i.e. tolerance for both the vegetative and resting phase) can be different from the temperature

Table 9
Geographic subdivision of Algal Vegetation

BOERGESEN and JÓNSSON	NIENBURG	SMITH, SETCHELL, HOYT	OKAMURA	FELDMANN
Arctic	Arctic	Boreal		
Subarctic	Arctic-French			
Boreal-Arctic	North-European	Upper Boreal		
Cold Boreal	West-European-Baltic			
Warm Boreal	Meridional-Norwegian	N. Temperate	Temperate	Boreal Atlantic
		N. Subtropical	Subtropical	Tropical Atlantic
		Tropical	Tropical	Pantropical
		S. Subtropical		
		S. Temperate		
		Upper Australian		
		Australian		
	Cosmopolitan		Cosmopolitan	Cosmopolitan
			Indo-Pacific	Indo-Pacific
	Meridional		Japanese-Okhotsk	Mediterranean

Table 10

Geographic Elements
(According to Chapman, 1946)

	Arctic		Antarctic
	Subarctic		Subantarctic
	Boreal-Arctic		Australian-Antarctic
Cold-Boreal-Atlantic	Cold-Boreal-Pacific	Cold-Australian-Atlantic	Cold-Australian-Pacific
Boreal-Atlantic	Boreal-Pacific	Australian-Atlantic	Australian-Pacific
North-Subtropical-Atlantic	North-Subtropical-Pacific	South-Subtropical-Atlantic	South-Subtropical-Pacific
	Tropical-Atlantic		Tropical-Pacific
	Pantropical		Cosmopolitan
	Caribbean		Arabican
	Indo-Pacific		

Table 11

Geographic Distribution According to Temperature
(According to Steemann Nielsen, 1944)

I. Species not dependent on temperature *(Pylaiella littoralis)*

II. Species inhabiting the cold regions (oligothermic)
Perennials and annuals (summer-annuals only)

III. Species inhabiting cold and cold-temperature regions (oligothermic and meso-α-thermic)
  1. Perennials *(Phyllophora brodiaei)*
  2. Annuals
     a. Occurring in summer as well as in winter, (in Danish waters: *Elachista fucicola)*
     b. Occurring in spring only (in Danish waters: *Monostroma grevillei)*

IV. Species inhabiting cold and cold-temperate and warm-temperate regions (oligothermic, meso-α-thermic, meso-β-thermic)
  1. Perennials *(Ascophyllum nodosum)*
  2. Annuals *(Enteromorpha intestinalis)*

V. Species inhabiting cold-temperate regions (meso-α-thermic).
  1. Perennials *(Membranoptera alata)*
  2. Annuals *(Sphacelaria cirrhosa)*
  This group excludes species with a limited period of vegetation in either the cold or warm season. Such species can be held to have a wider distribution.

VI. Species inhabiting cold-temperate or warm-temperate regions (meso-α-thermic and meso-β-thermic)
  1. Perennials *(Halidrys siliquosa)*
  2. Annuals *(Chaetomorpha aerea)*

VII. Species inhabiting warm-temperate regions (meso-β-thermic)

VIII. Species inhabiting warm-temperate and warm regions (meso-β-thermic and poly-thermic)

IX. Species inhabiting warm regions (poly-thermic)

tolerance of the vegetative one alone. Many benthic algae reveal quite similar features. The two aspects of temperature tolerance are identical for perennials and many annuals, which unlike the plankton, are non-motile. However, there are many species which develop their vegetation (and become fertile) only for a period of a few months, and they live in a state of rest (in different water temperatures) for the remainder of the year; with these species the absolute temperature tolerance differs from the vegetative temperature tolerance.

There are stenothermic and eurythermic species among the benthic algae, just as among the plankton (SETCHELL 1920). The stenothermic species have a small temperature tolerance and are therefore found in regions with slight variations in temperature, while the eurythermic species have a wide range of tolerance.

On the basis of such facts STEEMANN NIELSEN (1944) developed a system (cf. Table 11) which is based, above all, on the absolute temperature tolerance but which also takes account of the tolerance of the vegetative stage.

The Vertical Distribution of Algal Vegetation

Attempts to categorize algal vegetation according to its regional distribution and occurrence had already been made by J. G. AGARDH (1836) and OERSTED (1844). However, the first investigator to consider the ecological factors in greater detail was KJELLMAN. In his work on the algal vegetation in the White Sea (1877) he divided the vegetation into that of littoral, sublittoral, and elittoral zones. The littoral zone ranges from the highest level of high tide to the lowest level of low tide. The sublittoral zone ranges from the littoral to a depth of 20 fathoms, while lower parts of the sea bottom covered with algae were referred to the elittoral zone. In his work on algal zonation in the eastern Skagerrak (1878) KJELLMAN introduced the term "Region" for the term "Gebiet" formerly used. KJELLMAN's terminology has on the whole remained unchanged, and, it is still customary, therefore, to speak of littoral, sublittoral, and elittoral zones.

Several somewhat different concepts became widespread sometime later regarding the definition and limitation of the zones.

Zonation along coasts subject to tidal influence will first be considered, since conditions along this type of coast can be regarded as "normal". Areas with little or no tidal activity will be dealt with exclusively as more or less special cases.

The upper limit of the littoral zone has been fixed by ROSEVINGE (1898) as being where the vegetation begins, a line which in Greenland coincides with the lowest high tide level. BOERGESEN (1905) similarly defines the upper limit for the littoral zone, which, however, at the Faroe Islands in sheltered locations coincides with the highest tide level, and in exposed places may even exceed this level by 30 metres. The upper limit for the littoral zone is also similarly defined by KYLIN (1907), JONSSON (1912, and COTTON (1922). SERNANDER (cf. 1917, p. 90), in contrast, defines this limit as the high water level, and above the littoral zone there follows the supra-littoral zone in his categorization (LORENZ 1863, p. 193). Thus it can be said that he drew the line purely in view of physical factors, a limitation which obviously is not natural. The upper limit of the littoral zone does not only depend on the high-water mark, for other factors, such as waves and desiccation caused by insolation, also have an influence upon it. As KYLIN (1918, p. 66) stressed, they are these three factors, – the position of the high-water mark, the waves, and then insolation – which determine the upper limit of vegetation. He called this limit the "physiological high-water line" and defines it as the upper limit of the littoral zone, which seems logical. It is the same line which SJÖSTEDT (1928, p. 4) calls "Litus-Line". In nature this line is clearly marked. It roughly coincides with the lowermost limit of the *Verrucaria maura* association.

It is a striking characteristic of this zone, which actually is always found above the water surface, that *Verrucaria maura*-like lichens and *Cyanophyceae* spread over it, which give the rock a black colour. This phenomenon can be observed on the shores of almost any part of the world, and was simply called "the Black Zone" by STEPHENSON (1949).

With regard to the drawing of the border line between the littoral and supra-littoral zones, the lowermost limit of *Verrucaria maura*, mentioned above, may sometimes be misleading, because of the occurrence of other species of *Verrucaria*.

Another line which can be applied in many parts of the world is that one constituted by the upper limit of the *Balanus balanoides* association (or other Balanidae). Normally, this border line is clearly distinct (LEVRING 1937, STEPHENSON 1949, WENNBERG, unpubl.), and for practical purposes can very well be taken as the border line between the supra-littoral and the littoral zones.

The upper limit of the littoral zone is thus clearly marked, whereas the lower is not so clearly discernible. According to KJELLMAN (1877) the lower limit is set by the lowest tide level. Like the upper limit, the lower was later modified as to correspond roughly to level of low water at neap tides (cf. ROSENVINGE 1898; BÖRGESEN 1905; SERNANDER l.c.; PRINTZ 1926 etc.). This line, however, also is not distinct: according to WENNBERG (op. cit.) in sheltered places the lower limit is determined by the tide level, while in exposed places it lies above that level.

Below the littoral zone follows the sublittoral zone which reaches down to the elittoral region yet lower. The border line was established by KJELLMAN in his work referred to above as being at a depth of 20 fathoms. ROSENVINGE (1898 p. 237), however, holds that the lower limit of the sublittoral zone should coincide with the limit of the algal vegetation as such, a view which seems logical, and which has been recognized by subsequent authors. Thus the elittoral zone consists of that part of the sea bottom which is void of any vegetation. This limit is to be found in the case of the North-European coasts at a depth of approximately 25 to 40 metres.

In subtropic and tropical regions, where the water is much clearer than in other parts of the world, vegetation penetrates into much greater depths, viz. down to at least 100 metres. LUND (1957) traced several species down to a depth of 80 metres on the east coast of Greenland. WILCE (1967) has found a well developed algal vegetation under the ice down to at least 50 metres (occasionally even below 100 m) on the west coast of Greenland and arctic Canada. According to ZANEVELD (1966) benthic algae have been found in the Ross Sea (Antarctic) down to the remarkable depth of 150–300 metres. Also records of more than 600 metres are mentioned.

The conditions prevailing on coasts subject to tidal influences, as they have been recognized and described by several authors who had carried on investigations mainly in Northern Europe will become apparent from the descriptions given above. One, therefore, finds a littoral zone which is designated by an intermittent process of desiccation and which, above all, is limited by the high-tide and the low-tide levels. However, the upper and the lower limit of the littoral zone are not taken as two purely physical lines, but rather as two natural physiological boundaries which are influenced also by various other factors. Above the littoral begins the supralittoral zone which is never submerged; and below it begins the sublittoral zone which reaches down to the cessation of vegetation and which is always submerged. This kind of division into zones of algal vegetation may, at any rate, be made for all coasts subject to tidal influences (i.e. shores that are normally twice daily covered and uncovered by the tides).

As for some algologists it seems impossible to agree on the terms of supralittoral, littoral, and sublittoral, STEEMANN NIELSEN (1944) has suggested a different terminology:

I) The phytalic region, which ranges as far as vegetation occurs.

   1) The adythalic zone which regularly will be reached only by the crests or spray of the waves.

   2) The hemidythalic zone which is alternately submerged (either regularly or more or less irregularly) and corresponds to the tidal zone.

   3) The eudythalic zone which is always submerged.

II) The aphytalic region, the region without any chances of life for autotrophic plants.

Some authors have tried to establish more or less complicated general systems. It is obvious that this would be very difficult to achieve, because the vegetation in different areas is so highly varied in its composition. Such systems have been based entirely on hydrographical zones or on vegetation zones or both. The basic terminology originates mainly from earlier Scandinavian workers. Certain changes in definitions etc. have been neces-

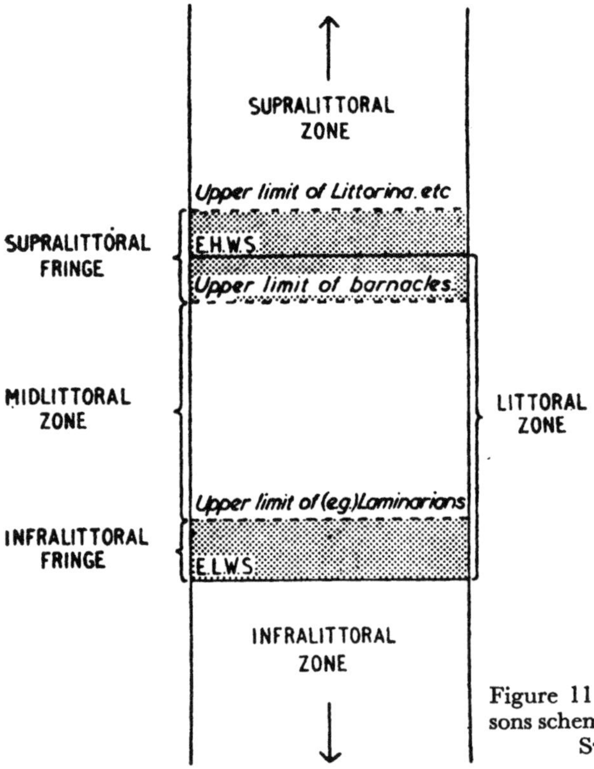

Figure 11: Diagram illustrating the Stephensons scheme of zonation and terminology (from STEPHENSON & STEPHENSON, 1949)

sary. In the attempts to create general systems where the vegetation is divided into zones and communities of higher and lower rank one, unfortunately, often gets lost in nomenclatorial discussions. Difficulties and misunderstandings have also been raised, when trying to get uniformity between the terminologies used by land and marine ecologists, and when terms have been translated from one language to another. A closer discussion of these questions and problems would be of minor interest to this study. The author feels that any system of this kind, after all, has to follow factors given by nature and has to be based upon careful investigations of the vegetation in the field.

As shown by STEPHENSON and STEPHENSON (1949 and other publications) and later LEWIS (1964) a system of classification of the shore organisms as well animals, algae, lichens, and other plants have to be considered at the same time. The Stephensons tried to propose a world wide system based upon the contention that certain types of organisms characterize approximately the same levels on all rocky shores. These provide a framework within which local features can be accommodated without obscuring those underlying features by which shores throughout the world can be compared. The Stephensons scheme of zonation appears in Fig. 11. It will be noted that the two authors in a later paper returned to the term "sublittoral" which is synonymous with "infralittoral" and which ought to be more commonly used by marine ecologists. The Stephensons thus recognised the following zones and fringes:

1) Supralittoral zone: The uppermost region of the shore characterized by landlichens or flowering plants.
2) Supralittoral fringe: A marginal belt characterized by small snails *(Littorina* etc.), Isopods *(Ligia)* and by blackening organisms *(Verrucaria* and Cyanophyta as *Calothrix)*. High water of spring tides invades at least the lower part of this fringe.
3) Midlittoral zone: The main tidal belt which is regularly covered and uncovered every day. It extends from the upper limit of barnacles (in quantity) and is dominated by barnacles, mussels, limpets, and a great variety of algae (the major quantity of which are different Fucaceae in the North Atlantic etc.).
4) Sublittoral fringe: The lowest zone of the shore uncovered only at the major tides and often only in calm weather. It extends from the upper limit of the Laminarias in cold temperate water, ascidians and certain Rhodophyta in warmer waters and corals in the tropics down to the lowest level of the water.
5) Sublittoral zone: The organisms living in this zone are always submerged.

In his recent study of rocky shore ecology of the British Isles LEWIS (1964) supports many of the ideas mentioned above. He is suggesting a modified terminology (Fig. 12) based upon the view "that the zones are biological entities which can only be defined by biological means". Physical definitions are thus rejected as there is no real coincidence between biological and tidal levels, and because the physical conditions are best revealed by the organisms. The term supralittoral is discarded as being entirely superfluous and illogical for a belt of marine organisms. Instead the littoral zone is divided into a littoral fringe (dominated by *Littorina* and *Verrucaria*) and an eulittoral zone (dominated by barnacles and fucoids). In his view to extend the littoral zone upwards to include all the marine organisms up to the lowest belt of true terrestrial vegetation, LEWIS comes very close to the early ideas of BÖRGESEN (1908) and COTTON (1912). The scheme of zonation

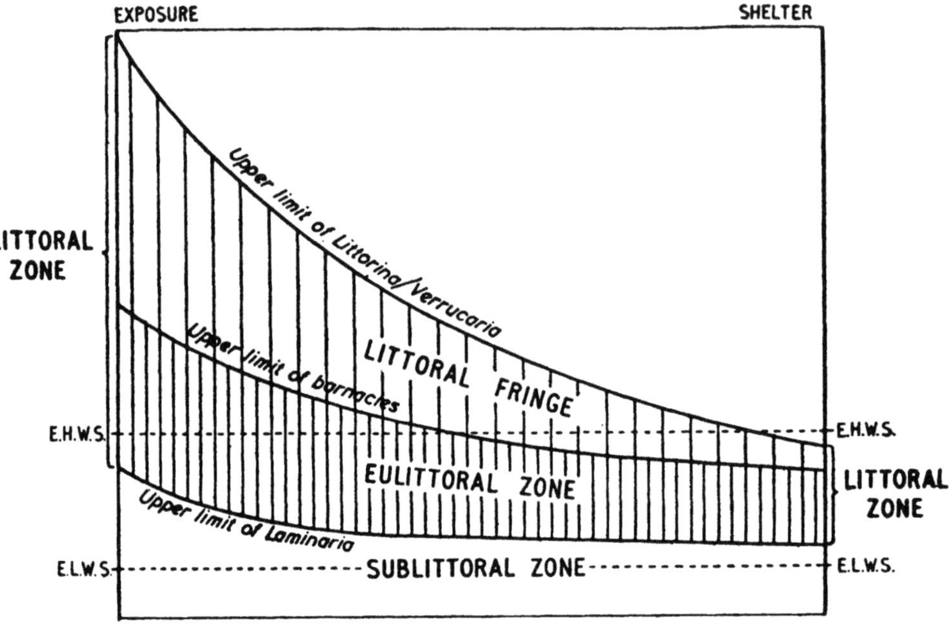

Figure 12: Scheme of zonation and terminology according to LEWIS (1964)

used by LEWIS appears in Fig. 12. It is obvious how very different the zonations are in exposed and sheltered habitats.

Of great value are the descriptions of the vegetation of various parts of the world, in which for different plant communities the subdivision into zones were stated. For the vegetation of western Europe, above all, are to be mentioned in this connection studies by BÖRGESEN (1905) on the Faeroe Islands, by JÓNSSON (1912) on Iceland, and by NIENBURG (1930) who gives a general description of the plant communities of the North European coastal area (Table 12, Fig. 13, Pl. II: 4, V: 9).

DAVY DE VIRVILLE (1940) subdivides the vegetation of the Atlantic coasts in Europe into a number of horizontal "basic belts" (cf. fig. 14).

There are belts of the yellowish-orange lichens *Xanthoria parietina*, *Caloplaca marina* throughout the supralittoral zone and a black belt of *Verrucaria maura* in the lower part. In the littoral zone follow belts of *Pelvetia canaliculata*, *Fucus spiralis*, *Fucus vesiculosus*, and *Fucus serratus;* further down and exposed only by the lowest tides grows *Laminaria digitata*, and below that *Laminaria hyperborea (cloustonii)*.

According to the same author there are in addition to these basic belts a number of facultative ones: In the upper shore there are lichen belts *(Lichina confinis* and *L.pygmaea)* and a belt of *Rivularia bullata*. On the same level as *Fucus vesiculosus* in sheltered situations there is a belt of *Ascophyllum nodosum*. On a level with, or somewhat above *Laminaria*, there may be belts of *Himanthalia lorea* and *Bifurcaria tuberculata*. There is no correlation between tide level and the lichen belts mentioned above. There is also a certain variation between different localities, as CHAPMAN (1943, p. 243) among others has pointed out.

Table 12

Plant Associations

I. SUPRALITTORAL PLANT ASSOCIATIONS (closed formations only)
- a) Supralittoral crust formation
- b) Filamentous algal formation

II. LITTORAL PLANT ASSOCIATIONS
1. Closed formations
   - + On solid substratum
     - a) Higher laminar algal formation
     - b) Littoral ligulate algal formation
     - c) Lower laminar algal formation
     - d) Filamentous algal ("Schnuralgen") formation
   - ++ On soft substratum
     - a) *Salicornia* formation
     - b) *Posidonia* formation
2. Open formations. Littoral shrub-like algae
   - a) On solid substratum
   - b) On soft substratum

III. SUBLITTORAL PLANT ASSOCIATIONS
1. Closed formations
   - + On solid substratum
     - a) Calcareous crust formation
     - b) Leathery-algal formation
     - c) Calcareous nodule ("Kalkknolle") formation
   - ++ On soft substratum
     - a) Common seagrass formation
2. Open formations
   - a) Sublittoral ligulate algal formation
   - b) Sublittoral shrub-like algal formation

Results of thorough investigations of the algal vegetation on the Dutch coasts have recently been published by DEN HARTOG, and the investigations of BURROWS and LODGE (1951) and of BURROWS, CONWAY, LODGE, and POWEL (1954) of the algal vegetation on the shores of Great Britain should be mentioned. A very comprehensive account of the plant and animal communities of rocky shores is to be found in the recent monograph by LEWIS (1964).

The algal vegetation in the south-west of Norway has been investigated by LEVRING (1937) and more recently by WENNBERG (unpublished). Near Bergen the difference in tide levels amounts to approximately 90 (40 to 150) centimetres. Here different belts can be distinguished. WENNBERG names these belts according to the species which have the highest degree of coverage; on a sheltered rocky coast, for instance he names them as follows in descending order:

1) The Maura *(Verrucaria maura)*, 2) the Balanus-Hildenbrandia-mucosa *(Balanus balanoides, Hildenbrandia rosea, Verrucaria mucosa)*, 3) the Lithothamnion-Lenormandii, 4) the Laminaria belts. The *Maura* belt belongs to the supralittoral zone; belt 2 and the major part of belt 3 belong to the littoral zone, while the remainder of belt 3 and 4 belong to the sublittoral zone. A number of other species also

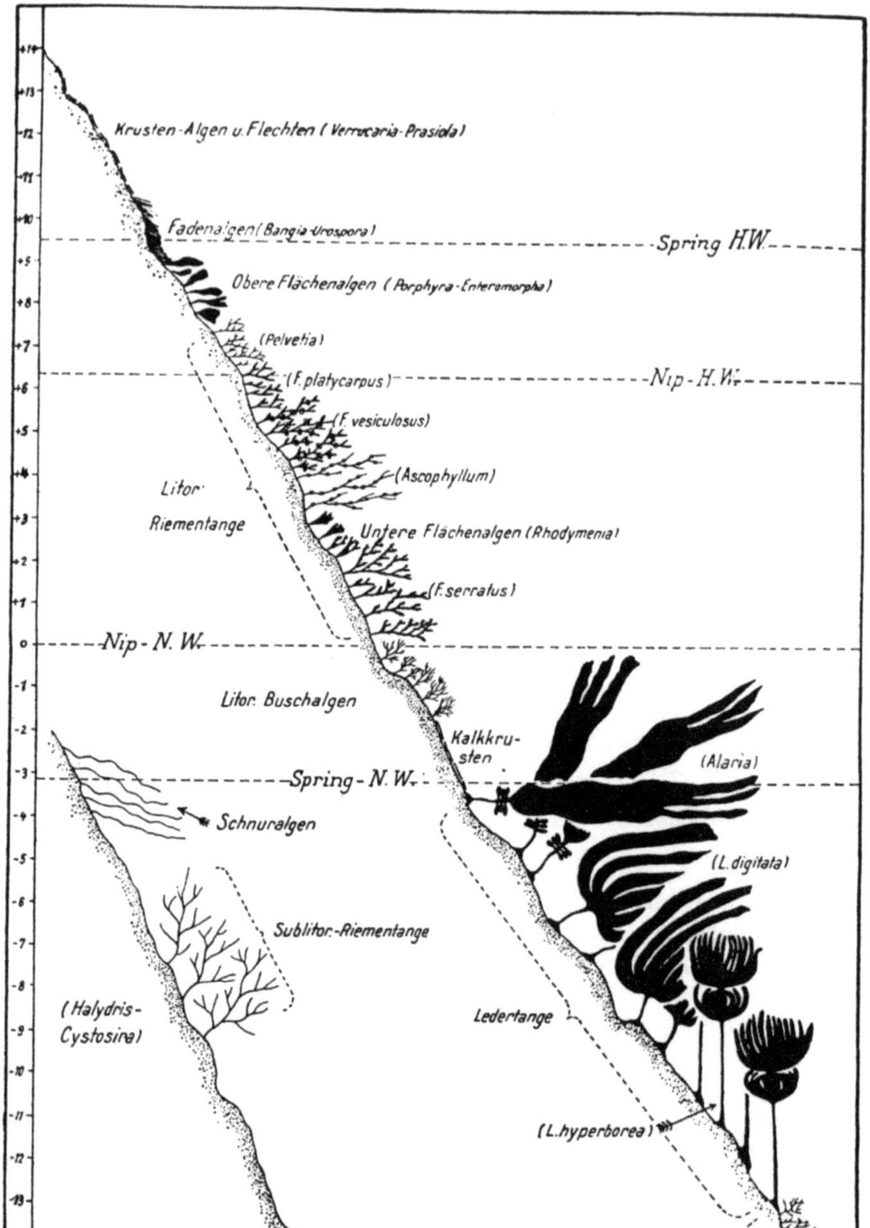

Figure 13: Diagram of the algal vegetation of the Channel coast (from NIENBURG)

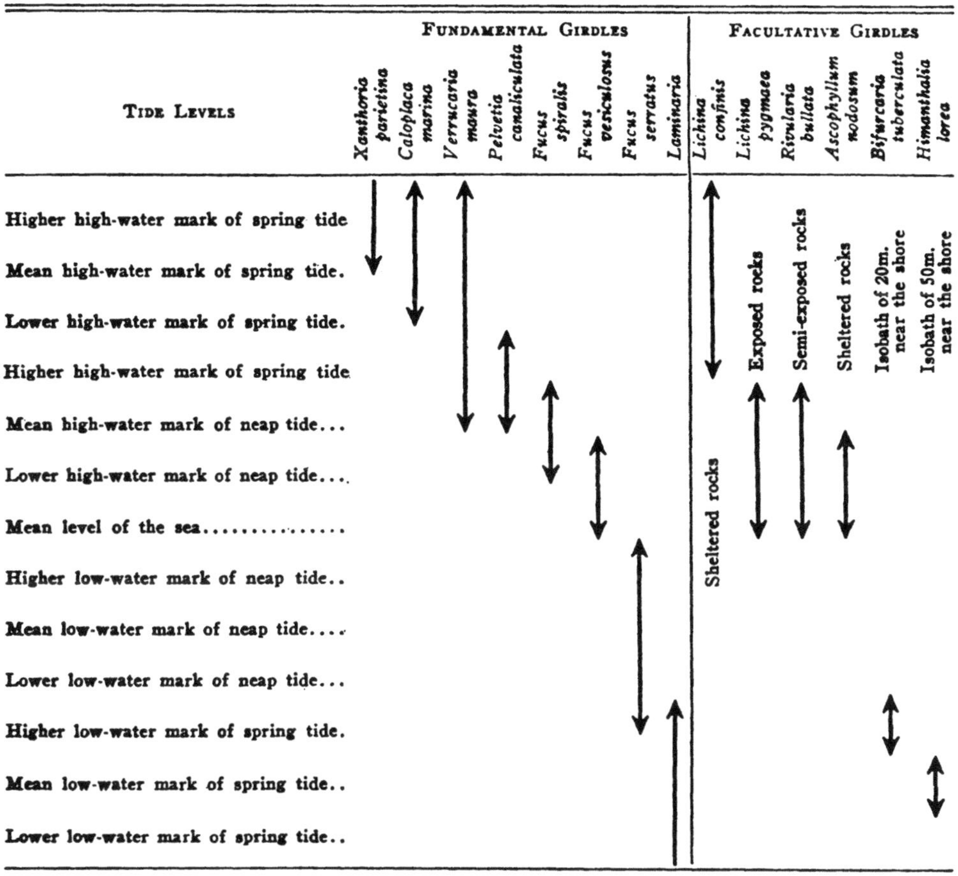

Figure 14: Table of the vegetation girdles on the Atlantic coasts of Europe in relation to tide levels (from A. Davy de Virville)

form belts, growing in more or less connected groupings within the belts mentioned above, such as *Pelvetia, Fucus spiralis, Fucus vesiculosus, Fucus serratus,* and *Ascophyllum.* According to studies of myself in various parts of north-west Europe (Scotland, Isle of Man, Iceland), where the differences between high-tide and low-tide levels are relatively great, the respective belts can be found to be more or less widespread in vertical extension.

The vegetation undergoes certain changes on exposed coasts; some species – e.g. most Fucaceae – disappear, while others, such as *Polysiphonia urceolata, P.brodiaei, Spongomorpha, Gigartina, Ceramium penicillatum, C.secundatum,* and *Corallina,* replace them. The littoral region is also extended in the vertical direction. According to Wennberg it occupies the same range on the Norwegian west coast and the southernmost point of Norway, the tidal range being nil in both areas. Other important, recent investigations of Norwegian algal vegetation are those by Printz (1926) of the Trondheim area and by Sundene (1953) of the Oslo area.

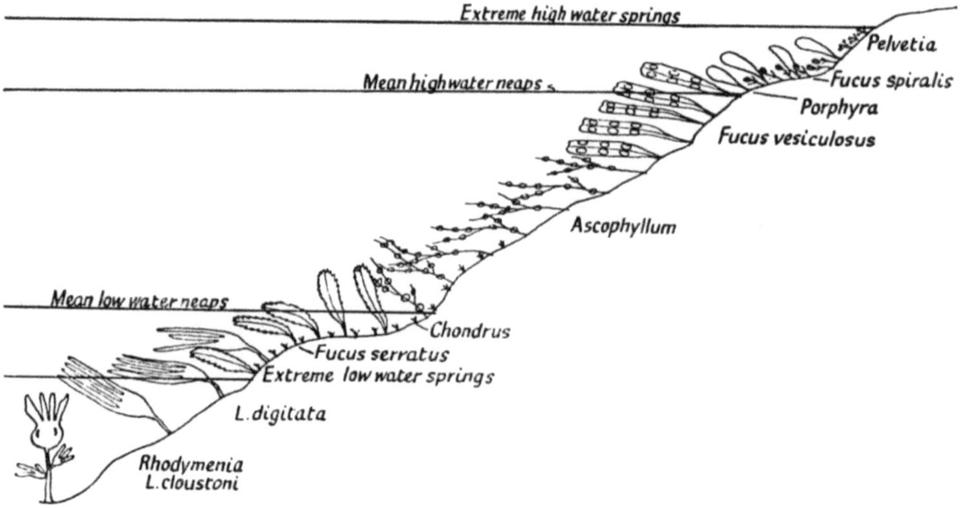

Figure 15: Diagram illustrating typical zonation on a European rocky shore (from CHAPMAN)

Roughly the same belts can be found in the arctic area of Norway (JAASUND 1965). On sheltered coasts there are *Pelvetia, Fucus spiralis, Fucus vesiculosus, Ascophyllum nodosum* (if present), *Fucus inflatus, Fucus serratus, Laminaria digitata* (normally), *Alaria esculenta* (occasionally). On exposed coasts: 1) *Porphyra, Bangia* (disappears in June and is replaced by *Blidingia*). 2) *Balanus;* 3) *Rhodymenia* and the small *Fucus distichus* or *Ascophyllum;* 4) *Acrosiphonia spinescens;* 5) *Polysiphonia urceolata, Gigartina* (occasionally together with *Monostroma, Spongomorpha, Rhodomela,* and in somewhat less exposed places also with *Dumontia, Scytosiphon* and *Himanthalia*); 6) *Alaria esculenta;* 7) *Laminaria digitata;* 8) *Laminaria hyperborea* down to a depth of approximately 10 metres, 9) *Lithothamnion* and others. A general review of Arctic algal vegetation is given by TAYLOR (1954).

As may be expected the belts of vegetation on the Atlantic coasts of North America resemble to those on the European coasts of the Atlantic. They have been studied particularly by PRAT (1933, 1940), BELL and MACFARLANE (1933), WILCE (1959), JOHNSTON and SKUTCH (1928), JOHNSTON and YORK (1917), TAYLOR (1957, 1960, 1961).

Descriptions of the subdivision into zones of marine vegetation on the North-American west coast can be found in the studies by GAIL (1918, 1920, 1922), MUENSCHER (1915), SCAGEL (1947), NEUSHUL (1965). GISLÉN (1943–1944) undertook comparative studies on the subdivision into zones of marine vegetation on the Californian and Japanese coasts, while DOTY (1946) described the division into vertical zones on the coasts of Oregon and California in relation to tides. Thus in the littoral zone there are belts of *Pelvetiopsis* and *Fucus,* similar to those in Europe and to those on the east coast of North America. The lowermost parts of the littoral zone in exposed localities are inhabited by communities of *Postelsia palmaeformis.*

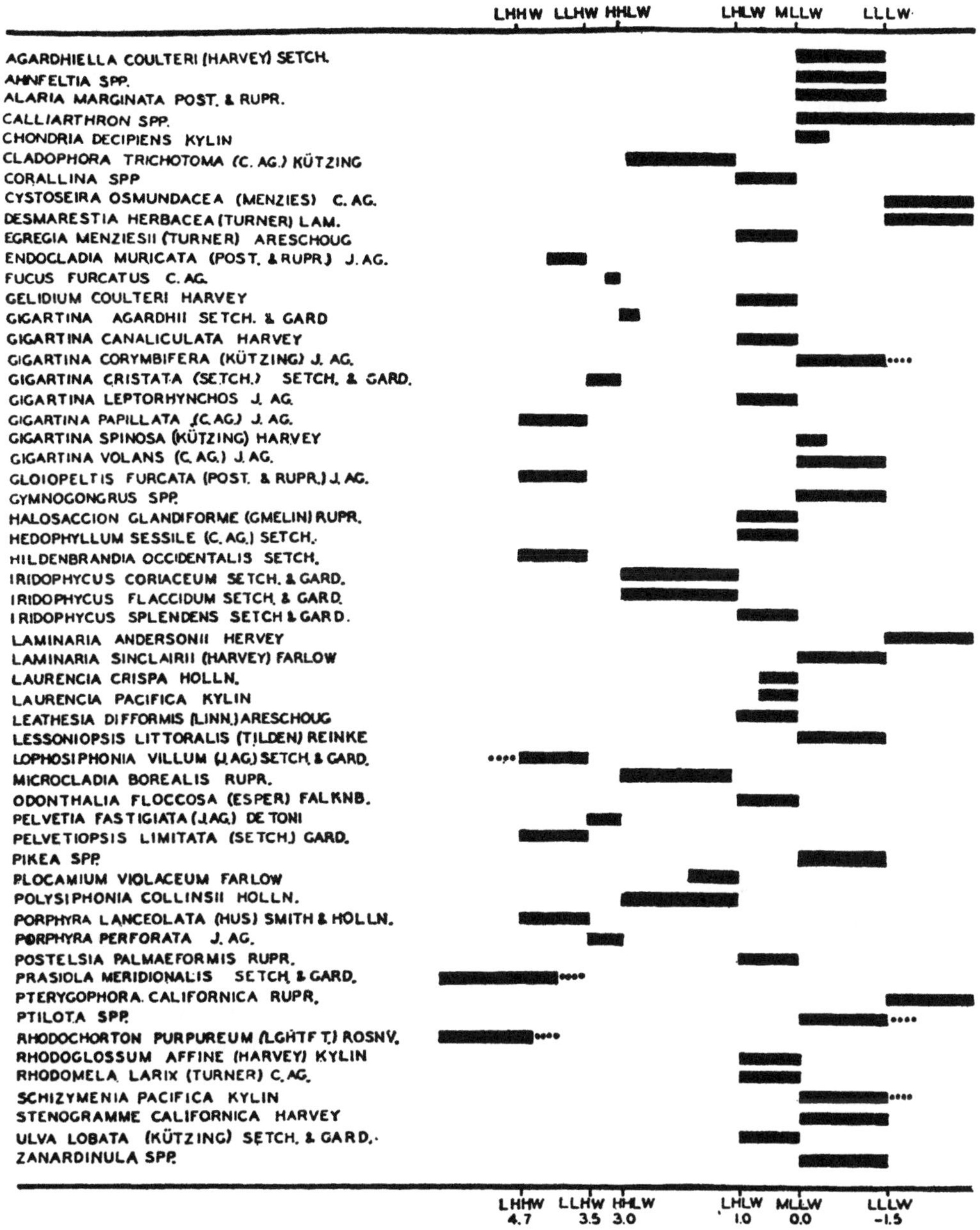

Figure 16: Chart to indicate vertical distribution of more conspicuous indicator algae along Central and Northern Pacific Coast of the United States. Corresponding tide heights (in feet above or below mean lower low water, MLLW) for San Francisco are given along lower axis as well as initials indicating critical tide factors correlated with breaks between zones along both horizontal axes (from Doty)

Various, very large Laminariaceae often occupy the sublittoral zone: *Laminaria, Alaria, Macrocystis, Nereocystis.*

SKOTTSBERG (1942) studied the zonation of algal vegetation in southern Chile, Tierra del Fuego, and the Falkland Islands. In these areas he distinguished the following zones:

1) Upper littoral Formation of algae resistant to desiccation and insolation, such as *Hildenbrandia lecannellieri, Bostrychia,* and *Porphyra.*
2) Intermediate-lower littoral Formation of surfresistant Rhodophyceae: *Iridea*-species and *Corallina.*
3) Intermediate-lower littoral Formation on less exposed, i.e. sheltered coasts: *Adenocystis* and Chlorophyceae. The main difference as from the equivalent zone in north Europe is the complete absence of Fucaceae.
4) Large Phaeophycean Surf-formation: *Durvillea antarctica.*
5) Large submerged Phaeophycean Formation: communities of *Lessonia,* often together with large species of *Desmarestia* which in colder, antarctic waters constitutes the main vegetation of the sublittoral zone together with *Phyllogigas, Ascoseira,* and *Cystosphaera.* Furthermore there are numerous medium-sized to small algae which form the ground vegetation, mainly consisting of Rhodophyceae.
6) Giant Phaeophycean Formation with floating thallus equipped with floats: *Macrocystis.*
7) Deep-water Rhodophycean Formation: often found as well-developed communities forming ground vegetation below the large Phaeophyceae.
8) Encrusting Corallinacean Formation: species of *Lithothamnion* etc. (above low-tide level, but down to the lower limit of vegetation).

A short, general review of antarctic marine algal vegetation is given by SKOTTSBERG (1964). The vegetation on the coasts of the Juan Fernandez Islands, however, is quite different. According to SKOTTSBERG (1941) the most important associations of the littoral zone are the following: The *Hydrocoleus lyngbyaceus, Hildenbrandia prototypus, Enteromorpha, Chondriella, Gelidium, Ulva rigida, Chaetomorpha firma, Splachnidium, Codium,* and *Litophyllum* associations and for the sublittoral zone the following, inter alia: the *Dictyotacean, Codium, Rhodomelacean,* and *Lithothamnion* associations.

Among recent ecological studies of the marine algae in Japan those by TANIGUTI (1962) have to be mentioned.

Amongst others CRIBB (1954), WOMERSLEY (1947, 1948, 1956), WOMERSLEY and EDMONDS (1958), SMITH (1952) have presented descriptions of vegetation on the Australian coasts. Except in tropical areas associations of *Hormosira banksii* (Pl. X), *Porphyra columbina, Corallina, Cystophora* species, *Ulva, Xiphophora* etc. are particularly striking in the littoral zone. The upper parts of the sublittoral zone are inhabited by species of *Cystophora* and *Sargassum,* by *Phyllospora comosa,* and *Ecklonia radiata,* and (in Victoria and Tasmania) by the giant algae *Sarcophycus potatorum* and *Macrocystis angustifolia.*

The vegetation of New Zealand resembles very much to the vegetation found in the south of South America (cf. OLIVER, 1923, CRANWELL and MOORE, 1938, and SKOTTSBERG, 1941). The supra-littoral zone contains associations of *Verrucaria* and lichens. In the littoral zone *Apophloea, Hormosira, Xiphophora, Carpophyllum, Splachnidium, Gigartina, Pachymenia, Durvillea* and others, and in the sublittoral zone *Lessonia, Ecklonia, Durvillea, Macrocystis,* and various Rhodophyceae should be mentioned in particular.

Zone	Western and Central Coasts					South-Eastern Coast
	Extreme to Strong Wave Action			Moderate Wave Action	Slight Wave Action	Extreme to Strong Wave Action
	Steeply Sloping Palaeozoic Rock	Horizontal Rock Platforms	Sand Beaches	Sheltered Coasts	Sandy or Muddy Flats	Horizontal Rock Platforms
Supralittoral	*Calothrix* *Melaraphe*	*Melaraphe* *Lichina* *Verrucaria*	*Talorchestia*	*Melaraphe* *Lichina*	Relatively bare when rocky Samphires	*Melaraphe* *Lichina* *Verrucaria*
Upper littoral	(*Chthamalus*) *Chamaesipho*	*Chthamalus* *Chamaesipho*	Isopods	*Chthamalus* *Chamaesipho*	Samphires, mangroves, or *Bembicium*	
Mid littoral	*Catophragmus* (Molluscs, *Galeolaria*, blue-green algae)	(*Catophragmus*) Molluscs, *Galeolaria*, blue-green algae *Brachyodontes rostratus*		Molluscs, *Galeolaria*, blue-green algae	*Bembicium*, *Modiolus inconstans*, mangroves, *Enteromorpha*	Molluscs, *Galeolaria*, blue-green algae
Lower littoral	*Balanus* – coralline mat	*Hormosira* Other algae	Bivalves	*Corallina* – *Gelidium*, *Hormosira*	*Hormosira* *Brachyodontes erosus*, *Katelysia*, *Pinna*	*Hormosira*
Sublittoral fringe	*Cystophora intermedia*	*Cystophora intermedia*		*Cystophora* sp. ↓	*Hypnea*, *Spyridia*, and other algae	*Durvillea* ↓
Upper sublittoral	Fucoid algae etc.	Fucoid algae, *Ecklonia*, red algae	Bivalves	Other algae	*Zostera* ↓ *Posidonia* ↓	*Macrocystis* ↓

Figure 17: Summary of intertidal zonation on South Australian coasts (from Womersley and Edmonds)

The algal vegetation of South Africa was analysed by LEVYNS (1929) and ISAAC (1937). In the littoral zone they found an association of *Porphyra capensis*, below which there is a bare belt, followed by *Chaetangium (C.saccatum* and *C.ornatum)*. In the lower part of the littoral zone there follow belts of *Iridaea capensis*, *Bifurcaria brassiciformis*, and finally *Champia lumbricalis*. In the sublittoral zone several Laminariaceae *(Ecklonia maxima*, *Macrocystis pyrifera* and *Laminaria pallida)* are especially prominent (Pl. V: 10).

Our knowledge of the algal vegetation on tropical coasts is still highly incomplete. In his works on Ceylon SVEDELIUS (1906) correlates the periodicity of the algal vegetation with the monsoon. In his studies of Samoha and Tahiti, SETCHELL (1924, 1926) deals especially with the vegetation of the coral reefs.

According to the author's own investigations on coral reefs on the east coast of Africa, the reefs, of which large parts are uncovered at low-tide, are covered with a relatively mixed vegetation: species of *Ulva*, *Chaetomorpha*, *Caulerpa*, and *Halimeda; Struvea anasto-*

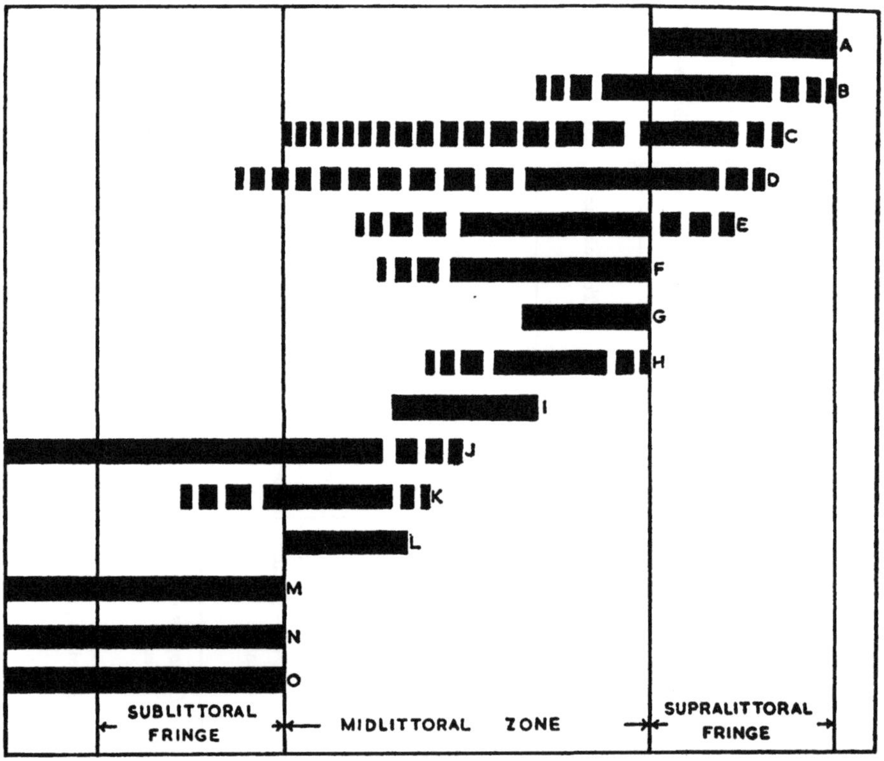

Figure 18: Diagram showing the zones occupied by some of the commoner intertidal animals and plants of the Sierra Leone Colony peninsula: A *Tactarius granosus*, B *Littorina punctata*, C *Bostrichia binderi*, D *Lyngbya confervoides*, E *Nerita senegalensis*, F *Chthamalus dentatus*, G *Bostrichia radicans*, H *Siphonaria pectinata*, I *Ostrea sp.*, J *Lithothamnia*, K *Gelidium pusillum*, L *Murrayella periclados*, M *Sargassum sp.* (?) *vulgare*, N *Caulerpa racemosa*, O *Padina sp.* (from LAWSON)

mosus, Boergesenia forbesii, species of *Sargassum, Turbinaria, Cystoseira,* and *Padina; Colpomenia sinuosa, Gracilaria, Hypnea, Gelidium,* Corallinaceae etc. The phanerogams *Cymodocea ciliata* and *Diplanthera uninervis* also occur there (Pl. VI: 11).

The vegetation of the African west coast has been investigated by LAWSON (1955, 1956, 1957 a–b). He mentions, *i. a.:* associations of *Bostrychia tenella* and of *Enteromorpha, Pylaiella fulvescens, Centroceras, Ulva, Lithothamnia, Gigartina acicularis, Hypnea, Gracilaria, Padina, Sargassum,* and *Dictyopteris.*

According to FELDMANN and LAMI (1937) for the West Indies (Guadeloupe) associations of *Bostrychia* and *Ralfsia expansa* are found in the supra-littoral and littoral zones. In the sublittoral zone in exposed situations associations of *Sargassum* and *Turbinaria* are predominant, while in more or less sheltered situations various *Dictyotales, Galaxaura* etc. are found. Corallinaceae are found generally everywhere. Of great interest are the descriptions given by TAYLOR (1960) on the composition of the vegetation in the lower parts of the sublittoral zone (Bermudas). In a depth below 50 metres the Chlorophyceae are predominant. Thus in a depth of 90 metres out of 23 species twelve were found to be Chlorophyceae, one to two are Phaeophyceae and nine Rhodophyceae. An explanation of this phenomenon seems to be the fact that the optical properties of these waters permit a high transmission of blue light, which can be utilized by the green algae.

A vegetation which is peculiar to tropical and near tropical regions is that of the mangrove swamps, composed of species of *Avicennia* and *Rhizophora.* The algae found on the roots and pneumatophores of the mangroves are especially characteristic: species of *Catenella, Caloglossa,* and *Bostrychia* and also *Chlorophyceae* (BÖRGESEN, 1911, FELDMANN and LAMI, 1936, TAYLOR, 1960).

Regions where tidal activity is absent or only negligible must be regarded as exceptional; as, for example, the coasts of the Kattegat and the Skagerrak, the Baltic Sea and the Mediterranean. Although certain changes in water level are recorded in such areas, they occur more or less irregularly and amount to no more than approximately 30 centimetres.

The vegetation zones of the Mediterranean have been described in detail by FELDMANN (1937). The supra-littoral zone contains belts of *Verrucaria maura* and certain Cyanophyceae amongst others, giving a black colour to the rocks. On exposed rocks one may find belts of. *Risoella verruculosa,* and below them often *Tenarea tortuosa.* Other littoral species tending to form associations are *Corallina mediterranea, Ralfsia verrucosa, Scytosiphon lomentaria, Enteromorpha intestinalis,* and *Cladophora rudolphiana.* These belts correspond to the Fucaceae on the west coasts of Europe. Noteworthy also is the occurrence of *Fucus virsoides* (probably identical with *Fucus spiralis)* in the central parts of the Adriatic Sea, where some tidal activity can be observed. As in the north-west of Europe, those algae grow in the upper parts of the littoral zone.

The upper part of the sublittoral zone in the Mediterranean is characterized by different species of *Cystoseira.* Other components of the vegetation worth mentioning are *Dictyopteris membranacea, Padina pavonia, Cladostephus verticillatus, Ulva* etc. The lower part of the sublittoral zone, reaching far down, is occupied by certain species of *Cystoseira* and *Sargassum,* calcarious algae etc. A vegetation consisting of the phanerogams *Zostera nana, Posidonia, Cymodocea nodosa* and of the Chlorophyceae *Caulerpa prolifera* occurs on soft bottom in the upper part of the sublittoral zone.

Ercegović (1932, 1934) distinguishes various belts of epilithic and endolithic Cyanophyceae growing on limestone on the Yugoslavian coast.

Changes in water level recorded on the Swedish west coast are approximately 50 centimetres. This zone, which is subject to more or less irregular exposure, is generally believed to correspond to the littoral zone on the Atlantic coasts of Europe. The vegetation and its zonation on the shores of the Skagerrak and the Kattegat has been described by several authors: Kylin (1907, 1918), Gran (1896), Sjoestedt (1928), Gislén (1930), Levring (1935, 1940, 1959) etc. Zonation of marine lichens has been reviewed by Eliasson (1965). All in all they are the same as found on the Atlantic coasts. However, they are more compressed and compact, and certain algae like *Pelvetia* do not occur there at all. The profiles given in Table 13 will convey an idea of these vegetation zones.

The algal vegetation of the Baltic was investigated particularly by Reinke (1889), Lakowitz (1907, 1929), Svedelius 1901), Du Rietz (1925, 1930, 1932), Levring (1940), and Waern (1952). It should be mentioned that there is a long period of low water in spring in that region. Therefore, the littoral zone contains no perennials, but only annuals which thrive during the summer months. Du Rietz, (l.c.) therefore, calls this zone the "Stufe der sommerannuellen Fadenalgen". In the upper part of the sublittoral zone there are associations of *Fucus vesiculosus* and *Pylaiella*. *Furcellaria*, *Polysiphonia nigrescens*, *Rhodomela subfusca* etc. are particularly characteristic of the lower parts of the sublittoral zone.

The vertical distribution and range of algal zones are, of course determined by several factors. Of these the tide (or changes in the water level due to other causes), exposure, influences caused by ice and temperature, and submarine illumination ought to be the main factors. Moreover there are still additional factors such as the salinity in areas where this is changing with different depths, and also environmental competition of various species of algae amongst themselves. It is quite obvious that there are differences in the precise boundaries between corresponding zones in different geographical areas. The differences in the terminology used often appears to be even greater. However, it seems to be fairly general to distinguish the following 4 biological zones:

1) Littorinoids and blackening organisms (*Verrucaria*, *Calothrix* etc.).
2) Fucoids of various kinds, barnacles, limpets, and other mid shore species.
3) Laminarians, certain Rhodophyta, corals etc.
4) Organisms, which are always submerged.

Table 13

Profiles

Profile No. 1

Norway, area around Bergen: Hetlö 16/7, 1935. Sheltered.

70– 25 cm *Verrucaria maura*
25– 0 cm *Pelvetia canaliculata, Verrucaria maura*
0– 10 cm No vegetation. *Balanus balanoides*, 0–90 cm
10– 30 cm *Fucus spiralis*
30– 70 cm *Fucus vesiculosus, Ascophyllum nodosum, Hildenbrandtia*
75–140 cm *Fucus serratus, Hildenbrandtia, Lithothamnion lenormandii.* From 90 cm: *Gigartina stellata* and *Cladophora rupestris*
140 cm *Laminaria digitata* with *Chorda filum, Halidrys siliquosa, Chordaria flagelliformis, Dictyosiphon foeniculaceus, Ectocarpus siliculosus, Rhodymenia palmata, Ceramium sp.*

Profile No. 2

Norway, area around Bergen: Ljösö 17/7, 1935. Exposed.

200– 0 cm *Verrucaria maura*
0– 10 cm No vegetation. *Balanus balanoides*
10– 20 cm *Fucus spiralis*
20– 70 cm *Fucus vesiculosus*
70–110 cm *Himanthalia lorea, Dumontia filiformis, Corallina officinalis, Gigartina stellata, Ceramium penicillatum, Polysiphonia brodiaei, Callithamnion arbuscula, Pylaiella rupincola, Scytosiphon lomentaria*
110–140 cm *Alaria esculenta*
140 cm *Laminaria digitata, Laminaria hyperborea*

Profile No. 3

Norway, area around Bergen: Agnö 1/8, 1935. Very exposed.

200– 0 cm *Verrucaria maura*
0– 80 cm *Balanus balanoides* with *Porphyra umbilicalis, Nemalion multifidum, Corallina officinalis, Chaetomorpha aerea, Blidingia minima*
80–115 cm *Corallina officinalis, Lithothamnion lenormandii, Nemalion multifidum, Dumontia filiformis, Gigartina stellata, Polysiphonia urceolata* and *brodiaei, Ceramium penicillatum* and *secundatum, Hildenbrandtia prototypus, Himanthalia lorea*
115–150 cm *Alaria esculenta*
150 cm *Laminaria digitata*

Profile No. 4

Swedish west coast, Kristineberg. Sheltered.

30– 0 cm *Verrucaria maura*
0– 5 cm No vegetation or *Hildenbrandtia*
5– 10 cm *Fucus spiralis, Hildenbrandtia*
10– 20 cm *Fucus vesiculosus, Hildenbrandtia*
20– 30 cm *Ascophyllum nodosum* with *Pylaiella litoralis, Hildenbrandtia, Cladophora rupestris, Enteromorpha clathrata, Ulva lactuca, Lithothamnion lenormandii*
30– 45 cm *Ascophyllum nodosum* and *Fucus serratus, Cladophora rupestris, Enteromorpha clathrata, Hildenbrandtia* and *Lithothamnion lenormandii*
45– 90 cm *Fucus serratus, Ceramium rubrum, Cladophora rupestris, Chondrus crispus, Corallina officinalis, Hildenbrandtia prototypus, Lithothamnion lenormandii, Furcellaria fastigiata*
90–140 cm *Halidrys siliquosa, Laminaria saccharina, Fucus serratus, Chondrus crispus, Corallina officinalis, Furcellaria fastigiata, Phyllophora membranifolia, Trailliella intricata*

ALEEM, A. A., Quantitative underwater study of benthic communities inhabiting kelp beds off California. – Science 123. 1956.
ANAND, P. L., An ecological study of the Algae of the British chalk-cliffs. – Journ. of Ecol., 25. 1937.
ATKINS, W. R. G., The influence upon algal cells of an alternation in the Hydrogen ion concentration of sea water. – Journ. Marine Biol. Ass. Plymouth, 12. 1922.
ATKINS, W. R. G., Quantitative considerations of some factors concerned in plant growth in water. Pt. I & II. – Journ. Cons. perm. pour l'Expl. de la Mer., 1. 1926.
ATKINS, W. R. G., Illumination in algal habitats. Bot. Notiser 1939.
BAKER, S. M., On the brown seaweeds of the salt-marsh. – Journ. Linn. Soc. London, Bot., 40. 1911.
BAKER, S. M. and BOHLING, M. H., On the brown algae of the salt marsh. Pt. II. – Journ. Linn. Soc. London, Bot., 43. 1915.
BEAUCHAMP, P. M. DE, Les grèves de Roscoff. Paris. 1914.
BELL, H. P., and MACFARLANE, C., The marine algae of the Maritimes Provinces of Canada. – Canad. Journ. of Research, 9. 1933.
BERGQUIST, P. L., Notes on the marine algal ecology of some exposed rocky shores of Northland, New Zealand. – Botanica Marina. 1. 1933
BERNER. L., Contribution a l'etude sociologique des Algues marines du golfe de Marseille. – Ann. Mus. Hist. nat. Marseille, 24. 1931.
BERNER, L., Sur l'épiphytisme chez Digenea simplex (Wulf.) Ag. Bull. Inst. Océanogr. Monaco. No. 606. 1932.
BERTHOLD, G., Über die Vertheilung der Algen im Golf von Neapel. – Mittheil. Zool. Stat. zu Neapel. 1882.
BIEBL, R., Ökologische und zellphysiologische Studien an Rotalgen der englischen Südküste. – Beih. Bot. Centralbl., 57A. 1937.
BIEBL, R., 1938. Trockenresistenz und osmotische Empfindlichkeit der Meeresalgen verschiedenen tiefer Standorte. – Jahrb. wiss. Bot. 86.
BIEBL, R., Temperatur und osmotische Resistenz von Meeresalgen der bretonischen Küste. – Protoplasma 50. 1958.
BÖRGESEN, F., A contribution to the knowledge of the marine algal vegetation on the coast of the Danish West Indian islands. – Bot. Tidsskr., 23. 1900.
BÖRGESEN, F., The algal vegetation of the Faeröese coasts. – Botany of the Faeröes, 3. 1905.
BÖRGESEN, F., The algal vegetation of the lagoons in the Danish West Indies. – Biolog. Arbejder Til. E. Warming. 1911.
BÖRGESEN, F., On Rosenvingea stellata, a new indian Alga, and on an interesting littoral algal vegetation in which this species is a characteristic constituent. – Dansk Bot. Ark., 5 (No. 2). 1928.
BÖRGESEN, F. and JÓNSSON, H., The distribution of the marine algae of the Arctic sea and the northern-most part of the Atlantic. – Botany of the Faeröes, Bd. 3, Appendix. 1905.
BROWN, L. B., Experiment with marine Algae in fresh-water. – Publ. Puget Sound Biol. Station, 1. 1915.
BURROWS, E. and LODGE, S., Autecology and the species problem in Fucus. – Journ. mar. biol. Ass. U. K., 30. 1951.
BURROWS, E., CONWAY, E., LODGE, S. and POWEL, H. T., The raising of intertidal algal zones on Fair Isle. – Journ. of Ecology, 42. 1954.
CHAPMAN, V. J., Studies in salt-marsh ecology. I–III. – Ibid. 26. 1938.
CHAPMAN, V. J., Studies in salt-marsh ecology. IV and V. – Ibid., 27. 1939.
CHAPMAN, V. J., Studies in salt-marsh ecology. VI and VII. Ibid. 28. 1940.
CHAPMAN, V. J., Studies in salt-marsh ecology. VIII. Ibid. 29. 1941.
CHAPMAN, V. J., Zonation of marine Algae on the sea-shore. – Proc. Linn. Soc. London. Sess., 154. 1943.
CHAPMANN, V. J., Marine Algal Ecology. – Bot. Review, 12. 1946.
COTTON, A. D., Clare Island survey. Pt. 15, Marine Algae. – Proc. Roy. Irish Acad., 31. 1912.
CRANWELL, L. M., and MOORE, L. B., Intertidal communities of Poor Knight Islands. – Trans. & Proc. Roy. Soc. New Zealand, 67. 1938.
CRIBB, A., The algal vegetation of Port Arthur, Tasmania. – Papers Proc. Roy. Soc. Tasmania 88. 1954.
DAMANT, G. C., Storage of oxygen in the bladders of the seaweeds Ascophyllum nodosum and their adaptation to hydrostatic pressure. – Journ. Exper. Biol., 14. 1937.
DAVIS, B. M., General characteristics of the algal vegetation of Buzzards Bay and Vineyard Sound in the vicinity of Woods Hole. – Bull. U. S. Bureau of Fisheries, 31, pt. 1. 1913.
DAVY DE VIRVILLE, A., Recherches écologiques sur la flore des flaques du littoral de l'Océan Atlantique et de la Manche. – Rev. gén. Bot., 46, 47. 1934–1935.
DAVY DE VIRVILLE, A., Les zones de végétation sur le littoral atlantique.– Soc. Biogéogr., 7. 1940.
DOTY, M. S., Critical tide factors that are correlated with the vertical distribution of marine Algae and other organisms along the Pacific coast. – Ecology, 27. 1946.
DOTY, M. S., Rocky intertidal surfaces. – Memoirs. Geol. Soc. America. 1. 1957.
DU RIETZ, E., Die Hauptzüge der Vegetation der Insel Jungfrun. Sv. Bot. Tidskr. 19. 1925.

Du Rietz, E., Algbälten och vattenståndsväxlingar vid svenska Östersjökusten. – Bot. Not. 1930.
Du Rietz, G. E., Zur Vegetationsökologie der ostschwedischen Küstenfelsen. – Beih. Bot. Cebtralbl., 49, Erg. Bd. 1932.
Du Rietz, G. E., Das limnologisch-thalassologische Vegetationsstufensystem. – Verhandl. int. Ver. f. teor. und angew. Limnologie, 9. 1940.
Emerson, R., The relation between maximum rate of photosynthesis and concentration of chlorophyll. – Journ. gener. physiol., 12, 1929.
Engelmann, Th. W., Über Assimilation von Haematococcus. – Bot. Zeitung 40. Leipzig 1882.
Engelmann, Th. W., Farbe und Assimilation. – Ibid. 41. 1883.
Engelmann, Th. W., Farbe und Assimilation. – quantitativen Beziehungen zwischen Absorbtion des Lichtes und Assimilation in Pflanzencellen. Ibid. 42. 1884.
Ercegović, A., Etudes écologiques et sociologiques des Cyanophycées lithophytes de la cote yougoslave de l'Adratique. – Bull. int. Acad. yougosl. des sc. et arts., 26. 1932.
Ercegović, A., Wellengang und Lithophytenzone an der ostadriatischen Küste. – Acta adriat. Inst. biol. ocean. Split., 3. 1934.
Feldmann, J., Sur quelques Cyanophycées vivant dans le tissu des Eponges de Banyuls. – Arch. Zool. exp. et gén., 75. 1933.
Feldmann, J., Recherches sur la végétation marine de la Méditerranée. La côte des Albères. – Rev. Algologique, 10. 1937.
Feldmann, J., La végétation benthique de la Méditerranée. – Soc. Biogéogr., 7. 1940.
Feldmann, J., and Davy de Virville, A., Les conditions physiques et la végétation des flaques littorales de la côte des Albères. – Rev. gèn. Bot., 45. 1933.
Feldmann, J., and Lami, R., Sur la végétation de la Mangrove à la Guadeloupe. – Compt. Rend. Acad. Sci. Paris, 203. 1936.
Feldmann, J., and Lami, R., Sur la végétation marine de la Guadeloupe. – Ibid., 204. 1937.
Fischer, Ed., Recherches de bionomie et d'océanographie littorale sur la Rance et le littoral de la Manche. – Ann. Inst. océanogr., N. S., 5. 1929.
Fischer-Piette, E., Sur quelques progrés récents et sur les méthodes et tendances actuelles en bionomie intercotidale. – Soc. Biogéogr., 7. 1940.
Föyn, B., Specific differences between northern and southern European populations of the green alga *Ulva lactuca* L. – Pubbl. Staz. Zool. Napoli, 27. 1955.
French, C. S. and Youngl V. K., The fluorescence spectra of red algae and the transfer of energy from phycoerythrin to phycocyanin and chlorophyll. – Journ. Gen. Physiol, 35. 1952.
Funk, G., Die Algenvegetation des Golfs von Neapel. Publ. della Staz, zool. die Napoli 7, Suppl. 1927.
Gabrielsen, E. K., Einfluß der Lichtfaktoren auf die Kohlensäureassimilation der Laubblätter. – Dansk bot. Archiv, 10. 1940.
Gaidukov, N., Über den Einfluß farbigen Lichts auf die Färbung lebender Oscillatorien. – Abhandl. preuß. Akad. d. Wissensch., 1902.
Gaidukov, N., Weitere Untersuchungen über den Einfluß farbigen Lichtes auf die Färbung der Oscillarien. – Ber. deutsch. bot Ges., 21. 1903.
Gaidukov, N., Die Farbenveränderung bei den Prozessen der komplementären chromatischen Adaption. – Ibid. 1903.
Gaidukov, N., Zur Farbenanalyse der Algen. – Ibid. 22. 1904.
Gaidukov, N., Die Farbe der Algen und des Wassers. – Hedwigia 43. 1904.
Gaidukov, N., Über komplementäre chromatische Adaption bei Porphyra und Phormidium. – Ber. deutsch. bot. Ges. 24. 1906.
Gail, F. W., Some experiments with Fucus to determine the factors controlling its vertical distribution. – Publ. Puget Sound Biol. Station, 2. 1918.
Gail, F. W., Hydrogen ion concentration and other factors affecting the distribution of Fucus. – Ibid., 2. 1920.
Gail, F. W., Photosynthesis in some of the red and brown Algae as related to depth and light. – Ibid., 3. 1922.
Gain, L., La flore algologique des régions antarctiques et subantarctiques. Deux. exped. antarct. francaise 1908–1910, 7. 1912.
Gislen, T., Epibioses of the Gullmar Fjord. Skriftserie K. Svensk. Vetenskapsakad. Kristinebergs zool. stat. 1877–1927, No. 3, 4. 1930.
Gislen, T., Physiographical and ecological investigations concerning the littoral of the Northern Pacific. I. A comparison between the life conditions on the littoral of central Japan and California. Lunds Univ. Årskr. N. F. 39 (No. 2). 1943.
Gislen, T., Ibid., 40 (No. 2). 1944.
Goor, A. C. J. van, Die holländischen Meeresalgen (Rhodophyceae und Chlorophyceae) insbesondere der Umgebung von Helder, des Wattenmeeres und der Zuidersee. – Verh. Kon. Akad. Wetensch. Amsterdam. 23, 2. 1923.
Gran, H. H., Kristianiafjordens algeflora. I. Rhodophyceae og Phaeophyceae. – Videnskabsselsk. Skrifter. I. Math. nat. Kl 2. 1896.
Grubb, V. M., Marine algal ecology and the exposure factor at Peveril Point, Dorset. – Journ. Ecology, 34. 1936.
Hamel, G., Sur les Chlorophycees de la côte basque francaise et leur repartition en ceintures. – Blumea, suppl. 2. 1942.
Hartog, C. den, The epilithic algal communities occurring along the coast of the Netherlands. Wentia I. 1959.

Harvey, H. W., The chemistry and fertility of sea waters. Cambridge. 1955.

Hatton, H., Essais de bionomie explicative sur quelques especes intercotidales d'Algues et d'Animaux. – Ann. Inst. Oceanogr., N. S., 17. 1938.

Hoefler, K., Hypotonietod und osmotische Resistenz einiger Rotalgen. – Österr. Bot. Zeitschr. 80. 1931.

Hoefler, K., Vergleichende Protoplasmatik. – Ber. d. Dt. Bot. Ges. 50. 1932.

Hoffmann, C., Die Atmung der Meeresalgen und ihre Beziehung zum Salzgehalt. – Jahrb. wissenschaftl. Bot. 71. 1929.

Howe, M. A., Plants that form reefs and islands. – Scient. Monthly, 36. 1933.

Hoyt, W. D., The periodic fruiting of Dictyota and its relation to the environment. – Amer. Jour. Bot. 14. 1927.

Hygen, G., Fotoperiodiske reaksjoner hos alger. – Blyttia 7. 1948.

Isaac, W. E., Some observations and experiments on the drought resistance of Pelvetia canaliculata. – Ann. Bot. 47. 1933.

Isaac, W. E., Studies of South African seaweed vegetation. I. West coast from Lambert Bay to the Cape of Good Hope. – Trans. Roy. Soc., South Africa, 25. 1937.

Jerlov, N. G., Optical studies of ocean waters. – Rep. Swedisch Deep-Sea Exp. 1947–48, Vol. III. Fasc. 1. 1951.

Johnson, D. S. and York, H. H., The relation of Plants to tide-levels. A study of factors affecting the distribution of marine Algae. – Carnegie Inst. of Washington. Publ. No. 206. 1915.

Johnson, D. S., and Skutch, A. F., Littoral vegetation on a headland of Desert Island, Maine. I. Submersible or strictly littoral vegetation. II. Tide pools. III. Adlittoral or non-sub-mersible vegetation. Ecology, 9. 1928.

Johnson, N. G., and Kullenberg, B., On radiant energy measurements in the sea. – Svenska hydr.- biol. komm. skr., Ser. 3. 1. 1946.

Jónsson, H., The marine algal vegetation of Iceland. -Botany of Iceland. Pt. I, Copenhagen. 1912.

Jorde, I., and Klavestad, N., The natural history of the Hardangerfjord. 4. The benthonic algal vegetation. – Sarsia 9. 1963.

Kain, J., Direct observations on some Manx sublitoral algae. – J. Mar. Biol. Ass. U.K. 39. 1960.

Kain, J., Aspects of the biology of Laminaria hyperborea 1. Vertical distribution. – Ibid. 42. 1962.

Kalle, K., Zum Probleme der Meerwasserfarbe. – Ann. d. Hydr. u. mar. Meteor. 66. 1938.

Kjellman, F. R., Bidrag till kännedomen af Kariska hafvets Algvegetation. Öfvers. af K. Vet. – Akad. Förh. 1877. No. 2. 1877.

Kylin, H., Studien über die Algenflora der schwedischen Westküste. Upsala. 1907.

Kylin, H., Über die Kälteresistenz der Meeresalgen. – Ber. Deutsch. bot. Ges., 35. 1917.

Kylin, H., Svenska västkustens algregioner. – Svensk Bot. Tidskr., 12. 1918.

Knight, M., and Parke, M. W., Manx Algae. – Mem. Liverpool Marine Biol. Comm., 30. 1931.

Lakowitz, K., Die Algenflora der Danziger Bucht. Ein Beitrag zur Kenntnis der Ostseeflora. Danzig 1907.

Lakowitz, K., Die Algenflora der gesamten Ostsee. – Danzig. 1929.

Lami, R., Le Fucus lutarius Kutz., dans ses stations francaise de la Manche occidentale. – Trav. crypt. déd. à L. Mangin, Paris. 1931.

Lami, R., Sur les conditions d'elairement et d'hygrometrie necessaires à quelques Algues cavernicoles dans les grottes de la region malouine. – Bull. Labor. marit. Dinard., 22. 1940.

Lawson, G. W., Seasonal variation of intertidal zonation on the coast of Ghana in relations to tidal factors. – Journ. Ecol. 45. 1957.

Lawson, G. W., Some features of the intertidal ecology of Sierra Leone. – Journ. West African Sci. Ass. 3. 1957.

Lewis, J. R., The ecology of rocky shores. – London. 1964.

Lewis, J. R., The litoral fringe on rocky coasts of Southern Norway and Western Sweden. – Botanica Gothoburgensia. 3. 1965.

Levring, T., Zur Kenntnis der Algenflora von Kullen an der schwedischen Westküste. – Lunds Univ. Årsskr. N. F. Avd. 2. 31. 1935.

Levring, T., Zur Kenntnis der Algenflora der Norwegischen Westküste. – Lunds Univ. Årskr. N. F., 33. 1937.

Levring, T., Studien über die Algenvegetation von Blekinge Südschweden. – Lund. 1940.

Levring, T., Some culture experiments with Ulva and artificial seawater. – Fysiogr. Sällsk. förhandl. Lund. 16. 1946.

Levring, T., Submarine daylight and photosynthesis of marine algae. – K. Vet. o. Vitt. S. handl. Göteborg IV Ser. B. 5. 1947.

Levring, T., Some modern aspects of growth and reproduction in marine algae in different regions. – Ann. Biol. 33. 1957.

Levring, T., Distribution verticale de la végétation algale marine specialement en Scandinavie. – Colloque Intern. C.N.R.S., 81, Paris. 1959.

Levring, T., Photosynthesis of some marine algae in clear, tropical oceanic water. – Botanica Marina. 1968.

Levyns, M. R., Sea weeds of the Cape Peninsula. Botanical features of the South Western Cape province. Capetown. 1929.

Lorenz, R., Physikalische Verhältnisse und Vertheilung der Organismen in Quarnerischen Golfe. Wien. 1863.

Lund, S., The marine algae of East Greenland I–II. Medd. om Grønland. 156. 1959.

LUBIMENKO, V., La biologie de la photosynthèse. – Rev. gen. bot. 40. 1928.

LUBIMENKO, V. et TIKHOVSKAIA, Z., Recherches sur la photosynthèse et l'adaptation chromatique chez les algues marines. – Trudy Sevastopol. biologic. stancii. 1. Leningrad. 1929.

MCFARLANE, W. N. and PRESCOTT, J., Standing crop, chlorophyll content and in situ metabolism of a giant kelp community in southern California. – Institute of Marine Science (Univ. Texas) 6. 1959.

MCLEAN, J. H., Sublittoral ecology of kelp beds of the open coast area near Carmel, California. – Biol. Bull. 122. 1962.

MASSART, J., Essai de geographie botanique des districts littoraux et alluviaux de la Belgique. – Rec. Inst. bot. Leo Errera, 7. 1908.

MOKYEVSKY, O. B., Geographical zonation of marine littoral types. – Limnology and Oceanography. Vol. 5. 1960.

MOLINIER, R. and PICARD, J., Parallelisme dans la repartition des peuplements terrestres et marine benthiques du bassin mediterraneen occidental. – Revue generale de botanique. 61. 1954.

MONTFORT, C., Die Trockenresistenz der Gezeitenplan und die Frage der Übereinstimmung von Standort und Vegetation. – Ber. Deutsch. Bot. Ges., 55. 1937.

MUENSCHER, W. L. C., A study of the algal association of San Juan Island. – Publ. Puget Sound Biol. Sta. 1. 1915.

MUENSCHER, W. L. C., Ability of sea-weeds to withstand desiccation. – Publ. Puget Sound Biol. Station, 1. 1915 A.

MUENSCHER, W. L. C., A study of the algal associations of San Juan Island. – Ibid. 1. 1915 B.

NEUSHUL, M., SCUBA diving studies of the vertical distribution of benthic marine plants. – Botanica Gothoburgensia. 3. 1965.

NIENBURG, W., Die festsitzenden Pflanzen der nordeuropäischen Meere. Handbuch der Seefischerei. I. Stuttgart. 1930.

NODDACK, I. and NODDACK, W., Die Häufigkeiten der Schwermetalle in Meerestieren. – Arkiv för zool. 32. 1939.

OERSTED, A. S., De regionibus marinis. København. 1844.

OLIVER, W. R., Marine litoral plant and animal communities in New Zealand. – Transact. New Zeal. Inst. 54. 1923.

OLLIVIER, G., Etude sur la flore marine de la cote d'Azur. – Ann. Inst. Oceanogr. nouv. ser., 7. 1929.

OLTMANNS, F., Über die Kultur- und Lebensbedingungen der Meeresalgen. – Jahrb. wiss. Bot. 23. 1892.

OLTMANNS, F., Morphologie und Biologie der Algen. Vol. 2. Jena. 1905.

OLTMANNS, F., Morphologie und Biologie der Algen. 2d Ed., Vol. 3. Jena. 1923.

PETERSON, O. and EKMAN, G., De hydrografiska förändringarna inom Nordsjöns och Östersjöns område under tiden 1893–1897. – K. V. A.s Handl. N. F., Bd. 29. No. 5. 1897.

POOLE, H. H. and ATKINS, W. R. G., On the penetration of light into sea-water. – Jour. Marine Biol. Assn. 41. 1926.

POST, E., Systematische und pflanzengeographische Notizen zur Bostrychia-Caloglossa Assoziation. – Rev. Algologique, 9. 1936.

PRAT, H., Les zones de vegetation et les facies des rivages de l'estuaire du Saint-Laurent, au voisinage de Trois-Pistolen. – Le Naturaliste canadien, 60. 1933.

PRAT, H., Observations bionomiques sur les rivages atlantiques de l'Amerique du Nord et les iles voisines. – Soc. Biogeogr., 7. 1940.

PROVASOLI, L., MCLAUGHLIN, J.J.A. and DROOP, M. R., The development of artificial media for marine algae. – Arch. Mikrobiol., 25. 1957.

PRUVOT, G., Essai sur les fonds et la faune de la Manche occidentale, compares a ceux du Golfe du Lion. – Arch. Zool. Exp. et Gen. 3e ser., 5. 1897.

PRINTZ, H., Die Algenvegetation des Trondhjemsfjordes. – Skr. utg. av Det Norske Vidensk. Akad. i Oslo. I. Matem. Naturv. Klasse. 1926.

RAUNKIAER, C., Types biologiques pour la géographie botanique. – Danske Videnssk. Selsk. Forhandl., 5. 1905.

REES, T. K., The marine Algae of Lough Ine. – Jour. Ecol., 23. 1935.

REES, T. K., Algal colonization at Mumbles Head. – Ibid. 28. 1940.

ROSENVINGE, L. KOLDERUP, Om Algvegetationen vid Grønlands kyster. Medd. om Grønland 20. 1898.

SAUVAGEAU, C., Sur deux Fucus recoltes à Arcachon. – Bull. Station Biol. Arcachon, 11. 1908.

SCAGEL, R. F., An investigation on marine plants near Hardy Bay, B. C. Victoria. 1947.

SCAGEL, R. F., An annotated list of the marine algae of British Columbia and northern Washington. – National Museum of Canada. Bulletin 150. 1957.

SCAGEL, R. F., The distribution of certain benthonic algae in Queen Charlotte Strait, British Columbia, in relation to some environmental factors. – Pacific Science. 15. 1961.

SERNANDER, R., De Nordeuropeiska Hafvens Växtregioner. – Svensk bot. Tidskr., 11. 1917.

SETCHELL, W. A., American Samoa. – Carnegie Inst. of Washington. Publ. No. 341. 1924.

SETCHELL, W. A., Phytogeographical notes on Tahiti. II. Marine vegetation. – Univ. of. Calif. Publ. Bot. 12. 1926.

SEYBOLD, A. and WEISSWEILER, A., Spektrophotometrische Messungen an grünen Pflanzen und an Chlorophyll-Lösungen. – Bot. Archiv, 43. 1942 a.

SEYBOLD, A. und WEISSWEILER, A., Weitere spektrophotometrische Messungen an Laubblättern und an Chlorophyll-Lösungen sowie an Meeresalgen. – Ibid. 44. 1942 b.

Shelford, V. E. and Gail, F. W., A study of light penetration into sea water made with the Kunz photoelectric cell with particular reference to the distribution of plants. – Publ. Puget Sound Biol. Sta. 3. 1922.

Sjöstedt, L. G., Litoral and supralitoral studies on the Scanian shore. – Lunds Univ. Årskr. N. F. 24 (No. 7). 1928.

Skottsberg, C., Observations on the vegetation of the Antarctic sea. – Bot. Stud. till. F. R. Kjellman. Uppsala. 1906.

Skottsberg, C., Communities of Algae in subantarctic and antarctic waters. – Kgl. Svensk. Vetensk. Ak. Handl. 19. 1941.

Skottsberg, C., Remarks on the plant geography of the southern cold temperate zone. – Proc. Roy. soc., B., 152. 1960.

Skottsberg, C., Antarctic phycology. 1964.

Smith, G. M., On the reproduction of some Pacific coast species of Ulva. – Amer. Journ. of Botany, 34. 1947.

Smith, G. M., Marine algae of the Monterey Peninsula. California Stanford. 1944.

Southward, A. J., The zonation of plants and animals on rocky sea shores. – Biol. rev., 33. 1958.

Steemann Nielsen, E., Havets Planteverden i ekologisk og produktionsbiologisk belysning. – Skr. udg. af Kommissionen for Danmarks Fiskeri- og Havundersøgelser No. 13, København. 1944.

Stephenson, T. A., and Anne, The universal features of zonation between tide-marks on rocky coasts. – Journ. Ecology 37. 1949.

Stephenson, T. A. and Anne, Life between tide marks in North America. – Ibid. 49. 1961.

Sundene, O., The algal vegetation of Oslofjord. – Skr. Norske Vidensk.-Akad. Oslo, I Mat.-Naturv. Kl. 1953.

Svedelius, N., Studier öfver Östersjöns Hafsalgflora. – Uppsala. 1901.

Svedelius, N., Über die Algenvegetation eines ceylonischen Korallenriffes mit besonderer Rücksicht auf ihre Periodizität. – Bot. Stud. till. F. R. Kjellman. Uppsala. 1906.

Svedelius, N., On the discontinuous geographical distribution of some tropical and subtropical algae. – Ark. f. bot. 19. 1924.

Sverdrup, H. U., Johnson, M. W. and Fleming, R. H., The oceans. New York. 1946.

Söderström, J., Equivalent algal zones i Western Sweden and the British Isles. – Botanica Gothoburgensia, 3. 1965.

Talling, J. F., Comparative laboratory and field studies of photosynthesis by a marine plankton diatom.–Limnology and Oceanography. 5. 1960.

Taniguti, M., Phytosociological study of marine algae in Japan. 1962.

Taylor, W. R., The marine Algae of Florida with special reference to the Dry Tortugas. – Carnegie Inst. of Washington. Publ. 379, 1928.

Taylor, W. R., Algae (in Cryptogamic flora of the Arctic). – Bot. rev., 20. 1954.

Taylor, W. R., Marine algae of the Northeastern coast of North America. – Ann. Arbor. 1957.

Taylor, W. R., Marine algae of the eastern tropical and subtropical coasts of the Americas. Ann. Arbor. 1960.

Taylor, W. R., Distribution in depth of marine algae in the Caribbean and adjacent seas. – Rec. adv. in botany, Univ. of Toronto press. 1961.

Tortonese, E., Modern views on the distribution of the marine benthos. – Hidrobiologi arastirma enstitusu yayini. Publ. Univ. Istanbul. 5 b. 1960.

Waern, M., Rocky-shore algae in the Öregrund Archipelago. – Acta Phytogeographica Suecia. 30. 1952.

Wassnik, E. C. and Kersten, J. A. H., Observation sur le spectre d'absorption et sur le rôle des caroténoides dans la photosynthèse des diatomées. – Enzymologia 12. 1946.

Wennberg, T., The distribution of certain marine algae on the Norwegian west coast. – Medd. Göteborgs Bot. Trädgård, 18. 1950.

Wilce, R. T., The marine algae of the Labrador Peninsula and Northwest Newfoundland. – National mus. Canada, 1958. Ottawa. 1959.

Wilce, R. T., Heterotrophy in arctic sublittoral seaweeds: an hypothesis. Botanica Marina, Vol. X, 1967.

Womersley, H. B. S., A critical survey of the marine algae of Southern Australia. I. Chlorophyta. – Austr. Journ. Mar. Freshwater. Res. 7. 1956.

Womersley, H. B. S. and Edmonds, S. J., A general account of the intertidal ecology of South Australian coasts. – Australian Journ. Marine and Freshw.res., 9. 1958.

Wood, R. D., Adapting SCUBA to aquatic plant ecology. – Ecology. 44. 1963.

Zaneveld, J. S., The littoral zonation of some Fucaceae in relation to desiccation. – Jour. Ecology, 25. 1937.

Zaneveld, J. S., Vertical zonation of Antarctic and Subantarctic bentaic marine algae. – Antarctic journ. of the United States. 1966.

Zaneveld, J. S., The occurrence of benthic marine algae under shore fast-ice in the western Ross Sea, Antarctica. – Proc. 5th Intern. Seaweed Symposium Halifax, Canada. 1966.

Classification of the Algae

By Tore Levring

The algae constitute no uniform or systematic group. Rather, they comprise various divisions of primitive plants of a low order which have undergone a more or less parallel evolution. Formerly a distinction was made between flagellates and algae. Flagellates were then so defined as to include coloured unicellular organisms moving by means of cilia but lacking cell walls. However, several colourless species, too, were classified among the flagellates. Several of them are closely related to the coloured species: they have only lost their photosynthetically active pigments. The coloured flagellates resemble several motile stages of certain algae (zoospores, gametes).

Already KLEBS (1893) had drawn attention to the close relationship between coloured flagellates and algae. In this century many scientists have clarified this relationship with greater precision. The flagellate-like forms occur as primitive stages in various parallel lines of development.

In the older systematic classification of algae colour was not allowed for as a feature of distinction. HARVEY (1836) was the first to distinguish between these various groups: *Chlorospermeae* (green algae), *Melanospermeae* (brown algae), and *Rhodospermeae* (red algae). These names were subsequently altered into *Chlorophyceae* (KÜTZING, 1845), *Phaeophyceae* (KJELLMAN, 1891), and *Rhodophyceae* (RUPPRECHT, 1855), respectively. HARVEY had classified the blue-green algae amongst the CHLOROPHYCEAE. The blue-green algae were called MYXOPHYCEAE by STITZENBERGER (1860) and CYANOPHYCEAE by SACHS (1874). From the grass-green algae LUTHER (1890) separated the yellow-green species as *Heterokontae*. Later they were called *Xanthophyceae* by ALLORGE (1930). The different groups of algae, according to STRAIN (1944), contain special types of chlorophyll, carotene, and xanthophyll. Their metabolism, too, differs. This is also true of the final products of photosynthesis: the Chlorophyceae produce starch, the Phaeophyceae form laminarin, while the Cyanophyceae produce cyanophycean starch, and the Rhodophyceae form rhodophycean starch etc. The ciliary mechanism of the reproductive organs of the various groups are differently formed. Other features indicate that not all groups are interrelated: they rather belong to different lines of development.

Today the algae will, in general, be classified in seven divisions:

I. Chlorophyta:	1. Class: Chlorophyceae
	2. Class: Conjugatophyceae
	3. Class: Charophyceae
II. Euglenophyta:	1. Class: Euglenophyceae
III. Chrysophyta:	1. Class: Xanthophyceae
	2. Class: Chrysophyceae
	3. Class: Bacillariophyceae
IV. Pyrrophyta:	1. Class: Desmophyceae (Desmokontae)
	2. Class: Dinophyceae
	3. Class: Cryptophyceae

V. Phaeophyta: 1. Class: Phaeophyceae
VI. Cyanophyta: 1. Class: Cyanophyceae
VII. Rhodophyta: 1. Class: Rhodophyceae
1. Subclass: Bangioideae
2. Subclass: Florideae

In the following review the mainly marine, benthic groups constitute the essential part. The fresh water forms and the phytoplankton are treated only briefly.

I. DIVISION: CHLOROPHYTA

This large and clearly defined group comprises forms of highly diverse lines of development and phases of greatly varying habit. The thallus is a unicellular or multicellular structure with green chromatophores (chloroplasts). Pigments, identical with those in higher plants, are to be found in them, of which the following have been proved to be present: chlorophyll-a and small quantities of chlorophyll-b; different variants of xantophyll and carotene (α and β). As well as in higher plants starch is produced as the final product of photosynthesis.

Motile reproductive cells (zoospores and zoogametes), generally with two or four cilia, are produced. As a rule the cilia are equal in function, of equal length, and arise from the front end of the swarmer. Sexual and asexual reproduction is found in the group, with sexual reproduction being either isogamous, anisogamous, or oogamous. The gametes are always produced in unicellular gametangia.

The green algae inhabit, above all, fresh water regions, and subsist under most varying conditions. It has been assumed for a long time that the origin of the higher green plants *(Bryophyta, Pteridophyta, Anthophyta)* should be traced back to the Chlorophyta. This assumption is in particular based on the fact that they have the same kinds of pigments and identical metabolic products.

The Division can be systematized by subdivision into three parallel classes. Some authors, however, prefer to regard the last two classes merely as orders belonging to the Chlorophyceae.

1. Class: *Chlorophyceae* or Green Algae in the strict sense.
2. Class: *Conjugatophyceae*, Conjugates
3. Class: *Charophyceae*, Stoneworts

1. Class: Chlorophyceae

These are the green algae in the strict sense of the definition, which comprise numerous forms of greatly varying degrees of organization: monadal and coccoid to trichoid, heterotrichous, siphonocladous, and siphoneous types.

The most primitive type is the monad, a motile unicell *(Chlamydomonas)* which contains chloroplasts, pyrenoids, vacuoles etc. and has two (rarely four) cilia. Of this type all stages up to the motile colonial form (e.g. *Volvox* coenobium) are found to exist. The coccoid (coccal) type comprises unicellular, non-motile forms *(Chlorococcum, Pediastrum, Hydrodictyon)*. The trichoid forms are filamentous organisms, i.e. they consist of rows of cells joined into threads of a simple or ramified structure or into bodies of a flat shape

(Ulothrix, Oedogonium, Enteromorpha, Ulva). The heterotrichous forms differ from the trichoid, as they consist of a creeping (horizontal) and a vertical filamentous system.

The siphonocladous stage is marked by large multinuclear cells. The thallus of siphoneous forms is of tubelike shape, and there are no partition walls. The protoplast contains a large number of nuclei and chloroplasts.

A cell wall will be found to exist in almost all cases. This wall is composed of two layers, with the inner layer containing cellulose, the outer pectin-like substances. The submicroscopic structure of the cell wall has been described, above all, by NICOLAI and PRESTON (1952, 1953, 1959, NICOLAI 1960). It was found that representatives of the *Cladophorales* and *Siphonales* have a cell wall structure similar to that found in higher plants. According to LEVRING (1955) only pectin-like substances within the walls of the young *Ulva*-zygotes can initially be proved to be present; cellulose, however, appears later on as little as two days. Certain forms of flagellata and quite generally zoospores and gametes, have no cell wall.

The protoplast, as in higher plants, contains one (or several) large central vacuoles, often divided by plasmatic walls. Only certain primitive forms have very small vacuoles or none at all. *(Volvocaceae, Pleurococcus, Prasiola* etc.). Contractile vacuoles occur amongst green flagellata and the coccoid forms. Certain large-cell marine variants *(Valonia, Caulerpa, Acetabularia* etc.) are suitable objects for the study of protoplasm currents. The protoplast is pressed tightly to the cell membrane and externally surrounded by a thin plasma membrane as usual.

The chloroplasts are important components of the cells. Amongst the simplest forms (e.g. *Chlamydomonadaceae)* the cells contain one cup- or piston-shaped chloroplast. Axial, asteroid chloroplasts are found in *Prasiola, Asterococcus* and others. The filamentous forms regularly have chloroplasts located on the wall. The cells either contain a plate-shaped chloroplast *(Ulothrix, Chaetophora)*, a net-like and partly open chloroplast *(Cladophora, Chaetomorpha, Oedogonium)*, or several small disc-shaped chloroplasts *(Siphonales)*. Pyrenoids are found in the chloroplasts of most of the Chlorophyceae. They consist of a central protein body surrounded by starch material. The pyrenoids, for these reasons, are considered to be concentration centres for starch. The view that they are particles of reserve protein has been proposed by a number of authors. Starch, however, can also be produced independently of the pyrenoids. Forms with no pyrenoids at all, such as *Microspora*, form starch directly in the chloroplast. In some Siphonales *(Caulerpa, Dicotomosiphon, Derbesia neglecta, Udotea, Halimeda, Penicillus, Pseudochlorodesmis)* special leucoplasts are present in addition to the chloroplasts. These algae produce starch exclusively by means of their leucoplasts. Reproduction of the pyrenoids takes place by division; however, they can also be produced anew.

The chloroplasts of the small-celled types will generally contain only one pyrenoid. In contrast the large chloroplasts of the large-celled types contain a large number of pyrenoids. In some cases the number of pyrenoids in the cell is used as a means of distinguishing species (e.g. in *Ulothrix* and *Enteromorpha)*.

The nucleus is very similar to the nuclei of higher plants. In its rest phase it is surrounded by a distinct nuclear membrane and contains one or more nucleoli which can be intensely coloured by means of nuclear stains: As far as it has become known hitherto the mitotic and meiotic division of the nuclei takes place in a manner similar to that found in higher plants.

Cilia occur in motile forms and in zoospores and zoogametes. Most of the motile forms have two *(Chlamydomonas)*, rarely four *(Carteria, Pyramimonas)* or more cilia *(Polyblepharides)*. The zoospores of the higher green algae have two or four, the gametes generally two cilia. The zoospores of *Oedogonium* and *Derbesia* are noteworthy for the coronoid arrangement of their cilia at the anterior.

The cilia always are of equal length. Their fine structure has been examined particularly by MANTON during the past few years.

Eye-spots will generally be found in zoospores and gametes, as well as in the motile forms. They are of orange to reddish colour, and are regarded as organelles for the perception of light, i.e. they direct the motion of the cell in correlation with the direction of light. These eye-spots are of a more or less complicated structure.

In some cases (e.g. *Chlamydomonas)* the eye-spot consists of a biconvex lens, a colourless photosensitive region and a pigment-disc. In the swarmers of *Cladophora* and *Ulva* the lens is located within the pigment-body.

Vegetative reproduction by fragmentation of the thallus, i.e. the disintegration into small pieces which develop into new complete individuals, occurs in various types of algae, as, for instance, in certain species of Ulothricales *(Stichococcus, Hormidium)* and in *Caulerpa* whose reproduction, in all probability, is mainly due to this process.

Another type of vegetative reproduction is the multiplication by means of akinetes developing from vegetative cells by augmentation and thickening of their membranes. They often have the appearance of typical permanent cells which can tide over periods of unfavourable conditions, such as desiccation and cold spells. Akinetes are occasionally produced in *Ulothrix*, *Microspora*, and *Cladophora*, regularly, however, in *Cladophorella*, *Pithophora*, and *Prasiola*.

Asexual Reproduction

This kind of reproduction is generally effected by zoospores similar to *Chlamydomonas*, but without cell walls. The zoospores are produced in zoosporangia which are similar to ordinary vegetative cells. Special zoosporangia are very rare. In *Oedogonium* only a single zoospore is produced, in other algae the zoospores are produced in each sporangium in powers of two (i.e. 2, 4, 8, 16, 32, 64 etc.). In certain cases *(Cladophora)* the number may be very large indeed (up to several thousands).

In most instances zoospores are liberated by an opening in the cell wall. For *Ulva* and *Enteromorpha* it has been proved (LEVRING 1955) that the zoospores are surrounded by a coat of pectin-like material. In general the zoospores are positively phototactic. After some time they become less active in their motions and become negatively phototactic, and the sedentary forms come to rest on the substratum. Aplanospores are to be found in certain genera (e.g. *Chlorococcales)*. They are of globular shape and have no cilia. The Volvocales form so-called filial coenobia. In some of the higher forms the reproductive organs are developed in certain limited parts and sections of the thallus.

Sexual Reproduction

Sexual reproduction is the rule in all groups of multicellular green algae. The gametes are very similar to the zoospores, and are produced in the same manner, with the main difference between them that the gametes are mostly somewhat smaller, and that they are produced in the mother cell in larger quantities than zoospores. The gametes are normally naked and uninucleate.

Isogamy, i.e. copulation of two equal gametes, is the most common – and at the same time the most primitive – kind of sexual reproduction. Some forms have gametes of different sizes (anisogamy) – i.e. smaller male gametes and larger female gametes. Frequently some species of a genus are isogamous, other anisogamous (e.g. *Ulva thuretii* and *lactua;* species of *Enteromorpha).*

Oogamy is found in some cases. In these the female cell (ovum) is particularly large, motionless and mostly fertilized in the oogonium. The male cell, which is motile, is termed spermatozoid.

During the process of fertilization the gametes are very active in their swarming movements. If there is a large number of them, a more or less distinct formation of groups is discernable for a short time. The gametes, which are usually positively phototactic, conjugate in pairs to form zygotes which soon become negatively phototactic and gradually slow down in their movements. They come to rest and become enveloped by a membrane. As in germinating zoospores cellulose can only be proved after a certain period *(Ulva, Enteromorpha;* cf. LEVRING op. cit.). Frequently the gametes are able to germinate without fertilization (parthenogenetically).

Normally the two nuclei of the gametes fuse immediately after copulation, in exceptional cases only shortiy before the germination of the zygote. Meiosis (reduction division) will occur very often as the first division of the zygote nuclei *(Volvox, Oedogonium, Ulothrix, Chlamydomonas* etc.).

The Life-Cycle of Chlorophyceae

Among the green algae four different types of life-cycle can be distinguished: in the simplest type the vegetative plant is haploid, while only the zygote is diploid. The first division is one of meiosis. In this process four haploid zoospores will be produced which subsequently develop into new haploid plants. This means that a change in the phase of the nucleus takes place: the haploid phase constitutes the vegetative plant producing gametes; the diploid phase is that of the zygote. Such plants – unicellular or multicellular – are called haplonts according to SVEDELIUS (1931). In addition to reproduction by means of gametes, these plants will frequently multiply by means of zoospores which are, however, not related to the regular life-cycle (accessory reproduction). For green algae this type of life-cycle is the most frequent one.

In the second type of life-cycle the vegetative plant is diploid. Reduction division takes place in the gametangia. The haploid gametes fuse into a diploid zygote which gives rise to a new, diploid plant. Plants in which such a change of the nuclear phase takes place are called diplonts (SVEDELIUS, op. cit.). Among the green algae this type of life-cycle is rare, e.g. in various Siphonales *(Bryopsis, Caulerpa, Codium)* and Siphonocladiales *(Acetabularia, Valonia).*

In the third type an alternation of generations of two morphologically equal generations takes place. One of them is haploid (gametophyte) and produces gametes. The diploid zygote originating from the copulation of the gametes germinates directly and develops into a diploid plant (sporophyte) which produces haploid zoospores. Reduction division takes place in the sporangia while the zoospores are being produced. The zoospores develop into new haploid gamophytes.

Typical representatives of this alternation of generations which, according to FRITSCH (1935, 1942), is known as an isomorphic alternation of generations, because the two generations are morphologically identical, are *Ulva*, *Enteromorpha*, *Chaetomorpha*, and *Microdictyon*.

The fourth type of life-cycle differs from the third in that way that the two generations are morphologically different, i.e. a heteromorphic alternation of generations takes place (FRITSCH, op. cit.). Characteristic representatives of this type are *Derbesia*, in which formerly the gametophyte was known as *Halicystis* (KORNMANN, 1930), *Monostroma* (KUNIEDA, 1934, SUNESON, 1947), and *Urospora* in which the sporophyte was formerly regarded as certain species of *Codiolum* (JORDE, 1933).

Systematic Classification

Some 350 genera comprising 8,000 species have been recorded, of which only ten per cent thrive in the sea. These marine species partly belong to groups which have an exclusively marine character.

The borderlines between the different species, as well as the number of subdivisions have been drawn differently at certain points by the different authors. This problem, however, will not be discussed in this study.

Fossile green algae are known particularly from the families of Dasycladaceae and Codiaceae, which had their greatest and most multifarious development of shapes and structures during the Triassic and the Jurassic Periods. Codiaceae are known to have been growing already in the early Paleozoic Period (Ordovician Epoch). From the Permian Period and the Cretaceous Period Codiaceae are preserved in good condition. During the Triassic Period they were found in abundance in certain areas of the earth.

1. Order: Volvocales (Green Flagellata)

Individual cells or colonies of cells, by means of cilia actively motile. *Palmella*-stages occur often. Asexual reproduction by means of zoospores or through the formation of

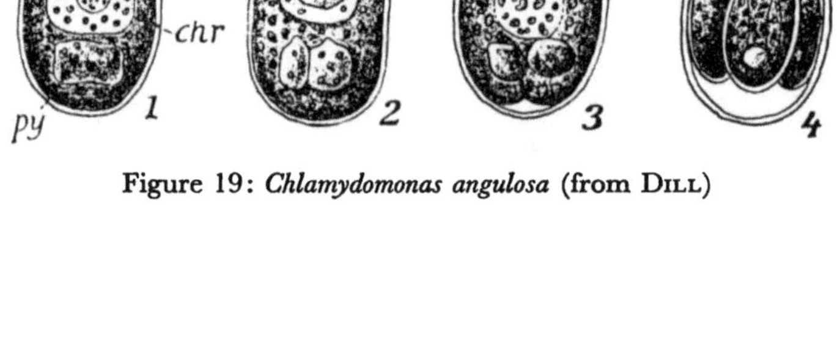

Figure 19: *Chlamydomonas angulosa* (from DILL)

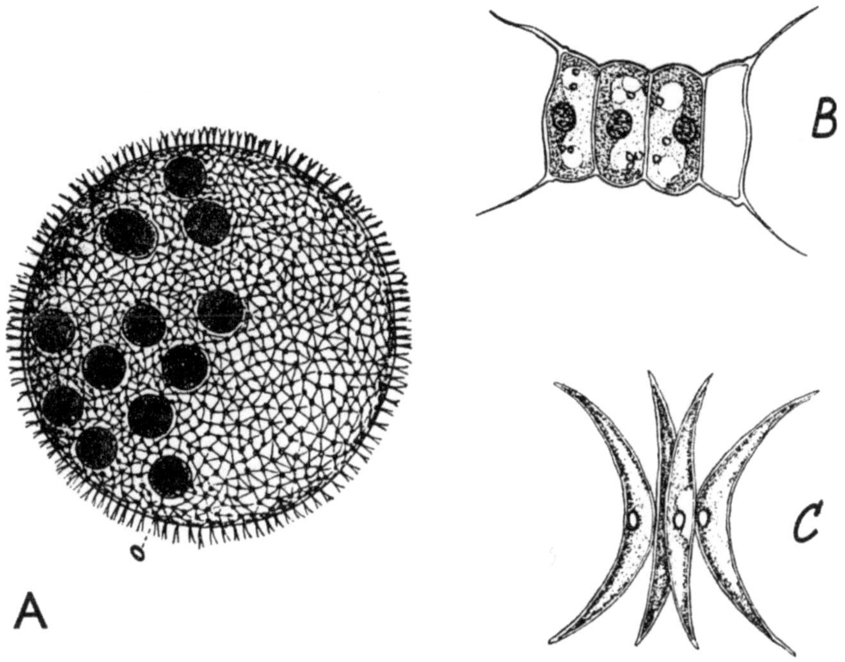

Figure 20: A. *Volvox* with oogonia (from KLEIN); B. *Scenedesmus quadricauda* and C. *S. acuminatus* (from FOTT)

motile filial colonies. Generally sexual reproduction (isogamy, anisogamy or oogamy). Approximately 90 genera are known, primarily thriving in fresh water.

Fam. CHLAMYDOMONADACEAE – forms living singly.

Fam. VOLVOCACEAE – Forming colonies.

2. Order: Tetrasporales

Normally the cells are non-motile and surrounded by a gelatinous substance. Most forms associate in amorphous colonies. Characteristic for most of them are so-called gelatine cilia (pseudo cilia), filamentous appendages which are inserted in the anterior end of the cell like the cilia serving for locomotion, but are non-motile and have no function.

The Tetrasporales constitute a group of their own, ranging between *Volvocales* and *Chlorococcales*.

1. Fam.: HYPNOMONADACEAE –
2. Fam.: CHLORANGIACEAE – *Prasinocladus* (marine species).
3. Fam.: CHAETOCHLORIDACEAE –
4. Fam.: TETRASPORACEAE – *Tetraspora, Gloeochaete.*
5. Fam.: CHARACIOSIPHONACEAE –

3. Order: Chlorococcales

Cells living solitary or in colonies, with distinct cell membrane. No cilia. Reproduction by means of zoospores or aplanospores. Sexual reproduction occurs rarely. The chloroplasts are mostly cup-shaped or mug-shaped, those of the higher types are thin, parietal and often subdivided into small integral parts.

The order comprises approximately ninety genera with seven hundred species thriving under greatly varying conditions, the majority of them, however, have their habitat in fresh water. They are completely absent in marine plankton. Many fresh-water species may be cultivated without any difficulties. Experiments with mass cultures of fast-growing species *(Chlorella, Scenedesmus)* have been made to produce technical raw materials, foods, and feeding materials. For obvious reasons these experiments have roused great interest.

1. Fam.: CHLOROCOCCACEAE – *Chlorococcum, Trebouxia, Bracteococcus.*
2. Fam.: CHLOROCHYTRIACEAE – Endophytically in marine algae, mosses, and angiosperms. Reproduction by means of 2– or 4–ciliated zoospores. *Chlorochytrium* (various marine species), *Phyllobium, Codiolum.*
3. Fam.: CHARACIACEAE – The majority thriving in fresh water. *Characium, Hydrianum, Aukyra.*
4. Fam.: HYDRODICTYACEAE – In fresh water. *Pediastrum, Sorastrum, Hydrodictyon.*
5. Fam.: MICRACTINIACEAE – *Micractinium, Acanthosphaera.*
6. Fam.: GLOECOCYSTIDACEAE – *Gloeocystis, Palmella, Palmodictyon.*
7. Fam.: RADIOCOCCACEAE – *Radiococcus.*
8. Fam.: BOTRYOCOCCACEAE – *Botryococcus.*
9. Fam.: DICTYOSPHAERIACEAE – *Dictyosphaerium.*
10. Fam.: OOCYSTACEAE – Some of Oocystaceae occur as lichen gonidia. *Chlorella, Lagerheimia, Oocystis, Glaucocystis, Tetraedron, Eremosphaera, Ankistrodesmus.*
11. Fam.: SCENEDESMACEAE – *Scenedesmus, Crucigenia, Acinastrum, Coelastrum.*
12. Fam.: PROTOSIPHONACEAE – According to various authors this family constitutes a link between the coccal and the siphonal green algae. *Protosiphon.*

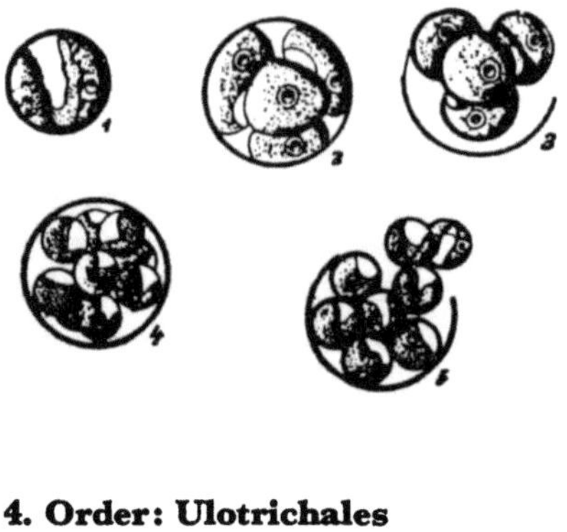

Figure 21: *Chlorella vulgaris* (from GRINTZESCO)

4. Order: Ulotrichales

Cells generally uninucleate are provided with one parietal plate-shaped chloroplast. In most of the cases cells form branched or simple filaments. Asexual reproduction by bi- or quadri-ciliate zoospores, but, aplanospores and akinetes also occur. Sexual reproduction is isogamous, anisogamous, or in certain genera also oogamous. Approximately eighty genera, comprising 450 different species have been recorded. The majority live in fresh water.

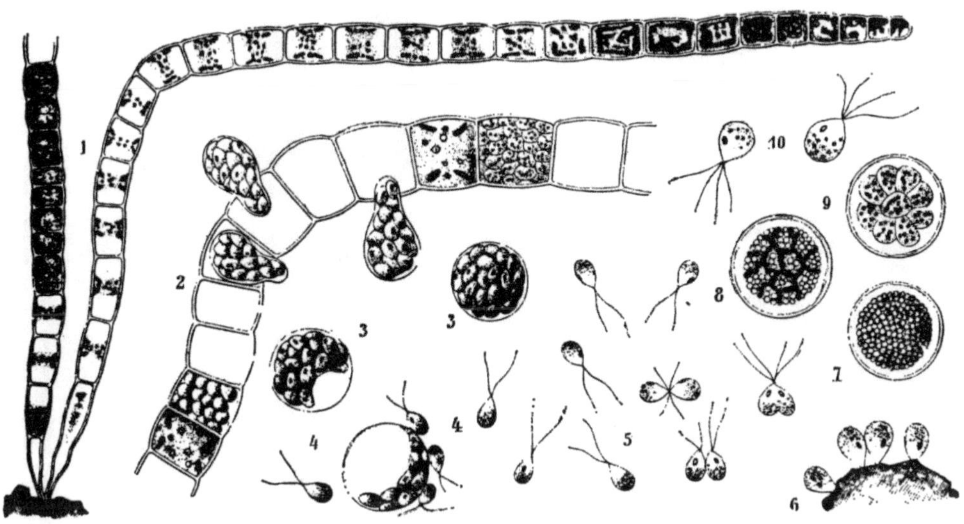

Figur 22: *Ulothrix zonata.* 1 Filaments; 2–4 formation and release of gametes; 5 gametes (before and after copulation); 6 young zygotes; 7–9 resting zygotes; 10 zoospores (from SKOTTSBERG)

1. Fam.: ULOTRICHACEAE – Unbranched, filamentous cells, uninucleate with belt like, parietal chloroplast. *Ulothrix* (some marine species, most of these species living in fresh water).
2. Fam.: MICROSPORACEAE – In fresh water. *Microspora.*
3. Fam.: CYLINDROCAPSACEAE – In fresh water only. *Cylindrocapsa.*
4. Fam.: SPHAEROPLEACEAE – *Sphaeroplea.*

5. Order: Ulvales

Cells are uninucleate and divided into two or three planes whereby a parenchymatous thallus is formed of a flat, leaf-like shape or the shape of a hollow or a solid cylinder. Asexual reproduction is effected by means of quadri- or (rarely) bi-ciliate zoospores. Sexual reproduction is isogamous or anisogamous and effected by bi-ciliate gametes. The majority of the species live in the sea, some, however, occur in brackish or fresh water.

1. Fam.: MONOSTROMACEAE – Flat thallus of monostratified structure, alternation of generations is heteromorphic (KUNIEDA, 1934; SUNESON, 1947). Gametes copulate (or in some species they develop parthenogenetically). Zygotes grow until they form quadri-ciliate zoospores developing into new gametophytes. They occur in the sea as well as in fresh water. *Monostroma.* Certain species of this family are cultivated in Japan (on nets of synthetic fibre). After being harvested they are dried and sold as a nutrient under the name of Aonori.
2. Fam.: ULVACEAE – The thallus is either flat and consists of two layers *(Ulva)* or cylindrical with a wall of one layer *(Enteromorpha, Blidingia, Capsosiphon)* or it is

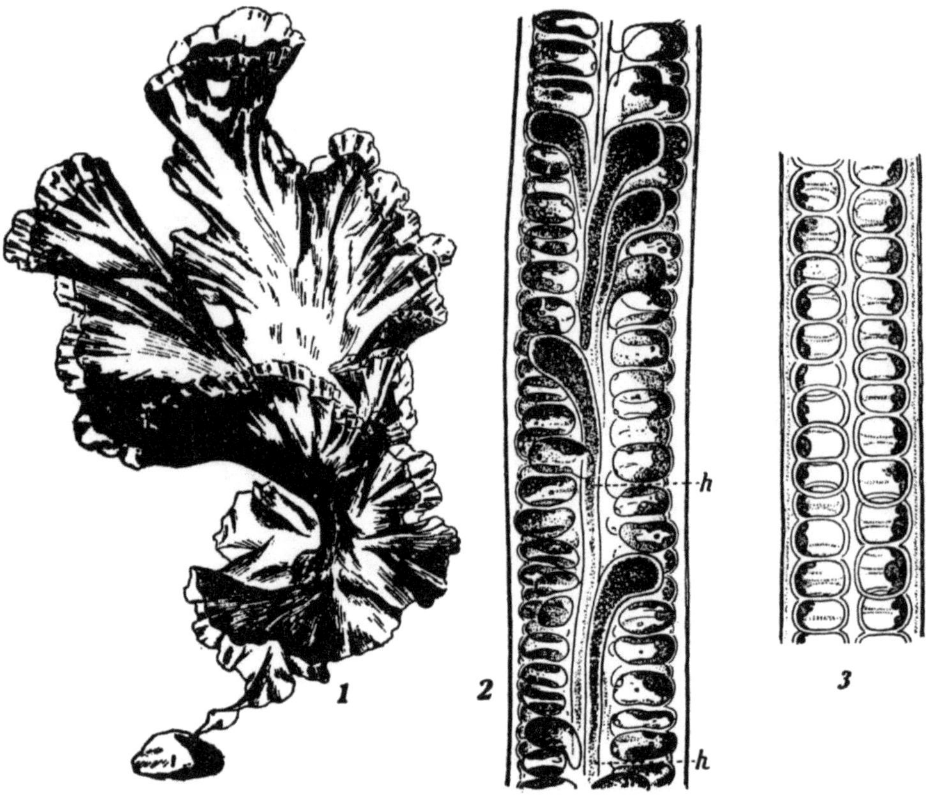

Figure 23: *Ulva lactuca*, 2–3 long. sections (from Thuret)

band-shaped, two or several cells wide *(Percursaria)*. The alternation of generations is isomorphic. *Ulva, Percursaria, Enteromorpha, Blidingia, Capsosiphon.*

3. Fam.: Schizomeridaceae – *Schizomeris* (in fresh water).

6. Order: Prasiolales

Thallus thread-like, either flat or shaped like a solid cylinder. The cells are provided with one stellate chloroplast containing one pyrenoid. Reproduction by means of akinetes or by division of cell complexes. No sexual reproduction.

Fam.: Prasiolaceae – *Prasiola* thrives on moist soils, rocks, and on sea shores.

7. Order: Chaetophorales

Thallus ramified and heterotrichous (i.e. consisting of creeping (horizontal) and vertical filamentous systems). Therefore many forms have the appearance of discs often provided with hairs. Sexual reproduction is in most of the cases isogamous, but anisogamy and oogamy also occur.

1. Fam.: CHAETOPHORACEAE – In fresh and in sea water, but also aerophytic. *Chaetophora, Draparnaldia, Stigeoclonium, Aphanochaete, Fritschiella, Protoderma, Microthamnion.*
2. Fam.: TRENTEPOHLIACEAE–Aerophytes, often of brown colour due to β-carotene. *Trentepohlia, Phycopeltis, Gongrosira.*
3. Fam.: COLEOCHAETACEAE – *Pilinia, Phaeophila, Acrochaete, Bolbocoleon, Ectochaete, Entocladia, Tellamia, Epicladia, Ochlochaete, Pringsheimiella, Pseudopringsheimia, Ulvella, Eugomontia.*
4. Fam.: PLEUROCOCCACEAE – *Pleurococcus* (typical aerophytes).

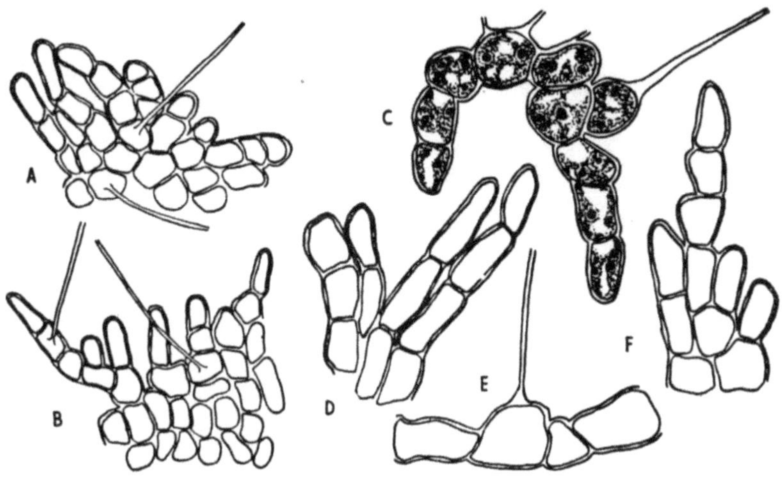

Figure 24: *Ectochaete leptochaete.* A–B x 395; C–F x 670 (from LEVRING)

8. Order: Cladophorales

Thallus thread-like, either branched or unbranched. Almost all cells are alike, large, with two nuclei or more, and a parietal, reticularly pierced chloroplast with many pyrenoids. Asexual reproduction usually by means of zoospores; sexual reproduction either isogamous or anisogamous. Normally alternation of generations is isomorphic. This order comprises only one family with approximately 350 species, some of which live in the sea, others in fresh water.

Fam.: CLADOPHORACEAE – *Urospora, Chaetomorpha, Cladophora, Aegagropila, Acrosiphonia, Spongomorpha, Pithophora, Rhizoclonium.*

9. Order: Oedogoniales

Thallus is thread-like, either branched or unbranched. Cells uninucleate, with a peculiar method of cell-division. Alternation of generations occurs. Sexual reproduction by means of oogamy. Certain species possess strongly reduced male gametophytes,

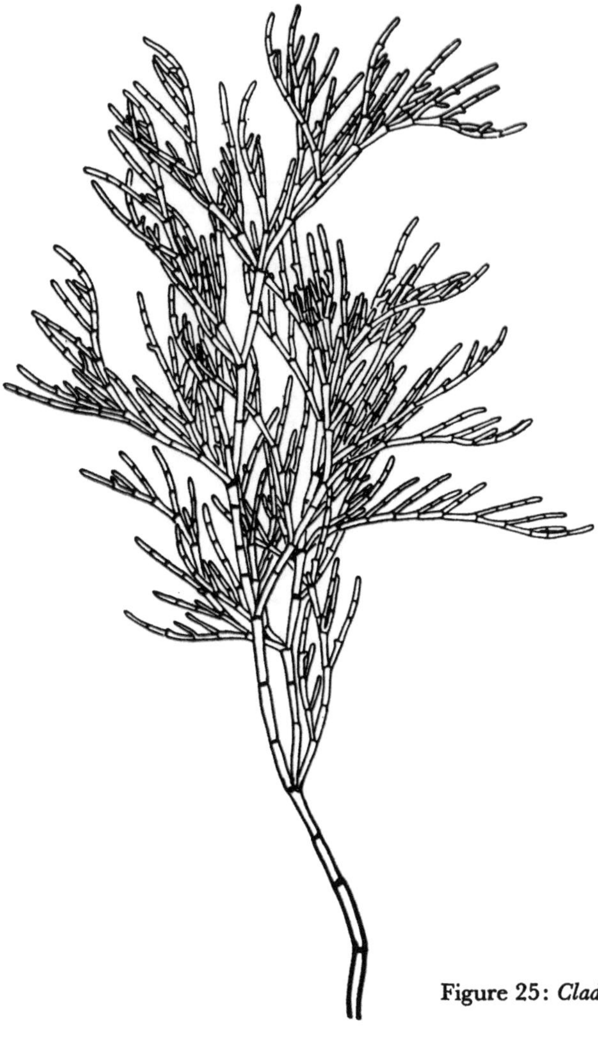

Figure 25: *Cladophora glomerata* (from SÖDERSTRÖM)

so-called dwarfed males. The zoospores and the gametes have their cilia on their anterior ends arranged in a ring.

Fam.: OEDOGONIACEAE – In fresh and brackish water: *Oedogonium, Bulbochaete, Oedocladium.*

10. Order: Siphonales

Greatly varying in shape. Thallus filamentous and tubelike (siphonal) without partition walls, so that the cell wall surrounds the whole protoplasmic mass containing many nuclei and chloroplasts (so-called coenocytes). In the most primitive cases the thallus is globular; generally, however, it consists of scarcely branched to abundantly ramified threads which very often form a kind of tomentum. Often a central pith system and a peripheral cortex system with assimilatory function can be distinguished. Reproduction by zoospores and aplanospores appears to be the exception rather than the rule. Alternation of generations is known only in isolated cases. Siphonales are generally diplonts. Sexual reproduction is either isogamous, anisogamous, or oogamous. About fifty genera comprising 500 species have been recorded, mainly from tropical and subtropical waters. Fossile forms have been found of times as early as the Ordovician Period.

1. Fam.: CHAETOSIPONACEAE – Thallus microscopic consisting of branched coenocytes. Provided with long hyaline hairs. *Blastophysa* (epiphytic on or endophytic in large marine algae or other marine plants).
2. Fam.: DERBESIACEAE – Sporophyte filamentous, simple or branched; reproduction by zoospores formed in large club-shaped zoosporangia. Zoospores provided with a ring of cilia. The sporophyte alternates with a spherical gametophyte which was

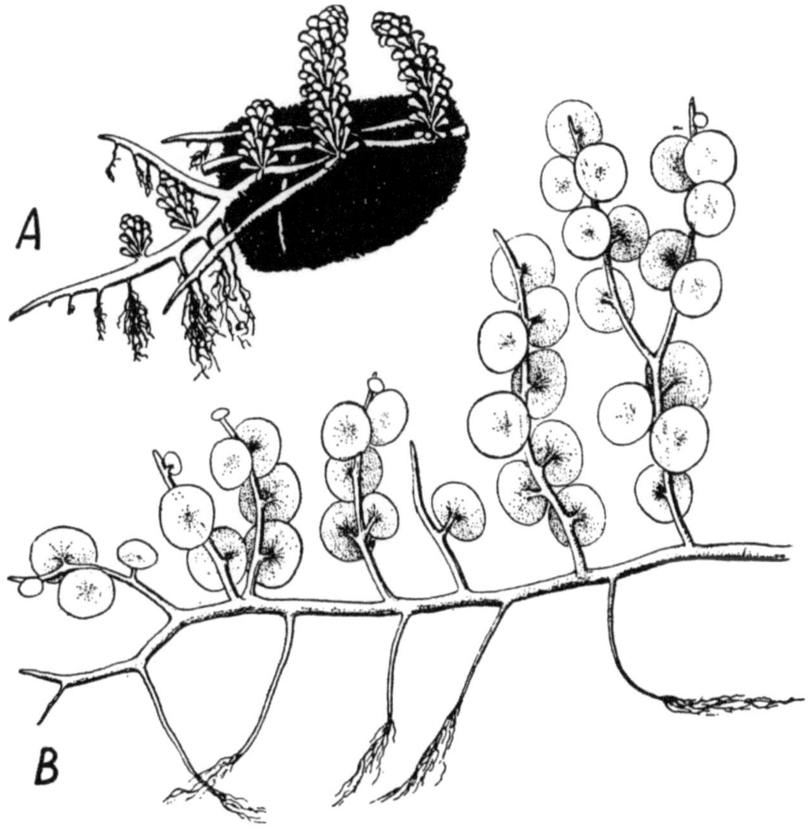

Figure 26: A. *Caulerpa racemosa;* B. *C. macrodisca* (A from BÖRGESEN; B from REINKE).

previously considered to be an independent alga, *Halicystis*. Its characteristics have been clarified in experiments carried through by KORNMANN (1938). The bi-ciliate gametes are anisogamous. *Derbesia* (approximately ten marine species).

3. Fam.: DICHOTOMOSIPHONACEAE – Thallus a *Vaucheria*-like, much bifurcated tube. Oogamous. Zoospores absent. A single genus with two species, in fresh water. *Dichotomosiphon*.
4. Fam.: CAULERPACEAE – Thallus morphologically divided into a rhizomelike main axis with basal rhizoids and erect, varied, often leaf-like assimilatory thallus lobes (assimilators). The thallus attains as a rule a length of several decimetres; it is strengthened inside by numerous cellulose cross-supports (trabeculae). In the cytoplasm there are not only chloroplasts but also leucoplasts secreting starch. Asexual reproduction occurs by fragmentation of the thallus, sexual reproduction by copulation of gametes. Reduction division takes place in gametangia. Caulerpaceae are also diplonts. Marine. A single genus, *Caulerpa*, with approximately sixty species, mostly in tropical waters.
5. Fam.: BRYOPSIDACEAE – Thallus divided into a rhizome-like, rhizoid-carrying part and into an erect, tubular main axis usually provided with numerous bilaterally or radially arranged branches. Diplobiontic; the gametangia are formed by transformation of side-branches. Gametes are bi-ciliate, and fertilization is anisogamous. Two marine genera, with about 30 species. *Bryopsis*, *Pseudobryopsis*.
6. Fam.: CODIACEAE – Thallus filamentous, much branched forming a dense, felty mass, giving the impression of a parenchymatic plant, differentiated into medulla and cortex. Some species encrusted with lime. Mostly diplobiontic. Bi-ciliate zoospores have been found in the *Halimeda*, but nothing is known of their germination and further development. Codiaceae are marine and occur mainly in warm and tropical waters. *Boodleopsis*, *Avrainvillea*, *Rhipalia*, *Cladocephalus*, *Udotea*, *Penicillus*, *Rhipocephalus*, *Halimeda*, *Pseudocodium*, *Codium*, *Chlorodesmis*.
7. Fam.: PHYLLOSIPHONACEAE – Thallus filamentous or vesicular, endophytic or endozooic. Reproduction by means of aplanospores. Three genera with some ten recorded species. *Phyllosiphon* and *Phytophysa* (endophytically in higher plants) *Ostreobium* (marine, habitat in old shells etc.).

11. Order: Siphonocladiales

Thallus, at first without partition walls (siphonal), later on subdivided into a number of multinucleate segments. Asexual reproduction by means of bi-ciliate gametes. Zoospores are found in several genera. Twenty-five genera with approximately 125 species are known. They are all marine and occur mainly in warm waters.

1. Fam.: VALONIACEAE – Thallus either vesicular or filamentous. All cells, except rhizoids, more or less similar. Some fifteen genera with approximately one hundred species are known. *Valonia*, *Valoniopsis*, *Ernodesmis*, *Siphonocladus*, *Chamaedoris*, *Dictyosphaeria*, *Petrosiphon*, *Cladophoropsis*, *Boodlea*, *Microdictyon*, *Struvea*, *Cystodictyon*, *Anadyomene*.
2. Fam.: DASYCLADACEAE – Plant consists of a long, axial cell anchored to substratum by rhizoids, bearing one or more whorls composed of simple or branched side-branches. Gametangia are formed on these branches. Often incrusted with lime. The

family is well known in the fossil record of the Ordovician Period (Lower Silurian) and also is well represented in the Triassic and Cretaceous. *Batophora, Dasycladus, Neomeris, Cymopolia, Chalmasia, Acetabularia, Acicularia.*

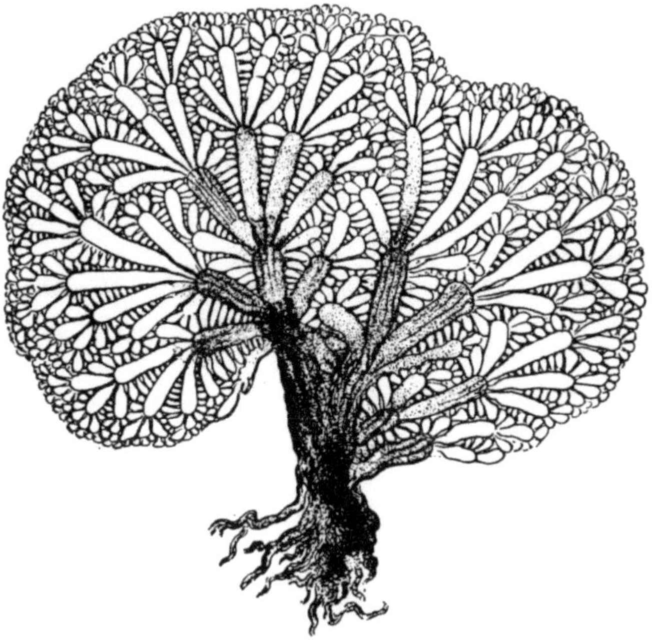

Figure 27: *Anadyomene flabellata* (from Oltmanns)

Figure 28: *Acetabularia mediterranea* (from Oltmanns)

2. Class: Conjugatophyceae

A special type of sexual reproduction, known as conjugation, distinguishes this class from that of the Chlorophyceae. In this sexual process the total content of cells is transformed into amoeboid gametes which emerge from the divided membrane and fuse in pairs or unite in copulation canals formed by the conjugating cells. Zygotes are surrounded by a thick wall for protection against low temperature, desiccation etc. Vegetative cells are haploid. The first division during the process of germination of zygotes is a meiosis. Most of the Conjugatae are unicellular with a characteristic habit that distinguishes them clearly from other green algae. Approximately four to six thousand species are known from fresh water. The class is subdivided into four orders.

1. Order: Mesotaeniales

Simple forms with cylindrical or spindle-shaped cells. *Mesotaenium*, *Cylindrocystis*, *Spirotaenia*.

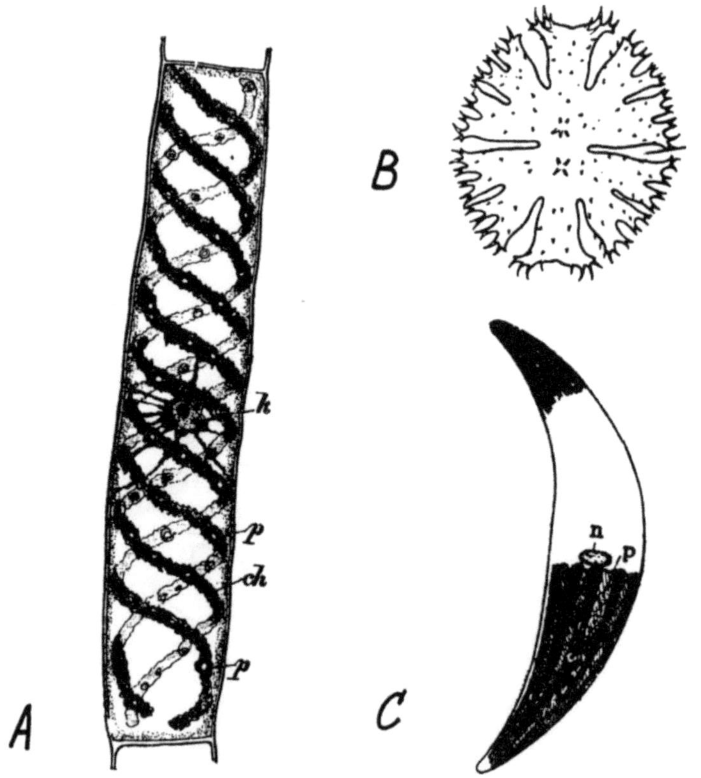

Figure 29: A *Spirogyra quinina* (from STRASBURGER); B *Micrasterias* and C *Closterium* (from FRITSCH)

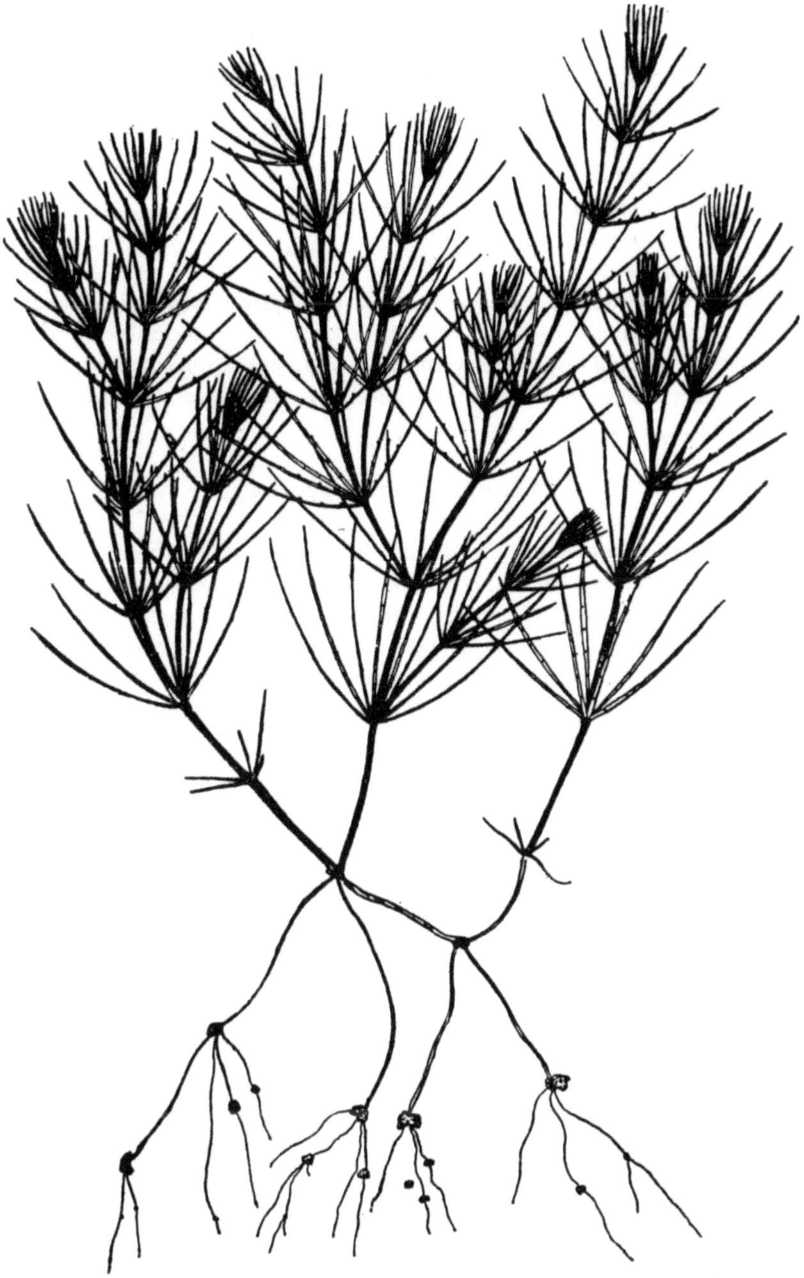

Figure 30: *Chara fragifera* (from MIGULA)

2. Order: Zygnematales

The Zygnematales consist of uniform, physiologically independent cells, forming unbranched filaments easily dissociating. The filaments are often surrounded by a gelatinous substance. *Mougeotia, Zygogonium, Spirogyra, Zygnema.*

3. Order: Gonatozygales

Cells living singly, provided with porous membrane.

4. Order: Desmidiales

Cell membrane divided into two parts, often of highly characteristic features. One separated half of the cell is older than its counterpart, due to the mode of cell division. There is only one family *(Desmidiaceae)* with twenty-six genera and approximately five thousand species. E.g., *Penium, Closterium, Pleurotaenium, Euastrum, Micrasterias, Cosmarium, Staurastrum, Desmidium.*

3. Class: Charophyceae

Thallus macroscopic, erect and, in form and size, comparable to horsetails. Characteristic for the thallus is the alternating structure of long internodal and short nodal cells. Division of the nodal cells gives rise to short vertical shoots, the so-called leaves. Cortical cells are produced by the division of the nodal cells. The thallus is often incrusted with lime. Sexual reproduction is peculiar and entirely different from that of other groups of algae. The female gametangium is bottle-shaped and contains one ovum, the oosphere. The male organ, the antheridium, is globular and of yellow to reddish colour. Spermatozoids have two cilia. The first division of zygote is a reduction division. Fossiles are known from the Devonian Period, and this group is well represented in the Jurassic and Cretacious Periods and in younger strata. Six genera with about three hundred species are on record; mainly in fresh, some in brackish water.

1. Order: Charales

Fam.: CHARACEAE – *Nitella, Tolypella, Chara.*

II. DIVISION: EUGLENOPHYTA

These have pure green chromatophores with pigmentation similar to that of Chlorophyta. Pyrenoid-like rounded bodies occur in several of these species. The final product of photosynthesis is paramylum, a polysaccharide related to starch, which occurs in the form of minute particles in the cells; oil is also produced. Almost all species are naked, unicellular flagellates with one, two or three cilia, terminally attached in canal-like recesses. A system of vacuoles and a large distinct cell nucleus is also to be found. Reproduction, in general, is asexual, simply by longitudinal cell division. Thick-walled resting stages, or cysts, are often formed. Sexual reproduction by fusion. Two nuclei of the same female cell (autogamy) have been proved in one group. Most species of this group live in fresh water.

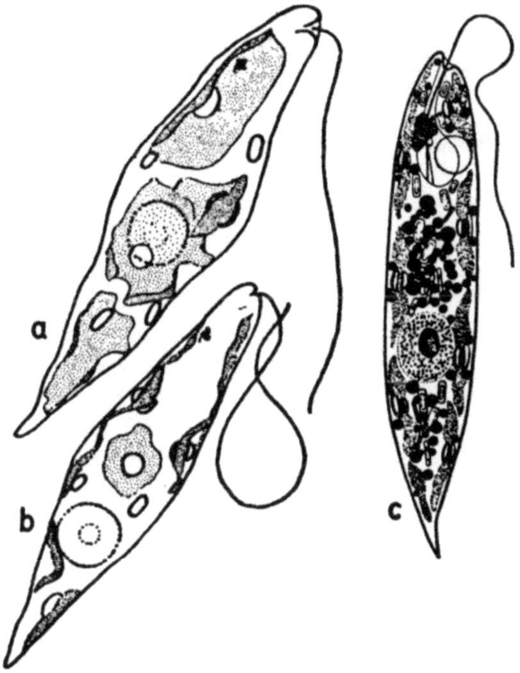

Figure 31: *Euglena gracilis* (X 858); (a–b from PRINGSHEIM, c from CHADEFAUD and PROVASOLI)

There is only one class:

1. Class: Euglenophyceae

1. Order: Euglenales
Phototrophic and mixotrophic, and also colourless species living saprophytically.

2. Order: Peranematales
Colourless species feeding osmotrophically or phagotrophically.

III. DIVISION: CHRYSOPHYTA

Their chromatophores are of yellow-greenish to yellow-brownish colour, as they contain much of xanthophyll and carotene. The products resulting from assimilation are oils, chrysose (leucosin), and carbohydrates of an unknown nature (not starch). Cell walls normally consist of two interlocked halves and often contain silicic acid. The cells may or may not possess cilia, they live isolated or in microscopic colonies. Reproduction is both sexual and asexual. Chrysophyta occur in all types of water over the entire globe. They are subdivided into three classes.

1. Class: Xanthophyceae (Heterokontae)

Chromatophores are yellow-greenish or green, containing chlorophyll-a and chlorophyll-e, β-carotene, and xanthophyll. The presence of xanthophyll causes the frequent yellowish colour of these algae. Exterior appearance of Xanthophyceae is similar to that of Chlorophyceae. However, their actual differences are great: they are differently pigmentated and have different assimilation products. The characteristic of the Xanthophyceae is the structure of their cell walls consisting of two interlocking parts. The most highly developed forms are filamentous. Motile, vegetative cells and zoospores are provided with two terminal cilia, each of different length. The shorter cilium may be so short as to be practically invisible. Most of the Xanthophyceae thrive in fresh water (e.g. *Tribonema*, *Vaucheria*) or even on moist ground (*Botrydium*, Pl. VI: 12). *Halosphaera* plays an important rôle as marine plankton.

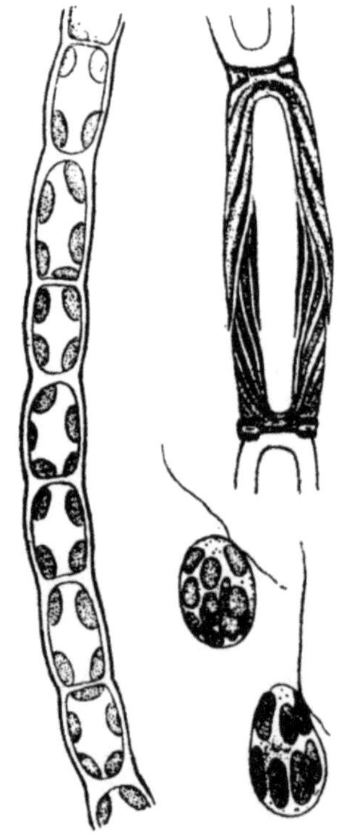

Figure 32: *Tribonema bombycina* (from GAY and LUTHER)

2. Class: Chrysophyceae

With yellow or brown chromatophores which contain various carotenes (often called phycochrysin). Cells usually contain one or two large parietal chromatophores, and frequently, naked pyrenoid-like bodies. Spherical, endogenically formed cysts with a silicified membrane are characteristics of the class.

Most of the Chrysophyceae are known in fresh water. Those occurring in the sea are above all the Calcareous flagellata *(Coccolithineae,* Pl. VI: 13, with an integument of calcarious bodies) and the Silico-flagellata *(Silicoflagelliineae,* Pl. VII: 14, with an internal siliceous skeleton). These are rather different from the many fresh water forms. The Coccolithophores, which are extremely small and delicate, obviously play an important rôle in the realm of marine plankton. They mostly occur in temperate or warm seas and may reach cell densities of some thousands to thirty millions per litre of sea water. They thrive even in depths of down to 1,000 to 4,000 metres.

3. Class: Bacillariophyceae (Diatoms)

Clearly differentiated from the other Chrysophyta. Unicellular or colonial organisms with a complicated and normally strongly silicified cell wall structure. The cell wall (frustule) always consists of two halves, of which the larger half (epitheca) fits over and

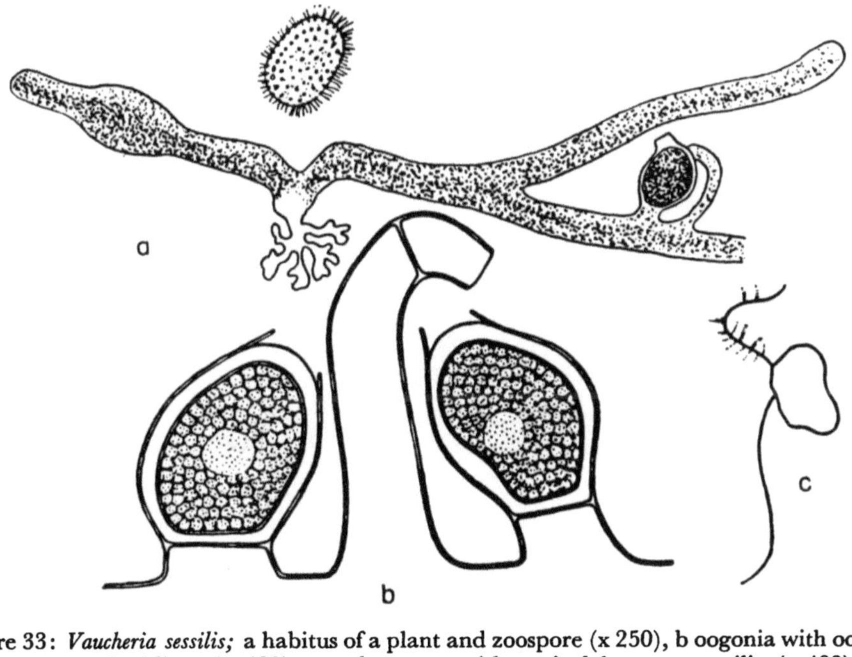

Figure 33: *Vaucheria sessilis;* a habitus of a plant and zoospore (x 250), b oogonia with oospores and empty antheridium (x 400), c androgamet with typical heterocont cilia (x 400) (from Wettstein, Fott and Koch)

covers the smaller one (hypotheca), as a lid fits over a tin. Bottom and roof of the wall are called valves; the two sides of the overlapping connecting-bands are called girdle-bands or pleurae. The valves often show a complicated structure of pores, ribs, moulds,

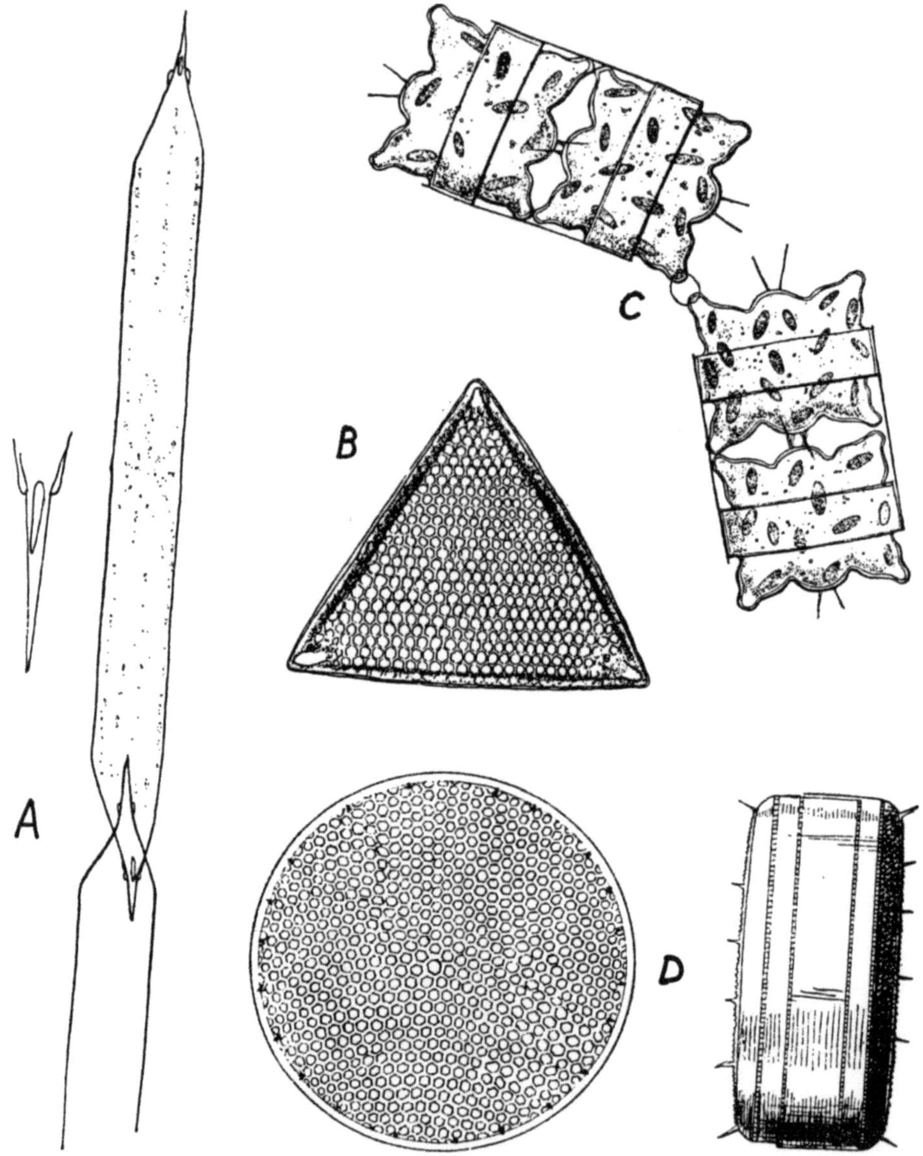

Figure 34: A *Rhizosolenia styliformis;* B *Triceratium favus;* C *Biddulphia aurita;* D *Coscinodiscus excentricus;* (A, C, D from LEBOUR, B from SMITH)

warts etc. A study of the structure of the valve-jacket of diatoms under the microscope permits the distinction to be made between valve side showing structure of valvae and the girdle side, showing structure of girdle.

Cells contain yellow or brown chromatophores which have pigments similar to those present in Phaeophyceae (chlorophyll-a and chlorophyll-c, β-carotene, fucoxanthin etc.). As assimilation products oils, volutin, and chrysose (at any rate, in some diatoms) are known to be formed.

Reproduction happens by simple division. The two daughter cells are always different

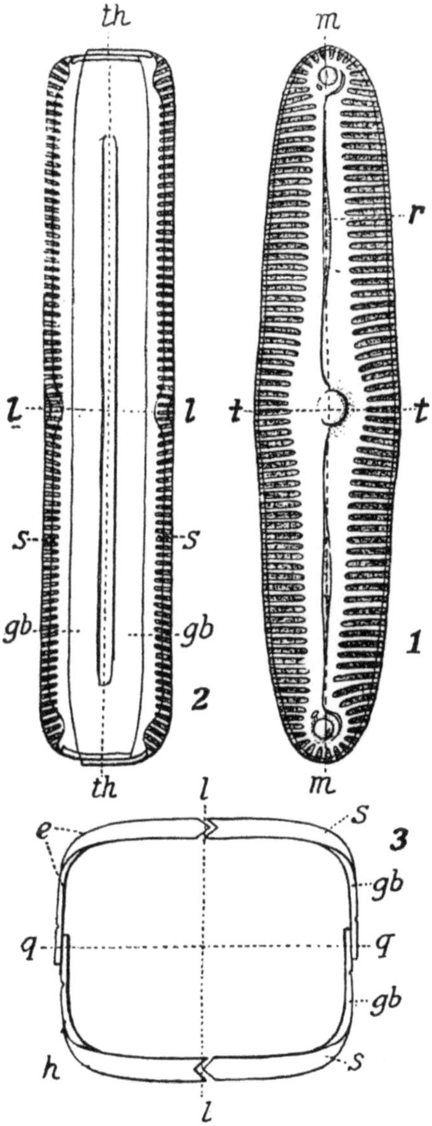

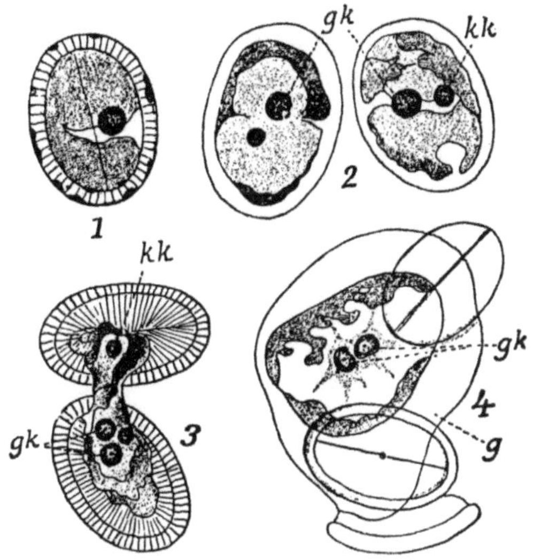

Figure 35: *Pinnularia viridis.* 1 valve view; 2 girdle view; 3 transverse section. S valve, gb connecting band, l longitudinal axis, q transverse axis, e epitheca, h hypotheca (from Pfitzer)

in size: One cell – the one with the epitheca of the mother cell – grows up to the size of the mother cell; the other – the one with the hypotheca – remains smaller than the mother cell. Thereby the greater part of the population becomes smaller on continuous cell division. The original size can be restored by the formation of so called auxospores, a kind of sexual process. In pennate diatoms two copulating cells will form two gametes which in turn bring forth two auxospores after copulation. Also several reduced processes for the formation of auxospores are known, in which the auxospores are created by automictic parthenogenesis (automixis) or without any fertilization (apomixis). In centric diatoms the formation of auxospores, according to von Stosch (1950, 1951, 1956, 1958), is a kind of oogamy with the ovum being impregnated by a motile spermatozoid. Certain centric diatoms form endogenous cysts as resting cells.

The diatoms are diploid organisms. The haploid phase is confined to the gametes. This is a characteristic feature which differentiates the diatoms from other algae. The fact that they are diplonts proves that they cannot be regarded to be primitive organisms. In all likelihood they have gradually separated from the other chrysophytes. But they have not developed beyond the unicellular stage.

1. Order: Centrales

Occurring almost exclusively as marine forms with centrically built valves, most of them are circular, some of them elliptic or triangular in shape. The structure of the valvae always concentrically arranged around a central point. The centric diatoms play a very important part in the realm of marine plankton. *Biddulphia, Chaetoceras, Coscinodiscus, Melosira, Rhizosolenia, Triceratium.*

2. Order: Pennales

The valves are not centrically shaped, but are either linear or somewhat less elongated, in some cases elliptic with a pinnate structure. The valves of most of the pennate forms have a longitudinal rift, the raphe, which is associated with their ability of spontaneous movement. Many species occur in marine habitats, though most of them are found in fresh water. *Cocconeis, Diploneis* (Pl. VII: 15), *Grammatophora, Licmophora, Nitzschia, Pinnularia, Pleurosigma, Thalassiothrix.*

IV. DIVISION: PYRROPHYTA

Chromatophores are of a yellow-greenish to dark brownish colour, due to their high content of β-carotene and xanthophyll. The assimilation product is starch as in Chlorophyceae. Most forms are flagellata of a very characteristic and specific morphology. They are provided with two cilia, normally of different length, shape, and of a different kind of motion. Particularly characteristic is the large nucleus, called dinokaryon, with chromatin nucleoli arranged like a string of beads. The Pyrrophyta, also their non-motile forms, multiply by means of highly characteristic zoospores. Sexual reproduction is rare.

1. Class: Desmophyceae (Desmokontae)

Simple flagellata, mainly occurring in the sea. *Prorocentrum, Exuviella, Desmomastix, Phalacroma* etc.

2. Class: Dinophyceae

Most of these are motile organisms with a characterizing cellulose valve showing a transverse and a longitudinal furrow on the surface of the cell. The two flagella arise from these furrows, the longitudinally directed flagellum from the longitudinal furrow and the transversely directed flagellum from the transverse furrow. They are found in marine as well as in fresh waters. Many of the species are important as marine plankton.

Dinophysis, *Gymnodinium*, *Noctiluca* (causing phosphorescence of the sea), *Peridinium*, *Gonyaulax* (*G. catenella* may produce a red colouring of the sea water, when it occurrs in great abundance. It contains a very strong poison, whereby certain mussels feeding on it can become lethal to man, if eaten), *Ceratium* (Pl. VII: 16, 17).

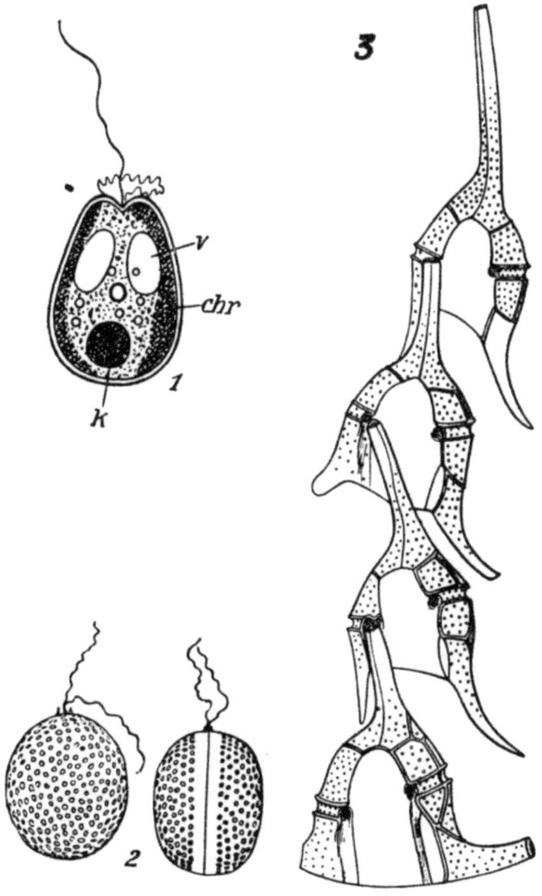

Figure 37: 1–2 *Exuviella marina* (after KLEBS); 3 *Ceratium tripos* (from KOFOID)

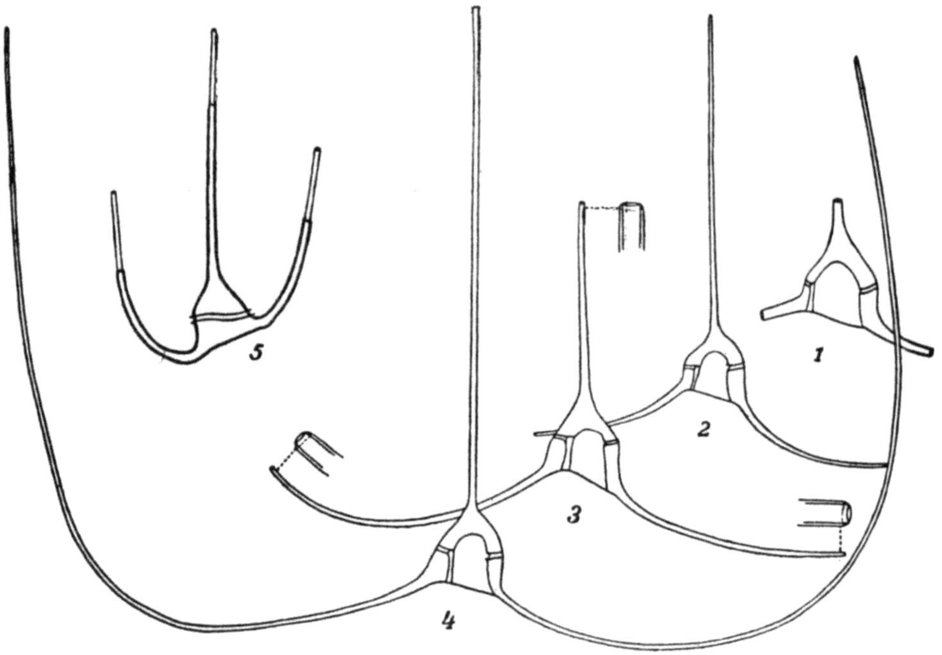

Figure 38: 1–4 *Ceratium trichoceras;* 5 *Ceratium Ostenfeldi* (from Kofoid)

Class: Cryptophyceae

A group of small flagellata occurring in the sea and fresh water, of undetermined systematic position. They are often held to belong to the Division of Pyrrophyta. However, they differ from the Pyrrophyta in their structure of the nucleus and the morphology of their cell. *Cryptomonas, Rhodomonas, Chilomonas.*

V. DIVISION: PHAEOPHYTA

The Phaeophyta or brown algae owe their characteristic shades of olive green to dark brown colour to the presence of a particular xanthophyll pigment, fucoxanthin (which also occurs in Diatoms and Chrysophyceae). Apart from fucoxanthin the Phaeophyta contain a number of pigments: chlorophyll-a and chlorophyll-c, xanthophylls different from fucoxanthin (violaxanthin, neoxanthin, flavoxanthin etc.), and carotene. Starch is not formed as end product of photosynthesis; the food reserves in the cells are the polysaccharide laminarin, the alcohol mannitol and fats.

Small vacuoles are very characteristic for the cells, the physodes, which contain fucosan, a substance showing many of the properties of tannin. The content of physodes produces an acid reaction contrary to that of normal large cell-vacuoles, which is alkaline. The number of physodes is generally especially high in assimilatory and reproductive cells.

Normally the cells of the Phaeophyta contain one single large nucleus with one or more nucleoli. In many cases mitosis is accompanied by the occurrence of centrosomes at the poles of the spindle, which always appear intranuclear.

In most of the cases the cells contain several, usually parietal, disc-shaped chromatophores; but in some cases there is a single plate-like chromatophore. The colour of the plants, which may be any shade between olive-green and dark brown, depends on the combination of the different pigments mentioned above.

Many species are able to accumulate iodine (as a water soluble organic compound) in the cells. Such algae have, therefore, been used as raw material for the extraction of iodine. Under certain conditions the algae are able to secrete free iodine from their living thallus.

The vegetative cell of the Phaeophyta has a distinct wall, differentiated into a firm inner portion and an outer gelatinous portion. The main constituent of the firm part is cellulose, that of the gelatinous one alginic acid (cf. p. 300ff), which in the membrane forms metal-compounds and protein-compounds. A third constituent of the wall is fucoidin, which is water soluble and makes the thallus more or less gelatinous, and prevents drying at occasional periods of air-exposure of the plant. Alginic acid especially is extracted commercially.

All Phaeophyta are multicellular. Unicellular or colonial forms are unknown. There is a great variation in size of the adult thallus from a microscopic cell-body of a few cells, to giant kelps, which may attain a length of up to 30 metres or even more. The Fucales, Sphacelariales, certain Dictyotales, and some other forms have terminal growth initiated by a single apical cell. In some Dictyotales (*Zonaria* etc.) there is a terminal meristem in form of a marginal row of apical cells. The majority of Dictyosiphonales show diffuse growth. A great number of Phaeophyta have an intercalary meristem. In the Laminariales it is situated in a zone between stipe and lamina. As the activity of the meristem – in many cases at least – is seasonal, a new lamina is formed between the old one and the stipe, which is perennial, viz. once a year. The Desmarestiales, most of the Chordariales, many Ectocarpales, and several others have an intercalary meristem which is located at the base of a terminal hair. At this so-called trichothallic growth new segments are cut off below, and to some extent also above the meristem. Growth in width of the thallus is effected in different ways. In Dictyotales, Sphacelariales and various other forms it is effected by longitudinal divisions of the segments formed first. In Chordariales and others certain radially directed filaments are formed, which gradually increase the width of the thallus. Many species have a special superficial layer of meristematical cells which effect the growth in width of the thallus through successive periclinal divisions.

The full-grown thallus of most forms shows various degrees of differentiation between external and internal regions. Superficial cells are smaller and more densely filled with chromatophores than the internal ones. This superficial layer thus forms a cortex encircling the medulla, which is composed of elongated, large, almost colourless cells. The cortex functions mainly as an assimilatory tissue, while the medulla works as a storage and communication tissue. The transition from cortex to medulla and also the degree of differentiation of tissues is very different in the various groups of the brown algae.

Normally all brown algae, except the Fucales, have an alternation of generations. The sporophyte forms either unilocular, plurilocular sporangia, or both. The one-celled unilocular sporangium begins to develop with an enlargement of an uninucleate cell. Meiosis takes place, and through several successive divisions a number of haploid nuclei is formed. There is then a cleavage into uninucleate protoplasts, not separated by walls, and a metamorphosis of each of such protoplasts into a biflagellate zoospore. In Dictyotales nonflagellated aplanospores are formed. The spores are released by a rupture of the sporangial wall. After germination these spores give rise to haploid sexual plants. The plurilocular sporangia are normally formed by a linear series of cells, which divide to form an elongate multicellular structure composed of many small cubical cells. The content of each cell will be transformed into a biflagellated zoospore of similar shape as those liberated from the unilocular sporangia. The spores from the plurilocular sporangia are formed without preceding meiosis. They are thus diploid and give rise to new generations of diploid sporophytes.

The sexual plants may be monoecious or dioecious, and either isogamous, anisogamous or oogamous. The gametangia of isogamous and anisogamous species display the same structure as the plurilocular sporangia. The same applies to the gamets, which are biflagellated and have the same shape as the zoospores. In oogamous species each oogonium produces one egg, or in the Fucales 1, 2, 4 or 8 eggs. Depending upon species each antheridium produces one or many spermatozoids.

The zoids – both sexual and asexual – are pear-shaped and usually contain an eye-spot. They are naked but seem to be surrounded by a thin gelatinous coat containing fucoidin (cf. Levring 1952). There are two lateral flagella (the Dictyotales apparently have only one), the anterior being longer than the posterior one. The Fucales are an exception, for their spermatozoids have a short anterior flagellum and a long posterior one. The eggs are spherical protoplasts without any cell-wall. As has been shown by Levring (op. cit.) they are surrounded by a gelatinous coat, tangentially stratified, and by a very thin socalled egg-membrane, which seems to play an important rôle as a kind of templete supporter at the wall-formation after fertilization.

Almost all Phaeophyta thrive in the sea. Only a few fresh-water genera are known (*Heribaudiella*, *Pleurocladia*, and *Bodanella*). Some marine species have been found to occur also in brackish water, and a number of Fucales constitute an important part of the salt-marsh vegetation in western Europe and also in other parts of the world. These salt-marsh forms are as a rule loose and entangled one with another, and with different phanerogams. As a rule they propagate by vegetative means only. One of the most striking examples of loose forms are the floating masses of *Sargassum* in the Sargasso Sea.

Most of the brown algae belong to the intertidal zone and the upper part of the sublittoral zone. In size, number of species, and quantities they attain their greatest development in colder waters. In these regions they often constitute the bulk of algal vegetation. Species belonging to Fucales and Laminariales will thus often occur in great quantities.

It seems very probable that the Phaeophyta represent a very old group among the plant kingdom. Fossils are known from periods as early as the triassic formation, some perhaps also from the palaeozoic period. The presence of motile reproductive cells indicates that Phaeophyta arose from unicellular flagellated ancestors. Of the forms living today the

Phaeophyta only seems to bear very slight relation to the Chrysophyta. This relationship is mainly based upon similarities in pigments and metabolic products (cf. CHADEFAUD, 1950; BOURRELLY, 1957).

The division Phaeophyta includes about 240 genera with altogether 1,500 species. In the following systematic review the arrangement proposed by KYLIN (1933) will be followed mainly. Thus the group is divided into three subclasses. There are several points

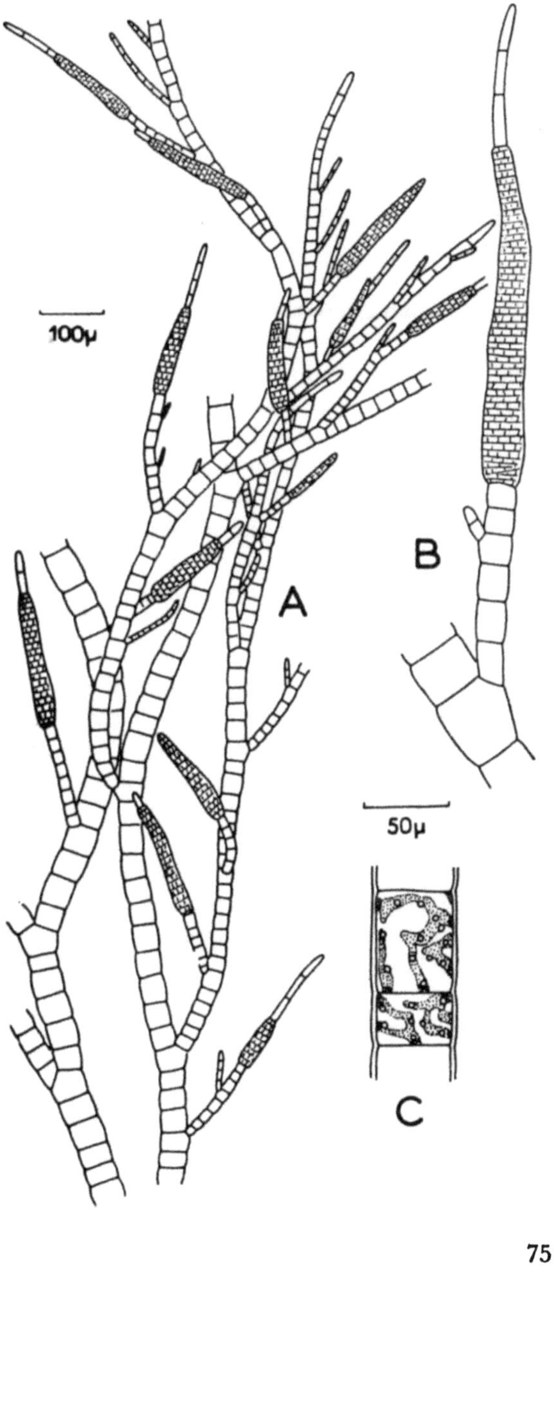

Figure 39: *Ectocarpus confervoides* var. *siliculosus* (from CARDINAL)

which still need further investigation, however, PAPENFUSS (1947), has combined the two orders Punctariales and Dictyosiphonales (as has been done in this review). A further division of the Punctariales into several new orders has also been suggested. It may also be pointed out that the limits between the Ectocarpales and the Chordariales still include several obscure points. The division contains only one class, Phaeophyceae, divided into 3 subsclasses.

1. Class: Phaeophyceae

1. Subclass: Isogeneratae

Life cycle with two alternating generations which are identical in vegetative structure. Growth of the thallus trichothallic, intercalary or apical. Thallus amorphous or of definite form and with or without an internal differentiation. The class is divided into five orders.

1. Order: Ectocarpales

The order includes the least specialized forms of the Phaeophyta. Many species are small, only a few centimetres or even microscopic. The thallus is filamentous with diffuse or trichothallic (rarely apical) growth. Normally the filaments are uniseriate (occasionally, as in *Pylaiella*, longitudinal divisions may occur), branched, either free (Fam. Ectocarpaceae) or adhere laterally to form pseudoparenchymatous, crustose bodies of more or less indefinite size (Fam. Ralfsiaceae). Cells contain one or more plate-like, stellate, ribbon-like, or numerous disc-shaped chromatophores. Terminal colourless hairs growing by means of basal meristems occur in many species or genera.

The Ectocarpales occur on rocks, wood or epiphytic on other algae. Even endophytic or endozoic species are known.

The life cycle has only been studied in a few cases. The sporophyte bears either unilocular or plurilocular sporangia, or both. There is no meiosis in the plurilocular sporangia, and the zoospores from them give rise only to other diploid plants. The first divisions in the unilocular sporangia are meiotic. The zoospores formed in these sporangia are thus haploid and give rise to new gametophytes. Parthenogenesis seems not to be uncommon.

Fam. ECTOCARPACEAE: filiform, generally uniseriate; subsimple or more or less branched. *Bodanella, Ectocarpus* (Pl. VIII: 18), *Feldmannia, Geminocarpus, Giffordia, Mikrosyphar, Phaeostroma, Spongonema, Streblonema.*

Fam. RALFSIACEAE: Crustose, composed of systems of radial Filaments from which arise erect series of cells forming a pseudoparenchymatous thick crust. *Heribaudiella, Lithoderma, Ralfsia.*

2. Order: Sphacelariales

Growth initiated by a single large apical cell; new segments are formed through transverse divisions. These primary segments are usually divided by both transverse and longitudinal septa. Older parts of the thallus, therefore, have a parenchymatous structure. A further increase in width occurs in some genera as the result of the formation of a kind of secondary cortex. Older parts often are also covered with a layer of rhizoid-like filaments.

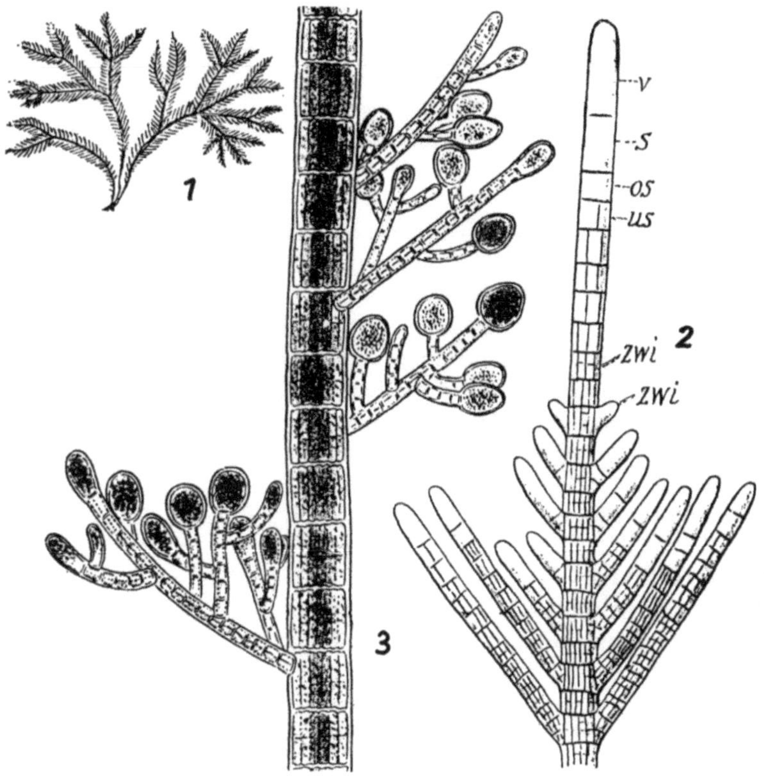

Figure 40: 1–2 *Sphacelaria plumigera*. 3 *S. racemosa* (from Reinke)

The thallus is filamentous, usually much branched and generally forming small tufts. The plants grow on rocks or on other algae.

Many species reproduce vegetatively by means of propagula, as a rule of a characteristic structure for a given species. They are often the only means of multiplication during a certain season of the year. Development of a propagulum begins in the same way as that of a lateral branch, but the apical cell is soon divided by longitudinal walls into two or three daughter cells. Each daughter cell is the beginning of a new branch. When separated from the mother-thallus the propagulum develops into a new plant after having been attached to a suitable substratum.

The life cycle of the Sphacelariales includes two isomorphic generations. Zoospores are formed in both plurilocular and unilocular sporangia. Both kinds or only one may occur in the same individual dependant on species. Meiosis takes place only in the young unilocular sporangia, and the haploid swarmers from this organ give rise to haploid sexual plants. The swarmers from the plurilocular sporangia are diploid and give rise to other diploid plants. The gametophytes as a rule form plurilocular gametangia and may be monoecious or dioecious. According to Moore (1946) some species of *Halopteris* are oogamous. Parthenogenesis has also been found among the Sphacelariales.

A monograph has been devoted to this order by SAUVAGEAU (1900–1914).

1. Fam.: SPHACELARIACEAE – *Chaetopteris, Sphacelaria*
2. Fam.: STYPOCUALACEAE – *Halopteris*
3. Fam.: CLADOSTEPHACEAE – *Cladostephus*
4. Fam.: CHORISTOCARPACEAE – *Choristocarpus*

3. Order: Cutleriales

Thallus flattened, blade or disc-like with trichothallic growth. The sporophytes produce unilocular sporangia only. The gametophytes are anisogamous and the biflagellated gametes are formed in plurilocular gametangia.

There are only two genera: *Zanardinia* has an alternation of two morphological identical (isomorphic) generations. In *Cutleria* they are somewhat different. The sporophyte consists of a prostrate, encrusting, lobed thallus. The gamatophyte is repeatedly forked, with the branches ribbon-shaped and ending in a tuft of hairs. The sporophytes were earlier considered to be a separate genus, *Aglaozonia:* they are still known as the Aglaozoniastage of *Cutleria.*

Fam.: CUTLERIACEAE – *Cutleria, Zanardinia*

4. Order: Tilopteridales

This order is rather imperfectly known and contains only a few small and very rare genera. The thallus is filamentous, freely branched with trichothallic growth. Upper parts

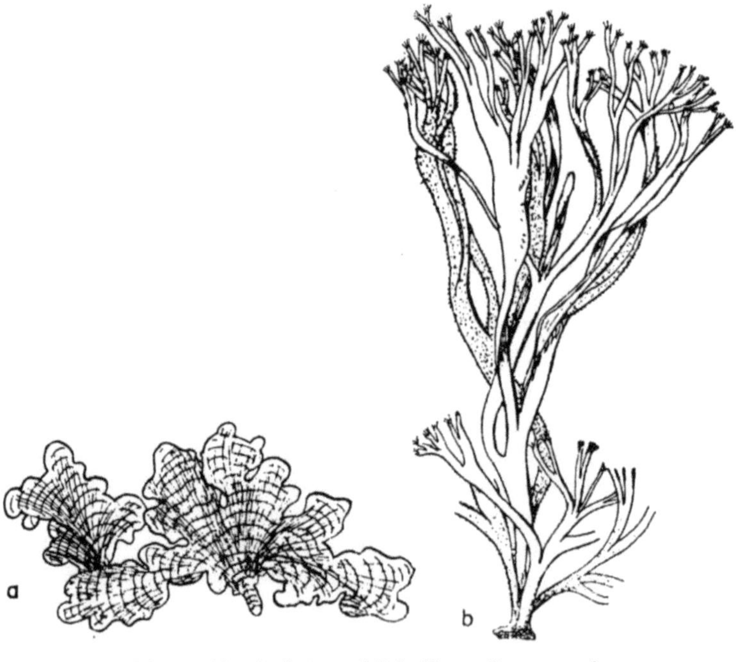

Figure 41: *Cutleria multifida* (from OLTMANNS)

are monosiphonous, *Ectocarpus*-like, lower parts polysiphonous, *Sphacelaria*-like. It seems that the life-cycle consists of two morphological similar generations. The sporophyte produces monolocular sporangia with one single aplanospore with four nuclei. The gametophyte seems to be oogamous.

Fam.: TILOPTERIDACEAE – *Tilopteris, Haplospora, Acinetospora*

5. Order: Dictyotales

The members of this order are quite distinct from other Phaeophyta. The thallus is usually erect, stipitate, foliose, ribbon-like or fan-shaped, and ramified, with the branches

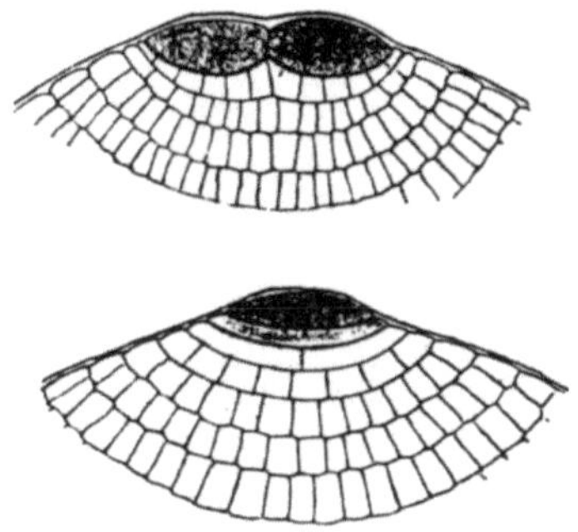

Figure 42: *Dictyota dichotoma;* terminal end of a branch showing initial cell and formation of dichotomy (from OLTMANNS)

usually in one plane. There is an attachment of rhizoids, which in some genera forms a thick attachment. In some genera – *Padina, Zonaria, Taonia* – concentric zones of hairs are formed on the thallus. Growth is initiated by a single apical cell *(Dictyota, Dilophus, Pachydictyon, Glossophora* etc.) or by a marginal row of apical cells *(Padina, Taonia, Dictyopteris, Zonaria* etc.).

The Dictyotales have a life-cycle with two morphological identical generations. The diploid sporophytes produce four or eight large aplanospores. Sexual reproduction is oogamous by eggs and spermatozoids provided with a single flagellum. Most species are dioecious. Both sporangia and oogonia as well as antheridia usually occur in definite sori.

The members of this order are most abundant in the warmer waters, especially in the tropics.

Fam.: DICTYOTACEAE – *Chlanidophora, Dictyopteris, Dictyota, Dilophus, Glossophora, Pachydictyon, Padina, Spathoglossum, Syringoderma, Taonia, Zonaria.*

2. Subclass: Heterogeneratae

Life cycle with two alternating generations which are of different sizes and of different vegetative structure. The sporophyte is always larger, generally of macroscopic size, with well differentiated anatomical structure and definite form. The sporophytes reproduce with zoospores from unilocular or plurilocular sporangia. The gametophyte is filamentous, microscopic. Reproduction of the gametophytes is isogamous, anisogamous or oogamous.

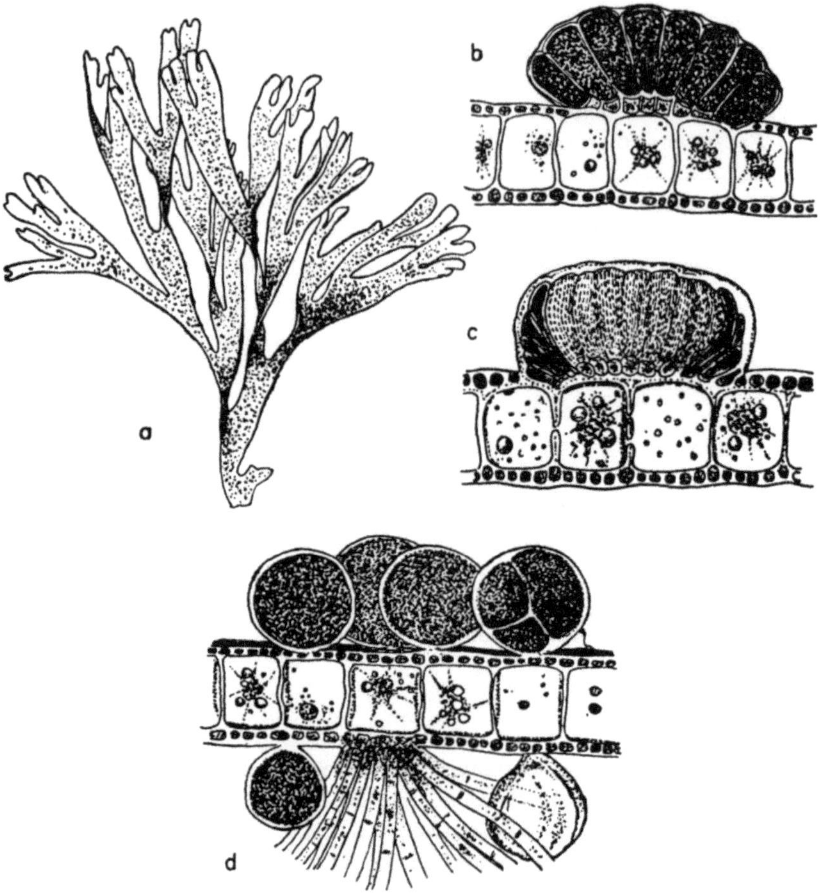

Figure 43: *Dictyota dichotoma;* a habitus; b transvers section with oogonia; c antheridia; d tetrasporangia; (from THURET)

I: Haplostichineae

Sporophytes with trichothallic growth and with a thallus composed of filaments, more or less separated, interwoven, or densely united to a pseudoparenchymatous structure. Intercalary longitudinal walls are not formed in the filaments. The gametophytes are always microscopic. The subclass is divided into three orders.

1. Order: Chordariales

Sporophytes of a filamentous structure. The most primitive representatives of the order (Fam. MYRIONEMATACEAE, ELACHISTACEAE, and CORYNOPHLAEACEAE) form minute crusts, small pulvinate or gelatinous, cushion-like growths. Forms belonging to other families are terete, branched and small to fairly large in size. The thallus is cylindrical (or more or less flattened) and differentiated into a central, large-celled medulla and a peripheral, small-celled photosynthetic cortex. Very often a transitional region can be distinguished between these two layers. The central part is either of uniaxial or multiaxial structure. Branching of the central filaments are either monopodial or sympodial and their growth either trichothallic or by means of a single apical cell.

1. Fam.: MYRIONEMATACEAE – Minute crusts growing epiphytically on other algae. – *Ascocyclus, Myrionema, Pleurocladia.*
2. Fam.: ELACHISTACEAE – Plants pulvinate, small, with a cushion-like basal portion and free, erect filaments, which usually are branched only at the base. *Elachista, Halothrix, Leptonema.*

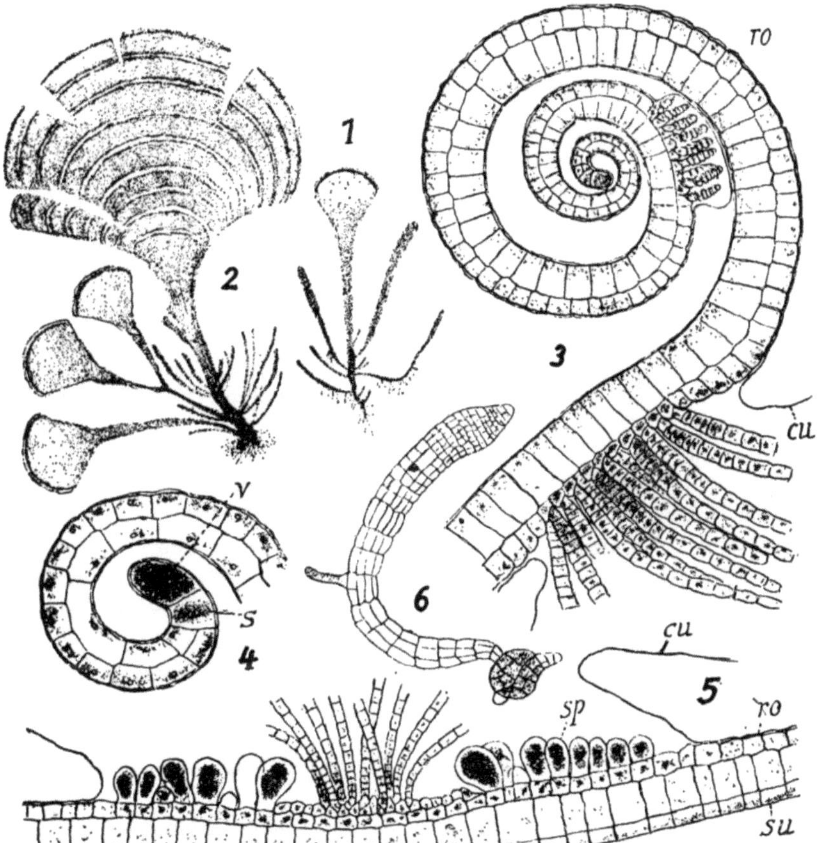

Figure 44: *Padina pavonia;* 1–2 habitus; 3–4 transvesse section of margin; 5 sorus 6 germeling (from OLTMANNS)

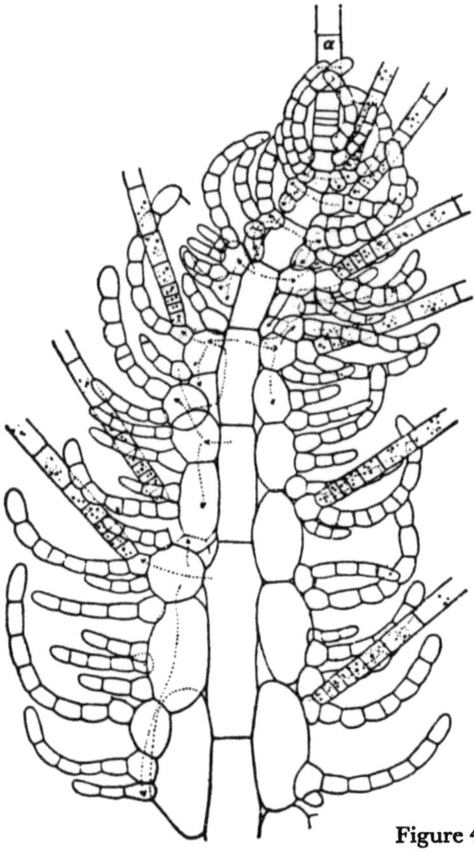

Figure 45: *Acrothrix gracilis*, apex x 400 (from Kuckuck)

3. Fam.: Corynophloeaceae – Thallus cushion-like of more or less irregular shape, gelatinous, up to 5 cm in size. – *Corynophloea, Leathesia.*
4. Fam.: Chordariaceae – Thallus terete, filamentous, more or less branched; one or numerous central filaments of monopodial or sympodial structure. Growth trichothallic. – *Chordaria, Cladosiphon, Castagnea, Eudesme* (Pl. VIII: 19), *Haplogloia, Levringia, Mesogloia, Myriogloia, Papenfussiella, Sphaerotrichia.*
5. Fam.: Acrothrichaceae – Terete, branched, with one single central filament ending with a hair. There is an intercalary meristematic zone between hair and central filament. – *Acrothrix.*
6. Fam.: Spermatochnaceae – Terete, branched, with apical growth. – *Spermatochnus, Stilophora.*
7. Fam.: Chordariopsidaceae – Terete, pseudoparenchymatous, with one single central filament with one apical cell. Paraphyses lacking when sterile. – *Chordariopsis.*
8. Fam.: Splachnidiaceae – Terete, cylindrical, branched. Paraphyses only in the top region. Growth through intercalary divisions at the base of the paraphyses. – *Splachnidium.*

2. Order: Sporochnales

The sporophyte, which in most of the species reaches a length of about 40 cm, is of trichothallic growth. The branches terminate with a tuft of hairs. The intercalary cell divisions take place at the base of the hairs. Unilocular sporangia are usually borne terminally and in dense clusters. The microscopic gametophytes are oogamous.

Fam.: SPOROCHNACEAE – *Carpomitra, Nereia, Sporochnus.*

3. Order: Desmarestiales

Members of this order are mainly found in cold waters in both hemispheres. Especially in the southern hemisphere they constitute an important part of the sublittoral vegetation. The thallus is cylindrical or in most of the species flat. All species have pinnate branching, with opposite or alternate branches. In addition to the ordinary branches there are uniseriate, hair-like branches, which in a charateristic way form tufts of hairs on the thallus of

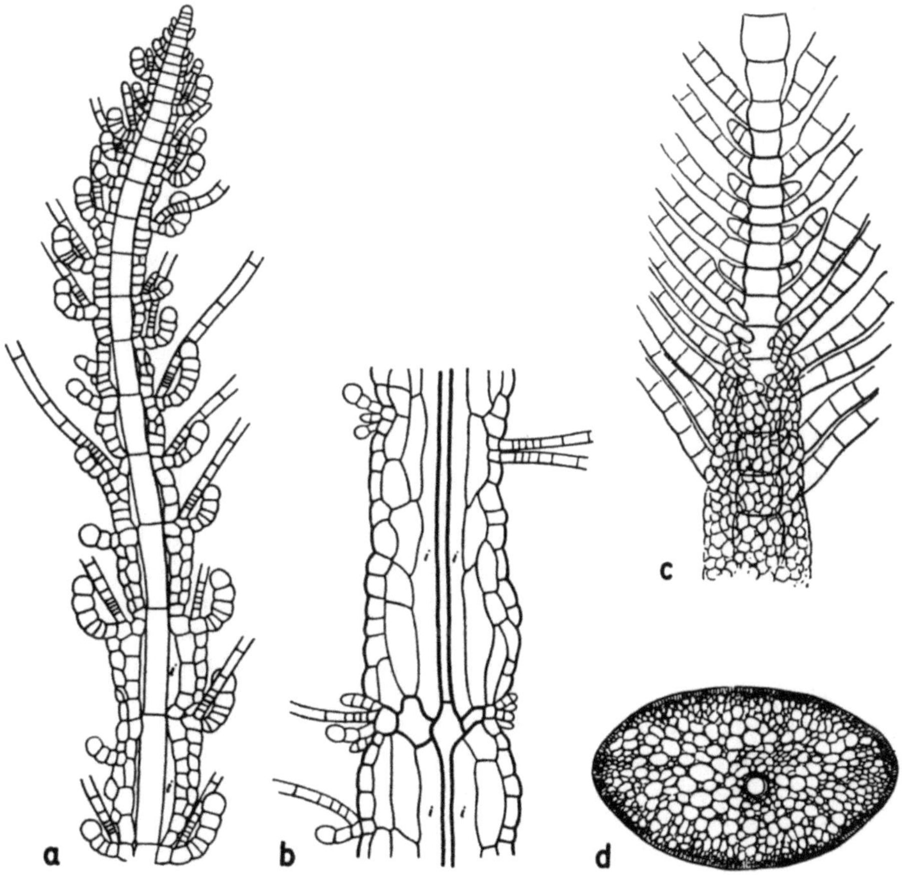

Figure 46: *Spermatochnus paradoxus;* a long. section of an apex, b mature thallus X 280. *Desmarestia ligulata;* c apex, d transv. section (from REINKE)

Desmarestia and *Arthrocladia* and a hairy covering on the thallus of *Phaeurus*. The hairs are persistent in *Phaeurus* and are shed in *Desmarestia* and *Arthrocladia* after the growing season.

The anatomical structure of the thallus is uniaxial with apical, trichothallic growth, and with an often considerable cortication. As far as it is known the sporophyte produces only unilocular sporangia. The microscopic gametophyte is oogamous.

1. Fam.: ARTHROCLADIACEAE – *Arthrocladia*
2. Fam.: DESMARESTIACEAE – *Desmarestia*, *Phaeurus*

II: Polystichineae

Sporophytes with parenchymatous thallus are produced by vertical and transverse divisions of intercalary cells. The sporophyte is always larger than the microscopic gametophyte, and of a well differentiated anatomical structure. The sporophyte produces zoospores in unilocular or plurilocular sporangia. The gametophytes are isogamous, anisogamous or oogamous.

Different opinions have been advanced on the most expedient division of the subclass into orders. In his contribution to a new system of classification of Phaeophyta KYLIN (1933) divided the Polystichineae into three orders: Punctariales, Dictyosiphonales, and Laminariales. The Laminariales represent a very well established group. As pointed out by KYLIN himself the Punctariales represent a somewhat ill-defined assemblage, which is in great need of monographic treatment. Since the borderline between the Punctariales and the Dictyosiphonales seems to be highly blurred, as pointed out by PAPENFUSS (1947), this author suggests the merging of the two orders under the name of Dictyosiphonales. But other authors have suggested the division of Punctariales in several new orders. One or two of these orders will then be included among the Isogeneratae (cf. FELDMANN, 1949; LUND, 1959). The discussion of this issue, however, is beyond the scope for the present work.

1. Order: Dictyosiphonales

The order is taken to be rather comprehensive, and includes small-sized to medium-sized plants living on other algae or on rocks etc. The gametophytes are microscopic, filamentous. The thalli of the sporophytes are of various shapes: solid, cylindrical or tubular to saccate, and simple or branched, flat and ribbon-like to foliaceous.

With the exception of *Colpomenia*, which has a hollow, irregular cushion-like thallus (Pl. VIII: 20), and a few others, the young plants are in the form of erect filaments with diffuse intercalary growth or apical growth. As a result of additional longitudinal cell divisions, which commence early, a thallus with true parenchymatous structure is formed. The increase in length of elder thalli of most members of the order is the result of diffuse intercalary growth. In *Chnoospora* and *Scytothamnus* (cf. LEVRING, 1941) growth is localized mainly to the subapical region, in *Scytosiphon* to the proximal part and in *Dictyosiphon* to the apical part of the thallus. At least young thalli often end in a terminal hair. Some genera have solitary or tufts of hairs scattered over the surface of the thallus. In some cases (e.g. *Colpomenia, Chnoospora)* they develop from special conceptacles.

The sporophytes produce either unilocular or plurilocular sporangia or both. The

Figure 47: *Asperococcus bullosus* (from THURET)

plurilocular sporangia serve as accessory reproductive organs. The zoospores formed there are diploid. The first division in the young unilocular sporangium is meiotic. The zoospores deriving from these organs are thus haploid. However, cases are known, where meiosis is suppressed (cf. SAUVAGEAU, 1929, KYLIN, 1934, FÖYN, 1934). Zoospores from such sporangia give rise to new sporophytes. In some genera (e.g. *Scytosiphon, Chnoospora, Colpomenia)* only plurilocular sporangia are known. Consequently such genera lack the sexual generation.

The gametophytes are microscopic, filamentous. They are, as a rule, isogamous, but cases of anisogamy are known, too.

1. Fam.: STRIARIACEAE – *Isthmoplea, Stictyosiphon, Striaria.*
2. Fam.: GIRAUDYACEAE – *Giraudya*
3. Fam.: MYRIOTHRICHIACEAE – *Myriotrichia*
4. Fam.: PUNCTARIACEAE – *Adenocystis, Asperococcus, Colpomenia, Desmotrichum, Endarachne, Hydroclathrus, Litosiphon, Petalonia, Punctaria, Soranthera, Utriculidium*
5. Fam.: SCYTOSIPHONACEAE – *Scytosiphon*
6. Fam.: CHNOOSPORACEAE – *Chnoospora, Scytothamnus*
7. Fam.: DICTYOSIPHONACEAE – *Dictyosiphon*

2. Order: Laminariales

The algae belonging to this order have a highly differentiated sporophytic generation. In size they exceed all other algae, and they also have a rather complex anatomical structure. With the exception of *Chorda*, which has a simple cylindrical, whip-like thallus, the sporophytes of the Laminariales are differentiated into hold fast, stipe and one or more

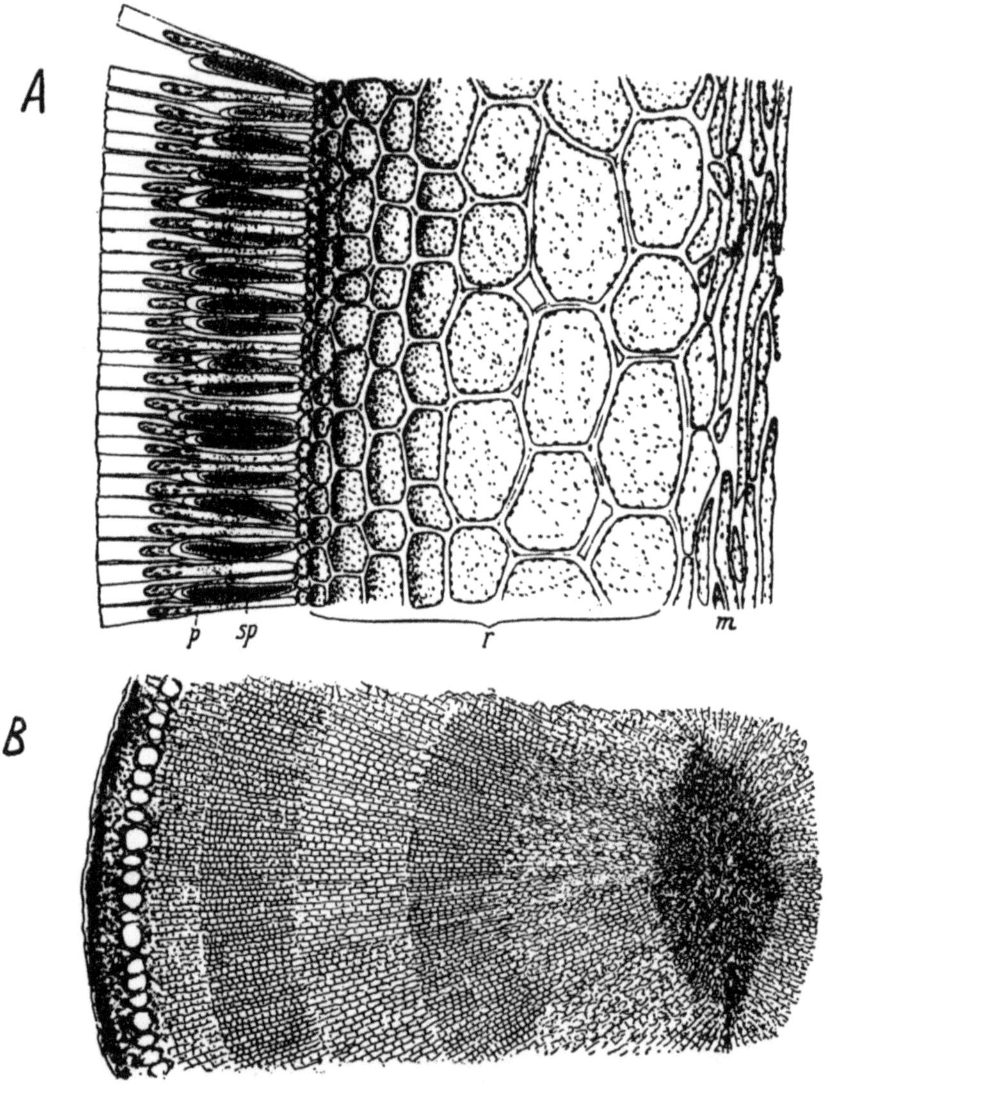

Figure 48: A. *Laminaria saccharina*. Long. section of thallus with sporangia (sp.), paraphysis (p), cortex (r) and medulla (m) (after KUCKUCK). B. Transv. section of old *Laminaria*-stipe (from OLTMANNS)

Figure 49: *Laminaria saccharina.* New thallus develops between stipe and old thallus (from OLTMANNS)

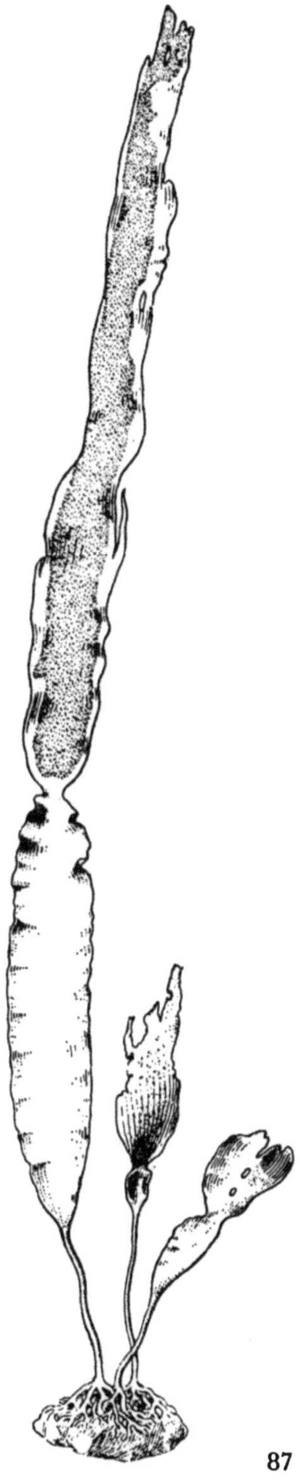

blades. The hold fast is in some forms a disc-like organ, in others it consists of a system of branched root-like haptera. In some cases, e.g. *Nereocystis luetkeana, Ecklonia maxima,* the stipe terminates with a gas bladder, and in the branched *Macrocystis* there is a bladder at the end of each stalk (Pl. IX).

Growth is due to an intercalary meristematic region usually situated between stipe and blade. Due to the activity of this meristem the stipe continuously increases in length. Blades of most of the species persist for about one year. They disintegrate at the end of the growing season after having discharged the zoospores. A new blade is formed between the meristem and the old blade.

The anatomical structure of stipe and blade is fundamentally the same; but due to the surface enlargement of the blade it is often difficult to follow the relationship between different cell-tissues there. The stipe is differentiated into a central medulla and cortex, each containing distinctive elements. The cortex and medulla of the stipe are continuous with those of the blade. The transition from one region to another is not abrupt. In early stages the medulla consists of longitudinal, parallel, unbranched filaments, hyphae, which later on are connected through cross-growing connecting filaments, so that a reticulate system of medullar filaments is formed. The medulla of a mature plant consists of transformed cortical cells, cross-connecting filaments, and hyphae. Some filaments are called trumpet hyphae because of their greater breadth near the transverse walls. Their nature seems to be rather like the sieve tubes of the vascular plants. There are numerous pores in their transverse walls. These "sieve plates" often become blocked off by callus pards when the trumpet hyphae become old. There is certain evidence for the fact that these tubes may function in the transport of food.

Normally, there is a continuous increase in diameter at the periphery of the cortex. Cells formed at the beginning of a growing season are bigger than those formed at the end, therefore the cortex contains a number of concentric rings corresponding to the age of the stipe.

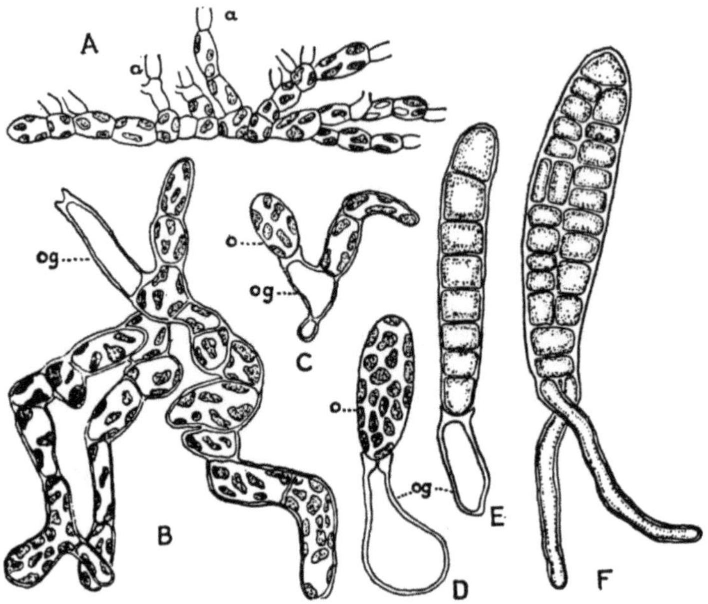

Figure 50: *Laminaria digitata.* A male, B–D female gametophytes. E–F young sporophytes; a antheridia, og oogonia, o egg. A X 600; B X 290; C and E X 320; D X 625; F X 390 (from KYLIN)

In many species there is a network of mucilage ducts in the outer cortex of both stipe and blade. The mucilage is formed by groups of secretory cells along the inner face of the ducts.

The sporophytes produce only unilocular sporangia. They are generally formed at a specific season, e.g. in spring or summer. The sporangia are formed in extensive superficial sori together with unicellular paraphyses on both surfaces of ordinary blades or on special sporophylls, e.g. *Alaria,* or in *Chorda* directly on the whip-like thallus. Species growing in the intertidal zone discharge their zoospores at the time of reflooding by incoming tides. The zoospores are haploid for the first division of the sporangium is meiotic.

The zoospores swarm for a time, then become attached to a suitable substratum, secrete a wall, and germination follows very soon. The gametophytes are filamentous, freely branched, in some cases only consisting of a few cells. They always seem to be dioecious, the male gametophyte having smaller cells and being more profusely branched than the female ones. The sexual reproduction is oogamous.

The order consists of about 30 genera with 100 species. The genus *Laminaria* is widely spread and has given the order name. Some forms from the Pacific coast of North and South America and from waters of the southern hemisphere are notable for their size and external form. The most common giant kelp is *Macrocystis,* with a repeatedly branch

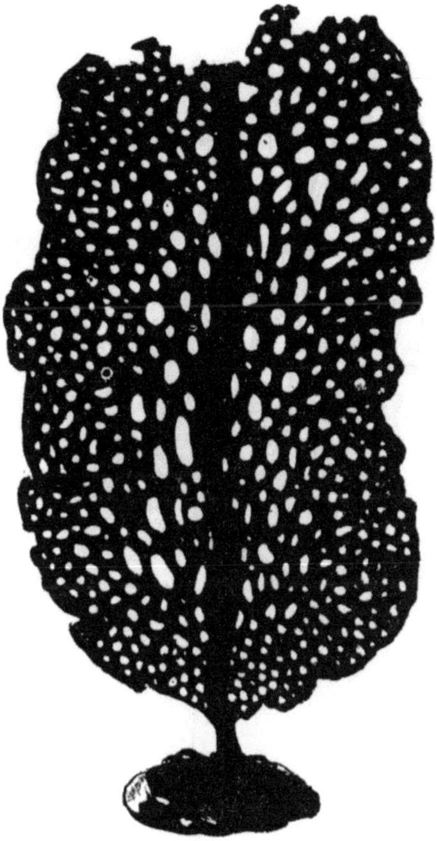

Figure 51: *Agarum Turneri* (from OLTMANNS)

stipe 30 to 50 metres long or more. There is a continuous formation of new blades at the tips. Each blade has a gas-bladder at its base. Another giant-kelp is *Nereocystis luetkeana* with an unbranched stipe, 20–25 metres long, terminating with a single large gas-bladder carrying numerous single blades, 3–5 metres long. A very strange form is the "sea-palm", *Postelsia palmaeformis*, which grows on exposed rocks in the intertidal zone and with a habit reminding of a palm.

Several species are of great commercial value.

1. Fam.: CHORDACEAE – *Chorda*
2. Fam.: LAMINARIACEAE – *Agarum, Costaria, Hedophyllum, Laminaria, Phyllaria, Saccorhiza*
3. Fam.: LESSONIACEAE – *Lessonia, Macrocystis, Nereocystis, Pelagophycus, Postelsia*
4. Fam.: ALARIACEAE – *Alaria, Ecklonia, Egregia, Eisenia, Undaria*

Figure 52: *Laminaria hyperborea* (from OLTMANNS)

3. Subclass: Cyclosporae

The members of this group have no alternation of generations. The plants are diploid. Meiosis takes place in the gametangia. Sexual reproduction is always oogamous. Growth

Figure 53: *Lessonia flavicans* (from POSTELS and RUPRECHT)

in length is initiated by a single cell or – in some cases – more apical cells. Growth in width is due to a meristematic activity of the surface layers of the cells, the meristoderm. The medullary layer consists of parallel longitudinal filaments laterally separated from one another by gelatinous material. At intervals adjacent cells retain local contact, and at these points the walls are usually provided with pits. Hyphae are produced from cells

Figure 54: *Alaria oblonga* (from KJELLMAN)

in the inner part of the cortex which grow into the medulla.

Like the Laminariales the members of Cyclosporae are characterized by marked morphological differentiation and anatomical complexity. The thallus is usually divided into holdfast, stipe, and frond. It is of parenchymatous structure. In a transverse section a peripheral cortex consisting of small cells with numerous chromatophores and a central medulla can be distinguished.

The branches of many genera contain sterile cavities, cryptostomata, where small groups of hairs projecting outwards through an ostiole are produced. When the plants become fertile such cavities are producing gametangia between the hairs. Such fertile cavities are called conceptacles. They may be scattered over the whole surface of the thallus, or more frequently, e.g. *Fucus,* they are limited to terminal parts of the branches, known as receptacles. In some genera, e.g. in *Sargassum,* the receptacles are special small branchlets which are produced when the plants are becoming fertile.

Depending on species the plants are either monoecious or dioecious. In monoecious forms oogonia and antheridia are produced in the same or in different conceptacles. The antheridia are usually formed on paraphyses. After meiosis and some mitotic divisions each antheridium usually contains 64 spermatozoid nuclei. At liberation a mass of spermatozoids is extruded. When exposed to water they become free. They are pear-shaped, biflagellate, and usually posses an eye-spot. Contrary to other Phaeophyta the posterior flagellum is the longer one of the two.

The first two divisions of the oogonium are meiotic. The subsequent development is slightly different in different genera. In *Fucus* and *Notheia* each oogonium produces eight, in *Ascophyllum, Bifurcariopsis, Durvillea, Hormosira,* and *Xiphophora* four, in *Pelvetia* two and in the remaining genera one egg. The formation of eight or four eggs appears to be very primitive. The others are modifications in different ways. After fertilization the zygote secrets a wall and begins to germinate within one or two days.

1. Order: Fucales

The order includes all members of the Cyclosporae. It is a large order with about 30 genera and 325 species distributed all over the world. No species attains the same high dimensions as some members of the Laminariales. Most of the species grow on rocks in the intertidal zone or in the upper part of the sublittoral zone. They attain their greatest development in colder waters. Many of the genera are distinguished by their external form and appearance (Pl. I-IV, X, XI).

Several species are of great commercial importance.

1. Fam.: Ascoseiraceae – *Ascoseira*
2. Fam.: Durvilleaceae – *Durvillea*, subantarctic, may reach a length of up to 10 metres.
3. Fam.: Hormosiraceae – *Hormosira*
4. Fam.: Fucaceae – *Ascophyllum, Axillaria, Cystosphaera, Fucus, Hesperophycus, Marginariella, Myriodesma, Pelvetia, Pelvetiopsis, Xiphophora*
5. Fam.: Seirococcaceae – *Seirococcus, Phyllospora, Scytothalia*
6. Fam.: Himanthaliaceae – *Himanthalia*
7. Fam.: Cystoseiraceae – *Bifurcaria, Bifurcariopsis, Carpoglossum, Cystophora, Cystophyllum, Cystoseira, Halidrys, Hormophysa, Landsburgia, Neoplatylobium, Platythalia*
8. Fam.: Sargassaceae – *Acystis, Carpophyllum, Coccophora, Sargassum, Scaberia, Scaeonophora, Turbinaria*

VI. DIVISION: CYANOPHYTA

On account of their colour the algae of the division Cyanophyta are also called blue-green algae. They mostly form macroscopic clusters or cushions on the ground and on rocks etc. Others form more or less irregularly shaped gelatinous conglomerations. Many are microscopic. Occasionally they reproduce so rapidly that the water is coloured red. Their cell structure is simple; reproduction takes place by division.

Cells are mostly cylindrical or spherical. The protoplast is subdivided into two plasma regions, the chromoplasma and centroplasma. The peripheral chromoplasma contains the pigments: phycocyanin and phycoerythrin, which are the specific colour pigments of these algae; they also contain chlorophyll-a, carotene, and xantophyll. Due to different combinations of pigments the Cyanophyta cells have many different shades of colour: yellow, brown, green, bluish-green, blue, red, black etc. The colour of even one and the same species may change due to the colour of ambient light. The first product of photosynthesis that can be proved is glycogen. It appears to be converted immediately into glycoproteides, which occur in the cells in the form of grains.

The centroplasma is the central pigment-free part of the cell. There is no actual cell nucleus, but the centroplasma contains variously formed nucleary substances.

The cell membranes contain pectines and pectine-like cellulose materials (cf. p. 374). Most of the species have gelatinous sheaths, which either have no structure at all or are laminated and of particular form, e.g. coloured tubes, bladder-like bags etc.

The cells will rarely grow individually. The most widespread forms are the colonial (Chroococcales) and the filaments. There are two types of filament: the pleurocapsan consisting of bundles of short filaments of thickwalled cells, and the hormogonal con-

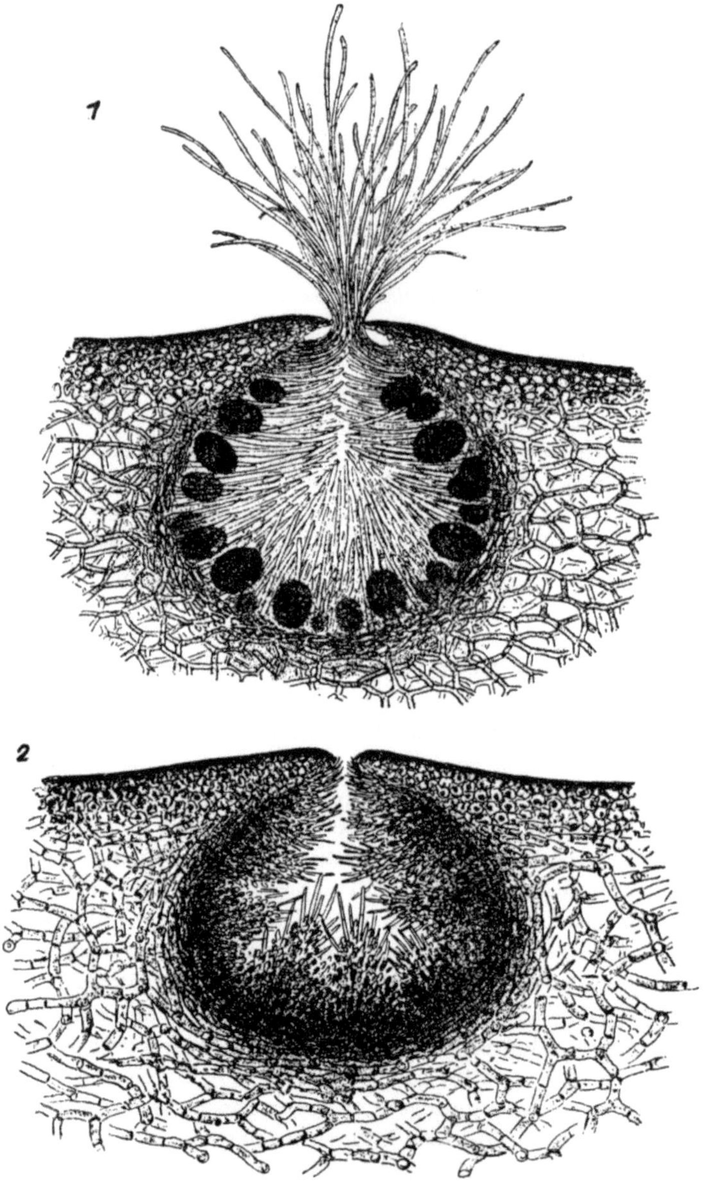

Figure 55: 1 Conceptacle of *Fucus spiralis* with oogonia and spermatangia; 2 male conceptacle of *F. vesiculosus* (from THURET)

sisting of trichomes of thin-walled cells, which form one physiological unit. The trichomes often are surrounded by a tubelike gelatinous sheath.

Heterocyst is a term used to describe certain cells which are formed individually from vegetative cells of the trichome. They are thick-walled and, due to the absence of pigments of their own, are only very slightly coloured. Their significance is as yet unknown.

The Cyanophyta reproduce by division of their cells into two. The young cells grow up and get the shape of the parental cells. In many chroococcal species there are so-called nannocysts which are accumulations of minute spores. Endospores arise through the enlargement of parental cells and the division of the contents thereof into several thin-walled younger cells. Exospores are formed from the parental cells by constriction. The reproductive bodies of filamentous Cyanophyta are described as hormogonia; they are formed as short filament-segments in the trichomes. Sexual reproduction is not known of this class.

The Cyanophyta are generally widespread in nature; the majority are fresh-water organisms. They grow in many places where other plants do not longer find the necessary environmental conditions to live. Several species are known to fix nitrogen.

The Cyanophyta are one of the oldest plant groups of the earth. Residual traces have been found which can be considered as fossile Cyanophyta in formations as old as the Precambrian and Palaeozoic. They are a very isolated group and have certain similarities in developmental history only with the Rhodophyta.

Approximately 150 families, amounting to 2,000 species, are known. They can be subdivided into four orders:

1. Order: Chroococcales

Unicellular, generally forming colonies. Reproduction by division, sometimes by nannocysts. *Microcystis, Aphanothece, Chroococcus, Gloeocapsa, Merismopedia, Enthophysalis.*

2. Order: Pleurocapsales

Regularly forming nests of short filaments of thick-walled cells. *Pleurocapsa, Hydrococcus, Hyella.*

3. Order: Dermocapsales

Unicellular, stationary, no vegetative cell division. Reproduction by endospores and exospores. *Dermocarpa, Chamaesiphon.*

4. Order: Oscillatoriales (Hormogonales)

Truly filamentous, with trichomes, often sheathed. Heterocysts frequent. Reproduction by hormogonia.

Stigonema, Scytonema, Calothrix, Rivularia, Gloeotrichia, Nostoc, Anabaena, Aphanizomenon, Oscillatoria, Spirulina, Phormidium, Lyngbya, Trichodesmium (*T. erythraeum* is held to cause the red colour of the Red Sea), *Beggiatoa.*

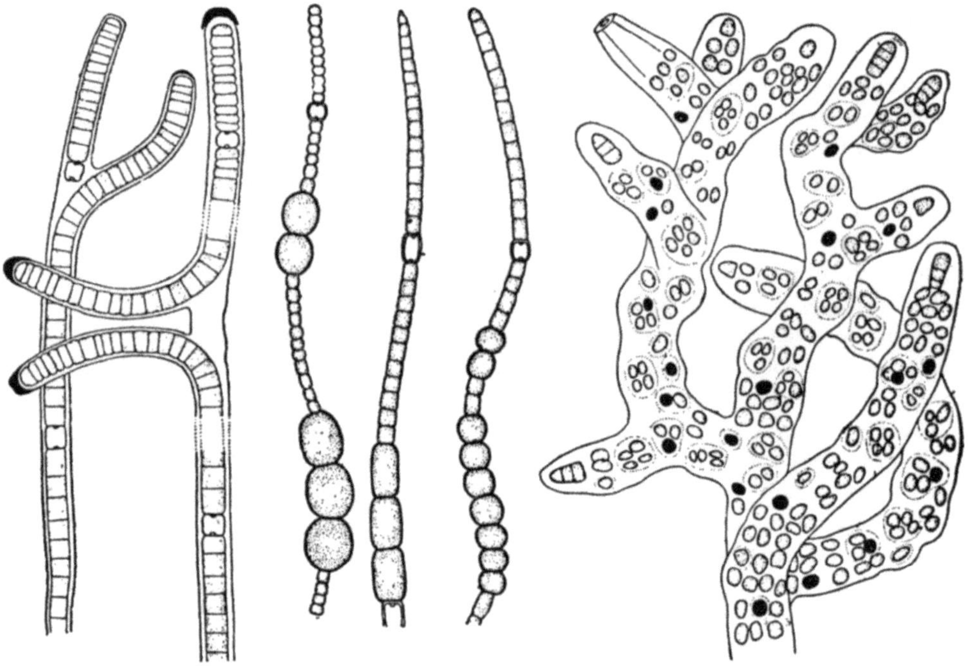

Figure 56: a *Scytonema evanescens;* b *Anabaena sp.;* c *Stigonema ramosissima* (from GARDNER)

VII. DIVISION: RHODOPHYTA

The Rhodophyta or Red algae are a large group of small to medium-sized plants, most of them characterized by comparatively elaborate thalli of more or less red colour due to the presence of the pigments phycoerythrin and phycocyanin. Compared with other algal groups the red algae show remarkable uniformity and may represent a very long line of evolutionary history.

Morphology

The cells of most of the Rhodophyta of simple structure are uninucleate, but those of a great number are multinucleate. The division of the nucleus appears to follow the normal course. The cells usually contain a large vacuole. The innermost layers of the cell wall consist of cellulose, whereas the outermost layers contain pectic substances. Dried aqueous extracts of such substances are commercially used under such names as agar, carrageenin etc. If hydrolysed with dilute acids they all produce galactose.

In some genera cell walls are incrusted with lime, e.g. *Corallina, Lithophyllum, Lithothamnion.* The thalli of such forms are often hard and stony, and especially in tropical seas, where they often occur in abundance they contribute in a large measure to the development of coral reefs. The cells of simpler types of red algae generally contain a

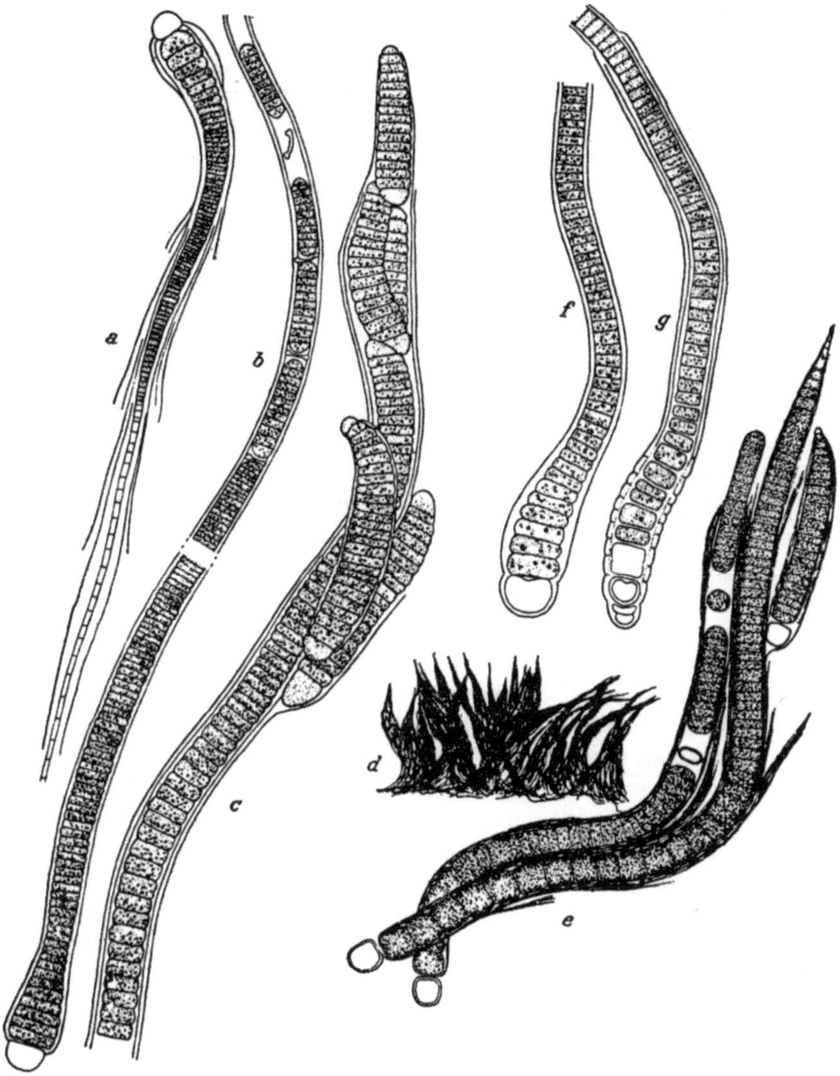

Figure 57: a, b *Calothrix parasitica*, c–e *C. pulvinata*, f–g *C. scopulorum*; (a–c, f–g from Bornet-Thuret, d–e from Fremy)

central, starshaped chromatophore with a pyrenoid. In most of the cases the chromatophores are band-shaped or discoid.

A few red algae are irradescent, probably due to the occurrence of yellowish protein bodies which reflect the shorter (blue) wavelengths of light.

Very few Rhodophyta are unicellular. The majority have a multicellular, filamentous, cylindrical, and simple or branched and flattened, foliaceous or disc-shaped thallus of a more or less complicated anatomical structure.

Pigmentation

The principal pigments are subject to little variation. The principal green pigment is chlorophyll-a, but small quantities of chlorophyll-d often are present. The principal carotenoids are ß-carotene and lutein, but several others may also occur in small quantities. Particularly characteristic for the group is the presence of phycobilins: phycoerythrin and phycocyanin. Phycoerythrin is usually present in such abundance that it will mask the chlorophylls and carotenes, and gives the plant a range of shades of red which accounts for their common name, Red algae. Phycocyanin is also present, sometimes in greater quantities giving a bluish to almost black colour to the plant. The variation in the amounts of these different pigments accounts for the wide diversity of colours to be found in the group. There is also a close correlation between the habitat and colour. The deeper occurring marine species, living in deeper ranges which are never exposed by the tides, have the most characteristic pink or bright red colour, whereas those thriving in the upper tidal area or in fresh water, rarely have the characteristic red colour, but usually various display shades of olive, brown, dullgreen or a bluish to black colour.

The red algae may be found in greater depth than other algae. The phycoerythrin can therefore be expected to play an important rôle (cf. p. 17).

As photosynthetic product a carbohydrate of the general nature of glycogen known as floridean starch is formed, which gives a dark-red stain with iodine-potassium iodine solution.

Reproduction

The reproductive process of this group is quite different from that of other algae. There are no flagellated reproductive cells. The femal reproductive organ, corresponding to the oogonium in other algae, consists of a swollen basal portion known as a carpogonium, containing a single egg, and a long slender projection, the trichogyne. The spermatangium, corresponding in function to the antheridium of other plants, is a unicellular structure discharging a single non motile male cell called spermatium. Since the naked male cells or spermatia have no mobility of their own, they are carried by the movements of the water. When, by chance, a spermatium becomes attached to a trichogyne, the spermatial and trichogyne walls break down, and the spermatial nucleus enters the trichogyne; it migrates down into the carpogonium where it fuses with the carpogonial nucleus, thus forming a diploid zygote from which asexual carpospores are formed in a more or less complicated way.

Occurrence

Most of the Rhodophyta occur in the sea. They are found in particularly great abundance in Australasia and in the warmer waters. They grow on rocks, stones etc. or as epiphytes on other algae or seagrasses, from the sea level down to the lower limits of the vegetation at a depth of 30–70 metres, in very clear water at about 200 metres. They usually occur in great abundance in lower intertidal and sublittoral zones where closed communities of one or more species may exist. Endophytes and a small number of partial parasites are also known among the Rhodophyta. Only about 50 species are known to grow in fresh water.

Relationship with other algae

The highly specialized and peculiar sexual reproductive process and the absence of motile reproductive cells in the Rhodophyta are generally considered as evidence of their very long evolutionary history. There seem to be no facts which could afford a relationship to any flagellate types probable. Chiefly for biochemical reasons an affinity with the Cyanophyta is now generally assumed. There are similarities of the pigments – phycoerythrin and phycocyanin – of the starches and of the mucilages of the cell walls, which strongly support this view. But there are also many fundamental points of difference between the Cyanophyta and the Rhodophyta. They probably represent two evolutionary lines from some common ancestor in a very distant geological period.

Classification

The division Rhodophyta contains about 560 genera with 3.800 species. There is only one class divided into two subclasses and 11 orders.

1. Class: Rhodophyceae

This is the only class of the Rhodophyta. It includes two distinct subclasses: Bangioideae and Florideae.

1. Subclass: Bangioideae

Compared with the other subclass, Florideae, this is a small well separated group of simple types. There are unicellular, filamentous or membranaceous forms with intercalary or diffuse growth and without pit-connections between the cells.

Asexual reproduction by gonidia or non-motile monospores. Sexual reproduction is only known in some genera. There is a very simple carpogonium which divides into carpospores. Five orders are recognized among the Bangioideae: Porphyridiales, Goniotrichales, Bangiales, Compsogonales, and Rhodochaetales.

1. Order: Porphyridiales

Unicellular organisms, living isolated or forming colonies.
Fam. PORPHYRIDIACEAE – *Porphyridium, Chroothece, Rhodospora, Rhodosorus.*

2. Order: Goniotrichales

Multicellular, filamentous, near microscopic forms, simple or branched. Cells with stellate or parietal chromatophores.

1. Fam.: GONIOTRICHACEAE – *Asterocytis, Goniotrichum.*
2. Fam.: PHRAGMONEMATACEAE.

3. Order: Bangiales

Filamentous *(Erythrotrichia, Bangia)*, disc-shaped *(Erythrocladia)* and membranous expanded forms *(Porphyra)*. Cells with stellate or parietal chromatophores. Sexual reproduction is known in some cases. The fertilized carpogones are directly divided into carpospores. Asexual reproduction with monospores.

1. Fam.: ERYTHROPELTIDACEAE – *Erythrocladia, Erythrotrichia, Erythropeltis, Porphyropsis.*
2. Fam.: BANGIACEAE – *Bangia, Porphyra* used as human food: "laver", "nori" in Japan, where it is cultivated for commercial purposes.

4. Order: Compsogonales

Filamentous, polysiphon with central cells.

Fam. COMPSOPOGANACEAE – Mainly tropical freshwater algae. *Compsopogon.*

5. Order: Rhodochaetales

Systematical position not quite clear, only one species is known.

Fam. RHODOCHAETACEAE – *Rhodochaete.*

2. Subclass: Florideae

Compared with the first subclass, Bangiodeae, from which it is well separated, it is more advanced, and includes the majority of the red algae. Whereas few are microscopic, the majority of the Florideae are small to medium-size algae. There is a great variety of forms: filamentous, cylindrical-branched or unbranched, gelatinous or cartilaginous, solid or hollow – flattened expansions of different size, texture, and branching. External appearance is no guide to internal construction, for each type of anatomical structure produces thalli of quite different forms and sizes. The cellwalls are generally composed of two layers. The inner, firmer walls consist mainly of cellulose, the outer walls of pectic substances. It is from this pectic layer that agar and carrageenin are extracted. Very characteristic of Floridean cells are the pit connections. It has been shown that the anatomical construction of the thalli of all Florideae is fundamentally filamentous. This is not only the case with monosiphonous types, but also with more elaborate forms, which have a parenchymatous structure in the adult parts of the plant, where usually some indication of the filamentous structure can be distinguished in the apical regions. Two main types of such filamentous constructions are known: the uniaxial type, where the thallus results from the activity of one single-branched central filament; and the multiaxial or the fountain type, where several such filaments can be distinguished. The Florideae are generally divided into 6 orders: Nemalionales, Gelidiales, Cryptonemiales, Gigartinales, Rhodymeniales, Ceramiales. This classification is based mainly on reproductive details.

After fusion of the spermatial and the carpogonial nucleus, the resulting diploid zygote soon undergoes, a nuclear as well as a cell division. In *Nemalion* it develops directly into a special cellular body, called gonimoblast, which finally forms asexual spores, known as carpospores. The gonimoblast thus represents a special generation living as a kind of parasite on the gametophyte. In many cases this gonimoblast is provided with a cover developed from neighbouring vegetative cells of the gametophyte. The entire formation of the gonimoblast and cover will then be called cystocarp. In most of the Nemalionales and all Gelidiales the carposporophyte originates directly from the fertilized carpogonium, but in other Florideae the zygote nucleus (or its derivative) passes into special nutritive cells, auxiliary cells from which the gonimoblast is formed.

The position of the auxiliary cells varies from order to order. In many cases carpogonial branches and auxiliary cells (or auxiliary mother cells) occur as a unit and are called procarps after SCHMITZ.

The term "auxiliary cell" originally was introduced by SCHMITZ (1883). He considered the function of these cells to be mainly nutritive. Different authors have used this term with more or less different meanings, which have caused much confusion (cf. DIXON 1961). According to KYLIN (1923, 1928, 1930, 1935, 1937, 1956) the use of the term "typical auxiliary cells" was restricted to such cells as served for the origin of the gonimoblast. They could not be cells of the carpogonial branch. For such nutritive cells belonging to the carpogonial branch the term "carpogon auxiliary cells" was introduced by WILKE and ZIEGENSPECK (1929 cf. also KYLIN 1956).

According to KYLIN four types of typical auxiliary cells can be distinguished:

1) The *Dumontia*-type: the auxiliary cells occur on certain branches (auxiliary cell branches) which are formed before fertilization.

2) The *Platoma*-type: an intercalary, vegetative cell serves as auxiliary cell.

3) The *Rhodymenia*-type: the auxiliary cell is cut off before fertilization from a daughter cell of the supporting cell of the carpogonial branch.

4) The *Ceramium*-type: the auxiliary cell is cut off after fertilization from the supporting cell of the carpogonial branch.

A contribution to the discussion of the term "auxiliary cell" was made recently by DREW (1954), who proposes the following definition: "An auxiliary cell is a specified cell of the gametophyte, with which the carpogonium fuses before the formation of gonimoblasts, or a cell with which the primary gonimoblast fuses. Auxiliary cells have a purely nutritive function in those cases where no nucleus is transferred, or combine nutritive and generative functions in those cases where the fertilization nucleus or its derivative is transferred to that cell and there initiates the development of secondary gonimoblasts."

The life cycle of most Florideae includes three generations: the gametophyte, with the dependent carposporophyte, and the tetrasporophyte. Meiosis takes place in the tetrasporangia. Such forms are called diplobiontic. Concerning Nemalionales many forms have two generations only: the gametophyte with the dependent carposporophyte. Meiosis takes place at the first division of the zygote or in the gonimoblasts at the formation of carpospores. Such forms are called haplobiontic.

1. Order: Nemalionales

From the anatomical point of view there are uni-axial as well as multi-axial types. Most of the Nemalionales are haplobionts, but also diplobionts are known. There is a wide range in complexity of structure of both the carpogonial branch and the carposporophyte. In the simpler forms gonimoblast filaments develop directly from the undivided, fertilized carpogonium (e.g. *Kylinia, Batrachospermum*). In *Nemalion* the carpogonium divides by a transverse wall, while the gonimoblast is developed from the upper one, and the lower fuses with the nearest cell of the growing gonimoblast. Other genera show further advances. Thus the gonimoblast filaments develop from certain cells in the carpogonial branch. In some cases a compound nutritive system is formed. In many cases meiosis takes place before formation of the carpospores (haplobiontic forms). In some cases there is an alternation of generation with the meiosis at the formation of tetraspores in the tetrasporophyte (diplobiontic forms).

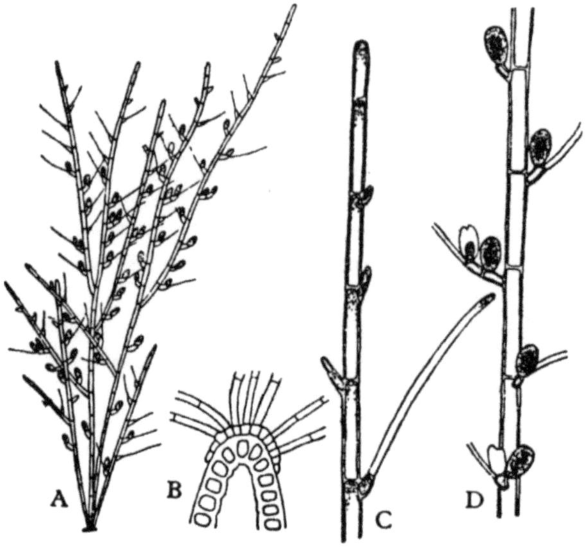

Figure 58: *Acrochaetium virgatulum* (from KYLIN)

The Nemalionales are divided into eight families, most of them well separated from each other. Their relationship to other Florideae – especially Gelidiales – is rather obscure. There are recent proposals either to divide Nemalionales into several orders or to include Gelidiales in Nemalionales.

1. Fam.: CHANTRANSIACEAE – Thallus of simple-branched filaments, microscopic to ca. 25 mm high. – *Kylinia, Acrochaetium, Rhodothamniella, Rhodochorton.*
2. Fam.: BATRACHOSPERMACEAE – Thallus cylindrical, branched unaxial. Only in fresh water – *Batrachospermum.*
3. Fam.: LEMANEACEAE – Thallus cylindrical, branched or simple, rigid uniaxial. In fresh water – *Lemanea.*
4. Fam.: THOREACEAE – Thallus cylindrical, branched, multiaxial. In fresh water – *Thorea.*
5. Fam.: HELMINTHOCLADIACEAE – Thallus cylindrical, branched gelatinous, multiaxial. Marine – *Nemalion, Helminthora, Helminthocladia, Liagora* (more or less incrusted with lime), *Cumagloia.*
6. Fam.: CHAETANGIACEAE – Thallus cylindrical or flattened, generally repeatedly dichotomously branched, texture firm – gelatinous, and of multiaxial structure. Most of the genera are haplobionts, but diplobionts exist *(Galaxaura)* – *Scinaia, Gloiophloea, Pseudoscinaia, Pseudogloiophloea, Chaetangium, Actinotrichia, Galaxaura.*
7. Fam.: NACCARIACEAE – Thallus cylindrical, branched, gelatinous, uniaxial – *Atractophora, Neoardissonia, Naccaria.*

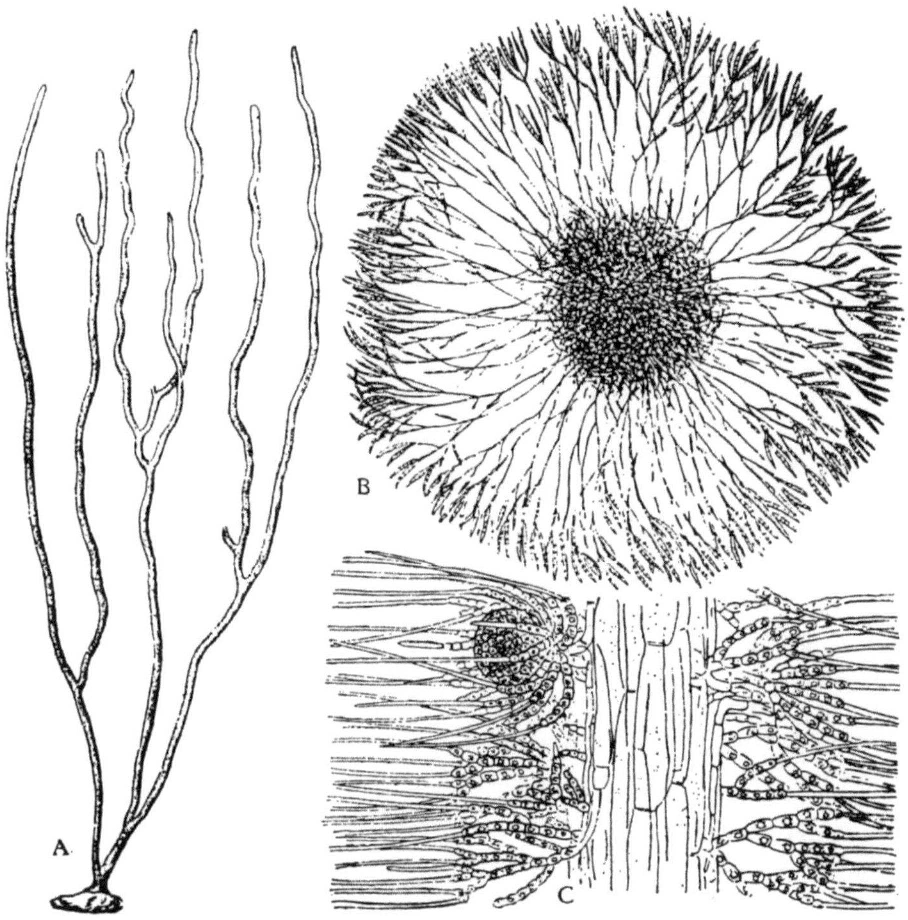

Figure 59: A–B *Nemalion multifidum*, habitus and transv. section (from OLTMANNS). C *Helminthora divariata*, long. section with gonimoblast (from THURET and BORNET)

8. Fam.: BONNEMAISONIACEAE – Thallus slenderly branched, with an evident axis, structure uniaxial – *Bonnemaissonia, Asparagopsis, Delisea, Ptilonia, Leptophyllis.*

2. Order: Gelidiales

Small or medium-sized plants, often with a more or less pinnate branching. From the anatomical point of view they are uniaxial, i.e. there is a central axis throughout the thallus. Typical auxiliary cells are absent. There is an alternation of generations with three phases as well as in most of the Florideae. The Gelidiales were originally

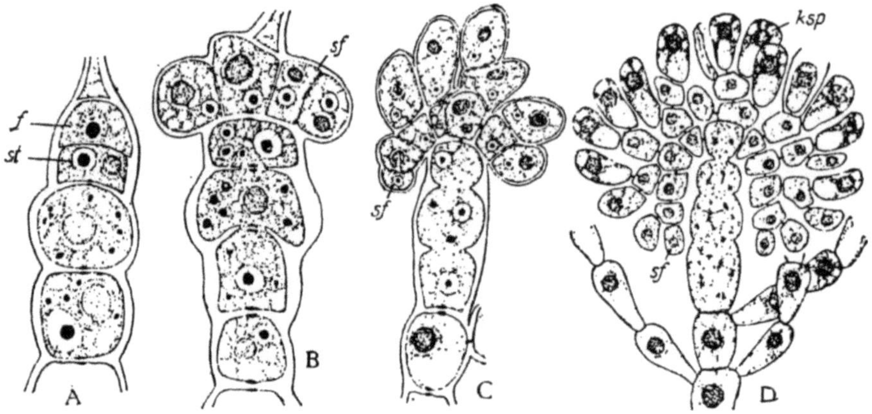

Figure 60: *Nemalion multifidum*, development of gonimoblast (from Kylin)

separated from the Nemalionales. Since the taxonomic position of Gelidiales according to recent investigations seems to be obscure, there are proposals to replace the order in the Nemalionales.

About 70 species, mainly in tropical and subtropical seas. Some species are of great economic importance as raw material for the agar-industry.

Fam.: Gelidiaceae – *Gelidiella, Gelidium, Pterocladia, Suhria* (Pl. XII: 28, 29).

3. Order: Cryptonemiales

Plants with very different forms – filiform, cylindrical, membranous, disc-shaped etc. Thallus corticated with either uniaxial or multiaxial structure.

The carpogonial branches always occur on special accessory branches, sometimes aggregated into sori, nemathecia or conceptacles. It is characteristic that the auxiliary cells are developed before fertilization on accessory branches, either close to or remote from the carpogonial branch. The majority are diplobionts. The order contains 12 families with about 100 genera and 700 species.

1. Fam.: Dumontiaceae – Uniaxial, carpogonial branches scattered, separated from the auxiliary cell branches. Cystocarps scattered – *Dudresnaya, Dumontia, Cryptosiphonia, Leptocladia, Dilsea, Constantinea.*
2. Fam.: Gloiosiphoniaceae – Uniaxial, carpogonial and auxiliary cell branches not separated – *Thuretella, Gloiosiphonia, Schimmelmannia.*
3. Fam.: Endocladiaceae – Cylindrical, branched, uniaxial – *Endocladia, Gloiopeltis.*
4. Fam.: Tichocarpaceae – Flattened, multiaxial – *Tichocarpus.*
5. Fam.: Rhizophyllidaceae – Thallus creeping and dorsiventral or erect. Multiaxial – *Rhizophyllis, Desmia, Ochtodes.*
6. Fam.: Polyideaceae – Cylindrical, repeatedly dichotomously branched or flattened with prolifications. Multiaxial – *Polyides, Rhodopeltis.*

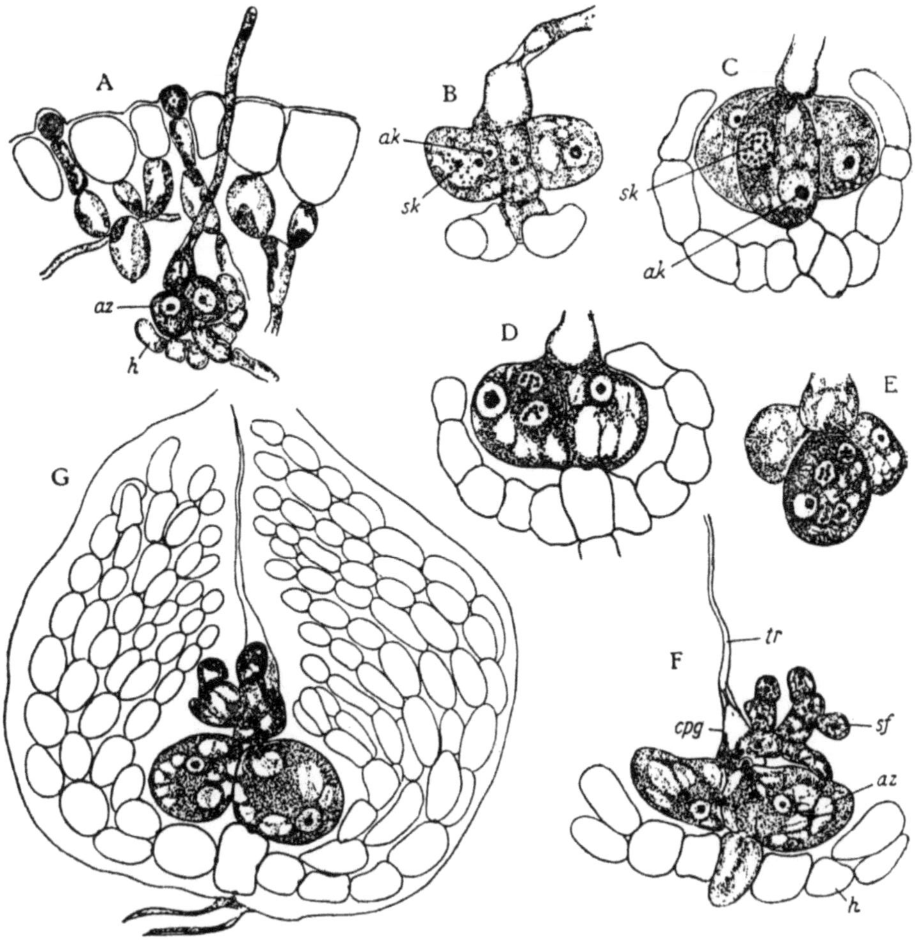

Figure 61: *Scinaia furcellata*, development of gominoblast; az auxilliary cell with haploid nucleus (ak); cpg carpogonium, h cover, sf gonimoblast filaments sk diploid nuclei transferred from the fertilized carpogonium, tr trichogyne (from SVEDELIUS)

7. Fam.: SQUAMARIACEAE – Encrusting, horizontally spreading – *Peyssonnelia, Rhododermis.*
8. Fam.: HILDENBRANDIACEAE – Encrusting, the lower side strongly adherent – *Hildenbrandia.*
9. Fam.: CORALLINACEAE – Thallus incrusted with lime, sometimes hard as stone, Multiaxial – *Sporolithon, Phymatolithon, Lithothamnion, Epilithon, Lithophyllum, Tenarea, Porolithon, Dermatolithon, Melobesia, Amphiroa, Corallina, Jania.*
10. Fam.: GRATELOUPIACEAE – Thallus differently shaped, multiaxial – *Halymenia, Gelinaria, Grateloupia, Cryptonemia, Aeodes, Pachymenia, Prionitis.*

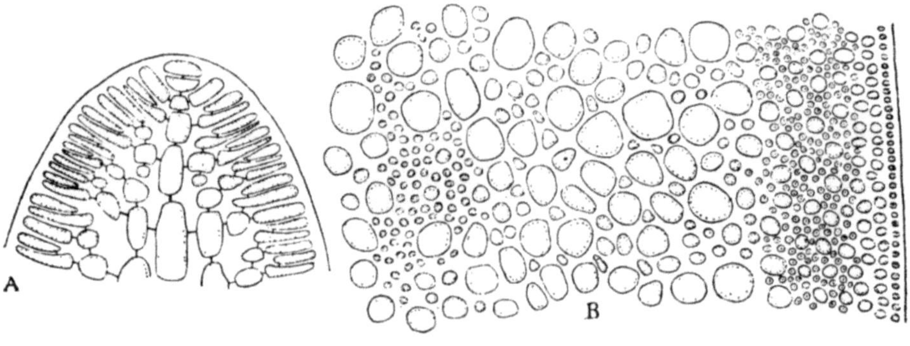

Figure 62: *Gelidium cartilagineum.* A long. section of apex; B transv. section (from KYLIN)

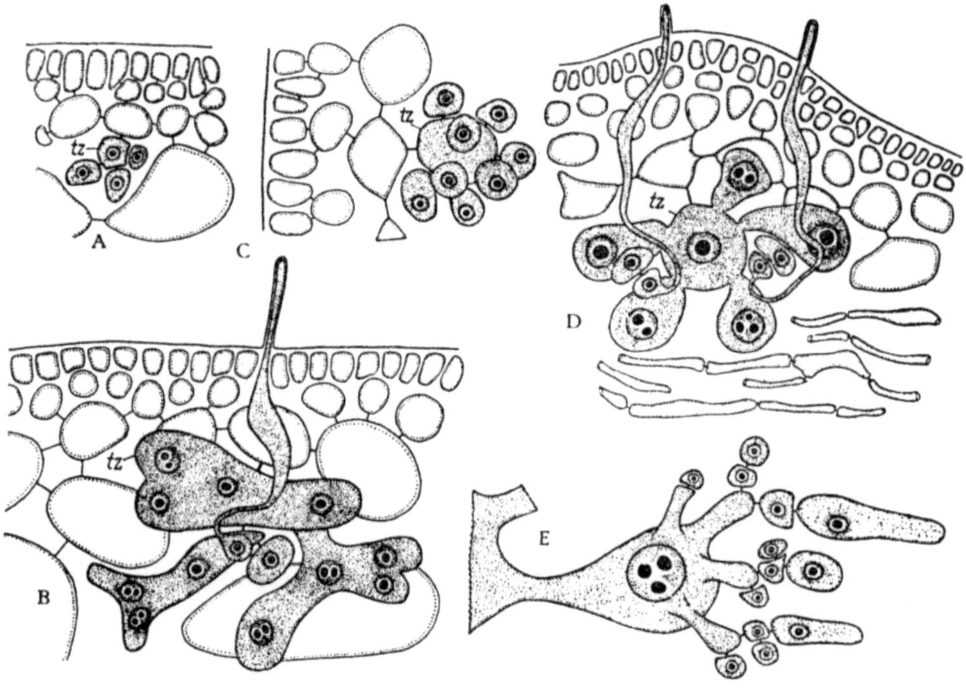

Figure 63: A–B *Callophyllis obtusifolia,* development of procarps. C–E *Callymenia reniformis,* development of procarps and gonimoblast filaments (from KYLIN)

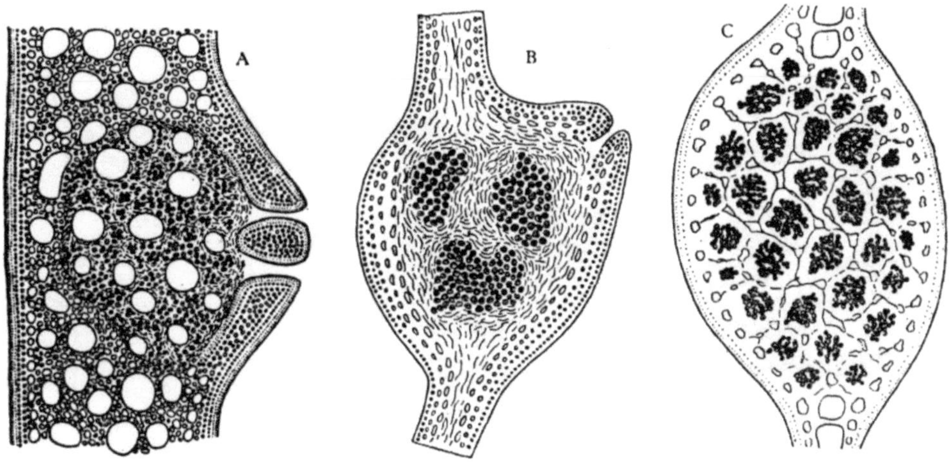

Figure 64: Transv. sections of cystocarps: A *Callymenia obtusifolia;* B *Callymenia reniformis;* C *Rhizopogonia asperata* (from KYLIN)

11. Fam.: CALLYMENIACEAE – Thallus foliaceous or branched, often soft. Uniaxial – *Pugetia, Callophyllis, Euthora, Callymenia.*
12. Fam.: CHOREOCOLACEAE – Minute, cushion-shaped. Parasites on different Rhodomelaceae – *Choreocolax, Harveyella.*

4. Order: Gigartinales

Plants showing various forms from filiform, cylindrical to fleshy – membranaceous or encrusting. Thallus either uniaxial or multiaxial. The carpogonial branches are formed by ordinary vegetative cells, and intercalary vegetative cells serve as auxiliary cells. Normally diplobionts. The order contains 21 families with about 80 genera and 900 species. Some of them are of great economic importance, i.e. *Gracilaria* (Pl. XII: 30), *Eucheuma,* and *Furcellaria* for making agar, *Gigartina* and *Chondrus* ("Irish moss") for the production of carrageen.

1. Fam.: CRUORIACEAE – Plants encrusting. Gonimoblasts develop from connecting filaments – *Cruoria, Petrocelis.*
2. Fam.: CALOSIPHONIACEAE – Cylindrical, branched, with central filament; gelatinous. Sporangia tetrapartite – *Calosiphonia.*
3. Fam.: NEMASTOMACEAE – Cylindrical or foliaceous, soft or gelatinous; multiaxial. Sporangia tetrapartite – *Nemastoma, Platoma, Schizymenia.*
4. Fam.: SEBDENIACEAE – Foliaceous, multiaxial. Gonimoblast with special protective envelope. Sporangia tetrapartite – *Sebdenia.*
5. Fam.: GRACILARIACEAE – Cylindrical or flattened, branched, no central filament. Structure cellular, firm, and often cartilagenous. Cystocarps projecting, hemispherical,

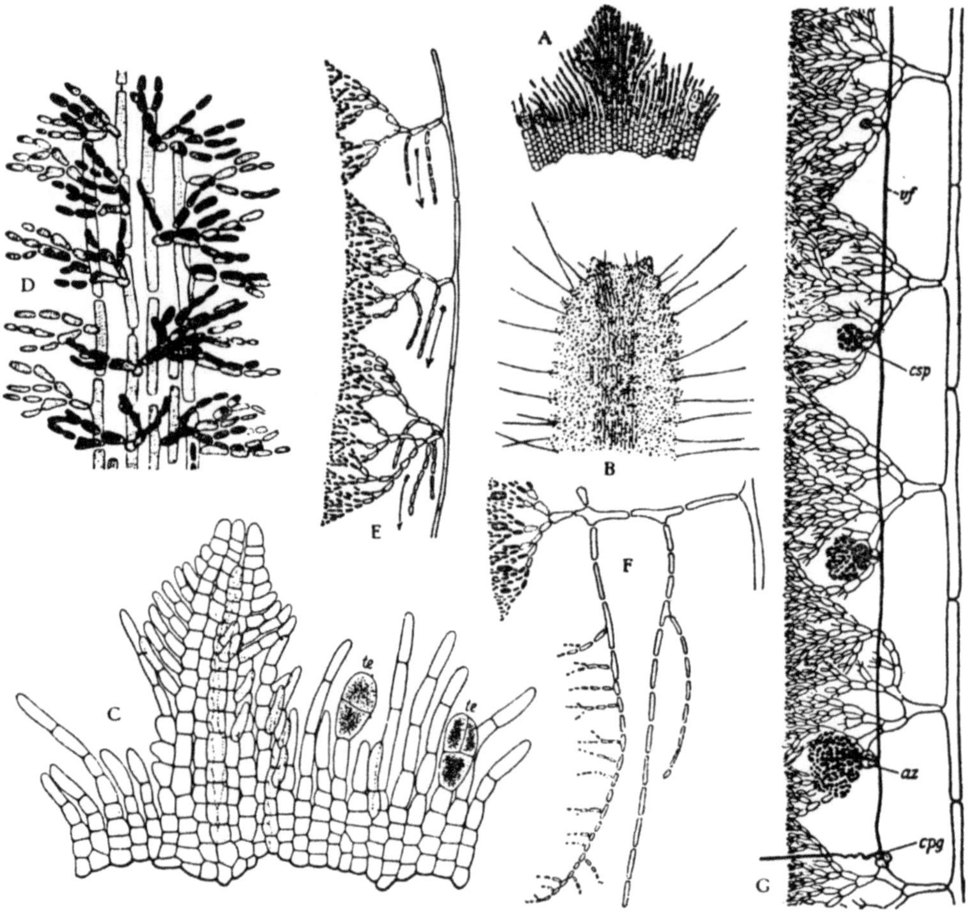

Figure 65: *Platoma Baiardii.* A and C young plant; B branching; D–F morphology of thallus; G development of gonimoplast. – az auxiliary cell, cpg carpogonium, csp carpospores, vf connecting filaments (from Kuckuck)

discharging carpospores through a pore. Sporangia tetrapartite – *Gracilaria, Melanthalia, Tylotus, Curdiea, Gelidiopsis.*

6. Fam.: Plocamiaceae – Thallus flattened, sympodial. Sporangia zonate – *Plocamium.*
7. Fam.: Sphaerococcaceae – Cylindrical or flattened, branched, with central filament. Cystocarps projecting. Sporangia zonate – *Heringia, Caulacanthus, Phacelocarpus, Sphaerococcus, Stenocladia.*
8. Fam.: Stictosporaceae – *Stictosporium.*

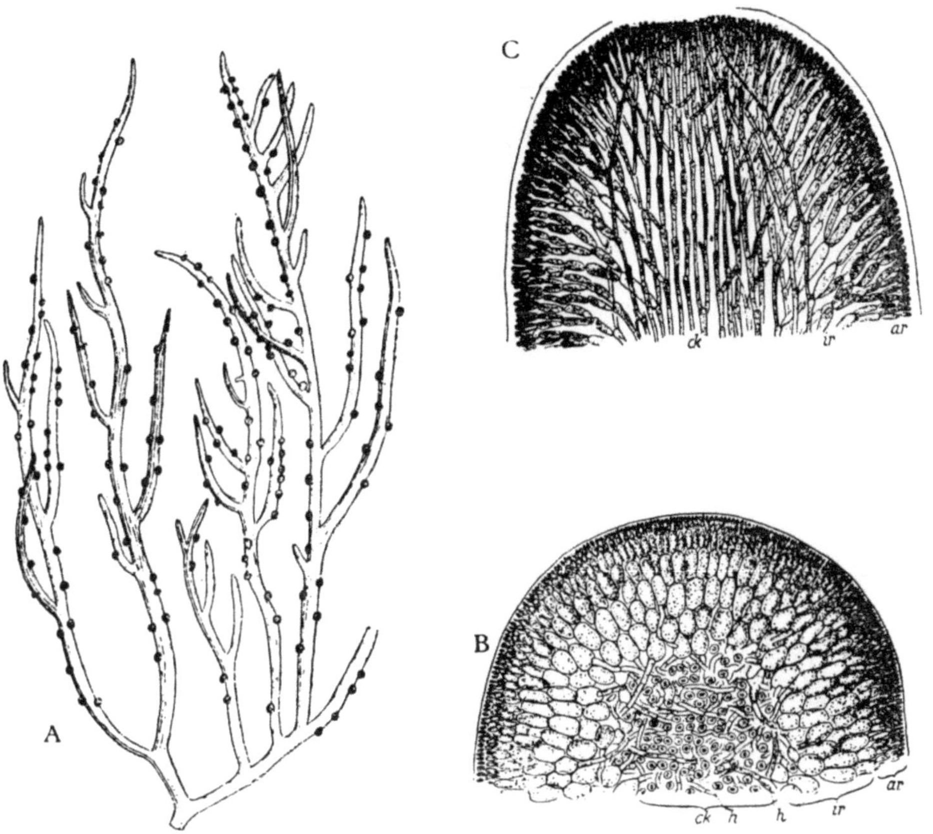

Figure 66: A *Gracilaria compressa* (from HAUCK). B–C *Furcellaria fastigiata*, long. and transv. sections (from OLTMANNS)

9. Fam.: SARCODIACEAE – Cylindrical or flattened, multiaxial, cystocarps projecting; sporangia zonate – *Trematocarpus, Sarcodia.*
10. Fam.: FURCELLARIACEAE – Cylindrical or flattened, multiaxial. Gonimoblasts immersed; sporangia zonate – *Halarachnion, Furcellaria.*
11. Fam.: SOLIERIACEAE – Cylindrical or flattened, with filamentous structure without central axis. Cystocarps with pore, more or less immersed; sporangia zonate – *Thysanocladia, Turnerella, Agardhiella, Solieria, Eucheuma, Meristotheca.*
12. Fam.: RISSOELLACEAE – *Rissoella.*
13. Fam.: RHABDONIACEAE – Cylindrical or flattened, branched. Structure filamentous with central axis. Cystocarps immersed; sporangia zonate – *Catenella, Rhabdonia.*
14. Fam.: RHODOPHYLLIDACEAE – Cylindrical or flattened. Structure filamentous or more or less cellular. Cystocarps projecting; sporangia zonate – *Cystoclonium, Rhodophyllis, Calliblepharis.*

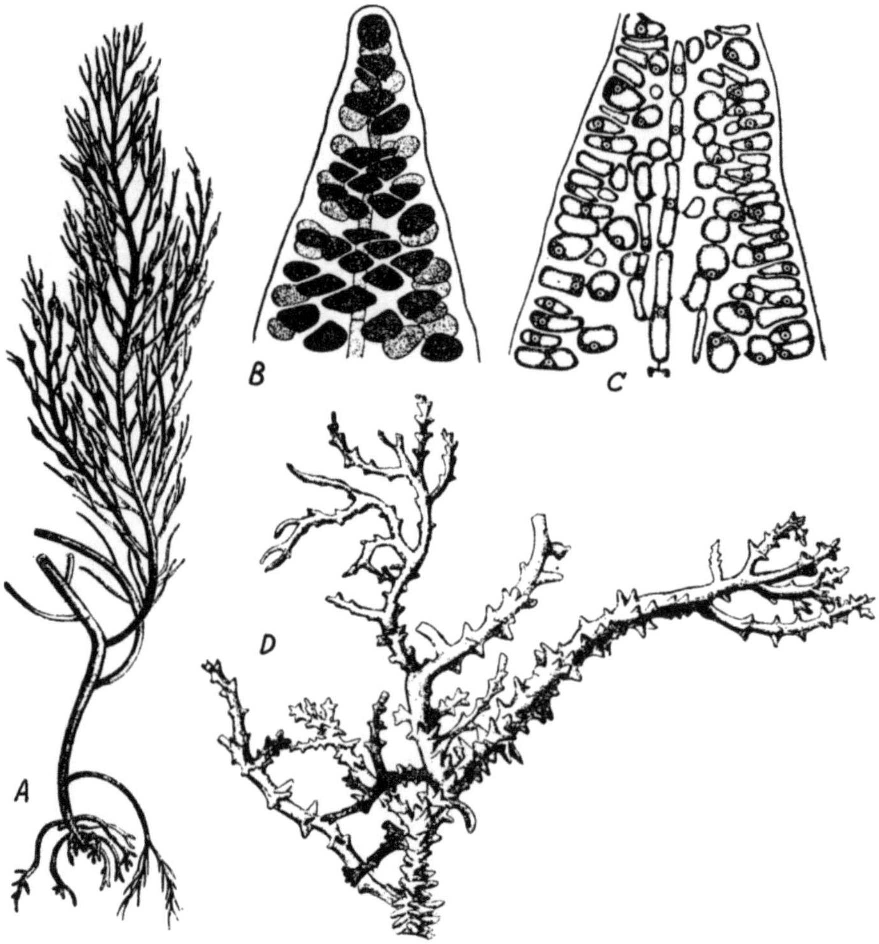

Figure 67: *Cystoclonium purpurescens;* A habitus; B–C long. sections. D *Eucheuma muricatum* (A and D from Kützing, B–C from Kylin)

15. Fam.: Hypneaceae – Cylindrical, branched, branch tips often hooked. Structure cellular; axes near tips showing a central filament. Cystocarps projecting; sporangia zonate – *Hypnea.*
16. Fam.: Mychodeaceae – Cylindrical or flattened; with central filament. Sporangia zonate – *Mychodea.*
17. Fam.: Dicranemaceae – *Dicranema.*
18. Fam.: Acrotylaceae – *Acrotylus.*
19. Fam.: Phyllophoraceae – Cylindrical or flat, dichotomously branched, often with prolifications. Multiaxial, structure cellular. Gonimoblasts immersed or

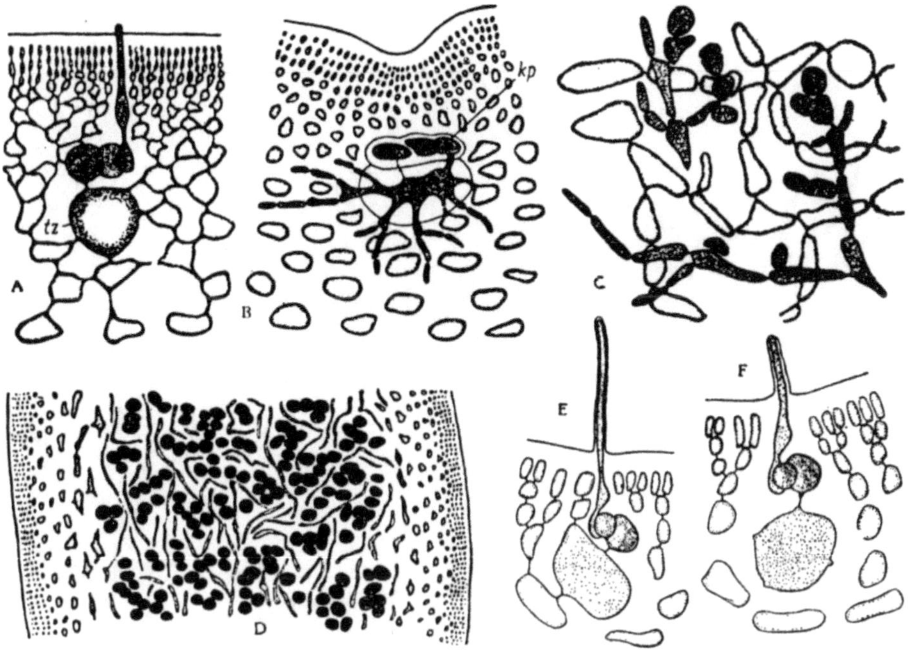

Figure 68: A–D *Chondrus crispus*, development of gonimoblast, E–F *Iridaea cordata*, procarps (from KYLIN)

projecting. Sporangia in superficial nemathecia, tetrapartite, seriate. Sporangial and gametangial reproduction greatly modified in some species – *Phyllophora, Stenogramme, Gymnogongrus, Ahnfeltia.*

20. Fam.: GIGARTINACEAE – Thallus generally flattened, above flabellate to foliaceous. Multiaxial, structure more or less filamentous. Cystocarps immersed or projecting. Sporangia in immersed sori, tetrapartite, seriate – *Chondrus, Iridaea, Rhodoglossum, Gigartina.*
21. Fam.: CHONDRIELLACEAE – *Chondriella.*

5. Order: Rhodymeniales

Thallus of various forms from filiform to fleshy-membranaceous, sometimes hollow, corticated. Structure of modified multiaxial type, commonly appearing parenchymatous, filamentous or cellular. It is characteristic that the auxiliary cell, which is formed before fertilization, is cut off from a daughter cell of the supporting cell of the carpogonial branch. Cystocarps projecting, enveloped by a pericarpium. Sporangia in sori or scattered over the thallus. Diplobionts. There are only two families with 35 genera and about 185 species.

1. Fam.: RHODYMENIACEAE – Thallus foliaceousy, or, if bushy with a distinct main

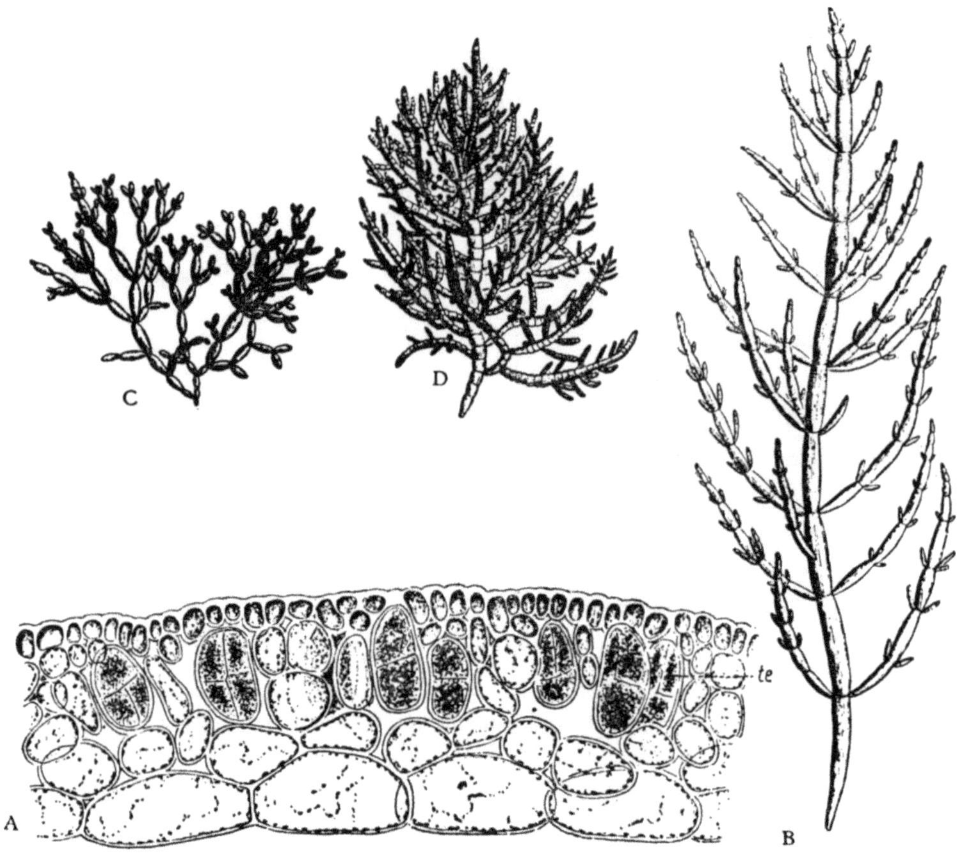

Figure 69: A *Rhodymenia pertusa*, section with tetrasporangia (after KYLIN); B *Chylocladia kaliformis;* C *Gastroclonium ovale;* D *Champia parvula* (from KÜTZING)

axis, solid or hollow. Growth from an apical meristem. Surface cells small, often in short radical series and containing chromatophores. Innermost cells large. Sporangia tetrapartite, rarely tetrahedral – *Fauchea, Gloioderma, Chrysymenia, Erythrocolon, Fryeela, Botryocladia, Gloiosaccion, Halosaccion, Rhodymenia* (Pl. XIII: 31), *Dendrymenia, Epymenia, Hymenocladia.*

2. Fam.: LOMENTARIACEAE – Thallus cylindrical or compressed, multiaxial, solid or hollow. Sporangia tetrahedral. Cystocarps projecting – *Lomentaria, Champia, Chylocladia, Gastroclonium.*

6. Order: Ceramiales

Thallus cylindrical, often filamentous, in a few genera compressed or membranaceous; branched, naked, or corticated, from the anatomical point of view with uniaxial structure.

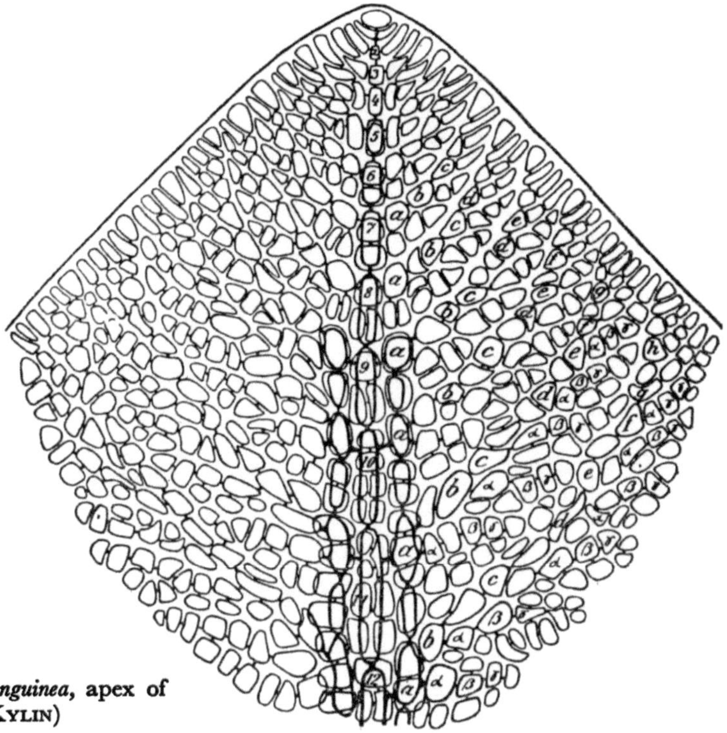

Figure 70: ***Delesseria sanguinea*****, apex of thallus (from KYLIN)**

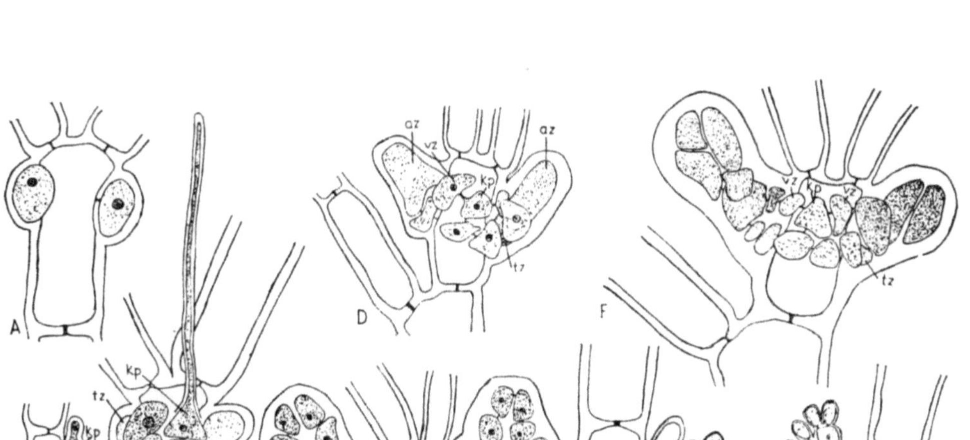

Figure 71: *Callithamnion bipinnatum;* A–C development of procarps; D–F development of gonimoblast; G–H spermatangia. – az auxiliary cell; kp carpogonium; tz supporting cell; vz connecting cell (x 435) (from LEVRING)

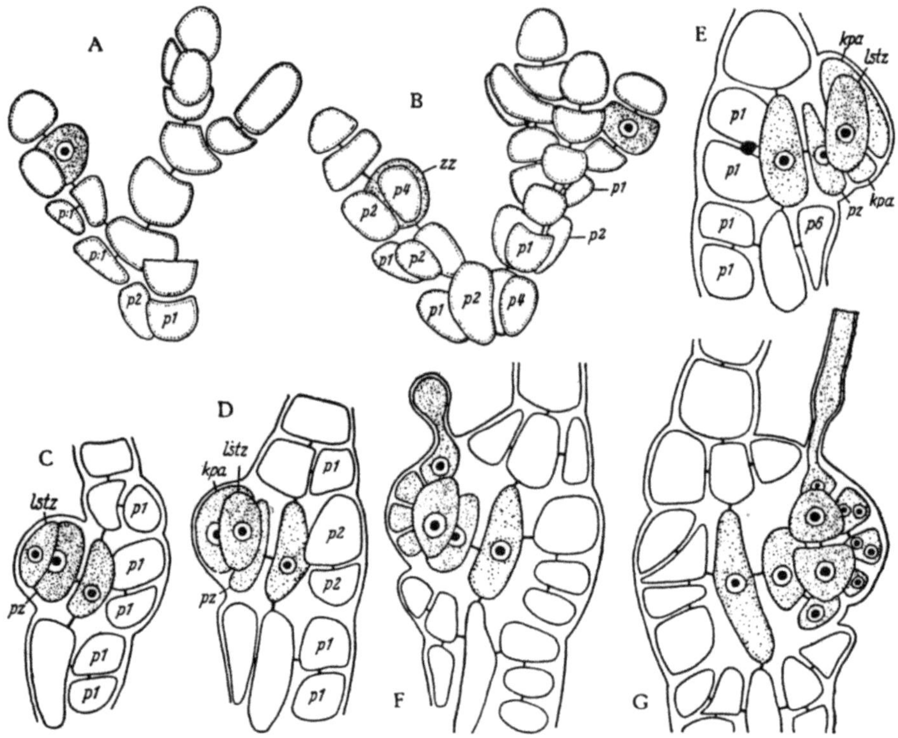

Figure 72: *Rhodomela virgata;* A apex with right spiral; B apex with left spiral; C–G development of procarp. – az auxiliary cell; kpa carpogonial branch; lstz lateral sterile cell; p 1, p 2, p 3 and p 4 first to fourth pericentral cell; zz fertile central cell (from Kylin)

Growth from an apical cell. The carpogonial branch is always four-celled and born on a pericentral cell or its equivalent. The auxiliary cells are formed first after fertilization, one or two cut off from the supporting cell of the carpogonial branch or a homologous pericentral cell. Gonimoblasts naked or enveloped by a pericarpium. Plants generally diplobionts.

There are four families with about 250 genera and 1,300 species.

1. Fam.: Ceramiaceae – Thallus cylindrical, occasionally flattened, often filamentous. Gonimoblasts naked or enveloped by branchlets. Sporangia tetrapartite or tetrahedral – *Crouania, Antithamnion, Ballia, Ceramium* (Pl. XIII: 32), *Centroceras, Spyridia, Callithamnion, Compsothamnion, Pleonosporium, Griffithsia, Spermothamnion, Bornetia, Plumaria, Ptilota, Dasyphila.*
2. Fam.: Delesseriaceae – Thallus usually foliaceous, simple or branched. Structure primarily uniaxial with growth from an apical cell. Sporangia tetrahedral,

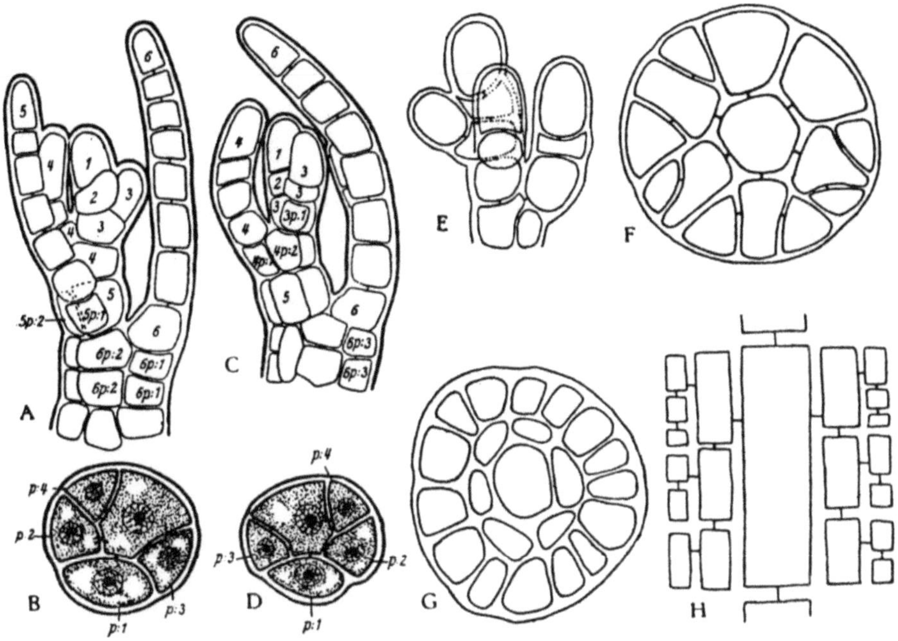

Figure 73: *Rhodomela virgata;* A, B and E apex with right spiral and transv. section of the same; C–D same of apex with left spiral; F–G transv. sections; H long. section (diagr.). – p pericentral cell; 6p: 1 = first pericentral cell of the sixth segment etc. (A–D from ROSENBERG, E–H from KYLIN)

usually in superficial sori. Gonimoblasts with ostiolate pericarpium – *Hypoglossum, Membranoptera, Grinnellia, Apoglossum, Delesseria* (Pl. XIII: 33), *Claudea, Polyneura, Phycodrys, Myriogramme, Schizoseris, Platyclinia, Nitophyllum, Hymenena, Cryptopleura, Martenisa.*

3. Fam.: DASYACEAE – Thallus cylindrical or flattened with sympodial growth. Sporangia tetrahedral, produced in special stichidia. Cystocarp with ostiolate pericarp – *Dasya, Heterosiphonia.*
4. Fam.: RHODOMELACEAE – Thallus cylindrical, flattened or rarely membranaceous, branches, often bushy, with monopodial growth from persisting apical cells producing an axial cell row. Branched colourless hairs, trichoblasts, often present. Axial cells generally surrounded by a number of pericentral cells cut off from them by longitudinal walls, producing a characteristic polysiphonous structure. Cortication frequent. Sporangia tetrahedral; cystocarps with ostiolate pericarp – *Polysiphonia, Lophurella, Brongniartella, Pterosiphonia, Rhodomela, Odonthalia, Bostrychia, Streblocladia, Herposiphonia, Polyzonia, Lophosiphonia, Rytiphlaea, Amansia, Vidalia, Lenormandia, Chondria, Laurencia.*

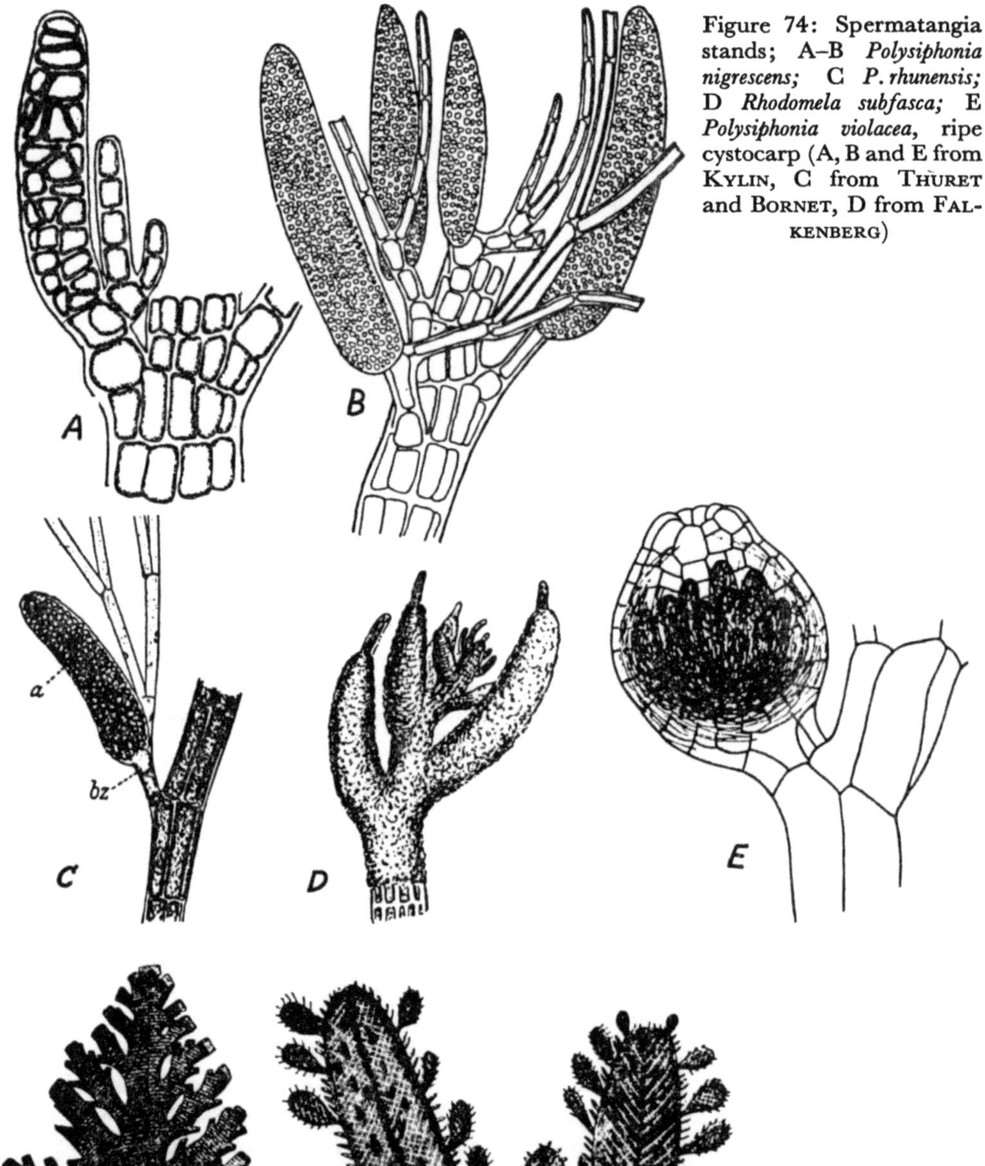

Figure 74: Spermatangia stands; A–B *Polysiphonia nigrescens;* C *P. rhunensis;* D *Rhodomela subfasca;* E *Polysiphonia violacea,* ripe cystocarp (A, B and E from KYLIN, C from THURET and BORNET, D from FALKENBERG)

Figure 75: A *Amansia multifida* (x 2,5); B *Lenormandia marginata* (nat. size) (A from FALKENBERG, B from HAEVEY)

AGARDH, C. A., Systema algarum. Lund. 1824.

AGARDH, J. G., Algae maris Mediterranei et Adriatici. Paris. 1842.

ALLEN, W. E., „Red Water" in La Jolla Bay in 1945. – Trans. Amer. Microsc. Soc. 65. 1946.

ALLISON, F. E., and HOOVER, S. R., Conditions which favour nitrogen fixation by a blue-green alga. – Trans. Third Int. Congr. Soil Sci. 1. 1935.

ALLORGE, P., Héterocontes ou Xanthophycées? – Rev. Algol., 5. 1930.

ARESCHOUG, J. E., Observationes phycologicae. I–V. Uppsala. 1866–84.

BARKER, H. A., The culture and physiology of the marine Dinoflagellates. – Arch. f. Mikrobiol. 6 (2). 1935.

BERTHOLD, G., Die geschlechtliche Fortpflanzung der eigentlichen Phaeosporeen. – Mitt. Zool. Stat. Neapel. 2 (3). 1881.

BERTHOLD, G., Bangiaceen des Golfes von Neapel und der angrenzenden Meeres-Abschnitte. – Fauna u. Flora des Golfes von Neapel. 8. 1882.

BITTER, G., Zur Anatomie und Physiologie von Padina pavonia. – Ber. Deutsch. Bot. Ges. 17. 1899.

BLIDING, C., Über Sexualität und Entwicklung bei der Gattung Enteromorpha. – Svensk. Bot. Tidskr. 27. 1933.

BLIDING, C., Studien über Entwicklung und Systematik in der Gattung Enteromorpha. II. – Bot. Notiser. 1939.

BLIDING, C., Zur Systematik der schwedischen Enteromorphen. – Bot. Notiser. 1944.

BÖRGESEN, F., Contributions à la connaissance du genre Siphonocladus Schmitz. – Overs. Dansk. Vidensk. Selsk. Forhandl. 1905.

BÖRGESEN, F., The marine Algae of the Danish West Indies. I. – Dansk Bot. Arkiv. 1, 2. 1913–1914.

BÖRGESEN, F., The marine algae of the Danish West Indies. 2. Phaeophyceae. – Dansk Bot. Arkiv 2 (2). 1914.

BÖRGESEN, F., Marine algae from Easter Island. – In Skottsberg, C. The natural history of Juan Fernandez and Easter Island 2 (9). Uppsala. 1924.

BÖRGESEN, F., Some Indian green and brown algae, especially from the shores of the presidency of Bombay. I. – Journ. Indian Bot. Soc. 9. 1930.

BÖRGESEN, F., Marine Algae of the Iranian Gulf. – Danish Scientific Investigations in Iran. Part I. Copenhagen. 1939.

BORNET, E., Note sur quelques Ectocarpus. – Bull. Soc. Bot. France 38. 1891.

BORNET, E., & FLAHAULT, C., Revision des Nostocacées hétérocystées contenues dans les principaux herbiers de France. Ann. Sci. Nat. Bot. Sér. 7, 3, 4, 5, 7. 1885–1887.

BOURELLY, P., Recherches sur les Chrysophycées. – Rev. Algol., Mém. hors-série 1. 1957.

BRAUN, ALEXANDER, Betrachtungen über Erscheinung der Verjüngung in der Natur insbesondere in der Lebens- und Bildungsgeschichte der Pflanze. – Leipzig. 1851.

BRAUN, ALEXANDER, Über die Richtungsverhältnisse der Saftströme in den Zellen der Characeen. – Monatb. Akad. Wiss. Berlin. 1852.

BRAUN, ALEXANDER, Algarum unicellularum genera nova vel minus cognita, praemissis observationibus de algis unicellularibus in genere. Leipzig 1855.

BREBNER, G., On the classification of the Tilopteridaceae. – Proc. Bristol Nat. Soc. Ser. 3, 8. 1897.

BURKHOLDER, P. R., Movement in the Cyanophyceae. – Quart. Rev. Biol. 9. 1934.

BUTCHER, R. W., An introductory account of the smaller algae of British coastal waters. Part 1: Introduction and Chlorophyceae. – Min. of Agric. Fish. and Food, Fish. Invest. Ser. 4. 1959.

CARTER, P. W., The life-history of Padina pavonia. I. The structure and cytology of the tetrasporangial plant. – Ann. Bot. 41. 1927.

CHADEFAUD, M., Les pyrenoides des algues. – Ann. Sci. Nat. Botanique 11th ser. 11. 1941.

CHADEFAUD, M., Les végétaux non vasculaires. – Chadefaud et Emberger: Traité de botanique. Tome 1. Paris. 1960.

CLELAND, R. E., The cytology and life-history of Nemalion multifidum Ag. – Ann. Bot. 33. 1919.

CLEVE-EULER, A., Die Diatomeen von Schweden und Finnland. – Kgl. Svenska Vetenskapsakad. Handl. Ser. 4, 2–5. 1951–1955.

CLINT, H. B., The life history and cytology of Sphacelaria bipinnata Sauv.-Publ. Hartley Bot. Lab. Univ. Liverpool, No. 3. 1927.

COHN, F., Untersuchungen über die Entwicklungsgeschichte der mikroskopischen Algen und Pilze. Nova Acta Acad. Caes. Leop., 24. 1853.

CZURDA, V., Morphologie und Physiologie des Algenstärkekornes. – Beih. Bot. Centralbl. 45, 1. 1928.

DAMMANN, H., Entwicklungsgeschichtliche und zytologische Untersuchungen an Helgoländer Meeresalgen. – Wiss. Meeresuntersuch. Abt. Helgoland, N. F. 18 (4). 1930.

DAMMANN, H., Entwicklungsgeschichtliche und zytologische Untersuchungen an Helgoländer Meeresalgen. – Wiss. Meeresunters. Abt. Helgoland N. F. 18. 1930–1932.

DANGEARD, P., Recherches sur les Porphyra. – Le Botaniste. 18. 1927.

DANGEARD, P., Le vacuome des Algues et sa transmission par les zoospores. – Compt. Rend. Acad. Sci. Paris 194. 1932.

Dangeard, P. A., Mémoire sur la famille des Péridiniens. – Le Botaniste 29. 1938.

Dangeard, P. A. & Dangeard, P., Recherches sur le vacuome des Algues inférieures. – Ibid. 118. 1924.

Deflandre, G., Strombomonas, nouveau genre d'Euglénacées (Trachelomonas Ehrb. pro parte). Arch. Protistenk. 69. 1930.

Deflandre, G., Contribution à l'étude des Silicoflagellidés actuels et fossiles. – Microscopie 2, 2–4. 1950.

Desikachary, T. V., Cyanophyta. – Ind. Council Agric. Res. Monogr. on Algae. New Delhi. 1959.

De Toni, J. B., Sylloge Algarum. Vol. 1–6. Padova. 1889–1924.

Dillwyn, L. W., British Confervae. London. 1802–1809.

Dixon, P. S., Studies on marine algae of the British Isles; the genus Ceramium. – J. mar. biol. Ass. U. K. 39. 1960.

Dixon, P. S., On the classification of the Florideae with particular reference to the position of the Gelidiaceae. – Bot. Marina, 3. 1961.

Dixon, P. S., The Rhodophyta; 1 some aspects of their biology. – Oceanogr. mar. Biol., Ann. Rev. 1. 1963.

Dixon, P. S., Perennation, vegetative propagation and algal life histories, with special reference to Asparagopsis and other Rhodophyta. – Botanica Gothoburgensia 3. 1965.

Drew, K. M., Contributions to the cytology of Spermothamnion Turneri (Mert.) Aresch. 1. The diploid generation. – Ann. Bot. 48. 1934.

Drew, K. M., The life-history of Rhodochorton violaceum (Kütz.) comb. nov. (Chantransia violacea. Kütz.). – Ibid. 49. 1935.

Drew, K. M., Spermothamnion Snyderae Farlow, a Floridean alga bearing polysporangia. – Ibid. N. S. 1. 1937.

Drew, K. M., An investigation of Plumaria elegans (Bonnem.) Schmitz with special reference to triploid plants bearing parasporangia. – Ibid. N. S. 3. 1939.

Drew, K. M., Contributions to the cytology of Spermothamnion Turneri (Mert.) Aresch. 2. The haploid and triploid generations. – Ibid. N. S. 7. 1943.

Drew, K. M., Nuclear and somatic phases in the Florideae. – Biol. Rev. 19. 1944.

Drew, K. M., Conchocelis-Phase in the Life-History of Porphyra umbilicalis (L.) Kütz. – Nature 164. 1949.

Drew, K. M., The organisation and inter-relationships of the carposporophytes of living Florideae. – Phytomorph. 4. 1954.

Drouet, F. and Daily, W. A., Revision of the coccoid Myxophyceae. – Butler Univ. Bot. Studies 12. 1956.

Elenkin, A. A., Monographia algarum Cyanophycearum aquidulcium et terrestrium in finibus URSS inventarum. Pars specialis (systematica). Fasc. 1. Moscow, Leningrad. 1938.

Engler, A., Syllabus der Pflanzenfamilien. 2 Ed. Berlin. 1898.

Falkenberg, P., Die Befruchtung und der Generationswechsel von Cutleria. – Mitt. Zool. Stat. Neapel 1. 1879.

Falkenberg, P., Die Rhodomelacéen des Golfes von Neapel und der angrenzenden Meeres-Abschnitte. – Fauna u. Flora d. Golfes von Neapel. 26. 1901.

Farmer, J. B. & Williams, J. L., Contributions to our knowledge of the Fucaceae: their life-history and cytology. – Phil. Trans. Roy. Soc. London. Ser. B 190. 1898.

Feldmann, J., Les algues marines de la côte des Albères, I–III. – Revue algol. 9. 1937.

Feldmann, J., Sur la classification de l'ordre des Siphonocladales. Rev. génér. Bot. 50. 1938.

Feldmann, J., Sur l'héteroplastie de certaines Siphonales et leur classification. Compt. Rend. Acad. Sc. Paris. 222. 1946.

Feldmann, J. & G., Recherches sur les Bonnemaisoniacées et leur alternance de générations. – Ann. Sci. Nat. 11 série Bot. 3. 1942.

Feldmann-Mazoyer, G., Recherches sur les Céramiacées de la Méditerranée occidentale. – Alger. 1940.

Forti, A., Sylloge Myxophycearum. In J. B. De Toni, Sylloge algarum omnium hucusque cognitarum 5. 1907.

Forti, A., Contribuzioni diatomologiche, XII. Atti del Reale Istituto Veneto di Sci., Lettere ed Arti, 71 (2). 1912.

Fott, B., Algenkunde. Jena. 1959.

Føyn, B., Vorläufige Mitteilungen über die Sexualität und den Generationswechsel von Cladophora und Ulva. – Ber. Deutsch. Bot. Ges. 47. 1929.

Føyn, B. R., Über den Lebenszyklus einiger Braunalgen. Vorläufige Mitteilung. – Bergens Mus. Årsbok 1934.

Føyn, B., Lebenszyklus und Sexualität der Chlorophycee Ulva lactuca L. – Arch. Protistenk. 83. 1934.

Føyn, B., Lebenszyklus, Cytologie und Sexualität der Chlorophycee Cladophora Suhriana Kützing. – Ibid. 83. 1934 A.

Frémy, P., Cyanophycées des côtes d'Europe. – Mém. Soc. Nat. Math. Cherbourg 41. 1933.

Fritsch, F. E., The structure and reproduction of the algae. 1–2. Cambridge. 1935–1945.

Fritsch, F. E., The interrelations and classification of the Myxophyceae (Cyanophyceae). New Phytol., 41. 1942.

Fritsch, F. E., Studies in the comparative morphology of the algae. I. Heterotrichy and juvenile stages. Ann. Bot. n. s., 6. Phaeophyceae. Ibid., 7. 1942–1943.

Fritsch, F. E., Present-day classification of algae. – Bot. Rev., 10. 1944.

Geitler, L., Cyanophyceae. – Rabenhorst: Kryptogamen-Flora. 14. Leipzig. 1932.

Geitler, L., Schizophyta: Klasse Schizophyceae.

– Engler und Prantl: Die natürlichen Pflanzenfamilien. 2. Aufl. 1b. 1942.

GEITLER, L., Schizophyceen. – Linsbauer: Handbuch der Pflanzenanatomie. Bd. 6. Teil 1. 2. Aufl. Berlin. 1960.

GEMEINHARDT, K., Oedogoniales. – Rabenhorst: Kryptogamen-Flora. Bd. 12. Abt. 4. Leipzig. 1938–1939.

GERLOFF, J., Beiträge zur Kenntnis der Variabilität und Systematik der Gattung Chlamydomonas. – Arch. Protistenk. 94. 1940.

GOJDICS, M., The genus Euglena. Madison. 1953.

GOMONT, M., Recherches sur les enveloppes cellulaires des Nostocacées filamenteuses. – Bull. Soc. Bot. France. 35. 1888.

GOMONT, M., Monographie des Oscillariées (Nostocacées homocystées). – Ann. Sci. Nat. 7. Bot. 15. 1892.

GMELIN, S. G., Historia Fucorum. – Petropol. 1768.

GRAHAM, H. W., Studies in the morphology, taxonomy, and ecology of the Peridiniales. – Sci. Res. Cruise 7 of Carnegie. Biol. 3. Carnegie Inst. Wash. Publ. 542. 1942.

GREVILLE, R. K., Algae Britannicae. Edinburgh. 1830.

GROVES, J. and BULLOCK-WEBSTER, G. R., The British Charophyta. Vol. 1–2. London. 1920–1924.

HAMEL, G., Chlorophycées des côtes françaises. Paris. 1931.

HAMEL, G., Phéophycées de France. Paris. 1931–1939.

HANSTEEN, B., Studien zur Anatomie und Physiologie der Fucoiden. – Jahrb. Wiss. Bot. 24. 1892.

HARTMANN, M., Untersuchungen über relative Sexualität. I. Versuche an Ectocarpus siliculosus. – Biol. Zentralbl. 45. 1925.

HARTMANN, M., Untersuchungen über die Sexualität von Ectocarpus siliculosus. – Arch. Protistenk. 83. 1934.

HARTMANN, M., Ergänzende Untersuchungen über die Sexualität von Ectocarpus siliculosus. – Ibid. 89. 1937.

HARVEY, W. H., Algae. In: J. T. Mackay. Flora Hibernica. Dublin. 1836.

HARVEY, W. H., Phycologia Britannica. London. 1846–1851.

HARVEY, W. H., Nereis Boreali Americana. Part I. Melanospermeae. Smiths. Contrib. to Knowledge, 3 (Art. 4): Part II. Rhodospermeae. – Ibid., 4 (Art. 5): Part III. Chlorospermeae. – Ibid., 10 (Art. 2): 1852–1858.

HEERING, W., Chlorophyceae 3. Ulothrichales, Microsporales, Oedogoniáles Pascher: Süßwasser-Flora. H. 6. Jena. 1914.

HEURCK, H. VAN, A treatise on the Diatomaceae. London. 1896.

HIRANO, M., Flora Desmidiarum Japonicarum. – Contr. Biol. Lab. Kyoto Univ. 1, 2, 3, 5, 7, 9, 11. 1955–1960.

HOLLENBERG, G. J., Culture studies of marine algae. I. Eisenia arborea. – Amer. Jour. Bot. 26. 1939.

HOLLENBERG, G. J., Culture studies of marine algae. II. Hapterophycus canaliculatus S. & G. – Ibid. 28. 1941.

HOYT, W. D., Periodicity in the production of the sexual cells of Dictyota dichotoma. – Bot. Gaz. 43. 1907.

HOYT, W. D., The periodic fruiting of Dictyota and its relation to the environment. – Amer. Jour. Bot. 14. 1937.

HOYT, W. D., The periodic fruiting of Dictyota and its relation to the environment. – Proc. Internat. Congr. Plant. Sci. Ithaca 1. 1929.

HUBER-PESTALOZZI, G., Das Phytoplankton des Süßwassers. Teil 1–5. – Die Binnengewässer 16. 1938–1961.

HUSTEDT, F., Bacillariophyta. – Pascher: Süßwasser-Flora. H. 10. 2. Aufl. Jena. 1930 A.

HUSTEDT, F., Die Kieselalgen. Rabenhorst's Kryptogamenflora. 7, I. 1930 B.

HYGEN, G., Über den Lebenszyklus und die Entwicklungsgeschichte der Phaeosporeen. Versuche an Nemacystus divaricatus (Ag.) Kuck. Nyt. Mag. Naturvidensk. 74. 1934.

ISRAELSON, G., The freshwater Florideae of Sweden. – Symb. Bot. Upsal. 6. 1942.

IYENGAR, M. O. P., On the formation of gametes in a Caulerpa. (Preliminary note). Jour. Indian Bot. Soc. 12. 1933.

JACOBSEN, J. P., Aperçu systématique et critique sur les Desmidiacées du Danemark. – Bot. Tidsskr. 8. 1874.

JAHN, T. L., The euglenoid flagellates. Quart. Rev. Biol. 21. 1946.

JÖNSSON, B., Zur Kenntnis des Baues und der Entwicklung des Thallus bei den Desmarestieen. – Lunds Univ. Årsskr. Avd. 2, 37 (6). 1901.

JORDE, I., Untersuchungen über den Lebenszyklus von Urospora Aresch. und Codolium A. Braun. – Nyt, Mag. Naturvidenskab. 73. 1933.

JAASUND, E., Aspects of the marine algal vegetation of North Norway. – Botanica Gothoburgensia, 4. 1965.

KANDA, T., On the gamethophytes of some Japanese species of Laminariales. Sci. Pap. Inst. Algol. Res. Fac. Sci. Hokkaido Imp. Univ. 1 (2). 1936.

KANDA, T., On the gametophytes of some Japanese species of Laminariales. II. – Ibid. 2 (1). 1938.

KANDA, T., On the gametophytes of some Japanese species of Laminariales. III. – Ibid. 2 (2). 1941.

KANDA, T., On the gametophytes of some Japanese species of Laminariales, IV. Ibid. 2 (2). 1941 A.

KARSTEN, G., Bacillariophyta (Diatomeae). In A. Engler and K. Prantl. Die Natürlichen Pflanzenfamilien. Leipzig. 1928.

KJELLMAN, F. R., Phaeophyceae (Fucoideae). In: Engler & Prantl, Die natürlichen Pflanzen-Familien, 1, Abt. 2. Leipzig. 1897.
KLEBS, G., Flagellatenstudien. II. – Ztschr. Wiss. Zool. 55. 1893.
KLEBS, G., Über Flagellaten und algenähnliche Peridineen Verh. d. Naturh.-Mediz. Vereins zu Heidelberg 11. 1912.
KNEBEL, G., Monographie der Algenreihe der Prasiolales, insbesondere von Prasiola crispa. – Hedwigia 75. 1935.
KNIGHT, M., Studies in the Ectocarpaceae. I. The life-history and cytology of Pylaiella litoralis Kjellm. – Trans. Roy. Soc. Edinburgh 53. 1923.
KNIGHT, M., Studies in the Ectocarpaceae. II. The life-history and cytologi of Ectocarpus siliculosus, Dillw. – Ibid. 56. 1929.
KNIGHT, M. & PARKE, M. W., Manx algae, etc. Liverpool Marine Biol. Comm. Mem. 30. Liverpool. 1931.
KNIGHT, M., BLACKLER, M. C. H. & PARKE, M. W., Notes on the life-cycle of species of Asperococcus. – Trans. Liverpool Biol. Soc. 48. 1935.
KOFOID, C. A., On significance of assymmetry in Triposolenia. Univ. Calif. Publ. Zool. 3. 1906.
KOFOID, C. A., On Peridinium steini Jörgensen, with a note on the nomenclature of the skeleton of the Peridinidae. – Arch. Protistenk. 16. 1909.
KOFOID, C. A. and SKOGSBERG, T., The Dinophysoidae. The Dinoflagellata. – Scientific results of „Albatross" exped. 1904–1905. Cambridge. 1928.
KOFOID, C. A. and SWEZY, O., The free-living unarmored Dinoflagellata. Mem. Univ. Calif. 5. 1921.
KOLKWITZ, R. und KRIEGER, H., Zygnemales. – Rabenhorst; Kryptogamen-Flora. 2. Aufl. Bd. 13, Abt. 2. Leipzig. 1941–1944.
KOMAREK, J., Die taxonomische Revision der planktischen Blaualgen der Tschechoslowakei.– Komárek und Ettl: Algologische Studien. Prag. 1958.
KORNMANN, P., Zur Entwicklungsgeschichte von Derbesia und Halicystis. Planta 28. 1938.
KRICHENBAUER, H., Beitrag zur Kenntnis der Morphologie und Entwicklungsgeschichte der Gattung Euglena und Phacus. – Arch. Protistenk. 90. 1937.
KRIEGER, W., Die Desmidiaceen. – Rabenhorst: Kryptogamen-Flora. 2. Aufl. Bd. 13. Leipzig. 1933–1938.
KUCKUCK, P., Beiträge zur Kenntnis der Ectocarpus-Arten der Kieler Föhrde. – Bot. Centralbl. 48. 1891.
KUCKUCK, P., Über Schwärmsporenbildung bei den Tilopterideen und über Choristocarpus tenellus (Kütz.) Zan. – Jahrb. Wiss. Bot. 28. 1895.
KUCKUCK, P., Beiträge zur Kenntnis der Meeresalgen. 1. Über Rhododermis parasitica Batters. Wiss. Meeresunters., 2 (1). 1896.
KUCKUCK, P., Beiträge zur Kenntnis der Meeresalgen. 8. Compsonema, ein neues Genus der Phaeosporeen. – Ibid. 1899.
KUCKUCK, P., Beiträge zur Kenntnis der Meeresalgen. 9. Über den Generationswechsel von Cutleria multifida (Engl. Bot.) Grev. – Wiss. Meeresuntersuch., Abt. Helgoland, N. F. 3. 1899.
KUCKUCK, P., Die Fortpflanzung der Phaeosporeen. – Ibid. N. F. (Abt. Helgoland), 5. 1912.
KUCKUCK, P., Beiträge zur Kenntnis der Meeresalgen. 10. Neue Untersuchungen über Nemoderma Schousboe. Ibid. 5. 1912 A.
KUCKUCK, P., Beiträge zur Kenntnis der Meeresalgen. 11. Die Fortpflanzung der Phaeosporeen. Ibid. 5. 1912 B.
KUCKUCK, P., Fragmente einer Monographie der Phaeosporeen. – Ibid. 17. 1929.
KUNIEDA, H., On the life-history of Monostroma. – Proc. Imp. Acad. Tokyo 10. 1934.
KUNIEDA, H. & SUTO, S., The life-history of Colpomenia sinuosa (Scytosiphonaceae), with special reference to the conjugation of anisogametes. – Bot. Mag. Tokyo 52. 1938.
KÜTZING, F. T., Phycologia generalis, oder Anatomie, Physiologie und Systematik der Tange. 1843.
KÜTZING, F. T., Phycologia germanica. Nordhausen. 1845.
KÜTZING, F. T., Species algarum. Leipzig. 1849.
KYLIN, H., Studien über die Algenflora der schwedischen Westküste. – Upsala. 1907.
KYLIN, H., Studien über die Entwicklungsgeschichte von Rhodomela virgata Kjellm. – Svensk bot. Tidskr. 8. 1914.
KYLIN, H., Die Entwicklungsgeschichte von Griffithsia corallina (Lightf.) Ag. – Zeitschr. Bot. 8. 1916 A.
KYLIN, H., Über die Befruchtung und Reduktionsteilung bei Nemalion multifidum. – Ber. Deutsche Bot. Ges. 34. 1916 B.
KYLIN, H., Über die Entwicklungsgeschichte von Batrachospermum moniliforme. – Ibid. 35. 1917.
KYLIN, H., Studien über die Entwicklungsgeschichte der Phaeophyceen. – Svensk Bot. Tidskr. 13. 1918.
KYLIN, H., Studien über die Entwicklungsgeschichte der Florideen. – Kgl. Svenska Vetensk. Ak. Handl. 63, No. 11. 1923.
KYLIN, H., Über die karotinoiden Farbstoffe der Algen. – Zeitschr. für Physiol. Chemie 166. 1927.
KYLIN, H., Entwicklungsgeschichtliche Florideenstudien. – Lunds Univ. Årsskr., N. F. 24. 1928.
KYLIN, H., Über die Entwicklungsgeschichte der Florideen. – Ibid. 26. 1930.
KYLIN, H., Über die Entwicklungsgeschichte der Phaeophyceen. – Ibid. 29 (7). 1933.
KYLIN, H., Zur Kenntnis der Entwicklungsgeschichte einiger Phaeophyceen. Ibid. 30 (9). 1934.

KYLIN, H., Remarks on the life-history of the Rhodophyceae.– Bot. Rev. 1. 1935.
KYLIN, H., Anatomie der Rhodophyceen. Berlin. 1937.
KYLIN, H., Über den Bau der Florideentüpfel. Förhandl. Kgl. Fysiografiska Sällsk. i Lund. 10, No. 21. 1940.
KYLIN, H., Die Phaeophyceenordnung Chordariales. – Lunds Univ. Årsskr. N. F. Avd. 2. 43. 1940.
KYLIN, H., Verwandtschaftliche Beziehungen zwischen den Cyanophyceen und den Rhodophyceen. – Ibid. 13, No. 17. 1943.
KYLIN, H., Die Rhodophyceen der schwedischen Westküste. – Lunds Univ. Årsskr. N. F. Avd. 2 40. 1944.
KYLIN, H., Die Phaeophyceen der schwedischen Westküste – Lunds Univ. Årsskr. N. F. Avd. 2. 43 (4). 1947.
KYLIN, H., Die Chlorophyceen der schwedischen Westküste. – Lunds Univ. Årsskr. N. F. Avd. 2. 45. 1949.
KYLIN, H., Die Gattungen der Rhodophyceen. Lund. 1956.
LACKEY, J. B., Studies in the life histories of Euglenida. IV. A comparison of the structure and division of Distigma proteus Ehrenberg and Astasia dangeardi Lemm. A study in phylogeny. – Biol. Bull. 67. 1934.
LAMOUROUX, J. V. F., Dissertations sur plusieurs espèces de Fucus peu connues ou nouvelles. Agen. 1805.
LEBOUR, M. V., The Dinoflagellates of Northern Seas. Plymouth. 1925.
LEFEVRE, M., Les Péridinites des Barbades. Ann. Crypt. Exot. 6. 1933.
LEMMERMANN, E., BRUNNTHALER, J. und PASCHER, A., Chlorophyceae 2. Tetrasporales, Protococcales, einzellige Gattungen unsicherer Stellung, – Pascher: Süßwasser-Flora. H. 5. Jena. 1915.
LEVRING, T., Zur Kenntnis der Algenflora von Kullen an der schwedischen Westküste. – Lunds Univ. Årsskr. N. F. Avd. 2. 1935.
LEVRING, T., Zur Kenntnis der Algenflora der norwegischen Westküste. – Lunds Univ. Årsskr. N. F. Avd. 2, 33. 1937.
LEVRING, T., Studien über die Algenvegetation von Blekinge, Südschweden. – Lund. 1940 A.
LEVRING, T., Die Phaeophyceengattungen Chlanidophora, Distromium und Syringoderma. – Kungl. Fysiogr. Sällskap. Lund Förhandl. 10. 1940 B.
LEVRING, T., Die Meeresalgen der Juan Fernandez-Inseln. – In Skottsberg, C. The natural history of Juan Fernandez and Easter Island 2 (22): 601–670. 30 figs. Pls. 49–53. Uppsala. 1941.
LEVRING, T., Remarks on the submicroscopical structure of eggs and spermatozoids of Fucus. and related genera. – Physiologia plantarum, 5. 1952.
LEVRING, T., The marine algae of Australia I., – Ark. f. Bot. Ser. 2.2. 1953.
LEVRING, T., Some remarks on the structure of the gametes and the reproduction of Ulva lactuca. – Botaniska Notiser 108, Lund 1955 A.
LEVRING, T., Contributions to the marine algae of New Zealand I. – Ark. f. Bot. Ser. 2.3. 1955 B.
LEVRING, T., Contributions to the marine algal flora of Chile. – Lunds Univ. Årsskr. N. F. Avd. 2. 56. 1960.
LEWIS, I. F., Periodicity in Dictyota at Naples. – Bot. Gaz. 50. 1910.
LINDEMANN, E., Peridineae. In: A. Engler and K. Prantz, Die natürlichen Pflanzenfamilien. Ed. 2. 2, 1926.
LINDSTEDT, A., Die Flora der marinen Cyanophyceen der schwedischen Westküste. Lund. 1943.
LINNAEUS, C., Species plantarum. Vol. 1, 2. Stockholm. 1753.
LINNAEUS, C., Genera plantarum. Ed. 5. Stockholm. 1754.
LINNAEUS, C., Systema naturae. Stockholm. 1766–67.
LUND, S., The marine algae of East Greenland, I–II. – Medd. om Grønland, 156. 1959.
LYNGBYE, H. C., Tentamen Hydrophytologiae Danicae, e c. Copenhagen. 70. 1819.
MATHIAS, W. T., The life-history and cytology of Phloeospora brachiata Born. – Bubl. Hartley bot. lab. 13. 1935.
MATHIAS, W. T., Halopteris filicina Kütz. The cytology o the reproductive organs. Ibid. 1935 A.
MIGULA, W., Die Characeen. – Rabenhorst: Kryptogemen-Flora. 2. Aufl. Bd. 5. Leipzig. 1897.
MOORE, L. B., Oogamy in the brown alga Halopteris. – Nature 157. 1946.
MURBECK, S., Über den Bau und die Entwicklung von Dictyosiphon foeniculaceus (Huds.) Grev. Vidensk.-selsk. Skrift. Christiania. Math.-naturvid. Kl. 1900.
MYERS, M. E., The life-history of the brown alga, Egregia Menziesii. – Univ. Calif. Publ. Bot. 14. 1928.
NÄGELI, C., Die neueren Algensysteme und Versuch zur Begründung eines eigenen Systems der Algen und Florideen. Zürich. 1847.
NEWTON, L., A handbook of the British seaweeds. London. 1931.
NICOLAI, E., Untersuchungen über Zellwandstrukturen bei Algen. – Beih. z. Schweiz. Forstver. 30. 1960.
NICOLAI, E. and PRESTON, R. D., Cell-wall studies in the Chlorophyceae. I–III. – Proc. Roy. Soc. B. 140, 141, 151. 1952, 1953, 1959.
NIENBERG, W., Die Konzeptakelentwicklung bei den Fucaceen. – Zeitschr. Bot. 5. 1913.
NIENBERG, W., Zur Entwicklungsgeschichte der Helgoländer Haplospora. – Ber. Deutsch. Bot. Ges. 41. 1923.

NIENBERG, W., Die Entwicklung der Keimlinge von Fucus vesiculosus und ihre Bedeutung für die Phylogenie der Phaeophyceen. – Wiss. Meeresunters. Abt. Kiel, N. F. 21. 1931.

NIZAMUDDIN, M., Classification and distribution of the Fucales. – Bot. Marina, 4. 1962.

OLIVER, F. W., On the obliteration of the sievetubes in Laminariae. Ann. Bot. 1. 1887.

OLSEN, S., Danish Charophyta. – Dan. Vid. Selsk. Biol. Skr. 3. 1944.

OLTMANNS, F., Morphologie und Biologie der Algen, ed. 2. Jena. 1922–23.

PAPENFUSS, G. F., Note on the life-cycle of Ectocarpus siliculosus Dillw. – Science 77. 1933.

PAPENFUSS, G. F., Alternation of generations in Sphacelaria bipinnata Sauv. – Bot. Notiser 1934.

PAPENFUSS, G. F., Alternation of generations in Ectocarpus siliculosus. – Bot. Gaz. 96. 1935.

PAPENFUSS, G. F., The development of the gametophyte of Spermatochnus paradoxus. – Kungl. Fysiogr. Sällskap. Lund Förhandl. 5 (20). 1935.

PAPENFUSS, G. F., Studies of South African Phaeophyceae. I. Ecklonia maxima, Laminaria pallida, Macrocystis pyrifera. – Amer. Jour. Bot. 29. 1942.

PAPENFUSS, G. F., Notes on algal nomenclature. II. Gymnosorus J. Agardh. – Ibid. 30. 1943.

PAPENFUSS, G. F., Extension of the brown algal order Dictyosiphonales to include the Punctariales. – Bull. Torrey Bot. Club 74. 1947.

PARKE, M., A contribution to knowledge of the Mesogloiaceae and associated families. – Publ. Hartley Bot. Lab. Univ. Liverpool, No. 9. 1933.

PASCHER, A., Über die Beziehungen der Cryptomonaden zu den Algen. – Ber. Deutsch. Bot. Ges. 29. 1911.

PASCHER, A., Zur Gliederung der Heterokonten. – Hedwigia, 53. 1913.

PASCHER, A., Über Flagellaten und Algen. – Ber. d. Deutsch. Bot. Ges., 32. 1914.

PASCHER, A., Studien über die rhizopodiale Entwicklung der Flagellaten. II. Über eine neue Amöbe – Dinamoeba varians mit dinoflagellatenartigen Schwärmern. – Arch. Protistenk. 36. 1915.

PASCHER, A., Über die Übereinstimmungen zwischen den Diatomeen, Heterokonten und Chrysomonaden. Ibid. 39. 1921.

PASCHER, A., Die braune Algenreihe der Chrysophyceen. Arch. Protistenk., 52. 1925.

PASCHER, A., Die braune Algenreihe aus der Verwandtschaft der Dinoflagellaten (Dinophyceen). – Ibid. 58. 1927.

PASCHER, A., Von einer neuen Dinococcale (Cystodinium phaseolus) mit zwei verschiedenen Schwärmertypen. – Ibid. 63. 1928.

PASCHER, A., Zur Kenntnis der heterokonten Algen. – Ibid. 69. 1930.

PASCHER, A., Zur Verwandtschaft der Monadaceae mit den Chrysomonaden, etc. Ann. Protistol. 2. 1930 A.

PASCHER, A., Systematische Übersicht über die mit Flagellaten in Zusammenhang stehenden Algenreihen, etc. – Beihefte z. Bot. Centralbl. 48. 1931.

PASCHER, A., Heterokonten. – Rabenhorst: Kryptogemen-Flora. 2. Aufl. Bd. 11. Leipzig. 1938–1939.

PERAGALLO, H. & M., Diatomées marines de France et des districts maritimes voisins. – Grez-sur-Loing. 1897–1908.

PETERS, N., Peridinea. – Die Tierwelt der Nord- und Ostsee. Bd. 1. Teil 2 d. Leipzig. 1930.

PRESCOTT, G. W., A suggested modification of terminology for the names of phyla and classes in the plant kingdom. 1950.

PRINGSHEIM, E. G., Some aspects of taxonomy in the Cryptophyceae. New Phytol. 43 (2). 1944.

PRINGSHEIM, E. G., Contributions towards a Monograph of the genus Euglena. – Nova Acta Leopoldina, N. F. 18. 1956.

PRINTZ, H., Chlorophyceae (nebst Conjugatae, Heterocontae und Charophyta). In: Engler & Prantl. Die natürlichen Pflanzenfamilien. 3. Leipzig. 1927.

PRINTZ, H., Vorarbeiten zu einer Monographie der Trentepohliaceen. – Nytt Mag. Naturvidensk. 80. 1939.

RANDHAWA, M. S., Zygnemaceae. – Ind. Council Agric. Res. Monogr. on Algae 1. New Dehli. 1959.

RAYSS, T., Sur les Caulerpes de la cote Palestinienne. – Palestine Jour. of Botany, Jerusalem Series, 2. 1941.

REINKE, J., Atlas deutscher Meeresalgen. – Berlin. 1889–1892.

ROSENVINGE, L. K., The marine algae of Denmark. Part 1–4. – K. Danske Vid. Selsk. Skr. 7. Raekke, Natv. og Math. Afd. 7. 1–4. 1909–1931.

ROSENVINGE, L. K., Remarques sur les Protoflorideae. – Nuov. Notarisia. 36. 1925.

ROSENVINGE, L. K., Phyllophora Brodiaei and Actinococcus subcutaneus. – K. Danske Vid. Selsk. Biol. Meddel. 8, No. 4. 1929

ROSENVINGE, L. K. & Lund, S., The marine algae of Denmark. Vol. 2. Phaeophyceae. – K. Danske Vid. Selsk. Biol. Skr. 1–6. 1941–1950.

RUPRECHT, F. J., Über das System der Rhodophyceae. Acad. Imp. Sci. St. Pétersbourg, Mémoire, 9. 1855.

SACHS, J., Lehrbuch der Botanik. – 4 Ed. Leipzig. 1874.

SAUVAGEAU, C., Sur la conjugaison des zoospores de l'Ectocarpus siliculosus. – Compt. Rend. Acad. Sci. Paris 123. 1896 A.

SAUVAGEAU, C., Observations relatives à la sexualité des Phéosporées. – Jour. de Bot. 10. 1896 B.

SAUVAGEAU, C., La copulation isogamique de l'Ectocarpus siliculosus est-elle apparente ou

réelle? – Mém. Soc. Nat. Sci. Nat. et Math. Cherbourg 30. 1897 A.
SAUVAGEAU, C., Sur quelques Myrionémacées. Ann. Sci. Nat. Bot. Sér. 8, 5. 1897 B.
SAUVAGEAU, C., Les Cutlériacées et leur alternance de générations. Ibid. 8, 10. 1899 A.
SAUVAGEAU, C., Les Acinetospora et la sexualité des Tiloptéridacées. – Jour. de Bot. 13. 1899 B.
SAUVAGEAU, C., Remarques sur les Sphacélariacées. – Bordeaux. 1900–1914.
SAUVAGEAU, C., Sur la sexualité hétérogamique d'une Laminaire (Saccorhiza bulbosa). – Compt. Rend. Acad. Sci. Paris 161. 1915.
SAUVAGEAU, C., Sur développement et la biologie d'une Laminaire (Saccorhiza bulbosa). – Compt. Rend. Acad. Sci. Paris, 160. 1915 A.
SAUVAGEAU, C., Sur les débuts du développement d'une Laminaire (Saccorhiza bulbosa) – Ibid., 161. 1915 B.
SAUVAGEAU, C., Recherches sur les Laminaires des côtes de France. – Mém. Acad. Sci. Paris 56. 1918.
SAUVAGEAU, C., Sur le développement de quelques Phéosporées. – Ibid. 26. 1929.
SCHILLER, J., Über Fortpflanzung, geißellose Gattungen und die Nomenklatur der Coccolithophoraceen, nebst Mitteilung über Copulation bei Dinobryon. Arch. Protistenk. 53. 1926.
SCHILLER, J., Coccolithineae. Rabenhorst's Kryptogamenflora. 2nd edit. 10, II. 1930.
SCHILLER, J., Dinoflagellata in L. Rabenhorst's Kryptogamen-Flora. 10 (3). 1931–1937.
SCHMIDT, J., Danmarks blaagrønne Alger. I. Hormogoneae. – Bot. Tidsskr. 22. 1899.
SCHMIDT, O. C., Beiträge zur Systematik der Phaeophyten. I. – Hedwigia 77. 1938.
SCHMITZ, F., Die Chromatophoren der Algen. Bonn. 1882.
SCHMITZ, F., Untersuchungen über die Befruchtung der Floridéen. – Sitzb. Akad. Wiss. Berlin. 1883.
SCHMITZ, F., Systematische Übersicht der bisher bekannter Gattungen der Florideen. Flora. 72. 1889.
SCHMITZ, F. & HAUPTFLEISCH, P., Rhodophyceae. In A. Engler and K. Prantl, Die natürlichen Pflanzenfamilien. Leipzig. Teil 1, Abt. 2. 1896–1897.
SCHREIBER, E., Untersuchungen über Parthenogenesis, Geschlechtsbestimmung und Bastardierungsvermögen bei Laminarien. – Planta 12. 1930.
SCHREIBER, E., Über die geschlechtliche Fortpflanzung der Sphacelariales. – Ber. Deutsch. Bot. Ges. 49. 1931.
SCHREIBER, E., Über die Entwicklungsgeschichte und die systematische Stellung der Desmarestiaceen. – Zeitschr. Bot. 25. 1932.
SCHREIBER, E., Über Kultur und Geschlechtsbestimmung von Dictyota dichotoma. – Planta 24. 1935.
SCHUSSNIG, B., Zur Entwicklungsgeschichte der Siphoneen. – Ber. Deutsch. Bot. Ges. 46. 1928 A.
SCHUSSNIG, B., Die Reduktionsteilung bei Cladophora glomerata. – Österr. Bot. Zeitschr. 77. 1928 B.
SCHUSSNIG, B., Zur Entwicklungsgeschichte der Siphonen. II. – Ber. Deutsch. Bot. Ges. 47. 1929.
SCHUSSNIG, B., Der Generations- und Phasenwechsel bei den Chlorophyceen. – Österr. Bot. Zeitschr. 79. 1930.
SCHUSSNIG, B., Der Generations- und Phasenwechsel bei den Chlorophyceen. III. – Österr. Bot. Zeitschr. 81. 1932.
SCHUSSNIG, B., Handbuch der Protophytenkunde. Bd. 1–2. Jena 1953–1960.
SCHÜTT, F., Peridiniales. In: A. Engler and K. Prantl. Die natürlichen Pflanzenfamilien. Teil. 1. Abt. 1 b.
SCHÜTT, F., Über Peridineenfarbstoffe. – Ber. Deutsch. Bot. Ges. 8. 1890.
SCHÜTT, F., Die Peridineen der Plankton-Expedition. I. Theil. Studien über die Zellen der Peridineen. – Ergebn. Plankton Exped. 1895.
SETCHELL, W. A. & GARDNER, N. L., The marine algae of the Pacific coast of North America. I. Myxophyceae. Univ. Calif. Publ. Bot., 8. II. Chlorophyceae. Ibid., 8. III. Melanophyceae. Ibid., 8. 1919, 1920, 1925.
SETCHELL, W. A. & GARDNER, N. L., Marine algae of the Revillagigedo Islands Expedition in 1925. – Proc. Calif. Acad. Sci. Ser. 4, 19. 1930.
SIRODOT, M. S., Étude anatomique, organogénique et physiologique sur les algues d'eau douce de la famille des Lémanéacées. – Ann. Sci. Nat. Bot. Ser. 5, 16. 1872.
SIRODOT, M. S., Les Batrachospermes. Organisation, fonctions developpement, classification. Paris. 1884.
SKOTTSBERG, C., Zur Kenntnis der subantarktischen und antarktischen Meeresalgen. I. Phaeophyceen. – Wiss. Ergebn. Schwed. Südpolar-Exped. 1901–1903, 4 (6), Stockholm. 1907.
SKOTTSBERG, C., Botanische Ergebnisse der schwedischen Expedition nach Patagonien und dem Feuerlande 1907–1909. VIII. Marine algae 1. Phaeophyceae. – K. Svenska Vet.-Akad. Handl. 61 (11). 1921.
SKUJA, H., Versuch einer systematischen Einteilung der Bangioideen oder Protoflorideen. – Acta Horti Bot. Univ. Latviensis. 11–12. 1939.
SMITH, G. M., The fresh-water algae of the United States. New York. 1933.
SMITH, G. M., Cryptogamic botany. Vol. 1. Algae and fungi. New York. 1938.
SMITH, G. M. (ed.), Manual of phycology. Waltham. 1951.
STACKHOUSE, J., Nereis Britannica. Bath. 1801.

STARR, R. C., A comparative study of Chlorococcum Meneghini and other spherical zoospore-producing genera of the Chlorococcales. – Indiana Univ. Publ. Sc. Ser. No. 20. 1955.

STOSCH, H. A. VON, Oogamy in a centric diatome. – Nature 165. 1950.

STOSCH, H. A. VON, Entwicklungsgeschichtliche Untersuchungen an zentrischen Diatomeen I–II. – Arch. Mikrobiol. 16, 1951 and 23, 1956.

STOSCH, H. A. VON, Kann die oogame Araphide Rhabdonema adriaticum als Bindeglied zwischen den beiden großen Diatomeengruppen angesehen werden? – Ber. deutsch. bot. Ges., 71. 1958.

STRAIN, H. H., Chloroplast pigments. – Ann. Rev. Biochem. 1944.

SUNESON, S., Notes on the life-history of Monostroma. – Svensk Botanisk Tidsskrift 41. 1947.

SVEDELIUS, N. E., Über den Generationswechsel bei Delesseria sanguinea. – Svensk bot. Tidskr. 5. 1911.

SVEDELIUS, N. E., Über die Tetradenteilung in den vielkernigen Tetrasporangiumanlagen bei Nitophyllum punctatum. – Ber. Deutsch. Bot. Ges. 32. 1914 A.

SVEDELIUS, N. E., Über Sporen an Geschlechtspflanzen von Nitophyllum punctatum, ein Beitrag zur Frage des Generationswechsels der Florideen. – Ibid. 32. 1914 B.

SVEDELIUS, N. E., Zytologisch-Entwicklungsgeschichtliche Studien über Scinaia furcellata. – Nova Acta Reg. Soc. Sci. Upsaliensis. Ser. 4, 4. 1915.

SVEDELIUS, N. E., Über die sogenannten Süßwasser-Lithodermen. – Zeitschr. Bot. 23. 1930.

SVEDELIUS, N. E., Nuclear phases and alternation in the Rhodophyceae. – Beih. Bot. Centralbl. 48. 1931.

SVEDELIUS, N. E., On the development of Asparagopsis armata Harv. and Bonnemaisonia asparagoides (Woodw.) Ag. – Ibid. 1933.

SVEDELIUS, N. E., The apomeiotic tetrad division in Lomentaria rosea in comparison with the normal development in Lomentaria clavellosa. – Symbolae Bot. Upsaliensis 2. 1937.

SVEDELIUS, N. E., Zytologisch-Entwicklungsgeschichtliche Studien über Galaxaura, eine diplobiontische Nemalionales Gattung. – Nova Acta Reg. Soc. Upsaliensis, Ser. 4, 13. 1942.

SYKES, M. G., Anatomy and histology of Macrocystis pyrifera and Laminaria saccharina. – Ann. Bot. 22. 1908.

SÖDERSTRÖM, J., Studies in Cladophora. – Botanica Gothoburgensia, I. 1963.

TAI, LI-SUN and SKOGSBERG, T., Studies on the Dinophysoidae, marine armored Dinoflagellates, of Monterery Bay, California. – Arch. Protistenk. 82 (3). 1934.

TAYLOR, W. R., Marine algae of the northeastern coast of North America. – Ann. Arbor. 1937.

THURET, G., Recherches sur les organes locomoteurs des spores des algues. – Ann. Sci. Nat. Bot. Ser. 2, 19. 1843.

THURET, G., Recherches sur la fécondation des Fucacées, suivies d'observations sur les antheridies des algues. – Ann. Sci. Nat. Bot. Ser. 4, 2. 1854.

THURET, G., Recherches sur la fécondation des Fucacées et les antheridies des algues. II. – Ibid. Ser. 4, 3. 1855.

THURET, G. & BORNET, E., Études phycologiques. Paris. 1878.

TIFFANY, L. H., The Oedogoniaceae. Columbus. 1930.

TILDEN, J. E., The algae and their life relations. – Minneapolis. 1935.

TORREY, H. B., An unusual occurence of Dinoflagellata on the California coast. – Amer. Nat. 36. 1902.

TRANSEAU, E. N., The Zygnemataceae. – Columbus. 1951.

TURNER, DAWSON, A synopsis of the British Fuci. – London. 1802.

TURNER, DAWSON, Fuci sive plantarum Fucorum generi a botanicis ascriptarum icones, descriptiones et historia. Vol. 1–4. – London. 1808 bis 1819.

VENKATARAMAN, G. S., Vaucheriaceae. – Ind. Council Agric. Res. Monogr. on Algae. New Dehli. 1962.

WETTSEIN, R. VON, Handbuch der Systematischen Botanik. Bd. 1. – Leipzig. 1901.

WILCE, R. T., A revision of the North Atlantic species of the Simplices Section of Laminaria. – Botanica Gothoburgensia 3. 1965.

WILLE, N., Über die Befruchtung bei Nemalion multifidum. (Web. et Mohr) J. Ag. – Ber. Deutsch. Bot. Ges. 12. 1894.

WILLE, N., Beiträge zur physiologischen Anatomie der Laminariaceen. Univ. Festskr. til H. Maj. Kong Oscar II, Regjeringsjubilaeet. Christiania. 1897.

WILLE, J. N. F., Conjugatae. Chlorophyceae. In: Engler & Prantl. Die natürlichen Pflanzenfamilien. Teil 1, Abt. 2. Leipzig. 1897.

WILLE, J. N. F., Conjugatae. Chlorophyceae. Ibid. Nachtr. Teil 1, Abt. 2. Leipzig. 1911.

WILLIAMS, J. L., The antherozoids of Dictyota and Taonia. Ann. Bot. 11. 1897.

WILLIAMS, J. L., Studies in the Dictyotaceae. I. The cytology of the tetrasporangium and the germinating tetraspore. – Ibid. 18. 141–160. Pls. 9, 10. 1904 A.

WILLIAMS, J. L., Studies in the Dictyotaceae. II. The cytology of the gametophyte generation. – Ibid. 18. Pls. 12–14. 1904 B.

WILLIAMS, J. L., Studies in Dictyotaceae. III. The periodicity of the sexual cells in Dictyota dichotoma. – Ibid. 19. 1905.

VISCHER, W., Zur Kenntnis der Gattung Botrydium Wallroth. – Ber. Schweiz. Bot. Ges. 48. 1938.

Yamanouchi, S., The life-history of Polysiphonia violacea. – Bot. Gaz. 42. 1906.
Yamanouchi, S., The life-history of Cutleria. – Ibid. 54. 1912.
Yamanouchi, S., The life-history of Zanardinia. – Ibid. 56. 1913.
Yamanouchi, S., The life-history of Corallina officinalis var. mediterranea. – Ibid. 72. 1921.

Marine Algae as Raw Material

By Heinz A. Hoppe

The developments of the last 20 years have rendered it necessary to give special attention to the different species of Chlorophyceae, Phaeophyceae, and Rhodophyceae in their capacity as basic raw materials for industry.

A short historical survey clearly shows the trends in seaweed research and utilization during the course of the centuries.

The earliest particulars on seaweed utilization originate from a Chinese herbal, compiled according to the directions of the Emperor Shen Nung 2700 B.C. It seems that already at that time seaweed played a certain rôle among the peoples of the Far East as medicaments and victuals. But real economic utilization was initiated by the Chinese and Japanese around 1670.

Roman women, living at the time of Virgil and Horace, used a red dye made from algae for cosmetics.

One of the earliest uses of marine algae was agricultural fertilizing, a practice already known in antiquity in the Far East. Not until the 12th century, however, fertilizers are mentioned from algae in European records. Growth-territories are the coasts of France, Ireland, and Scotland. During the 17th century regulations have been published by the French Government regarding harvesting and working up of seaweed; also since that time seaweed was used on the Normandy and Britanny coasts in Kelp-kilns, in which a mixture of soda and potash was obtained for glass making by ashing the seaweed. The French Kelp-industry was very important during the reign of Louis XIV. In 1720 Kelp-kilns were also erected in Great Britain, and by the end of the century Scotland alone produced more than 20.000 tons p.a., for which some 400.000 tons of fresh seaweed were necessary.

The Kelp-kilns lost their importance, when Leblanc discovered a new method of soda-manufacture in 1791 and with a cheaper way of making glass.

After Bernard Courtois' discovery of iodine, there began a new period of Kelp-industry. At that time seaweed was the most important raw material for the extraction of iodine, although up to 50% of the iodine were lost in the incinerating process.

The extraction of iodine from seaweed lost its importance after 1873 with the discovery of the Chilean saltpetre-deposits. To-day iodine is also obtained from other materials, nevertheless seaweed remains an important raw material in several territories, e.g. Japan and the U.S.S.R.

Japan did not start iodine-extraction on an industrial scale until the time of 1900. At the begin of the First World War about 100 tons iodine-salt have been produced by Japan, and during the war the quantity produced increased by a factor of 2½, but felt again afterwards.

In the Far East other products were also obtained from marine algae, of which the first to be known in Europe was agar-agar from various red algae.

Japan was the main producer of agar up to 1939. The somewhat mysterious production of "kanten", the old Japanese name for agar, has been carried on by traditional methods for centuries. To-day modern methods are also used in the Far East.

Agar was produced in California for the first time in 1919, but the North American agar industry developed from these beginnings after 1923, while by 1939 industries had also developed in other countries, in Australia, South Africa, Spain, Morocco etc.

"Irish Moss" or carrageen from the European coasts had been in use for several centuries as thickener for sweets and similar products. A modern industry has been developed during the last decades for manufacturing Irish Moss extract, carrageenan and its derivates, and importance and utilization of these products continue to grow.

A number of brown algae are of great economic value for the manufacture of alginic acid and alginates.

The discoverer of alginic acid was the English pharmacist Stanford, born 1837 in Sussex. In 1883 Stanford published the discovery of a new substance in seaweed, which he called "Algin". He believed this to be a protein, until the Norwegian Krefting 1896 produced a pure alginic acid, which he called "Tang Acid". Stanford died 1899 without having experienced the rise of the industry, for which he had made the basic discovery. The structure of alginic acid has only been clarified in recent years.

The U.S.A. have started producing alginic acid and alginates in 1929, and now alginate-industries exist in Great Britain, France, Norway, and other countries.

During the course of time further industries have been based on seaweed, e.g. the production of furcellaran in Denmark and of phyllophoran in the U.S.S.R.

For centuries certain seaweeds have been used as human and animal food, and are still used to-day on the basis of the old experience, but now with exact scientific knowledges and modern methods of cultivation, preparation, packing etc.

Investigations of numerous marine algae have led to the discovery of new active medicaments and antibiotical substances.

These developments have not come to an end, on the contrary they stand at the threshold of a new era, an era for which modern technology will largely be responsible. Many details remain to be completed, especially since the coasts which not yet have been investigated hold an unknown potential of raw materials for the seaweed industry.

Chlorophyceae

Acetabularia

More than 20 species are recorded from tropical and subtropical oceans. The algae have an umbrella-shaped appearance. Development covers a period of three years. The bestknown species *A. mediterranea* occurs in the Mediterranean and on the neighbouring Atlantic coast. It is often incrusted with calcium carbonate. Investigations have been made of the chemical composition of the cell walls[1]).

A. major. Coasts of Thailand, Indonesia (Java), Japan, the Philippines. The alga is applied in South-East Asian areas in medicine as a treatment of gallstone and other stone developments.

Literature

1) Iriki, T. and Miwa, T., Nature (London) 1960. 185. 178

General Literature

Fott, B., 1959; Fritsch, F. E., 1956; Zaneveld, J. S., Ec. Bot. 13 (2): 89. 1959; see Index of Literature.

Caulerpa

Approximately 60 species are recorded, especially in the warmer oceans. The thallus of some species is very small, in other cases it is up to one metre long. The most common species is *C. prolifera* (Forsk.) Lam., which is abundant in the Mediterranean areas

C. prolifera. The alga flourishes on the Mediterranean coast in depths of up to 15 metre. and more. Growth has been investigated experimentally[1]). Gessner and Hammer have investigated the production capacity of *C. prolifera* in the Mediterranean *Caulerpa-Cymodocea* colonies[2]).

C. filiformis. The constituents of this alga have been investigated in detail[3–6]). It contains xylane and a polysaccharide esterified with sulphuric acid, which, when hydrolysed, yields galactose, glucose, mannose, rhamnose, xylose, and uronic acid[7]) [8]). The constitution of xylane has been analysed [9–11]).

The polysaccharides of *C. racemosa* and *C. sertularioides* have likewise been investigated. They are similar to those of *C. filiformis*[6]) [11]).

Some *C.*-varieties are used for human consumption. They are salten when being fresh.

C. ancepas. The xylane, the main cell-wall constituent, has been investigated [12]).

C. clavifera. Known as "limu fuafua", occurring in Pacific areas. The alga is used as a relish because of its strong spicy flavour[13]) [14]).

C. fergusoni. Area of Singapore. The alga is used for human consumption.

C. peltata var. *macrodisca.* Inhabitant of the coasts of Ceylon, Indonesia (Bangka, Celebes), the Philippines, Polynesia. This variety is used as a food product on Bangka and eaten as a salad in Luzon[13–15]).

Plate I

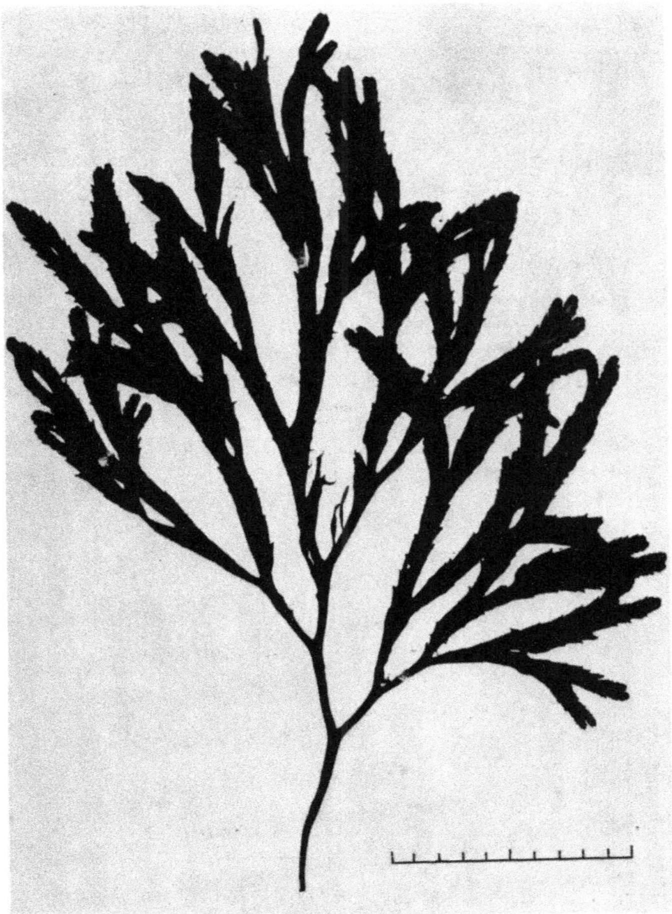

1. *Fucus serratus*, exposed (West Coast of Sweden). (Photo LEVRING)

2. *Fucus serratus*, from a sheltered locality with very strong currents (West Coast of Sweden). (Photo LEVRING)

Plate II

3. Bay of Fundy (Canada) at lowtide: Vegetation of mainly *Ascophyllum nodosum* (Photo LEVRING)

4. *Ascophyllum* (centre) and *Fucus*. Scotland (Photo LEVRING)

5. *Chondrus*, *Himanthalia*, *Fucus serratus*. Scotland (Photo LEVRING)

6. *Fucus spiralis*. West Coast of Sweden (Photo LEVRING)

Plate IV

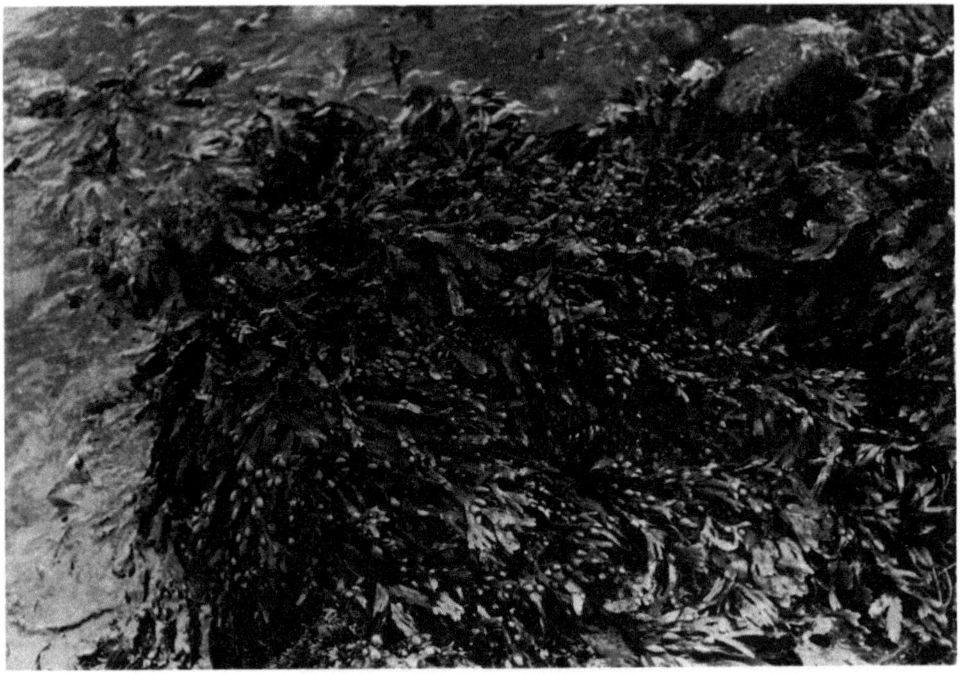

7. *Fucus vesiculosus*. Scotland (Photo Levring)

8. *Laminaria hyperborea*. Isle of Man (Photo Levring)

9. *Laminariae.* Ireland (Photo Hoppe)

10. Kelp (*Macrocystis* and *Ecklonia*). Cape Town, South Africa (Photo Levring)

Plate VI

11. Reef near Mombaza, East Africa. Vegetation of *Sargassum*, *Turbinaria* etc. (Photo LEVRING)

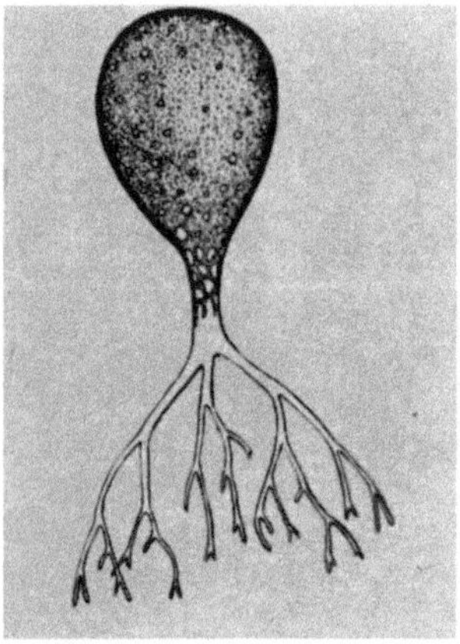

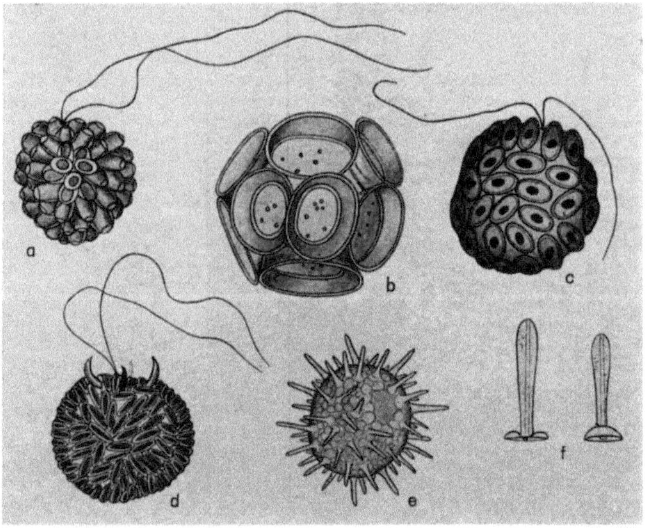

12. *Botrydium granulatum*. From VORONIN

13. *Coccolithineae* (x1500–1900). a. *Lohmannosphaera adriatica*. b. *Pontosphaera syracusana*. c. *Calyptrosphaera insignis*. d. *Syracophaera quadricornu*. e. *Rhabdosphaera multistylis*. f. Coccoliths from *Rhabdosphaera claviger*. (From SCHILLER)

14

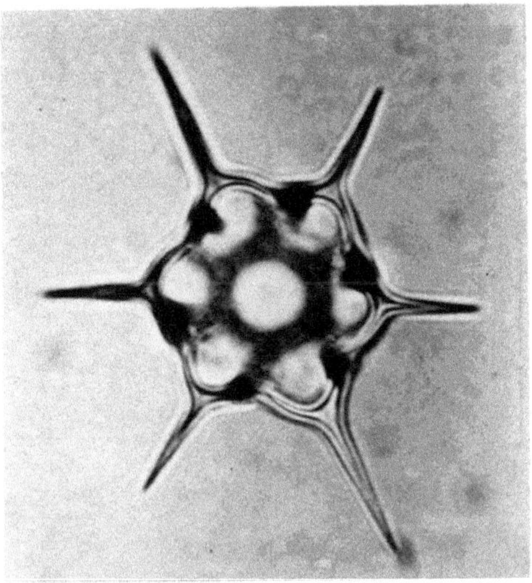

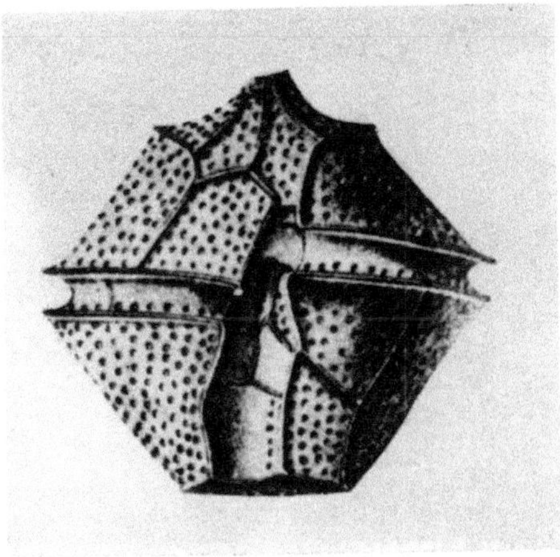

16

15

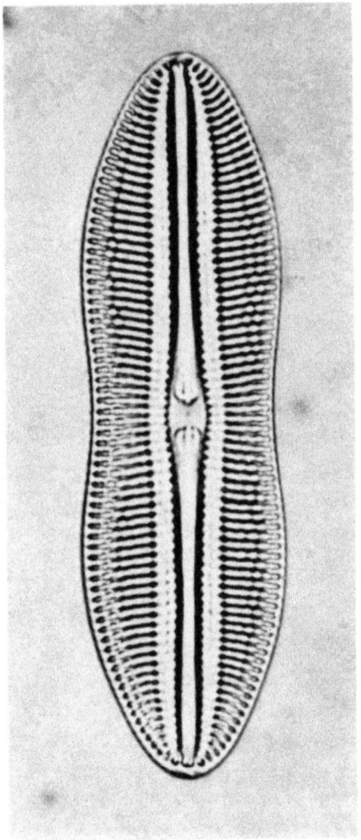

14. Silicoflagellate: *Distephanum speculum* (x 1300) (Photo EDSBAGGE)
15. *Diploneis* sp. (x 750) (Photo EDSBAGGE)
16. *Goniaulax polyedra* (From KOFOID)
17. *Noctiluca miliaris* (From DOFLEIN-REICHENOW)

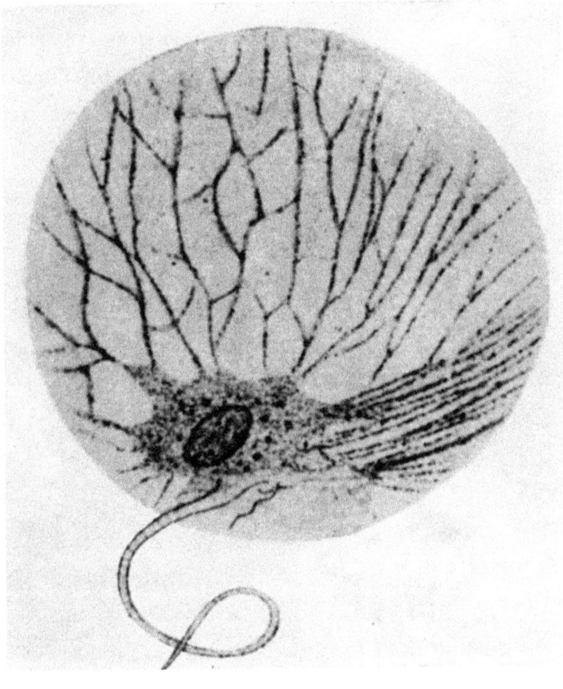

17

Plate VIII

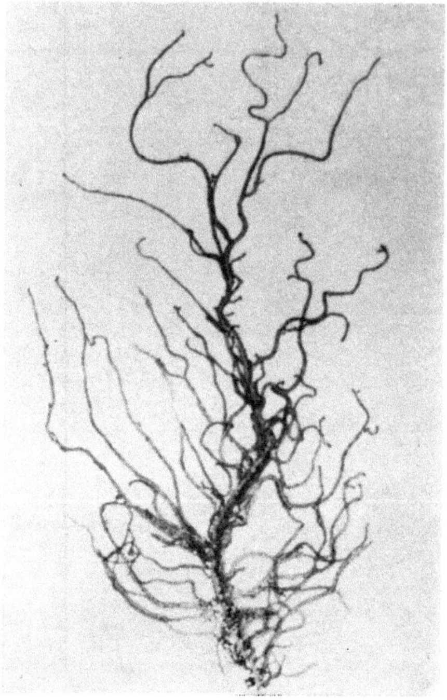

18. *Ectocarpus siliculosus* (x 0,6) (Photo LEVRING)

19. *Eudesme virescens* (x 0,5) (Photo LEVRING)

20. *Colpomenia sinuosa*. Natural size (From OLTMANNS)

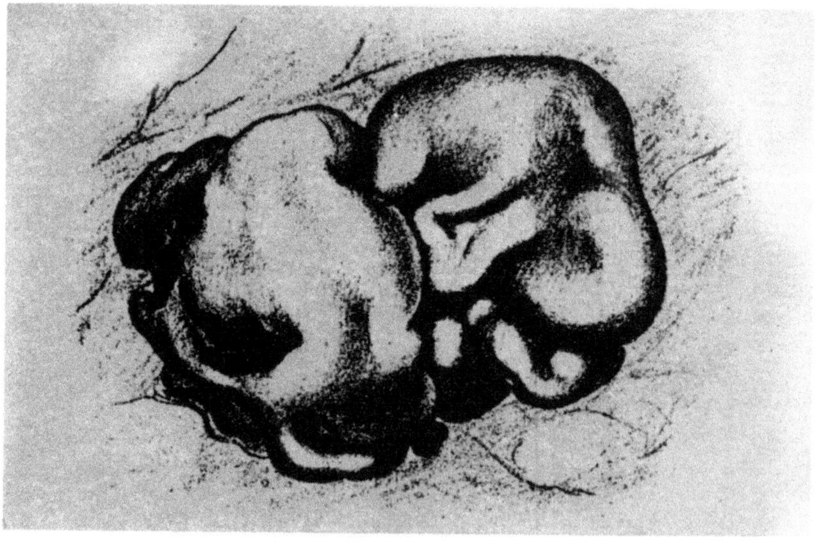

21. *Nereocystis luetkeana*. Pacific Coast, USA (Photo Skottsberg)

22. *Macrocystis angustifolia*. Australia (Photo Otto J. Schmid)

23

25

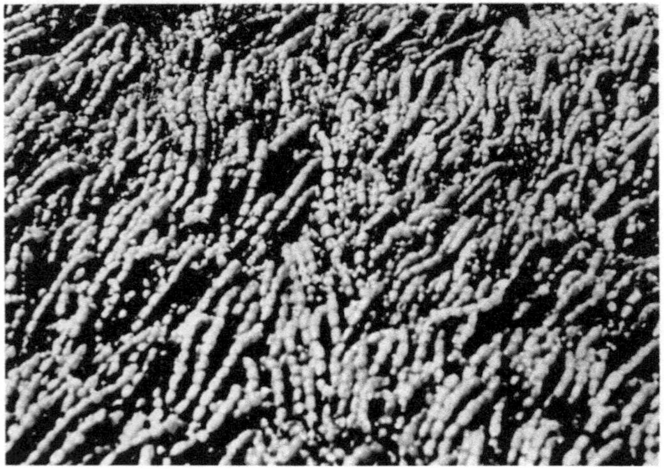

24

23. *Hormosira banksii.* New South Wales, Australia (Photo LEVRING)

24. *Hormosira banksii.* Littoral. New South Wales, Australia (Photo LEVRING)

25. *Hormosira banksii f. labillardieri.* On mudflats, Kangaroo Island, South Australia (Photo LEVRING)

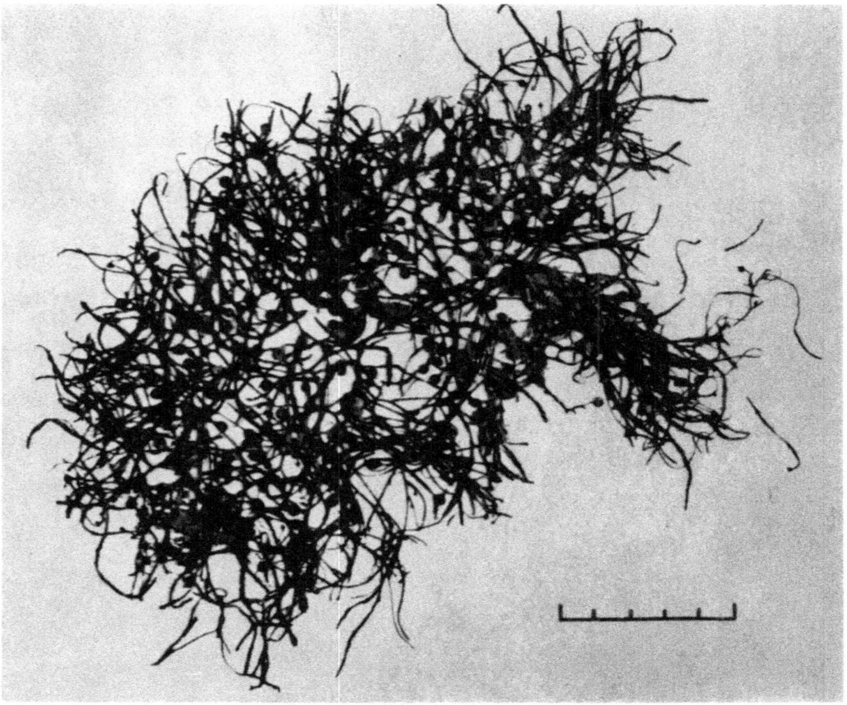

27. *Sargassum natans*. Sargasso-Sea (Photo LEVRING)

26. *Ascophyllum nodosum* (Photo LEVRING)

Plate XII

28. *Gelidium cartilagineum*. California. (x 0,25) (Photo LEVRING)

29. *Pterocladia lucida*. New Zealand. (x 0,25) (Photo LEVRING)

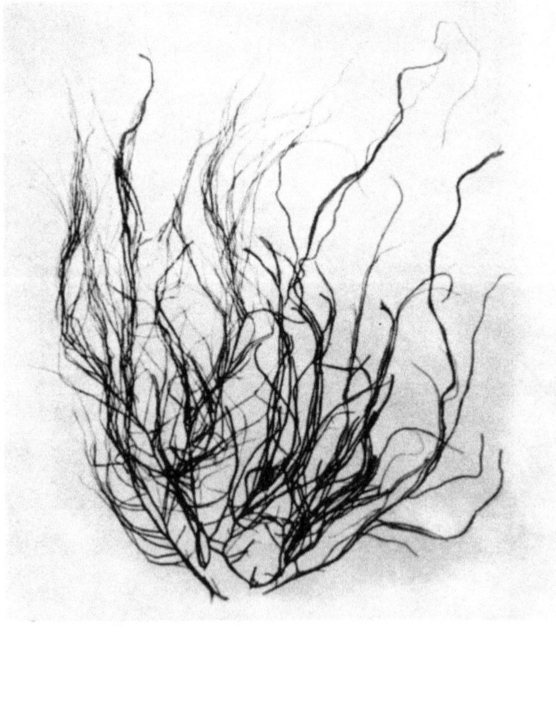

30. *Gracilaria confervoides*. (x 0,38) (Photo LEVRING)

31

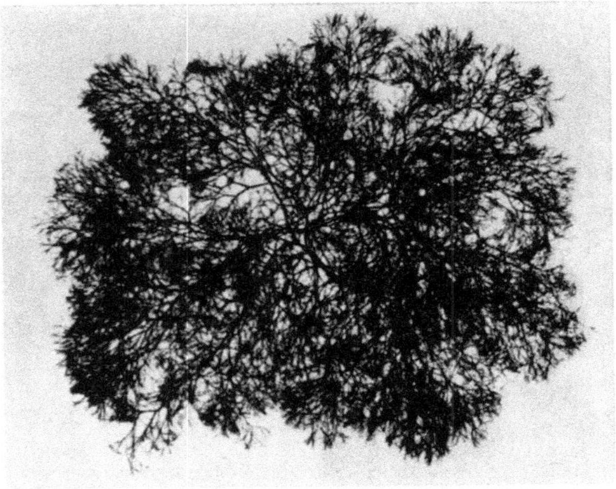

32

33

31. *Rhodymenia palmata.* (x 0,37) (Photo LEVRING)

32. *Ceramium rubrum.* West Coast of Sweden. (x 0,4) (Photo LEVRING)

33. *Delesseria sanguinea.* West Coast of Sweden. (x 0,36) (Photo LEVRING)

34. The culture ground of *Monostroma* in low tide
(YONEZU; Sept. 18, 1959)

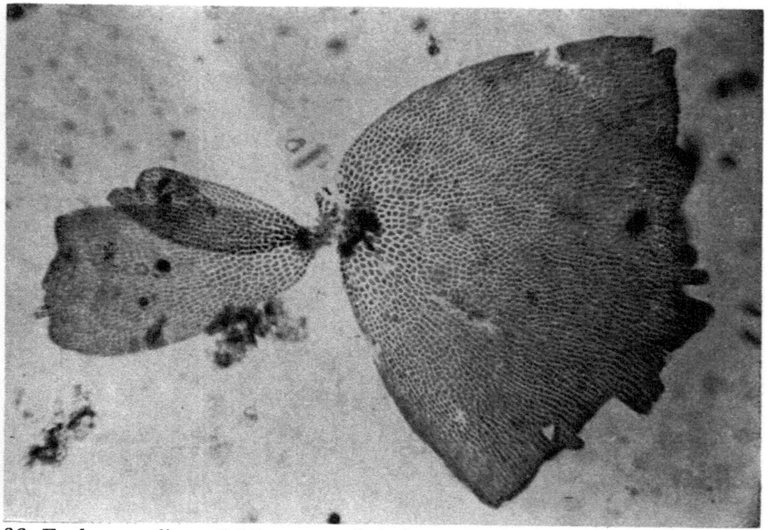

36. Early germling of *Monostroma latissimum* (KUTZ) WITTROCK (?),
from about 40 days after germination of spores. (x 40) (Oct. 26, 1959)

38

39

40

38. Dried *Laminaria* stipes for the manufacturing of alginic acid and alginates. Ireland (Photo H. A. Hoppe)

39. Device for dredging *Laminaria hyperborea*. Scotland (Photo Levring)

40. *Laminaria hyperborea* stipes with *Rhodymenia* and other epiphytes (Photo Levring)

41

42

43

41. Drying of *Chondrus* in Nova Scotia, Canada (Photo HOPPE)

42. Drying of *Chondrus* in Ireland (Photo HOPPE)

43. Drying of *Rhodymenia* in Nova Scotia, Canada (Photo HOPPE)

C. racemosa. Inhabitant of the coasts of Ceylon, India (Pamban), Malaya, Indonesia, the Philippines, widespread also on the Atlantic coast. The alga is used for human consumption in Malaya and plays a certain rôle for nutrition on the Philippines. It is cultivated in large containers and sold immediately after harvest while still being fresh[13–15]).

C. racemosa var. *clavifera.* Occuring on the coasts of Ceylon and Malaya. The alga is used as a dessert for rice dishes. On Celebes it is sold on the markets[13]) [15]).

C. racemosa var. *laetevirens.* Recorded on the coasts of Ceylon, Indonesia (Bali, Ceram), the Philippines, Australia, West Indies. The alga is used for human consumption on southern Bali[13–15]).

C. racemosa var. *uvifera.* Inhabitant of the coasts of the Red Sea, Ceylon, Indonesia (Celebes, Buru, Rotti Islands), the Philippines, Japan, Australia, West Indies, the Atlantic coast of America. The alga is eaten in the raw state, mostly as a salad[15]).

C. serrulata (= C. freycinetii). Inhabitant of the coasts of Singapore, the Philippines, Indonesia (west coast of Sumatra, east coast of Borneo, Celebes, Timor, Molukken), New Guinea, and more or less of all tropical oceans. The alga is used for human consumption in the area of Singapore[14]) [15]).

C. sertularioides. Particularly on the coasts of Indonesia, the Philippines, and in all tropical oceans. The alga is consumed in a raw state and as salad[15]).

The following species are to be found on the Pakistani coast (Karachi): *C. fastigiata, C. peltata, C. chemnitzia, C. racemosa, C. sertularioides, C. scalpelliformis, C. taxifolia, C. faredii, C. manorensis, C. gureshii.* Taxonomy, ecology, and distribution have been studied. The latter three species are a sp. nov.[16]).

C. charoides. A new record for the Indian Ocean region[17]).

C. veravalensis. Luxuriant growth, at low-tide level off the reef at Jaleshwar, Verawal, India[18]).

C. verticillata. A new record for India[19]). The amino acids of the Indian *C.*-species have been studied[20]).

The following are known to occur on the Japanese coast: *C. ambigua, C. brachypus, C. fergusoni, C. Okamurai, C. paroifolia, C. peltata, C. racemosa* varieties, *C. scalpelliformis, C. serrulata* and others.

C. scalpelliformis var. *denticulata* occurs in a number of localities on the Japanese coast[21]).

Records of *Caulerpa* from south-eastern Queensland have been described[22]). *C.* varieties are on record to occur also on the coast of Columbia and Yucatan[23]).

Literature

1) Jacobs, W. P., Third Int. Seaweed Symp., 1958
2) Gessner, F. and Hammer, L., Bot. Mar. II (1/2): 157–163. 1960
3) Institute of Seaweed Research. Annual Report 1957
4) ibid. 1958
5) ibid. 1959
6) ibid. 1960
7) Hirst, E. L., Third Int. Seaweed Symp., 1958
8) Hegnauer, R., Chemotaxonomie der Pflanzen. Birkhäuser Verlag, Stuttgart, 1962
9) Mackie, J. M. and Percival, E., J. Chem. Soc. 1151. 1959
10) Mackie, J. M. and Percival, E., ibid. 2381. 1960
11) Mackie, J. M. and Percival, E., ibid. 3010. 1961
12) Iriki, T. et al., Nature (London) 1960. 187.82
13) Chapman, V. J., Seaweeds and their Uses. Methuen & Co. Ltd., London, 1950
14) Uphof, J. C. Th., Dictionary of Economic Plants. H. R. Engelmann (J. Cramer), Weinheim, 1959

15) ZANEVELD, J. S., Ec. Bot. 13 (2): 89. 1959
16) NIZAMUDDIN, M., Bot. Mar. VI (3/4): 204–223. 1964
17) THIVY, F. and VISALAKSHMI, V., ibid. V (4): 101–104. 1963
18) THIVY, F. and CHAUHAN, V. D., ibid. V (4): 97–100. 1963
19) THIVY, F. and VISALAKSHMI, V., Bot. Mar. V (1): 29–32. 1963
20) LEWIS, E. J. and GONZALVES, E. A., J. Mar. Biol. Ass-India 1 (2): 203–205. 1960
21) NOMURA, Y., Bull. Jap. Soc. Phyc. 5 (1): 25. 1957
22) CRIBB. A. B., Univ. d. Pap. Dept. Bot. 3 (23): 209–220. 1958
23) TAYLOR, W. R., Bot. Ser. Field Museum of Natural History 20 (4): 1941

General Literature

CHAPMAN, V. J., 1950; FOTT, B., 1959; FRITSCH, F. E., 1956; PAX, F., 1962; SEGAWA, S., 1957; see Index of Literature

Chaetomorpha

Approximately 60 species of *Ch.* in oceans, brackish water, and fresh water. Thallus mostly filamentous, not branched, and fixed by an elongated basal cell.

Ch. linum (MÜLLER) Kütz. and *Ch. aerea* Kütz. are the best-known marine species of this genus.

Ch. aerea. Known in German as "Ufer-Borstenhaar"; Baltic and North Sea coast. European Atlantic coast, Mediterranean, particularly in summer. Thallus tufts consist of individual stiff unbranched filaments, formed by large cells visible to the naked eye.

Ch. antennina. Coast of Indonesia, Vietnam, Hawaiian Islands, Mexico, Ecuador, West Indies, South Africa. This alga is eaten in Hawaii[1]).

Ch. crassa. Coasts of the Red Sea, Indonesia, the Philippines, Japan, Australia. This alga is made to a gelatinous dessert, and is also often eaten as a salad[1]).

Ch. javanica. Coasts of Indonesia (Java, Amboina), West Indies. Chinese population living in Indonesia gather the alga during the monsoon months, soak it in fresh water, dry it in the sun and use it as food, after boiling with[1]).

Ch. linum. This species is very widespread. Coast of the western Baltic, not in very large quantities, particularly on the fringes of mudbanks of the North Sea. The alga is often found as a tangled mat of filaments of a dark greenish colour on the mudbanks[2]). Morphology and cell-physiological investigations, correlations between respiration and turgor regulations have been reported by KESSELER[3]) [4]). The polysaccharides have been studied[5]) [6]).

Ch. media. Coast of India. The steroids of this species have been investigated[7]).

Ch. melagonium. Coast of Canada. The inorganic constituents have been analysed[8]). Genuine cellulose was found. The chemical composition of the cell wall has also been investigated[9]).

Ch. crassa, *Ch. media*, *Ch. moniligera*, *Ch. spiralis* are recorded from Japan.

For physiological and biochemical investigations on *Ch. aerea*, *Ch. linum*, *Ch. melagonium*, see LEWIN (1962).

Literature

1) ZANEVELD, J. S., Ec. Bot. 13 (2): 89. 1959
2) LEVRING, T., Studien über die Algenvegetation von Blekinge, Südschweden, Dissertation, Lund, 1940

3) KESSELER, H., Helg. Wiss. Meeresunters. 7 (3): 114–124. 1960
4) KESSELER, H., ibid. 8 (2): 243–256. 1962
5) Institute of Seaweed Research. Annual Report 1961
6) ibid. 1962
7) GOVINDACHARIA, T. R. et RAJAGOPALAN, S., First Int. Seaweed Symp., 1952
8) YOUNG, E. G. and LANGILLE, W. M., Publ. 4636 of the Nat. Res. Council, Halifax, Canada, 1958
9) PRESTON, R. D., Third Int. Seaweed Symp., 1958

General Literature

CHAPMAN, V. J., 1950; FOTT, B. 1959; FRITSCH, F. E., 1956; HEGNAUER, R., 1962; KUCKUCK, P., 1953; LEWIN, R. A., 1962; PAX, F., 1962; SEGAWA, S., 1957; see Index of Literature.

Cladophora

Approximately 150 *Ch.* species, mostly marine, are recorded. The thallus is bushy, ramified, mostly consisting of elongate cells. The alga is fixed with a rhizoid cell. The genus is widespread over the whole earth, with the exception of the polar regions. Some species are annuals, others are perennial.

C. rupestris (L.) Kütz. Baltic and North Sea coast. Ranging from the northern Arctic ocean to France. East coast of North America. In the Baltic in the deep water, on the North Sea coast (Heligoland) on substrata falling dry at low tide. This alga mainly grows on stones or rocks, generally more or less in open areas.

The constituents of *C. rupestris* have been studied in detail[1]). The chemical composition of the cell wall has been investigated[2]). The alga has a high content of cellulose and protein. The polysaccharides have been investigated[3]). SMITH-MONTGOMERY reported on the water-soluble polysaccharides. The alga produces an acid mucilage, which contains about 20 per cent of sulphuric acid. Hydrolysis yields arabinose, galactose, xylose, rhamnose, and glucose[5]) [6]). O'COLLA describes the mucilage as the galactan sulphate known as "cladophorin"[7]). Further investigations have been reported[8–14]). The structure of sulfated polysaccharides has been studied[15]), and the carbohydrates were reported by MANNERS at the FOURTH INTERNATIONAL SEAWEED SYMPOSIUM[16]). The amino acids of the alga have been described by BLACK[17]).

C. rupestris has been investigated for carotene and carotinoid content[18]). English scientists investigated the enzymes. The properties of amylase, maltase, laminarase, and salicinase have been studied[15]) [19–23]). Extracts of the alga contain polypeptides as well as protein[15]) [19]). The nitrogenous constituents have been investigated[20]) and the inorganic constituents analysed[24]). Genuine cellulose has been found[6]).

The following species are recorded as occuring on the Atlantic coasts, particularly on the North Sea and Baltic coast: *C. arcta*, *C. lanosa*, *C. seriacea*. *C. hutchinsiae* only occurs in the Atlantic, and apparently not in the Baltic at all. Chemical investigations on it have been reported by BLACK[1]).

The following species have been reported to occur on the Japanese coast: *C. densa*, *C. glaucescens*, *C. japonica*, *C. ohkuboana*, *C. rudolphiana*, *C. rugulosa*, *C. stumpsoni*, *C. wrightiana*.

The structure of *C. wrightiana* and the life history[26]) have been investigated[25]). A species of *Cladophora* from Japan and its vicinity has been described by SAKAI[30]).

An elaborate work, "Studies in *Cladophora*", has been published by SOEDERSTROEM in 1963[27]).

For physiological and chemical investigations on *Cladophora* see LEWIN (1962).

The sterols of various Indian species have been investigated[28]. Also some of the South African *C.* species have been investigated[29].

Literature

1) BLACK, W. A. P., Ann. Rep. Chem. Soc. 50. 1953
2) PRESTON, R. D., Third Int. Seaweed Symp., 1958
3) FISHER, I. S. and PERCIVAL, E. E., Second Int. Seaweed Symp., 1955
4) SMITH, F. and MONTGOMERY, R., The Chemistry of Plant Gums and Mucilages. Reinhold Publishing Corporation, New York, 1959
5) HIRST, E. L., Third Int. Seaweed Symp., 1958
6) HEGNAUER, R., Chemotaxonomie der Pflanzen. Birkhäuser Verlag, Stuttgart, 1962
7) LEWIN, R. A., Physiology and Biochemistry of Algae. Academic Press Inc., New York, 1962
8) Institute of Seaweed Research. Annual Report 1958
9) ibid. 1960
10) ibid. 1961
11) ibid. 1962
12) FISHER, I. S. and PERCIVAL, E., J. Chem. Soc. 1957. 2666–2675
13) FISHER, I. S. and PERCIVAL, E., ibid. 1958. (1): 49
14) O'DONNELL, J. J. and PERCIVAL, E., ibid. 1959. 1739
15) Institute of Seaweed Research. Annual Report 1956
16) MANNERS, D. J., Fourth Int. Seaweed Symp., 1961
17) BLACK, W. A. P., The Algae ex Altschul, Processed Plant Proteins Foodstuffs. Academic Press Inc., New York, 1958
18) OWEN, E. C., J. Sci. Food. Agric. 5 (9): 449–453. 1954
19) Institute of Seaweed Research. Annual Report 1955
20) ibid. 1957
21) DUNCAN, W. M. et al., Biochem. J. 63. 44–51. 1956
22) DUNCAN, W. M. and MANNERS, D. J., ibid. 69 (3): 343–348. 1958
23) DUNCAN, W. M. and al., ibid. 73: 295. 1959
24) YOUNG, E. G. and LANGILLE, W. M., Publ. 4636 of the Nat. Res. Council, Halifax, 1958
25) SAKAI, Y., Bull. Jap. Soc. Phyc. 8 (1): 1–4. 1960
26) CHIHARA, M., J. Jap. Bot. 35 (1): 1–11. 1960
27) SÖDERSTRÖM, J., Studies in Cladophora. Botanica Gothenburgensia I, 1963
28) GOVINDACHARI, T. R. and RAJAGOPALAN, S., First Int. Seaweed Symp., 1952
29) SIMONS, R. H., Bothalia 7 (2): 189–199. 1960
30) SAKAI, Y., The Species of Cladophora from Japan and its Vicinity. Inst. of Algological Research, Hokkaido Univ. V (1): 1964

General Literature

BRENNAN, A. T., 1950; FOTT, B., 1959; FRITSCH, F. E., 1956; KUCKUCK, P., 1953; LEVRING, T., 1940; LEWIN, R. A., 1962; SEGAWA, S., 1957; Tokida, J., 1957; TOKIDA, J., 1954; see Index of Literature.

Codium

Approximately 50 species are recorded for all oceans, except the Arctic area. The thallus of the *Codium* species has many different forms, spongy, cushion-like, globular, cylindrical, flat or branched. The chemical nature of the cell wall has been investigated[1].

C. amphibium. This alga has been described as a species of doubtful validity[2].

C. decorticatum. Tropical western Atlantic, west coast of Africa, South-West Africa, island groups in the Atlantic, Mediterranean region. The species has been described by SILVA[3].

C. dichotomum. Coasts of the Atlantic, Indian, and Pacific oceans. Coast of Formosa, Japan, Sakhalin, Korea, Siberia[4]. The reproduction of this species off the French Atlantic coast has been reported by DANGEARD[5]. Constituents have been investigated[6][7]. The alga is used for human consumption[4].

C. difforme. Inhabitant of the coasts of the Atlantic ocean, of the Mediterranean, the

Bahamas, Kerguelen, Island and Falkland Islands, of the central and south coast of Chile, the Malayan islands and the Indian ocean[8]).

C. dimorphum. Recorded for the coast of West Patagonia[8]).

C. distichum. A carbohydrate has been isolated, which on hydrolysis yielded arabinose, galactose, and mannose[9]).

C. divaricatum. Inhabitant of the Japanese coast. The alga grows near the low-water level, mostly in bays, and is harvested in April and May. This alga is dried or salted for human consumption. Japanese name "Kiru-miru"[10]).

C. duthiae. The polysaccharides have been investigated[11]) [12]).

C. fragile. Coasts of the Atlantic ocean. After the Second World War this alga was found on the port piers of Heligoland, where it had been rarely found in former times[13]). Coast of Japan. The thallus is up to 40 centimetres long, dark green, branching dichotomous. Branches 3–8 millimetres thick, round, spongy, felty. The alga is eaten in Japan. It is harvested in spring, dried or salt-cured. Japanese name "Miru"[10]) [14]). The plant contains a polysaccharide probably esterified with sulphuric acid, which on hydrolysis yields glucose, mannose, methylpentose, and xylose[9]). The polysaccharide has been described in the ANNUAL REPORTS[11]) [12]) [15]). Three polysaccharides have been isolated by British workers, a water-soluble sulphated arabogalactan, a starch type polysaccharide, and a mannane[16]). The ethereal oils have been investigated[17]) [9]). A subspecies *tomentosoides*, is recorded from the French coast[18]).

C. intertextum. Tropical western Atlantic (Bermudas, Caribbean Sea, Brazil). SILVA has described this species in detail[3]).

C. intricatum. Coast of Japan. The alga is used for human consumption. Japanese name "Motsure-miru". The iodine content of the dried alga is recorded as 0.13 to 0.16 per cent.

C. isthmocladum. Tropical western Atlantic. Coasts of Yucatan and Columbia[19]), Bermudas, North Carolina, Florida, Gulf of Mexico, Caribbean Sea, Brazil. SILVA has described this species in detail[3]).

C. latum. The finding of glucose by ENDO has been reported by BLACK[20]).

C. Lindenbergii. Recorded on coasts of Japan. This alga is used for human consumption either in a fresh, dried or in a salt-cured state[14]).

C. Muelleri. Occurs at the coasts of the Hawaii Islands, Australia, Tasmania, New Zealand. It is used for human consumption on the Hawaii Islands in several preparations[14]) [21]).

C. repens. Habitat in tropical western Atlantic (Bermudas, Florida, Guadeloupe, Barbados, Curaçao, Trinidad, Brazil), Canary Islands. This species was described by SILVA[3]).

C. spongiosum. Tropical western Atlantic and other tropical areas, coasts of Brazil, Natal, Mozambique, Mauritius, Australia. Coasts of various Pacific islands. *C. spongiosum* is described in detail by SILVA[3]).

C. Taylori. Tropical western Atlantic (Bermudas, east coast of Florida, Gulf of Mexico, Caribbean Sea, Brazil), also, Canary Islands, Cape Verde Islands, Ghana, Israel. This species has been described by SILVA[3]).

C. tenue. Coasts of the Red Sea and of the Indian Ocean, Indonesia, the Philippines, Japan, and South Africa. The alga is sold on the Philippines as a food material. It is consumed in a raw state as a salad[14]) [21]).

C. tomentosum. Widespread on the coasts of all oceans. In the Adriatic Sea. Thallus erect, up to 50 centimetres high, round stipiform, ramified. The alga is used in Indonesia, Malaya, Japan, and on the Hawaii Islands as human food, in soups or with soya sauce or vinegar[14] [21]. It contains Vitamin A[22] [10]. In Japan *C. tomentosum* is also used for art handicraft work.

C. vermilara. Coast of the Adriatic Sea[23]. In addition to *C. divaricatum*, *C. fragile*, and *C. tenue* listed above. SEGAWA mentions the following species occuring on the Japanese coasts: *C. adhaerens*, *C. coarctatum*, *C. contractum*, *C. cylindricum*, *C. intricatum*, *C. latum*, *C. mamillosum* var. *minus*, *C. pugniformis*, *C. saccatum*, *C. tomentosum*[24].

Similarly, SILVA also reports on *C. adhaerens*, *C. adhaerens* var. *arabicum*, *C. bursa*, *C. difforme*, *C. lineare*, *C. tomentosum* with various varieties[3].

The same author investigated *C.* species on the South-African coast[25], in South-Australian waters[26], at the coast of Great Britain[27] and in Scandinavia[28].

Codium – as far as we could see – is one of the genera of chlorophyceae occurring in many areas and investigated by numerous authors. Reference should be made of publications dealing with *C.* species on the French coast[29–31].

Physiological and biochemical investigations of *C. dichotomum*, *C. decorticatum*, *C. fragile*, *C. intricatum*, and *C. tomentosum* have been reported in LEWIN (1962). Also the raw fibre of certain *C.* species, such as *C. fragile*, *C. lanatum*, and *C. tomensosum* have been investigated[9].

Literature

1) IRIKI, T. and MIWA, T., Nature (London) 1960. 185. 178
2) SILVA, P. C. and IRVINE, D. E. G., J. Mar. Biol. Ass. U. K. 39 (3): 631–636. 1960
3) SILVA, P. C., Nova Hedwigia I (3/4): 497–536. 1959
4) TOKIDA, J., Fac. of Fish. Hokkaido Univ. 2 (1): 1954
5) DANGEARD, P., C. R. Acad. Sc. (Paris) 247 (18): 1425–1427. 1958
6) JANECKE, H. et KEHR, W., Pharmaz. Ztg. 105 (25): 684. 1960
7) O'COLLA ex LEWIN, R. A., Physiology and Biochemistry of Algae Academic Press Inc., New York, 1962
8) LEVRING, T., Contributions to the Marine Algal Flora of Chile. Lunds Univ. Arsskrift N. F. Avd. 2. 56 (10). 1960
9) HEGNAUER, R., Chemotaxonomie der Pflanzen. Birkhäuser Verlag, Stuttgart, 1962
10) CHAPMAN, V. J., Seaweeds and their Uses. Methuen & Co., Ltd., London, 1950
11) Institute of Seaweed Research. Annual Report. 1960
12) ibid. 1961
13) KUCKUCK, P., Der Strandwanderer. J. F. Lehmanns Verlag, München, 1953
14) UPHOF, J. C. TH., Dictionary of Economic Plants. H. R. ENGELMANN (J. CRAMER), Weinheim, 1959
15) Institute of Seaweed Research. Annual Report 1962
16) LOVE, J. and PERCIVAL, E., Biochem. J. 1962, 84
17) KATAYAMA, T., J. Fac. Fish. & Anim. Husbandry Hiroshima Univ. 2 (1): 67–77. 1958
18) LAMI, R., Bull. Lab. Marit. Dinard 43. 132. 1957
19) TAYLOR, W. R., Bot. Ser. Field Museum of Natural History 20 (4). 1941
20) BLACK, W. A. P., Ann. Rep. Chem. Soc. 1953, Vol. 50
21) ZANEVELD, J. S., Ec. Bot. 13 (2): 89. 1959
22) FOTT, B., Algenkunde. Gustav Fischer Verlag, Jena, 1959
23) VOUK, V., Thalassia Jugoslav. 1 (6/10): 29–39. 1959
24) SEGAWA, S., The Seaweeds of Japan. Hoikusha, Osaka, Japan, 1957
25) SILVA, P. C., J. South Afr. Bot. 25 (2): 103–165. 1959
26) SILVA, P. C., . . .
27) SILVA, P. C., . . .
28) SILVA, P. C., Svensk. Bot. Tidskr. 51 (1): 117–134. 1957
29) FELDMANN, J., Bull. Soc. Phyc. de France 1: 10. 1955
30) DANGEARD, P., ibid. 3: 4–6. 1957
31) DELEPINE, R., Rev. gén. Bot. (785): 366–394. 1959

General Literature

CHAPMAN, V. J., 1950; FOTT, B., 1959; FRITSCH, F. E., 1956; HEGNAUER, R. 1962; HOFFMANN, C., 1938; LEWIN, R. A., 1962; PAX, F., 1962; SEGAWA, S., 1957; TOKIDA, J., 1954; see Index of Literature.

Collinsiella

Collinsiella tuberculata S. et G. Pacific coast of North America. It is likely that it is a variety of *Enteromorpha intestinalis*. Investigations on the life history have been reported.

Literature: Scagel, R. F., Can. J. Bot. 38 (6): 969–983. 1960

Enteromorpha

Approximately 40 species are recorded to live in the sea or in brackish water, and some are recorded from salty land-locked bodies of water. The *Enteromorpha* is one of the most widespread genera of green algae. The individual species are very difficult to get distinguished. Thallus is inflated or tubular, crisped, contorted, or band-shaped, sometimes branched, sedentary in the early stage, often free-floating when mature. Species of *Enteromorpha* among the green algae are most studied by research workers. Topics investigated include submicroscopical and chemical structure of gamets[1]). Relation of photosynthesis transients to respiration[2]), the influence of various organic substances on the alga[3]), the effect of saccharose on the contents of pigments[4]) and the chemical composition of the cell wall[5]), the volatile constituents[6–8]) have been studied, and a number of terpenes have been found. The antibacterial substances have been investigated[9]) [10]). Vitamin C (15 milligramme p% g of fresh weight) and traces of Vitamin B_1 have been found[11]).

A detailed investigation of water-soluble green algal polysaccharides, specifically of some *Enteromorpha* species, has been made in Great Britain under the supervision of Percival[12–16]). For physiological and biochemical investigations, see Lewin (1962).

E. species are used for human consumption in many countries, particularly in East Asia, mostly as a salad. But they are also used as a feeding material for animals and for fertilizing purposes, inter alia in India.

E. compressa. Known as "Flacher Darmtang" in German, of world-wide distribution. Thallus linear and lanceolate, with narrowed base. On rocks and stones in the littoral zone exposed at low water; occuring during the whole year (Kuckuck).

Segawa has reported on the occurrence of this species in Japan. The species also occurs on the Malayan and Indonesian coasts, on the Philippines, where it is also used for human consumption[17]), on some of the Pacific islands and on the coasts of New Caledonia[18]).

The constituents of the alga have been investigated[14]) [20]) [21]), in particular the polysaccharides[22]), which contain rhamnose and uronic acid[23]), and saccharose[8]). Antimicrobial substances have been found in algae of Baltic provenance[9]).

Black, according to the Standard Tables of Food Composition, indicates, for Japan, provitamin A, vitamin B_1, B_2, niacin, and vitamin C as constituents[24]).

E. flexuosa. The constituents of this alga have been investigated[19]).

E. intestinalis. Known as "Darmtang" (German), coasts of India, Malaya, Indonesia (Java), the Philippines, Hawaiian Islands, Japan.

Dangeard reported on this alga in 1957[25]). The influence of anorganic ions for example aluminium, arsenic etc. has been studied[26]); chemical investigations have been reported[19]; proteins and amino acids[27] [28], carotene and carotenoid have been analysed[29]. A very high ash content, namely 45.4 per cent, has been found on mass examination of *E. intestinalis*[30].

E. intestinalis is eaten in the Philippines, mostly as a salad, and, owing to its popularity, it is imported to Malaya. It is likewise used for human consumption in Japan and Hawaii[17]. *E. intestinalis* is also used as a fertilizer.

E. linza. European coasts, from the western Baltic to the Mediterranean. East coast of North America, coast of Japan. Inhabits the littoral zone, all year round; thallus flat, mostly lanceolate, undulated with tubular stipe. Morphology, taxonomy, and reproduction[31–34], amino acids[24] [35] and sterols[36] [8] have been investigated.

E.prolifera. Coast of Japan. The alga is eaten on the Hawaiian Islands in numerous preparations[18]. The var. *tubulosa* exists on the Malayan peninsular and is also eaten[17].

A number of other *E.* species has been investigated, e.g. *E. flabellata*, on the coasts of Morocco[37], and the culture of this alga[38]; *E. intermedia* in the Bassin d'Archachon[40]; *E. lingulata*, in Texas[41]; and *E. marginata*[42], also found amongst the "Nori" cultures in Tokyo Bay[43]; *E. sancti joannis* at Saint Jean de Luz[45]; *E. flexuosa* eaten on the Hawaiian Islands as a salad; *E. Hendayensis* n. sp. described by Dangeard and Parriaud[39]. *E. nana* var. *minima* on the coasts of Japan, its ecology has been studied[44].

A number of *E.* species have been described by Levring (1940) from North European coasts[46]. Reports have been published on growth forms, and environment[47]. A number of new Mediterranean species[48] and species from the Atlantic coast of Morocco, have been described[49]. Numerous *E.* species have been described by Segawa and Tokida. *E. intestinalis*, *E. linza*, and *E. prolifera* have a certain commercial importance in East Asia[50]. Where ever they are of considerable abundance, *E.* species are used for human nutrition. In Japan they are described as "Awo-nori" or "Ohashi-nori", and are also grown in cultivated beds.

Literature

1) Levring, T., Coll. Int. du Centre Nat. de la Recherche Scientifique, Dinard, 1960
2) Blinks, L. R., Science 131 (3409): 1316. 1960
3) Litvinenko, L. G., Ukrain. Bot. Zhurn. 17 (4): 36–40. 1960; ibid. 17 (5): 32–37. 1960; Fiziol. Rast. 7 (4): 466–469. 1960; Nauck Ezhegod. Odessk. Univ. Biol. Fak. 2. 41–42. 1960
4) Litvinenko, L. G., Ukrain. Bot. Zhurn. 16 (6): 49–54. 1959
5) Preston, R. D., Third Int. Seaweed Symp., 1958
6) Katayama, T., Bull. Jap. Soc. Sci. Fish. 21 (6): 420–424. 1955
7) Katayama, T., ibid. 22 (4): 248–250. 1956
8) Hegnauer, R., Chemotaxonomie der Pflanzen. Birkhäuser Verlag, Stuttgart, 1962
9) Roos, H., Kieler Meeresforsch. XIII (1): 1957
10) Katayama, T., Bull. Jap. Soc. Sci. Fish. 22 (4): 248–250. 1956
11) Humm, H. J., The Red Algae of Economic Importance ex Tressler, D. K. et Lemon, J. Mc. W., Marine Products of Commerce. Reinhold Publishing Corporation, New York, 1951
12) Institute of Seaweed Research. Annual Report 1955
13) ibid. 1956
14) ibid. 1957
15) ibid. 1960
16) McKinnel, J. P. and Percival, E., J. chem. Soc. 1962, 3141
17) Zaneveld, J. S., Ec. Bot. 13 (2): 89. 1959
18) Uphof, J. C. Th., Dictionary of Economic Plants. H. R. Engelmann (J. Cramer), Weinheim, 1959
19) Black, W. A. P., Ann. Rep. Chem. Soc. 50. 1953
20) Mita, K., Bull. Jap. Soc. Sc. Fish. 26 (10): 1010–1015. 1960
21) Smith, F. and Montgomery, R., The Chemistry of Plant Gums and Mucilages. Reinhold Publishing Corporation, New York, 1959
22) Percival, E., Fourth Int. Seaweed Symp., 1961

23) O'Colla ex Lewin, R. A., Physiology and Biochemistry of Algae. Academic Press Inc., New York, 1962
24) Black, W. A. P., The Algae ex Altschul, Processed Plant Proteins Foodstuffs. Academic Press Inc., New York, 1958
25) Dangeard, P., Col. Int. du Centre Nat. de la Recherche Scientifique, Dinard, 1960
26) Peters, B., Meddel. f. Göteborgs Bot. Trädgard XVIII, 1948
27) Coulson, C. B., Chem. and Ind. 1953, 997–998
28) Coulson, C. B., J. Sci. Food Agric. 6. 674–682. 1955
29) Owen, E. C., ibid. 5 (9): 449–453. 1954
30) MacPherson, M. G. and Young, E. G., Canad. J. of Bot. 30. 67–77. 1952. CA 46. 5149. 1952
31) Dangeard, P., Botaniste 43 (1/6): 103–117. 1960
32) Arasaki, S. and Shihira, I., Jap. J. Bot. 17 (1): 92–100. 1959
33) Comps, B., C. R. Acad. Sci. (Paris) 251 (19): 2067–2069. 1960
34) Baudrimont, R., Botaniste 43 (1/6): 157–167. 1960
35) Takagi . . . .
36) Tsuda, K. and Sakai, K., Chem. and Pharm. Bull. 8 (6): 554–558. 1960
37) Dangeard, P., C. R. Acad. Sci. (Paris) 244 (12): 1589–1592. 1957
38) Gayral, P., Botaniste 43 (1/6): 85–100. 1960
39) Dangeard, P. and Parriaud, H., C. R. Acad. Sc. (Paris) 250 (18): 2972–2975. 1960
40) Parriaud, H., Fourth Int. Seaweed Symp., 1961
41) Taylor, W. R., Bot-Ser. Field Museum of Natural History 20 (4). 1941
42) Dangeard, P., C. R. Acad. Sci. (Paris) 246 (3): 347–351. 1958
43) Iwamoto, K., Bull. Jap. Soc. Phyc. 8 (1): 4–12. 1960
44) Akiyama, K., Bull. Tohoku Reg. Fish. Res. Lab. 13. 108–114. 1959
45) Dangeard, P., C. R. Acad. Sci. (Paris) 251 (16): 1603–1606. 1960
46) Levring, T., Studien über die Algenvegetation von Blekinge, Südschweden. Dissertation, Lund, 1940
47) Burrows, E. M., J. Linn. Soc. London. Bot. 56 (366): 204–206. 1959
48) Bliding, C., Bot. Not. 113 (2): 172–184. 1960
49) Gayral, P. et al.
50) Tokida, J., Fac. of Fish. Hokkaido Univ. 2 (1). 1954

General Literature

Chapman, V. J., 1950; Fott, B., 1959; Fritsch, F. E., 1956; Hegnauer, R., 1962; Hoffmann, C., 1938; Kuckuck, P., 1953; Lewin, R. A., 1962; Pax, F., 1962; Segawa, S., 1957; Tokida, J., 1954; see Index of Literature.

Halimeda

Records mention 15 species in tropical and temperate oceans. The thallus consists of individual flat plated links, often strongly incrusted.

H. opuntiae. Inhabitant of the coast of Puerto Rico. The antimicrobial substances of this species have been investigated.

General Literature

Fott, B., 1959; Fritsch, F. E., 1956; Olesen, P. E. et al., Bot. Mar. VI (3/4): 224–232. 1964; see Index of Literature.

Monostroma

M. Grevillei (Thur.) Wittr. Coasts of the northern Arctic oceans down to France, east coast of North America, individual specimens recorded from the Baltic. *M. Grevillei* consists of soft, mucilaginous, torn and ragged lobes. It is found particularly in spring on rocks and on other algae[1]). Kornmann has described the development of *M. Grevillei*[2]). Levring has found individual and very small specimens of this species on the Swedish south coast, and he has also described *M. latissimum* and *M. balticum* n. spp.[3]). *M.* species have also been recorded from the Irish coast[4]).

SEGAWA lists the following *M.* species for the Japanese coast: *M. angicava, M. pulchrum, M. zostericola, M. latissimum* (Pl. XIV: 36, 37), and *M. nitidum.* The three first species mentioned, together with *M. Grevillei* and *M. fuscum* are recorded from the coast of Oshoro[5]).

TOKIDA mentions the economic possibilities of *M. arcticum, M. angicava, M. crassidermum, M. fuscum* var. *typicum,* and *M. fuscum* var. *splendens* for southern Sakhalin, and mentions also *M. zostericola* and *M. undulatum* var. *Farlowii.* Several *M.* species are recorded for the Bay of Tokyo[6]).

Several Japanese authors have mentioned Japanese *M.* species[7]) [8]), particularly the characteristic of the thallus[9]) and life history[10]). In 1956 SEGI wrote of the development of *Monostroma* in the sea[11]).

KORNMANN and SAHLING have investigated the taxonomy and development of *Monostroma* species from Heligoland and revised the classification of the genus. *M. Grevillei, M. arcticum, M. undulatum,* and *M. leptodermum,* occuring on Heligoland, and also *M. angicava, M. nitidum, M. latissimum, M. zostericola,* and *M. Wittrockii* have been studied[12]). A revision of Ulotrichales, based on phylogenetic considerations, has been reported by KORNMANN in PHYCOLOGIA in 1963[13]).

M. latissimum and *M. nitidum* are cultivated on various parts of the Japanese coast (Pl. XIV: 34, 35), particularly in Ise, Mikawa and Tokyo Bays, in places, where *Porphyra* does not grow so well, or where *Porphyra* is out of season. The thalli of these algae are light green and very tender. Like *Porphyra, Monostroma* has been formerly cultivated on bamboo bundles, but in modern times nets have been used, either of palm, hemp palm, straw, and synthetic fibre yarns. The major cultivation area is Ise Bay, where in September the nets are cast in the shallow sea near the river mouth. Investigations on the *Monostroma* cultures have been reported by SEGI in 1958 on occasion of the THIRD INTERNATIONAL SEAWEED SYMPOSIUM[14]), and in 1961 this author has published a work on the correlation between "Distribution of early germlings of *Monostroma* and tidal current in the culture ground" in BOTANICA MARINA[15]).

In the breeding areas there are districts known as "Taneba" districts where the stationary spores and young germlings in early autumn are found in greater quantities than in other areas. Sometimes the location of these areas of abundance shifts, apparently on account of surface currents, since the swarm-spores for the most part are brought in from the open sea. This shift of "Taneba" can be predicted if surface currents and wind directions at the time of swarming will be observed.

The highly valued food "Aonori" is produced from *M. nitidum* and *M. latissimum,* but it is also made from species of *Enteromorpha.* The production of "Aonori" specialities has risen in recent years. In 1958 1,400 tons of dried *Monostroma* have been processed in Japan[15]). *Monostroma* is also used as a salad. The STANDARD TABLE OF FOOD COMPOSITION, JAPAN, indicates the following dates for *Monostroma:* 16 per cent protein, 0,2 per cent fat, 66,6 per cent sugar, 12,1 per cent ash[16]). The vitamin content can also be regarded as of particular importance. Physiological and biochemical investigations of *M. nitidum* have been reported by LEWIN (1962)[17]). *M. nitidum* was the source of the first triterpene isolated from algae; Haliclonasterol and Friedelin have been also isolated[18]) and investigated[19]). *M. angicava* showed a high catalase activity[20]).

M. quaternarium. Inhabitant of the coast of India (Naini Valley). An account of its occurence has been given[21]).

Literature

1) KUCKUCK, P., Der Strandwanderer. J. F. Lehmanns Verlag, München, 1953
2) KORNMANN, P., Helg. Wiss. Meeresunters. 8 (2): 195–202. 1962
3) LEVRING, T., Studien über die Algenvegetation von Blekinge, Südschweden. Dissertation, Lund, 1940
4) BRENNAN, A. T., Inst. f. Ind. Res. and Standards, Dublin, 1950
5) TOKIDA, J. and MASAKI, T., Fac. of Fish. Hokkaido Univ. 10 (3): 173–195. 1959
6) IWAMOTO, K., J. Tokyo Univ. Fish. 47 (1): 93–101. 1960
7) OGATA, E., Bull. Jap. Soc. Sci. Fish. 22 (7): 400–407. 1956
8) SHIHIRA, I., Bot. Mag. Tokyo 71 (845/846): 378–385. 1958
9) SHIHIRA, I., J. Jap. Bot. 33 (12): 372–375. 1958
10) YAMADA, Y and TATEWAKI, M., Proc. IX. Int. Bot. Congr. 2. 438. 1959
11) SEGI, T., Rept. Fac. Fish. Pref. Univ. Mie 2 (2): 312–316. 1956
12) KORNMANN, P. and SAHLING, P. H., Helg. Wiss. Meeresunters. 8 (3): 302–319. 1962
13) KORNMANN, P., Phycologia 3 (2): 60. 1963
14) SEGI, T., Third Int. Seaweed Symp., 1958
15) SEGI, T. and KIDA, W., Bot. Mar. II (3/4): 223. 1961
16) BLACK, W. A. P., The Algae ex Altschul, Processed Plant Proteins Foodstuffs. Academic Press Inc., New York, 1958
17) LEWIN, R. A., Physiology and Biochemistry of Algae. Academic Press Inc., New York, 1962
18) HEGNAUER, R., Chemotaxonomie der Pflanzen. Birkhäuser Verlag, Stuttgart, 1962
19) TSUDA, K. and SAKAI, K., Chem. and Pharm. Bull. 8 (6): 554–558. 1960
20) TAKAGI, M., Bull. Jap. Soc. Sc. Fish. 18. 483–487. 1953 CA 56021. 1955
21) SINGH, K. P. and BHARGAWA, K. S., Curr. Sc. 28 (5): 205. 1959

General Literature

CHAPMAN, V. J., 1950; FOTT, B., 1959; FRITSCH, F. E., 1956; HOFFMANN, C., 1938; SEGAWA, S., 1957; TOKIDA, J., 1954; see Index of Literature.

Prasiola

Some 20 species, fairly widespread, marine littoral, tenestial or fluvial.

P. japonica. This alga is eaten in Japan. BLACK has reported the presence of provitamin A and vitamins B_1 and B_2 in this alga[1]).

P. stipitata. This alga occurs on numerous sea coasts, for example in southern Sweden and, *inter alia*, on the Canadian coast near Halifax. LEWIN has cultivated this alga and has investigated its physiology[2]).

Literature

1) BLACK, W. A. P., The Algae ex Altschul, Processed Plant Proteins Foodstuffs. Academic Press Inc., New York, 1958
2) LEWIN, R. A., Canad, J. of Bot. 33. 5–10. 1955

General Literature

FOTT, B., 1959; FRITSCH, F. E., 1956; LEVRING, T., 1940; LEWIN, R. A., 1962; TOKIDA, J., 1954; see Index of Literature.

Spongomorpha

Sp. arcta (Dillw.) Kütz. *(= Acrosiphonia centralis)*. This alga has been reported from the Baltic and the Swedish south coast, but in fact it is widespread elsewhere. *Sp. arcta* is usually a bushy plant, several centimetres high. For physiological and biochemical investigations see LEWIN (1962), and polysaccharides of the alga have been investigated[1–6]. The thallus contains approximately 12 per cent of an acid water-soluble polysaccharide[7].

Green algae contain volatile, occasionally very malodorous, constituents. These are often sulphur-containing compounds, for example thetin, in *Sp. arcta*[7]. This alga also contains antimicrobial substances[1]. The inorganic constituents of *Sp. arcta* have also been analysed[8].

KORNMANN investigated *Sp. lanosa* and published a revision of the genus *Acrosiphonia*[9] [10]. The same author also included *Spongomorpha* and *Acrosiphonia* in his revision of the Ulotrichales[11]. Very interesting is the recent study of JÓNSSON[12].

Literature

1) JANECKE, H. and KEHR, W., Pharmaz. Ztg. 105 (25): 684. 1960
2) Institute of Seaweed Research. Annual Report 1957
3) ibid. 1958
4) O'DONNELL, J. J. and PERCIVAL, E., Third Int. Seaweed Symp. 1958
5) O'DONNELL, J. J. and PERCIVAL, E., J. chem. Soc. 2168. 1959
6) O'COLLA ex LEWIN, R. A., Physiology and Biochemistry of Algae. Academic Press Inc., New York, 1962
7) HEGNAUER, R., Chemotaxonomie der Pflanzen. Birkhäuser Verlag, Stuttgart, 1962
8) YOUNG, E. G. and LANGILLE, W. M., Publ. 4636 of the Nat. Res. Council, Halifax, Canada, 1958
9) KORNMANN, P., Helg. Wiss. Meeresunters. 7 (5): 195–205. 1961
10) KORNMANN, P., ibid. 8 (2): 219–242. 1962
11) KORNMANN, P., Phycologia 3 (2): 60. 1963
12) JÓNSSON, S., Recherches sur des Cladophoracées marines (Thèses), Paris, 1962

General Literature

FRITSCH, F. E., 1956; LEVRING, T., 1940; LEWIN, R. A., 1962; see Index of Literature.

Ulva

Approximately 30 species in marine and brackish waters, ubiquitous. Thallus mostly flat, consisting of tow cell-layers. PAPENFUSS has discussed the genera and the status of the order[1]; and LEVRING has written on the submicroscopical and chemical structure of gamets[2]. *Ulva* starch[3] and amino acids have been investigated[4] [5].

Ulva species are valuable salads or vegetables because of their vitamin content[6], and they are suitable for fertilizing purposes because of their high nitrogen content.

Culture experiments have been made by LEVRING[7].

U. lactuca L. Called "Meersalat" (in German), or "Sea Lettuce", "Green Laver" (in English), "Aosa" or "Awosa" (in Japanese). Several reports have been published on occurence at the European, north and south American coasts. There is a great number of specimens which have been referred to this species[8]. *U. lactuca* consists of irregular, membranaceous, light green lobes with jagged fringes. In the early stage the alga is sedentary on rock, or other algae, later on it is free floating in shallow water. Specific differences between northern and southern European populations have been reported[9]. *Ulva* grows particularly well in water which is rich in nutrients. The structure of gamets and the reproduction of *U. lactuca* have been investigated by LEVRING[10]. PROVASOLI has investigated the influence of plant hormones on the growth of this alga[11]. The following qualities have also been studied: the chemical composition of the cell wall[12]), photosyntheses and phototaxis on the gamets[13]), potassium accumulation and sodium secretion[14]), the reversible replacement of potassium by rubidium ion[15]), and the exchange of sodium ions[16]). For physiological and biochemical investigations see LEWIN (1962).

The polysaccharides of *U. lactuca* have been studied[17]) [18]). They amount to approximately 20 per cent of the plant[19–26]); also a report has been published on acid polysaccharides of this species[27]). The polysaccharide is known as ulvin contains rhamnose[28]). The amino acids have been analysed[29]) and the enzymes of carbohydrate and amino acid metabolism too[31–34]). Investigations have been carried out on sterols[35]), on vitamins C and B_1[6]), and antimicrobial substances found in the alga[36]), and on inorganic components[37]).

The alga is eaten as a salad in many areas, for example on the Scottish coast, the south-west coast of North America, and also in Peru where it is known as "lechuga de samba". In Chile this alga is used in soups and as a vegetable; in India, as well as in many East-Asian areas it is used as a fodder and fertilizing material. In some regions *U. lactuca* is also used in popular medicine for gout. In Malaya, Indonesia, and on the Philippines the alga is used as a salad, a vegetable, or as a soup constituent[38]).

The following additional species of *Ulva* have been investigated:

U. elegans. Inhabitant of the coast of Morocco. This species has been described and its biology studied[39]).

U. fasciata. Occurs in East-Asian coastal areas. The alga is used for human consumption, particularly on the Hawaiian Islands[40]). GAYRAL made cultivation experiments in the laboratory[41]).

U. latissima. The alga occurs in considerable populations of the coast of Chile where it is used for human consumption. The Chilean name is "luche". It occurs on the coast of Columbia and in the Singapore area too, where it grows in large beds in the estuaries of rivers. It is also used as a pig feed and feed for ducks.

U. linearis. Inhabitant of the coast of Morocco. Studies with cultural experiments have been made on this alga[42]).

U. linza (= Enteromorpha linza). Coast of Chile, widespread populations[8]); central California[36]). The alga has been proved to have antimicrobial activity[36]) [43]).

U. lobata. The alga has been examined[44]). Other physiological and biochemical investigations have been reported by LEWIN (1962).

U. mutabilis. FOEYN has reported on this species[45–47]).

U. nematoida. Inhabitant of the coasts of Chile, Peru, Juan Fernandez, growing on rocks or epiphytically[8]). This alga is also eaten in New Caledonia[40]).

U. penniformis. This alga is used as a medical preparation in some parts of Siberia[40]).

U. pertusa. Known also as "Sea Lettuce" (English) and "Ana-awosa" (Japanese). This species is existing at the coasts of Formosa, China, Japanese islands, Kurile Islands, Sakhalin, Korea, Malayan islands[48]). Studies with this species have been made on nuclear and cell division[49]), on frost resistance[50]), on chemical composition in detailed investigations[51]) [52]), when the polysaccharides have been found to contain rhamnose and uronic acid [17]) [28]), on the raw protein content, which amounts to 31.55 per cent of the dried alga[53]), on the amino acids[4]), on the volatile constituents, and on terpenes and other organic compounds[54]) [55]). The anthelmintic effects and the influence on the muscle contraction of earth worms has also been investigated[56] [22]). For physiological and biochemical studies see LEWIN (1962).

The alga is used in China as a drug for the reduction of excessive body temperature. It is also used for nutrition purposes[40]). The economic significance of this alga for East Asia has been studied by TOKIDA.

U. reticulata. This alga is collected in Singapore and mainly used as a pig feed[38]).

U. rigida. Occurrence and distribution of this alga have been investigated, and physiological and biochemical inquiries have been reported by LEWIN (1962).

U. taeniata. For physiological and biochemical investigations see LEWIN (1962).

A number of *U.* species occurring on the French Atlantic and Mediterranean coasts have been investigated[57–62]).

In addition to *U. pertusa* SEGAWA mentioned the following species as occurring on Japanese coasts: *U. conglobata, U. fasciata, U. reticulata, U. sublittoralis.*

Literature

1) PAPENFUSS, G. F., J. Linn. Soc. London Bot. 56 (367): 303–318. 1960
2) LEVRING, T., Coll. Int. du Centre Nat. de la Recherche Scient., Dinard, 1960
3) MEEUSE/KREGER
4) BLACK, W. A. P., The Algae ex Altschul, Processed Plant Proteins Foodstuffs. Academic Press Inc., New York, 1958
5) TAKAGI, M. and KURIYAMA, Bull. Fac. Fish. Hokkaido Imp. Univ. 10 (1): 72–76. 1959
6) HUMM, H. J., The Red Algae of Economic Importance ex Tressler, D. K. et Lemon, J. Mc. W., Marine Products of Commerce. Reinhold Publishing Corporation, New York, 1951
7) LEVRING, T., Sällskap Lund, Förhandl. 16 (7): 1–12. 1946
8) LEVRING, T., Contributions to the Marine Algal Flora of Chile. Lunds Univ. Arsskrift N. F. Avd. 2 56 (10): 1960
9) FÖYN, B., Publ. Staz. Zool. Napoli 27. 261–270. 1955
10) LEVRING, T., Bot. Notiser 108 (1): 40–45. 1955
11) PROVASOLI, L., Third Int. Seaweed Symp., 1958
12) PRESTON, R. D., Third Int. Seaweed Symp., 1958
13) HAXO, F. T. and CLENDENNING, K. A., Biol. Bull. 105 (1): 103–114. 1953
14) HAYWARD, H. R. and SCOTT, G. T., ibid. 105 (2): 366. 1953
15) DE VOE, R. et al., ibid. 111 (2): 296. 1956; ibid. 112 (2): 249. 1957
16) SCOTT, G. T. et al., Science 125 (3239): 160. 1957; CA 51. 10. 75180. 1957
17) SMITH, F. et MONTGOMERY, R., The Chemistry of Plant Gums and Mucilages. Reinhold Publishing Corporation, New York, 1959
18) BLACK, W. A. P., Ann. Rep. Chem. Soc. 50. 1953
19) BRADING, J. W. E. et al., J. Chem. Soc. 319–324. 1954. CA 49. 3. 1584g. 1955
20) JANECKE, H. and KEHR, W., Pharmaz. Ztg. 105 (25): 684. 1960
21) PERCIVAL, E., Fourth Int. Seaweed Symp., 1961
22) HEGNAUER, R., Chemotaxonomie der Pflanzen. Birkhäuser Verlag, Stuttgart, 1962
23) Institute of Seaweed Research. Annual Report 1958
24) ibid. 1960
25) ibid. 1961
26) ibid. 1962
27) McKINNELL, J. P. and PERCIVAL, E., Biochem. J. 1962, 84
28) O'COLLA ex LEWIN, R. A., Physiology and Biochemistry of Algae. Academic Press Inc., New York, 1962
29) SMITH, D. B. and YOUNG, E. G., J. of Biol. Chem. 217 (2): 845–853. 1955
31) JACOBI, G., Planta 49 (1): 1–10. 1957
32) Institute of Seaweed Research. Annual Report 1955
33) ibid. 1959
34) DUNCAN, W. A. M., Biochem. J. 73. 295–298. 1959
35) GOVINDACHARI, T. R. et RAJAGOPALAN, S., First Int. Seaweed Symp., 1952
36) ROOS, H., Kieler Meeresforsch. XIII (1). 1957
37) YOUNG, E. G. and LANGILLE, W. M., Publ. 4636 of the Nat. Res. Council, Halifax, Canada, 1958
38) ZANEVELD, J. S., Ec. Bot. 13 (2): 89. 1959
39) GAYRAL, P., C. R. Acad Sc. (Paris) 251 (5): 768–770. 1960

40) Uphof, J. C. Th., Dictionary of Economic Plants. H. R. Engelmann (J. Cramer), Weinheim, 1959
41) Gayral, P., Fourth Int. Seaweed Symp., 1961
42) Gayral, P., Botaniste 43 (1/6): 85–100. 1960
43) Pratt, R. et al., J. Am. Pharm. Ass. 40. 575–579. 1951
44) Blinks, L. R., Pl. Physiol. 34 (3): 200–203. 1959
45) Föyn, B., Biol. Bull. 118 (3): 407–411. 1960
46) Föyn, B., Bot. Mar. III (2): 60. 1961
47) Föyn, B., ibid. IV (1/2): 156. 1962
48) Tokida, J., Fac. of Fish. Hokkaido Univ. 2 (1). 1954
49) Yabu, H. and Tokida, J., Bot. Magazine Tokyo 73 (863): 182–185. 1960
50) Terumoto, J., Low Temp. Sc. 18 B. 35–38. 1960
51) Black, W. A. P., Ann. Rep. Chem. Soc. 50. 1953
52) Mita, K., Bull. Jap. Soc. Sc. Fish. 26 (10): 1010–1012. 1960; ibid. 26 (10): 1013–1015. 1960
53) Takagi, M., Bull. Fac. Fish. Hokkaido Univ. 1 (1): 35–43. 1950
54) Katayama, T. and Tomiyama, T., Bull. Jap. Soc. Sci. Fish. 17 (5): 122–127. 1952. CA 48. 20. 12013p. 1954
55) Katayama, T., ibid. 19 (7): 793–797. 1953. CA 49. 7. 4804f. 1955; ibid 21 (6): 412–415. 1955. CA 50. 18. 13184a. 1956; ibid. 21 (6): 416–419. 1955. CA 50. 18. 13184c. 1956
56) Katayama, T., ibid. 22: 251. 1956; ibid. 22: 253. 1956
57) Dangeard, P., Botaniste 43 (1/6): 119–155. 1960
58) Baudrimont, R., Cahiers de Biol. Mar. 1 (3): 251–258. 1960
59) Dangeard, P., Fourth Int. Seaweed Symp., 1961
60) Baudrimont, R., ibid.
61) Bliding, C., Bot. Not. 113 (2): 172–184. 1960
62) Dangeard, P., Coll. Int. du Centre Nat. de la Recherche Scient., Dinard, 1960

General Literature

Chapman, V. J., 1950; Fott, B., 1959; Fritsch, F. E., 1956; Hegnauer, R., 1962; Hoffmann, C., 1938; Kuckuck, P., 1953; Lewin, R. A., 1962; Pax, F., 1962; Segawa, S., 1957; Tokida, J., 1954; see Index of Literature.

Phaeophyceae

Agarum

Perennial species recorded from Northern Asia and America.

A. cribrosum Bory. From the coasts of Japan, particularly Northern Honshû, Hokkaido, moreover Kurile Islands, Sakhalin, Southern Korea, Siberian coast of the Japan Sea, coasts of Okhotsk Sea, Kamchatka, Bering Sea, Pacific coast of North America from Alaska to Washington, Atlantic coast of North America, Arctic regions. On rocks, mostly in deep water, perennial with short stipe, 30 to 90 centimetres long, 20 to 30 centimetres wide. Mature blades have strongly marked central ribs with numerous perforations.

Drifting fronds are collected, particularly on Japanese coasts, (the Japanese name "Aname") and form the raw material for the production of alginic acid and alginates.

A. fimbriatum. East-Asian coastal areas. This species has fewer perforations than *A. cribrosum*. It can be used in the same manner.

Literature

FRITSCH, F. E., 1952; MIYABE, K., 1957; SEGAWA, S., 1957; TOKIDA, J., 1954; TOKIDA, J. et MASAKI, T., 1959; see Index of Literature.

Alaria

A. esculenta (L.) Grev. Known also as "Honey Ware" (Orkney Islands), "Badderlocks" (Scotland), "Murlins" (Ireland), "Marinkjarni" (Iceland), "Kutare" (Norway).

Colder coastal regions of the Northern Atlantic, particularly coasts of Lofoten Islands, Iceland, Scotland, the Orkney Islands, and Ireland. Coasts of Japan, particularly Hokkaido.

A. esculenta is perennial and grows in exposed areas near or below low-water. The relatively short stipe gives place to a long blade with serrated edge but is continued in the very prominent central rib. During the summer months two rows of ligulate sporophylls form on the upper ends of the stipe. The growth of the alga has been studied[1]).

Seasonal variations in chemical composition have been investigated[2]) [3]), and constituent sugars have been analyzed[4]). Alginic acid content has been found to amount to approximately 30 to 35 per cent. The alga is rich in mannite and xylose, and methylpentose have been proved to be present[5]). Amino acids[6]), vitamins[6]) [7]), and variation in nitrate content have been examined[8]). There is a high niacin content[6]) [7]).

The alga is eaten in various European maritime countries, e.g. in Scotland and Ireland, particularly the parts known as "fingers" and "central rib"[9]). *A. esculenta* is also used as a fodder together with hay, mainly for cattle, sheep, and horses. Other *A.* species are used for human consumption and feeding purposes.

A. corrugata. Known as "Chishima-wakame" in Japan. Recorded from the coasts of Etorofu and Uruppu, Kurile Islands[10]).

A. crassifolia. Known as "Ezo-wakame" in Japan. Occurs in Hokkaido, on the coast from Muroran as far southward as Hakodate and the Saurika district in Northern Honshû[10]) [11]), Korea, southern offshore island Che Ju Do, and to a lesser extent also on the Southern Coast.

Reproduction of the alga has been studied[12]), the chemical constituents[13]), including alginic acid, laminaran, nitrogen compounds, proteins, and fatty acid have been investigated[14]) [19]).

A. dolichorhachus. Known also as "Enaga-wakame" (Japanese).f. *typica* from the Kurile Islands, Sakhalin, Kamchatka, the Bering Sea, the Aleutian Islands, and Alaska; f. *longipes* from the Kurile Islands and Sakhalin[15]).

A. fistulosa. Other names: "Oni-wakame" (Japanese); "Stringy Kelp" (English). Coasts of the Northern Pacific from Etorofu Island as far northward as the Bering Sea, West Coast of Sakhalin. From there the alga will often be tossed up on to the coast of Hokkaido. This species is abundant on the North American coast between Alaska and California[10]) [14]) [15]).

A. fistulosa, with thallus lobes of up to 25 metres length, and 2 metres breadth, is the largest species of *Alaria*. The alga is used for human consumption. The product has been described in Japan as "kausam" or "kauan"[15]).

A. macrophylla. Other name: "Kunashiri-wakame" (Japanese). Kurile Islands, eastern coast of Kunashiri Island[10]).

A. macroptera. This species is, perhaps, identical with *A. corrugata*. Coast of Hokkaido, Kurile Islands, Sakhalin, Kamchatka, and the Okhotsk Sea[15]).

A. marginata. Vancouver Island. The alga is abundant there[16]) and is used as a fertilizing material. Formerly *A. marginata* has been used in the kelp industry for the production of potassium salts.

A. ochotensis. Other name: "Karafuto-wakame" (Japanese). Coasts of Sakhalin and Kamchatka. In Sakhalin this kelp is harvested in early summer and sold on the markets as a foodstuff in Sakhalin[15]).

A. praelonga. Coasts of Japan. This species has been mentioned by SEGAWA[11]).

A. tenuifolia. Vancouver Island[16]). This alga is used like *A. marginata*.

A. valida. Alga has been investigated. Vitamin B and C have been shown to be present, 53 milligrams vitamin C have been found per 100 grammes fresh weight of alga[17]).

A. yezoensis. Known also as "Ainu-wakame" (Japanese). Occurs in Hokkaido, along the coasts from Hidaka north-eastward as far as Nemuro. The sporophylls of the alga are eaten raw by the fishermen of Nemuro[10]). *A.* species is also recorded from the Philippine Islands[18]).

Physiological and biochemical investigations of *A. crassifolia*, *A. esculenta*, and *A. valida* are reported in LEWIN (1962).

Literature

1) BAARDSETH, E., Second Int. Seaweed Symp., 1955
2) BAARDSETH, E. and HAUG, A., Norsk inst. f. tang-og tareforskning, Rep. 2, 1953
3) HAUG, A. and JENSEN, A., ibid. Rep. 4. 1954
4) JENSEN, A., ibid. Rep. 9. 1956
5) YOUNG, E. G., Carbohydrate Accumulation by lower Plants (Rhodophyta) ex W. Ruh-

land, Handbuch der Pflanzenphysiologie. Springer-Verlag, 1958
6) Ericson, L. E. and Carlson, B., Kungl. Svenska Vetenskaps. Akad. 6 (49): 1953
7) Larsen, B., Norsk inst. f. tang- og tareforskning, Rep. 19. 1958
8) Larsen, B., and Jensen, A., ibid. Rep. 15. 1957
9) Brennan, A. T., Inst. f. Ind. Res. and Stand., Dublin, 1950
10) Miyabe, K., J. of the Sapporo Agric. College I, 1–50. 1957
11) Segawa, S., The Seaweeds of Japan. Hoikusha, Osaka, 1957
12) Yabu, H., Bull. Fac. Fish. Hokkaido Univ. 8 (3): 185–189. 1957
13) Togasawa, Y. et al., Bull. Jap. Soc. Sci. Fish. 20 (3): 189–192. 1954; ibid. 20 (3): 193–195. 1954
14) Mautner, H. G., Ec. Bot. 8 (2): 174. 1954
15) Tokida, J., Fac. of Fish. Hokkaido Univ. 2 (1): 1954
16) Scagel, R. F., Prov. Dep. of Fish. Victoria, B. C. 1. 1947
17) Humm, H. J., The Brown Algae ex Tressler, D. K. et Lemon, J. Mc. W., Marine Products of Commerce. Reinhold Publishing Corporations, New York, 1951
18) Velasquez, G. T., Philippine Seaweeds Report, 1955
19) Hegnauer, R., Chemotaxonomie der Pflanzen. Birkhäuser Verlag, Stuttgart, 1962

General Literature

Chapman, V. J., 1950; Fott, B., 1959; Fritsch, F. E., 1952; Hoffmann, C., 1938; Hoppe, H. A., 1958; Hoppe, H. A. and Schmid, O. J., 1962; Lewin, R. A., 1962; Pax, F., 1962; Uphof, J. C. Th., 1959; see Index of Literature.

Arthrothamnus

This genus stands between *Laminaria* and *Alaria*.

A. kurilensis Ruprecht. Southern and Central Kurile Islands, Sakhalin. Coasts of China and Japan. On rocks in the sublittoral, near low-water, in cold-water regions. Often up on shore in quantity, particularly near Cape Nishinotoro and Yenchishi.

Holdfast composed of filiform hapteres, stipe erect, branching up to 6–7 times.

A. kurilensis has a high sugar content and can be used for nutrition purposes. The Japanese name is "Chishima-nekoashi-kombu". The alga is used as a raw material for kombu.

A. bifidus. Coasts of China, Japan and the Kurile Islands. Thallus lobes 1.5 to 5 metres in length and 6 to 7 centimetres in breadth, somewhat broadened toward base. Normally only tornup algae are harvested, because *A. bifidus* grows in depths of more than 7 metres.

The alga has a high sugar content and is rich in iodine. The Japanese name is "Nekoashi-kombu". *A. bifidus* is often sold for giving flavouring food. It is also utilized for kombu production.

General Literature

Chapman, V. J., 1950; Fritsch, F. E., 1952; Hoffmann, C., 1938; Miyabe, K., 1957; Segawa, S., 1957; Tokida, J., 1954; Uphof, J. C. Th., 1959; see Index of Literature.

Ascophyllum

A. nodosum (L.) Le Jolis. Also known as "Knotentang" (German), "Knobbed Wrack", "Yellow Tang", "Pigweed" (English), "Grisetang", "Knuppetang" (Norwegian) (Pl. II: 3, 4; Pl. XI: 26).

Rocky coasts of Northern Atlantic Ocean, particularly Iceland, Norway, Scotland,

Ireland, France; also Atlantic coast of the Iberian Peninsular. Atlantic and Pacific coasts of Canada[1–6]).

A. nodosum occurs abundantly, particularly in sheltered bays exposed at low-tide; sometimes it is found together with *Fucus* species, but often it covers large areas exclusively.

The alga has a holdfast; it is bushy, irregularly branched, margin bluntly serrated, of 30 to 150 centimetres in length, and approximately 1 centimetre in width. Central rib absent. The large air bladders are individually placed. The alga is of leathery consistency and yellow-brown to dark olive-green in colour. The stalked receptacles are globular or oval and arise in the axils of the marginal serrations. *A. nodosum* is a perennial alga and, like *Fucus*, numbered among the "rockweeds".

Floating *Ascophyllum* is recorded from Sargasso Sea[7]). The *A. nodosum* flora of the White Sea coast has been studied[8]).

BAARDSETH has made a statistical study of the structure of the *Ascophyllum* zone[9]). This author also has reported on the quantitative composition of the fucoid zone in Norway to the THIRD INTERNATIONAL SEAWEED SYMPOSIUM[10]). The free-living forms have been investigated[11]). The growth in relation to harvesting problems has been studied in detail[12–14]). This investigations, which have been made mainly on the Canadian, Norwegian, and Irish coasts, will be of great significance, in view of the quantity of *A. nodosum* potentially available as a raw material. Studies on reproduction and rate of growth in cropped areas are still being continued[15]) [16]). PRINTZ has reported to the SECOND INTERNATIONAL SEAWEED SYMPOSIUM on recuperation and recolonization, and on phenology of *A. nodosum*[17]) [18]). MUNDA has studied the influence of salinity of the medium on chemical composition of the alga in cultures[19]). Water and electrolyte exchange has also been studied by this author[20]).

Particular attention has been given to find the best method for drying the alga. Thus, through-circulation drying[21]) [22]), drying in steam-heated drum-dryer[23]) and non-thermal drying[24]) have been evaluated.

For physiological and biochemical investigations of *A. nodosum* see LEWIN (1962).

Seasonal variation and the influence of habitat in the chemical composition have been investigated[25]) [26]) [27]).

The alga is used directly for fertilizing and feeding purposes. It is also a raw material for alginic acid, though less suitable than *Laminaria*, *Macrocystis* or *Nereocystis*[28–33]). In Canada *A. nodosum* is used for sodium alginate production, and seasonal variations in the alginic acid content have been investigated[34]). *Ascophyllum*, according to MAASS, supplies alginates, which are suitable for semi-solid gels, elastic films, and electrolyte-resistant solutions.

The mucilage has been studied[35]), and the polysaccharide has been analyzed[36]) [37]) [38]). The alga contains fucoidan[39]), and a substance similar to fucoidan has been isolated from the reproductive branches[40]). Reducing compounds, polyphenols with tanning properties also have been found[41–44]).

Laminaran is present[45]); this substance has been reported in appreciable amounts and appears to be confined to the larger brown algae[46]).

The proteins and amino acids have been investigated[47–50]), a number of vitamins, including vitamin B_{12}, have been found[51]), also carotene and other carotinoids[52]) [53]). The vitamin C content is reported to be 62 milligrams per 100 grammes of fresh weight[54]).

Further investigations in this field have been reported by JENSEN at the FOURTH INTERNATIONAL SEAWEED SYMPOSIUM[55]). The influence of season, habitat, and age of tissue on niacin content has been analyzed[56]) and antibacterial activity of an extract investigated[57]).

YOUNG has studied the inorganic constituents of the alga[58]). The changes in nitrate content[59]) and iodine exchange[60]) have also been investigated. The iodine concentration in the thallus of *A. nodosum* is over 200 times of that of ocean water.

Total analysis of *A. nodosum* has been made by the INSTITUTE OF SEAWEED RESEARCH[61]). Since this tabulation can be regarded to be of guiding value, it has been inserted here:

Chemically, it is composed of carbohydrates, proteins, fats, vitamins, and mineral matter, the composition depends on the habitat and time of harvesting (BLACK, J. Soc. chem. Ind., Lond., 1948, 67, 355).

Ascophyllum also contains fucosterol 0.23% (BLACK & CORNHILL, J. Sci. Fd Agric., 1951, 2, 387) and chlorophyll 340 p.p.m. (OWEN, loc. cit.) and growth-promoting substances (ERICSON, Ark. Kemi, 1953, 6, 427).

Range in Composition during the Year
(Black, loc. cit.)

% Water			70–79
% on dry weight basis:		Carbohydrates:	
Ash	16–25	Alginic acid	24–28
Iodine	07–20	Mannitol	6–12
Crude protein	3–10	Fucoidan	12–14
Crude fat	2.5–4.3	Laminarin	3–7
Cellulose	2–3	Other carbohydrates	11–17
Nitrogen-free extractives	58–74		

Mineral Matter
(Typical analysis, meal as used in Edinburgh & East of Scotland pig feeding trial)

CaO	2.3	SO_3	4.10
K_2O	2.7	P_2O_5	0.22
Na_2O	3.1	Si_2	4.70
CI	2.9	MgO	1.56

Trace Elements in P.P.M. dry Matter
(Black & Mitchell, J. Mar. biol. Ass. U.K., 1952, 30, 575)

	May	January		May	January		May	January
Co	0.73	0.41	Sn	0.7	1.0	Ag	0.2	0.3
Ni	4.4	1.5	Zn	60	103	Rb	—	80
Mo	0.29	0.69	V	1.5	1.9	Li	—	4
Fe	283	168	Ti	26	9	Sr	700	2600
Pb	4	6	Cr	1.0	0.7	Ba	13	50
Mn	27	50	Cu	4	4			

Vitamins

Vitamin	Content of Vitamin in p.p.m. Dry Matter	Reference
β-carotene (vitamin A precursor)	16–25	OWEN, J. Sci. Fd Agric., 1945, 5, 449.
Fucoxanthin (vitamin A precursor?)	238	CARROLL (unpublished).
B_1	1.4–5.4*	GERDES, Arch. Mikrobiol., 1951, 16, 53.
B_2	7.5	CARROLL (unpublished).
Pantothenic Acid	0.2*	ERICSON, Ark. Kemi, 1953, 6, 511.
Niacin	12.3	
B_{12}	0.08	COATES & Kon. (unpublished).
C	2674	CREACH & BARAUD, C. R. Soc. Biol., Paris, 1954, 148, 105
D_3	0.01	COATES & Kon (unpublished).
E	156.298	BROWN, Chem. & Ind., 1953, 174.
K	14.2	DAM & GLAVIND, Biochem. J., 1938, 32, 485.

* *Fucus vesiculosus*

Amino Acids

Ascophyllum contains all the essential amino acids (COULSON, Chem. & Ind., 1953, 971) and is particularly high in methionine (963 p.p.m.) and cystine (284 p.p.m.) (GREGORY & YOUNG, unpublished).

The seaweed meal from *A. nodosum* has growing importance as a fodder additive. Investigations on feeding experiments have been carried out mainly in Great Britain and Norway[1] [62–66]).

Literature

1) PRINTZ, H., Tang og Tare. J. Grundt Tanum Forlag, Oslo, 1953
2) PRINTZ, H., Utilization of Seaweeds in Norway
3) MOORE, L. B., New Zealand Science Rev. 9. 180–184. 1951
4) BRENNAN, A. T., Inst. f. Ind. Res. and Stand., Dublin, 1950
5) CABRERO-GOMAZ, F., Estudio de las Algas marinas espanoles. C. S. I. C., Madrid, 1951
6) WORT, D. J., Canad. J. Bot. 33. 323–340. 1955
7) WINGE, Ö., Rep. on the Danish Oceanogr. Exp. 1908–1910. 1923
8) KUSNEKOV, V. V., Leningrad Izdatel'stva Akad. Nauk SSSR, Chapter 4, 1960
9) BAARDSETH, E., Norsk inst. f. tang- og tareforskning, Rep. 11, 1955
10) BAARDSETH, E., Third Int. Seaweed Symp., 1958
11) GIBB, D. C., J. Ecol. 45 (1): 49–83. 1957
12) MACFARLANE, C. I., Proc. Nova Scotian. Inst. Sci. 18 (2): 27–33. 1931/32
13) PRINTZ, H., Avh. Norske Vid. Akad. Oslo, 1. Mat. Nat. Kl. 4. 1950
14) BAARDSETH, E. ......
15) PRINTZ, H., Avh. Norske Vid. Akad. Oslo, Mat. Nat. Kl. 3. 1959; ibid. 4. 1959
16) BLACKLER, H., Second Int. Seaweed Symp., 1955
17) PRINTZ, H., Second Int. Seaweed Symp., 1955
18) PRINTZ, H., ibid.
19) MUNDA, I., Bot. Mar. V (2/3): 84. 1963
20) MUNDA, I., Bot. Mar. ...
21) MITCHELL, T. J. and POTTS, C. S., J. Sci. Food Agric. 402–412. 1955
22) MERRITT, J. H. and COSGRAVE, E. T., J. Sci. Food Agric. 9 (5): 300–308. 1958
23) BOOTH, E., ibid. ... 1956

24) REID, K. C. and JACKSON, PH., ibid. ... 1956
25) BLACK, W. A. P., J. Soc. Chem. Ind. 67. 355–357. 1948
26) BLACK, W. A. P., The Algae ex Altschul, Processed Plant Proteins Foodstuffs. Academic Press Inc., New York, 1958
27) LARSEN, B. and HAUG, A., Third Int. Seaweed Symp., 1958
28) VINCENT, D. L., Canad. J. of Technol. 34. 220–226. 1956
29) BASHFORD, L. A. et al., J. Soc. Chem. Ind. 69 (11): 337. 1950
30) BLACK, W. A. P., Ann. Rep. Chem. Soc. Vol. 50. 1953
31) VINCENT, D. L., Canad. J. of Technol. 34. 220–226. 1956
32) ROSE, R. C., ibid. 29. 19–28. 1951 CA 45. 4377. 1951
33) SMITH, F. and MONTGOMERY, R., The Chemistry of Plant Gums and Mucilages. Reinhold Publishing Corporation, New York, 1959
34) BARRY, C. and FLOOD, D. T., Third Int. Seaweed Symp., 1958
35) DILLON, T. et al., Proc. Roy Irish. Acad. 55. Sect. B, 9. 1953
36) JENSEN, A., Norsk inst. f. tang og tareforskning, Rep. 9. 1956
37) JANECKE, H. and KEHR, W., Pharmaz. Ztg. 105 (25): 684. 1960
38) LARSEN, B. and HAUG, A., Fourth Int. Seaweed Symp., 1961
39) BLACK, W. A. P. et al., J. Sci. Food Agric. ... 1952. WFA (5). 1953
40) O'COLLA ex LEWIN, R. A., Physiology and Biochemistry of Algae. Academic Press Inc., New York, 1962
41) LARSEN, B. and HAUG, A., Nature 181 (4617): 1224. 1958. CA 52. 19. 16484d. 1958
42) HAUG, A. and LARSEN, B., Acta Chem. Scand. 12 (4): 650–657. 1958
43) HAUG, A. and LARSEN, B., Third Int. Seaweed Symp., 1958
44) HEGNAUER, R., Chemotaxonomie der Pflanzen. Birkhäuser Verlag, Stuttgart, 1962
45) MAUTNER, H. G., Ec. Bot. 8 (2): 174. 1954
46) KREGER ex LEWIN, R. A., Physiology and Biochemistry of Algae. Academic Press Inc., New York, 1962
47) COULSON, C. B., Chem. and Ind. 1953. 997–998
48) COULSON, C. B., J. Sci. Food Agric. 6. 674–682. 1955
49) COULSON, C. B., Chem. and Ind. 1953, 971–972
50) SMITH, D. B. and YOUNG, E. G., Publ. 3731 of the Nat. Res. Council, Halifax, Canada, 1955
51) TEERI, A. E. and BIEBER, R. E. Science 127 (3313): 1500. 1958
52) HAUG, A. and LARSEN. B., Norsk inst. f. tang- og tareforskning, Rep. 16. 1957
53) OWEN, E. C., J. Sci. Food Agric. 5 (9): 449–453. 1954
54) HUMM, H. J., The Brown Algae ex Tressler, D. K. et Lemon, J. Mc. W., Reinhold Publishing Corporation, New York, 1951
55) JENSEN, A., Fourth Int. Seaweed Symp., 1961
56) LARSEN, B., Norsk inst. f. tang- og tareforskning, Rep. 19. 1958
57) VACCA, D. C., J. Am. Pharm. Ass. Sci. Ed. 43. 24–26. 1954
58) YOUNG, E. G., Publ. 4636 of the Nat. Res. Council, Halifax, Canada, 1958
59) LARSEN, B. and JENSEN, A., Norsk inst. f. tang- og tareforskning, Rep. 15. 1959
60) BAILEY, N. and KELLY, S., Biol. Bull. 109 (1): 13–21. 1955
61) Inst. of Seaweed Research, Inveresk, 1956
62) BLACK, W. A. P., Chem. and Ind. 1640. 1955
63) HIE, J. and SANNAN, F., Third Int. Seaweed Symp., 1958
64) JENSEN, A., ibid.
65) JENSEN, A., Norsk inst. f. tang- og tareforskning, Rep. 24. 1960
66) JENSEN, A. and MINSAAS, J., Tidsskr. Norske Landbruk. 67 (4): 81–96. 1960
67) STOLOFF, L., Ec. Bot. 16 (2): 86–94. 1962

General Literature

Annual Reports 1955–1962; CHAPMAN, V. J., 1950; FOTT, B., 1959; FRITSCH, F. E., 1952; HEGNAUER, R. 1962; HOFFMANN, C., 1938; HOPPE, H. A., 1958; HOPPE, H. A. and SCHMID, O. J., 1962; KUCKUCK, P., 1953; LEWIN, R. A., 1962; MAASS, H., 1959; PAX, F., 1962; UPHOF, J. C. TH., 1959; see Index of Literature.

Ascoseira

Antarctic, Half-Moon Island, eastern end of Livingston Island.

Ascoseira. Has a repeatedly forked stipe, each branch terminating in a long and rather narrow, leathery blade. A detailed description has been supplied by NEUSHUL[1]).

No work has been published to date on the industrial utilization of Antarctic kelps. An investigation of these facts, however, would be highly interesting.

Literature

1) Neushul, M., Bot. Mar. V (1). 19. 1963. General Literature: Fritsch, F. E., 1952; see Index of Literature.

Bifurcaria

B. tuberculata. Atlantic coasts of Europe and Africa. Material from the Irish coasts[1]) and French coasts (Dinard)[2]) has been investigated.

B. tuberculata has a much branched, perennial rhizome, attached to the substratum by small adhesive discs (hapteres). Fronds up to 25 centimetres long, branched yellow when fresh, blackish when dried.

B. brassiciformis. Coast of Cape Province. The alga serves as a raw material for the production of alginic acid and alginates. Seasonal variation in chemical composition has been studied[3]).

Further *B.* species have been investigated[4]), including *B. rotunda*[5]).

B. sisymbrioides is another South-African species on record.

B. laevigata, also from South Africa, has a discoid holdfast (hapteres), and differs from other *B.* species. It is now placed in *Bifurcariopsis*.

Literature

1) Brennan, A. T., Inst. f. Ind. Res. and Stand., Dublin, 1950
2) Lami, R., Bull. Lab. Marit. Dinard 44: 55, 1959
3) v. Holdt, M. M., et al., J. Sci. Food Agric. 193–196, 1955
4) Ross, R., J. Linn. Soc. Bot. 55 (633): 753–754, 1958
5) Borel, Soc. Phyc. de France, Bull. No. 2, 1956

General Literature
Fritsch, F. E., 1952; see Index of Literature.

Chnoospora

A small genus, from warmer seas, the tropics, particularly the Pacific districts.

Ch. pacifica J. J. Agardh (= *Ch. minima* (Her.) Papenfuss). Coasts of the Indian Ocean, particularly Ceylon, Indochina, Indonesia, also Japan, Australia, Hawaiian Islands, Mexico, and West Indies. The *Ch.* species have small, dichotomously branched thalli with numerous tufts of hair. *Ch. pacifica* is approximately 7.5 centimetres high. It is eaten as a salad in Indochina.

Ch. implexa. In South Asia and East Asia. This species is eaten in Hawaii.

It is very difficult to distinguish between *Ch.* species. *Ch. implexa*, is perhaps, identical with *Ch. pacifica*.

General Literature

Fritsch, F. E., 1952; Segawa, S., 1957; Zaneveld, J. S., 1959; see Index of Literature.

Chorda

Chorda filum (L.) Stackh. Known as "Gemeine Meersaite" (in German) and "Bootlace Weed" or "Seatwine" (in English) being used to make a kind of string.

Arctic Oceans to the French Atlantic Coast. Northern Japan. Often abundant in the sublittoral. Thallus unbranched, 2 to 4, sometimes up to 8, metres long, 2 to 6 millimetres thick, string-like, round, hollow inside, in the young stages densly covered with whitish hairs. Often several thalli grow from one small hapteron.

In Scandinavian countries this alga often is used during the winter months with *Laminaria* species as an additive for cattle fodder.

Material collected in the Baltic Sea has been found to contain antimicrobial substances[1]).

Ch. filum contains alginic acid and can be regarded as a raw material for the production of alginic acid and alginates[2]).

Japanese *Chorda* species have been mentioned in several publications[3]) [4]). *Ch. filum* has been described by Tokida[5]).

The results of physiological and biochemical investigations have been reported in Lewin[6]).

Ch. tomentosa. Also known as "Zottige Meersaite" (in German). It closely resembles to *Ch. filum*, but is always covered with a dense yellow-brown gelatinous pile. Occurs on the Atlantic Coast from Norway to France. *Ch. tomentosa* is by Jaasund[7]) regarded as the type species of a new genus *Halosiphon.*

Literature

1) Roos, H., Kieler Meeresforsch. XIII (1): 1957
2) Smith, F. and Montgomery, R., The Chemistry of Plant Gums and Mucilages. Reinhold Publishing Corporation, New York, 1959
3) Nishibayashi, T. and Inoh, S., Biol. J. Okayama Univ. 4 (1/2): 67–78, 1958
4) Inagaki, K. I., Sc. Pap. Inst. Alg. Res. Hokkaido Univ. 4 (2): 87–197, 1958
5) Tokida, J., Fac. of Fish. Hokkaido Univ. 2 (1): 1954
6) Lewin, R. A., Physiology and Biochemistry of Algae. Academic Press Inc., New York, 1962
7) Jaasund, E., Bot. Notiser 110, 1957

General Literature

Brennan, A. T., 1950; Chapman, V. J., 1950; Fott, B., 1959; Fritsch, F. E., 1952; Hoffmann, C., 1938; Kuckuck, P., 1953; Levring, T., 1940; Segawa, S., 1957; see Index of Literature.

Chordaria

Ch. flagelliformis (Müll.) Ag. Known in German as "Gemeiner Geisseltang". Arctic Ocean, North Atlantic.

Coasts of Northern Japan (Honshû, Hokkaido), Kurile Islands, Sakhalin, Siberia, Chile (Magellan Straits), between tidemarks.

Main axis about 3 millimetres thick, with branches which are longer than the main axis. The latter is of compact structure, elastic, with long hair, and with very gelatinous membranes.

The alga is eaten in East Asia. On Kamchatka it is known as "Nebbpett". It is recorded to have a minor anticoagulant effect.

Ch. magellanica. Occurs on coasts of Chile and Patagonia.

Ch. linearis. Recorded from Fuegia and the Falkland Islands.

Ch. chordaria. Coasts of Australia.

Ch. firma. Coasts of Japan, China, and Korea.

Ch. Nagaii. Coasts of Hokkaido, Kurile Islands, Sakhalin.

Physiological and biochemical investigations of *Ch.* species have been reported by LEWIN (1962).

General Literature

CHAPMAN, V. J., 1950; FRITSCH, F. E., 1952; HOFFMANN, C., 1938; KUCKUCK, P., 1953; LEVRING, T., 1960; LEWIN, R. A., 1962; SEGAWA, S., 1957; TOKIDA, J., 1954; see Index of Literature.

Cladostephus

C. spongiosus (Lightf.) Ag. Known in German as "Wolliger Seequirl". Arctic Ocean to the North-African Atlantic Coast, East-Coast of North America. Subantarctic America. This alga has a sparcely bifurcated thallus with dense, verticillate branches.

C. verticillatus. Known in German as "Echter Seequirl". North Atlantic Ocean to Mediterranean. This species is larger than *C. spongiosus*, and the verticillate branches are arranged very loosely.

C. setaceus. Found at the coasts of southern South America and the Falkland Islands.

C. antarcticus. Coasts of Patagonia, Fuegia, and the Falkland Islands. The carbohydrates of *C.* species have been investigated[1] [2]. For physiological and biochemical investigations of *C. spongiosus* see LEWIN[3].

Literature

1) FANSHAWE, R. S. and PERCIVAL, E., J. Sci. Food Agr. 9 (4): 241–243, 1958
2) JANECKE, H. and KEHR, W., Pharmaz. Ztg. 105 (25): 684, 1960
3) LEWIN, R. A., Physiology and Biochemistry of Algae. Academic Press Inc., New York, 1962

General Literature

FRITSCH, F. E., 1952; KUCKUCK, P., 1953; LEVRING, T., 1960; see Index of Literature.

Colpomenia

C. sinuosa (Mert.) Werb. et Sol. Widespread in the warmer seas, particularly in East Asia. Coasts of India (East Coast of Madras), China and Japan; European coasts (Scotland, Ireland, West Coast of Norway[1–4]); coast of Juan Fernandez[5]) (Pl. VIII: 20).

During the last decades the var. *peregrina* has been found at the European Atlantic coasts, where its large thalli often cause considerable damage to oyster banks (French Atlantic Coast)[6]).

C. sinuosa contains alginic acid and can be considered as a raw material for the production of alginic acid and alginates[7]). Material of Indian provenance contains approximately 16 to 17 per cent alginic acid.

C. bullosa (Saunders) Yamada. (= *C. sinuosa* (Roth) Derbes et Solier *f. deformis*). Coasts of Japan.

Literature

1) Blackler, H., Nature, 162 (4130): 1001, 1948
2) Blackler, H., Fourth Int. Seaweed Symp., 1961
3) Brennan, A. T., Inst. f. Ind. and Stand., Dublin, 1950
4) Wennberg, T., Meddelanden från Göteborgs Botaniska Trädgard XVIII, 1950
5) Levring, T., Lunds Univ. Årsskrift N. F. Avd. 2, 56 (10), 1960
6) Mendez, D. C., Vie et Milieu 8 (1): 92–98, 1957
7) Smith, F. and Montgomery, R., The Chemistry of Plant Gums and Mucilages Reinhold Publishing Corporation, New York, 1959

General Literature

Fritsch, F. E., 1952; Segawa, S., 1957; see Index of Literature.

Costaria

C. costata (Turn.) Saunders (= *C. Turneri* Grev.). In the Pacific at the coasts of North America (from Alaska to California), coasts of Asia (Northern Honshû, Hokkaido, Kurile Islands, Sakhalin, Korea, Siberia, Kamchatka).

The blade is traversed by five prominent longitudinal ribs, which project alternately on the two surfaces, three beeing visible on the one and two on the other. Between the ribs there are numerous puckerings. Holdfast of filiform hapteres. Stipe 2.5 to 15 centimetres. The blades will be 30 to 70 centimetres long, 2 to 3 metres in larger specimens. Japanese name "Zarame" or "Sujime". This alga often grows together with *Laminaria* species.

Young specimens of the alga are eaten by coastal population.

Physiological and biochemical investigations have been reported in Lewin (1962).

The alga could be used industrially, failing quantities of raw materials of other species. Tokida has described *f. cuneata*.

General Literature

Chapman, V. J., 1950; Fritsch, F. E., 1952; Lewin, R. A., 1962; Miyabe, K., 1957; Segawa, S, 1957; Tokida, J., 1954; see Index of Literature.

Cymathere

C. triplicata. Coasts of Japan[1]), Pacific coasts of North America (Vancouver Islands to Thomas Point) in thick patches, almost unmixed[2]). The alga has been studied and has been found to contain alginic acid, laminaran, and mannitol. The alga can be considered for industrial utilization[3]). It also contains antimicrobial substances[4]) [5]).

Literature

1) MIYABE, K., J. of the Sapporo Agric. College I: 1–50, 1957
2) SCAGEL, R. F., Prov. Dep. of Fish. Victoria, B. C., (1): 1947
3) CUTCHEON, S. Mc. et al., J. Am. Pharm. Ass. 38: 196–200, 1949
4) ROOS, H., Kieler Meeresforsch. XIII (1), 1957
5) MAUTNER, H. G., Ec. Bot. 8 (2): 174, 1954

General Literature

CHAPMAN, V. J., 1950; FRITSCH, F. E., 1952; MAASS, H., 1959; SEGAWA, S., 1957; see Index of Literature.

Cystophora

Southern Hemisphere. Australia, New Zealand. Genus still largely unknown. A description of Antarctic *C.* has been given by NEUSHUL[1]).

The *Cystophora* growing on the coastal rocks of Australia and New Zealand are eaten by sheep.

Literature

1) NEUSHUL, M., Bot Mar. V (1). 19. 1963. General Literature: CHAPMAN, V. J., 1950; FRITSCH F. E., 1952; see Index of Literature.

Cystophyllum

A large genus, widely distributed in Japanese waters. *C.* has the habit of a *Cystoseira*, and the two genera are doubtless closely allied. It is not easy to distinguish sharply between mature plants. The only general difference lies in the restriction of airbladders to the ultimate branchlets in *Cystophyllum.*

Recorded from the coasts of Japan, Korea, Sakhalin, and the Kurile Islands, viz.: *C. crassipes*, *C. geminatum*, *C. hakodatense*, *C. sisymbrioides*, and *C. turneri.*

C. hakodatense has been described in detail. This alga contains antimicrobial substances[1]). The extracted fat contains 22.9 per cent solid fatty acids and 76.8 per cent liquid fatty acids[2]) [3]).

SEGAWA[4]) records floating *Cystophyllum* in the Japanese waters, particularly in the regions of Tsuyazaki and Iki-Tusima.

Several Japanese *C.* species are used for human consumption.

C. muricatum. Coasts of Ceylon (Gulf of Mannar) and India (East-Coast of Madras)[5]). This alga contains approximately 20 per cent of alginic acid.

Physiological and biochemical investigations of *C. crassipes*, *C. hakodatense*, and *C. sisymbrioides* have been reported in LEWIN (1962).

Literature

1) ROOS, H., Kieler Meeresforsch. XIII (1), 1957
2) MAUTNER, H. G., Ec. Bot. 8 (2): 174, 1954
3) HEGNAUER, R., Chemotaxonomie der Pflanzen, Birkhäuser Verlag, Basel, 1962
4) SEGAWA, S., The Seaweeds of Japan, Hoikusha, Osaka, 1957
5) VALSAN, A. P., Current Science, India, 24 (10): 343–345, 1955

General Literature

CHAPMAN, V. J., 1950; FRITSCH, F. E., 1952; HOFFMANN, C., 1938; LEWIN, R. A., 1962 SEGAWA, S., 1957; TOKIDA, J., 1954; see Index of Literature.

Cystoseira

Approximately 60 species, in warmer waters, often forming large colonies. Perennial. Thallus mostly terete or flattened, with air-bladders on the axis. This is a polymorphic genus.

FRITSCH lists numerous species occurring on the coasts of Great Britain, and BRENNAN mentions several from Irish coasts[1]). *C.* species are especially characteristic of the Mediterranean.

C. abrotanifolia C. Ag. Can be regarded as a raw material for the production of alginic acid and alginates[2]).

C. barbata J. Ag. Coasts of the Adriatic and Black Seas. This species can also be regarded as an alginate raw material[2]). Physiological and biochemical investigations have been reported by LEWIN. The alga has antilipemic activity[9]).

C. mediterranea Ag. Sauvageau. Mediterranean coasts. Alginate raw material. This alga contains D-mannuronic acid and L-guluronic acid[2]).

C. prolifera. Japanese coasts.

C. stricta. Mediterranean coasts. A paper, "l'étude expérimentale de la réinstallation d'un peuplement" was published concerning this species[11]).

C. tamariscifolia. Coasts of Great Britain[3]).

C. species occur in many places at the coasts of the Mediterranean, Adriatic, Aegean and Black Seas, and above all, on the coasts of Turkey[4-8]) [10]).

Literature

1) BRENNAN, A. T., Inst. f. Ind. Res. and Stand., Dublin, 1950
2) SMITH, F. and MONTGOMERY, R., The Chemistry of Plant Gums and Mucilages. Reinhold Publishing Corporation, New York, 1959
3) NAYLOR, M., Third Int. Seaweed Symp., 1958
4) GUERN, M., C. R. Acad. Sci. (Paris), 249 (14): 1254–1256, 1959
5) GUERN, M., Fourth Int. Seaweed Symp., 1961
6) VOZZHIUSKAYA, V. B., Akad. Nauk SSSR Trudy Inst. Okeanol., 23: 168–184, 1957
7) REZNICHENKO, O. G., Akad. Nauk SSSR Trudy Inst. Okeanol. 23: 185–194, 1957
8) ALFIMOV, N. N., Bot. Zhurn. 45 (8): 1196–1198, 1960
9) GÜVEN, K. C. and AKTIN, E., Bot. Mar. VI 1964
10) MAKKAVEEVA, E. B., Trudy Sevastopol Biol. Stants. 12: 168–191, 1960
11) HUVÉ, P., Comm. Int. Explor. Médit. Rapp. et Proc. Verb. n. s. 15 (2) 121–125, 1960

General Literature

FOTT, B., 1959; FRITSCH, F. E., 1952; LEWIN, R. A., 1962; SEGAWA, S., 1957; see Index of Literature.

Desmarestia

D. aculeata (L.) Lamour. Known as "Stacheltang" in German. Arctic Ocean, North Atlantic, West Coast of North America (Bering Islands to southern California).

Perennial, up to 2 metres long, depressed, irregularly branched, in spring with yellow hairy tufts, in summer and autumn bare, with short spine-like branches.

The high acidity and the content of free sulphuric acid are noteworthy. The alga contains laminaran[1])[2]), mannitol has also been found[3]). The carbohydrates have been investigated by other authors[4]). *D. aculeata* contains antimicrobial substances[5]). The acid content of the extracted of the alga has been investigated for several times[6]). The pH is extremely low. This is mainly due to the sulphuric acid content, but malic, citric, and succinic acids are also present in smaller quantities[7])[8]).

D. viridis. Arctic Ocean, North Atlantic, coasts of Europe, Greenland, North America, and Japan.

Yellow-coloured alga, appearing in summer. Branching opposite, dense, thin and hairy in the upper parts. Thallus turning green when exposed to air from liberated H_2SO_4 which destroys the pigments.

This species has been investigated by several authors. No mannitol was found[2]). The life cycle has been described by Kornmann[11]).

D. ligulata. Widespread in the colder waters of the Northern and Southern Hemisphere. Coasts of the Atlantic, Japan, and New Zealand. Usually near low-tide level, often, however, at considerable depths in the sublittoral.

D. tabacoides. Japanese coasts.

D. herbacea. Pacific coasts of America from Alaska to the Subantarctic. The alga reaches 2,5 metres in length[9]).

D. Rossii. Subantarctic America, Falkland Islands, Kerguelen (Desolation Island)[9]).

D. compressa and *D. anceps* are recorded from the Antarctic. They are richly branched and often of considerable size. *D. aculeata*, *D. ligulata*, *D. munda*, *D. tabacoides*, and *D. viridis* have been studied physiologically and biochemically (Lewin)[10]).

Literature

1) Black, W. A. P. et al., J. Sci. Food Agric. 4: 58–64, 1953
2) Mautner, H. G., Ec. Bot. 8 (2): 174, 1954
3) Stoll and Jucker, Heteroside ex W. Ruhland (ed.), Handbuch der Pflanzenphysiologie, Springer Verlag, 1958
4) Bouveng, H. et Lindberg, B., Acta Chem. Scand. 9 (1): 168–169, 1955
5) Roos, H., Kieler Meeresforsch. XIII (1), 1957
6) Wirth, H. E. et Rigg, G. B., Amer. J. Bot. XXIV, 68–70, 1937
7) Meeuse, B. J. D., Biochem. et Biophys. Acta 19 (2): 372–374, 1956
8) Eppley, R. W. et Bovell, C. R., Biol. Bull. 115 (1): 101–106, 1958
9) Levring, T., Lunds Univ. Årsskrift N. F. Avd. 2, 56 (10), 1960
10) Lewin, R. A., Physiology and Biochemistry of Algae. Academic Press Inc., New York, 1962
11) Kornmann, P., Helg. Wiss. Meeresunters. 8 (3): 287–292, 1962

General Literature

Brennan, A. T., 1950; Chapman, V. J., 1950; Fritsch, F. E., 1952; Kuckuck, P. 1953; Levring, T., 1940; Segawa, S., 1957; see Index of Literature.

Dictyopteris

D. polypodioides (= *D. membranacea*). A sublittoral perennial seaweed of the British and Irish shores[1]), also widespread in warmer oceans, for example, in the Mediterranean. This alga grows in tufts and is light olive-green. It is edible and is used in popular medicine in Mediterranean countries. Alginic acid, D. mannuronic, and L. guluronic acid have been found in it[2]).

A number of *D.* species occur on the Japanese coasts; some of them have been studied in detail, for example, *D. divaricata*[3]) [4]), the volatile constituents of which have also been investigated[5]) [6]).

D. plagiogramma. Hawaii. This alga is highly regarded as a foodstuff.

D. justii. Puerto Rico. Antibiotics from this species have been investigated[7]).

Physiological and biochemical investigations of *D. polypodioides* and *D. divaricata* have been reported in LEWIN.

Literature

1) BRENNAN, A. T., Inst. f. Ind. Res. and Stand., Dublin, 1950
2) SMITH, F. and MONTGOMERY, R., The Chemistry of Plant Gums and Mucilages. Reinhold Publishing Corporation, New York, 1959
3) YABU, H., Bull. Fac. Fish. Hokkaido Univ. 8 (4): 290–296, 1958
4) NISHIBAYASHI, T. and INOH, S., Bot. Mag. Tokyo 852: 261–268, 1959
5) BLACK, W. A. P., Ann. Rep. Chem. Soc. 1953, Vol. 50
6) HEGNAUER, R., Chemotaxonomie der Pflanzen, Birkhäuser Verlag, Basel, 1962
7) OLESEN, P. E. et al., Bot. Mar. VI. 1964

General Literature

CHAPMAN, V. J., 1950; FRITSCH, F. E., 1952; LEWIN, R. A., 1962; SEGAWA, S., 1957; UPHOF, J. C. Th, 1959; see Index of Literature.

Dictyota

Approximately 30 species, Atlantic Ocean, from Norway to the Canary Islands, Mediterranean, Indian and Pacific Oceans (particularly Japan), widespread.

D. dichotoma (Huds.) Lamour. Known in German as "Gemeine Gabelzunge". Widespread in the areas mentioned above. Erect, bifurcatedly divided, and flat thallus, composed of two cell-layers, 10 to 20 centimetres high, golden yellow with greenish-blue iridescence, growing from a cylindrical rhizome. Mostly in tufts on the substratum. Morphogenesis has been investigated in this alga[7]).

Material from Japan has been analysed by various authors, and, *inter alia*, ethereal oils have been found[2-5]). This species is also reported to contain alginic acid[1]).

D. acutiloba. Hawaiian Islands. This alga is sold for food on the markets of Hawaii and is often cultivated in what are known as "algal gardens".

D. apiculata. Coasts and islands of the Pacific Ocean, near Malaya, Indonesia (Celebes), Australia, Tasmania. This alga occurs on rocky substrata in the sublittoral. It is eaten locally[6]).

Other *D.* species are used in a similar manner, some of them are known to have a somewhat bitter taste.

D. barbayresiana. Recorded for India (on the East-Coast of Madras), contains 17 to 18 per cent alginic acid.

D. dilatata. Recorded from Japan.

D. divaricata. Puerto Rico. This alga has antibiotic activity[8]).

D. linearis. This alga contains alginic acid[1]).

Physiological and biochemical investigations of *D. dichotoma* and *D. flabellata* have been reported in LEWIN (1962).

Literature

1) SMITH, F. and MONTGOMERY, R., The Chemistry of Plant Gums and Mucilages. Reinhold Publishing Corporation, New York, 1959
2) ANDO, Y., Bull. Jap. Soc. Sci. Fish. 19: 717–721, 1953
3) YABU, H., Bull. Fac. Fish. Hokkaido Univ. 8 (4): 290–296, 1958
4) NISHIBAYASHI, T. and INOH, S., Bot. Mag. Tokyo, 852: 261–268, 1959
5) HEGNAUER, R., Chemotaxonomie der Pflanzen, Birkhäuser Verlag, Basel, 1962
6) ZANEVELD, J. S., Ec. Bot. 13 (2): 89, 1959
7) KUMAYAGAE, N. and INOH, S., Kramosomo, 46/47: 1521–1530, 1960
8) ALMODOVAR, L. R., Bot. Mar. VI. 196

General Literature

BRENNAN, A. T., 1950; CHAPMAN, V. J., 1950; FOTT, B., 1959; FRITSCH, F. E., 1952; KUCKUCK, P., 1953; LEVRING, T., 1960; LEWIN, R. A., 1962; SEGAWA, S., 1957; UPHOF, J. C. TH., 1959; see Index of Literature.

Durvillea

D. species are characteristic of the subantarctic oceans.

D. antarctica (= *D. utilis* Bory). "Bull Kelp". Littoral zone of areas exposed to strong surf, where the plants are constantly sprayed by water. Coasts of Australia and New Zealand (abundantly), Chatham, Campbell, and Auckland Islands[1]). South America, Central Chile to Cape Horn, Island of Chiloé, Falkland Islands, Kerguelen. The circular attaching discs (hapteres) may be more than half a metre in diameter and bear several fronds which can reach a length of 10 metres.

The stipe is relatively short and grows thicker annually for about 7.5 centimetres. It is flattened distally where it passes into the leathery blade, which is divided into numerous narrow whip-like segments. The central tissue contains a number of gas-filled bladders.

This alga, being a raw material for alginic acid has been investigated several times[2-4]). The alginic-acid content has been valued to be approximately 30 per cent.

Durvillea is used by the Maoris as a drug against scabies named "rimuroa"[5]).

D. antarctica is eaten in Chile, after washing, drying and roasting. Considerable quantities of this material, known as "cochayugo", are used. The air-bladders of the alga are used by the natives of New Zealand and South America for the manufacture of bags, water containers etc.

Reports have been published by FRANCKI[6]) [8]) on the value of the New Zealand *D. antarctica* as a growth-promoting substance and fertilizer.

D. Harveyi. Subantarctic regions of South America, Falkland Islands, Kerguelen, Crozet Island. Often growing together with *D. antarctica*[1]).

The alga is smaller than the species described above, and the undivided part of the blade is more extensive. But specimens of 30–40 feet have been found with a blade area of 216 sq. feet.

D. caesestipes. Chatham Island. The distribution of this species has been described[7]).

D. willana. New Zealand. This alga has been investigated[4]).

Literature

1) LEVRING, T., Lunds Univ. Årsskrift N. F. Avd. 2. Bd. 56 (10), 1960

2) SANNIE, C., C. R. Acad. Sci. (Paris) 232, 2040–2041, 1951

3) MOORE, L. and CRIBB, A. B., Nature 169, 4314, 1100, 1952

4) MOSS, B. and NAYLOR, M., Trans. Roy. Soc. New Zealand 81 (4): 473–478, 1954

5) BROOKER, S. G. and COOPER, R. C., New Zealand Medic. Plants 15 (1): 1, 1961

6) FRANCKI, R. J. B., Third Int. Seaweed Symp. 1958

7) NAYLOR, M., Nature 173, 4414, 1099–1100, 1954

8) FRANCKI, R. J. B., Plant and Soil XII (4). 1960 (I, II)

General Literature

CHAPMAN, V. J., 1950; FRITSCH, F. E., 1952; HOPPE, H. A., 1958; HOPPE, H. A. and SCHMID, O. J., 1962; MAASS, H., 1959; PAX, F. 1962; UPHOF, J. C. TH., 1959; see Index of Literature.

Ecklonia

Species of *E.* occur mainly in the Southern Hemisphere, but also on the Japanese coasts, mostly sublittoral.

E. maxima (Osbeck) Papenf. (= *E. buccinalis*). Known also as "Bamboo Seaweed", "Sea Bamboo" (Pl. V: 10); West Coast of South Africa, abundant from Cape Agulhas to the West Coast. Prefers warmer waters. The quantities occurring are estimated to amount to thousands or millions of tons[1]). Coasts of Australia, Tasmania, New Zealand.

Younger plants resemble *Laminaria*. The alga reaches a length of 33 feet. The upper parts of the adult specimens have hollow stipes floating on the water surface. They reach a lenght up to 5 metres and thicken out on the upper end to form a bladder. The blade parts are small, often serrated.

The alga ist used industrially as a raw material for alginic acid. Content about 35 to 38 per cent. The seasonal variations in the chemical composition have been studied in detail[2]) [3]). The content of laminaran is considerable[2]) [3]). Physiological and biochemical investigations have been reported by LEWIN (1962).

The alga is used also as a fertilizer and feeding material. On Tasmania potassium salt was produced from *E. maxima* during the last war.

E. bicyclis (Kjellm.) Setch. Coasts of Japan. The alga is eaten. HOFFMANN gave an account of its composition. It is industrially utilized for the production of iodine.

E. cava. Coasts of Japan. Japanese name "Kajime". Morphogenetical studies have been reported[6]).

The fresh alga contains approximately 0.25 per cent of iodine. It is utilized in the Japanese kelp industry.

E. exasperata. East and South-East Coast of South Africa (cf. *E. richardiana*)[1]).

E. kurome. Coasts of China and Japan. This alga is eaten in these countries for human nutrition.

E. latifolia. Coasts of China and Japan. This alga is eaten in East Asia.

E. radiata. Coasts of South Africa, Australia and New Zealand[4] [5]. The Australian alga has been investigated in regard to its content of alginic acid, mannitol, laminaran, and fucoidan[4].

In South Africa the alga is used together with other *E.* species for fodder while on New Zealand it is used as a fertilizer material.

E. radicosa. Coasts of Japan. This alga is eaten.

E. richardiana. East and South-East Coast of South Africa. The alga is smaller than *E. maxima.* It is often thrown up on the shores with other algae after gales. Between Port Elizabeth and East London and near Bird Island in the Algoa Bay it occurs in abundance, mostly, however, in deeper waters.

The suitability of the alga as a raw material for the production of alginic acid requires investigation[1].

E. stolonifera. Coasts of Japan.

Literature

1) Shuttleworth, R. G., South Afric. Ind. Chem. 1951, 115–117
2) v. Holdt, M. M. and al., J. Sci. Food Agric. 6: 193–196, 1955
3) Cronk, et al., Food Ind. of South Africa, Ref. WFA (5), 41, 1951
4) Stewart, C. M. and Higgins, H. G., Nature 187 (4736): 511. 1960
5) Bergquist, P. L., Physiol. Pl. 11 (4): 760–770, 1958
6) Nishibayashi, T. and Inoh, S., Biol. J. Okayama Univ. 6 (1/2): 27–36, 1960

General Literature

Chapman, V. J., 1950; Fott, B., 1959; Fritsch, F. E., 1952; Hegnauer, R., 1962; Hoffmann, C., 1938; Hoppe, H. A., 1958; Hoppe, H. A. and Schmid, O. J., 1962; Lewin, R. A., 1962; Maass, H. 1959; Miyabe, K., 1957; Pax, F., 1962; Segawa, S., 1957; Uphof, J. C. Th., 1959; see Index of Literature.

Eckloniopsis

Coasts of Japan. This alga is used in Japan as a kind of New-Year decoration. Nothing is known about its industrial utilization.

E. radicosa (Kjellman) Okamura. Morphogenetical investigations have been made[1].

Literature

1) Nishibayashi, T. and Inoh, S., Biol. J. Okayama Univ. 6 (1/2): 27–36, 1960

General Literature

Chapman, V. J., 1950; Miyabe, K., 1957; Segawa, S., 1957; see Index of Literature.

Ectocarpus

Cosmopolitan, mostly in colder seas.

E. siliculosus (Dillw.) Lyngb. Coasts of Heligoland, Baltic, Mediterranean, North America. The alga forms a mucilaginous tuft of finely branched filaments of a yellow-brown colour. It grows both on stones and other algae.

Antimicrobial substances[1] and alginic acid[2] have been shown to be present.

E. tomentosus (Huds.) Lyngb. Coasts of the Northern Atlantic, mostly on *Fucus serratus,* a filiform alga of rather dark brown colour, often felty and twisted like rope.

E. breviarticulatus and *E. indicus* are recorded from East Asian coasts. LEVRING mentions *E. confervoides*, *E. exiguus*, *E. Constaniciae*[3]).

Physiological and biochemical studies of *E. confervoides*, *E. globifer*, *E. siliculosus*, and *E. tomentosus* have been given by LEWIN (1962).

Extracellular products in pure cultures of *E.* have been investigated[4]).

The development of a form of *E. confervoides* has been shown by KORNMANN[5]).

The *Ectocarpaceae* studies of KUCKUCK, published by KORNMANN, should be mentioned here[6]).

Literature

1) ROOS, H., Kieler Meeresforsch. 13 (1): 1957
2) SMITH, F. and MONTGOMERY, R., The Chemistry of Plant Gums and Mucilages. Reinhold Publishing Corporation, New York, 1959
3) LEVRING, T., Lunds Univ. Årsskrift N. F. Avd. 2, 56 (10), 1960
4) FOGG, G. E. and BOALCH, G. T., Nature 181 (4611): 789–790, 1958
5) KORNMANN, P., Pubbl. Stz. Zool. Napoli XXVIII: 32–43, 1956
6) KUCKUCK, P., Helg. Wiss. Meeresunters. Vol. I–VII, 1964

General Literature

FOTT, B., 1959; FRITSCH, F. E., 1952; KUCKUCK, P., 1953; LEWIN, R. A., 1962; SEGAWA, S., 1957; see Index of Literature.

Egregia

Extensive beds at and below low-water mark, and in rock pools, on the West Coast of North America, often in abundant colonies.

E. Menziesii (Turn.) Aresch. Also known as "Feather-Boa Kelp", "Ribbon-Kelp".

West Coast of North America (Vancouver Island, California)[1]). At an early stage short flat, ligulate segments sprout out from the lower margins of the blade and form the flattening stipe.

With older plants numerous perforations of the blade are recognizable. Thallus lobes are 4 to 8 metres long.

The alga is used as a fertilizer. It is said, in respect of nitrogen and phosphate content, to be more valuable than the Giant Kelps. The iodine content is 0.08 per cent, which is very low, as well as that of potassium, namely from 10 to 13 per cent.

An alginic acid raw material. Fresh algae contain about 3 per cent of alginic acid, and when dried about 18 to 20 per cent, laminaran and mannitol are present in the usual quantities[2]) [3]), extracts of this alga show antibiotic activity[4]). The proteins have been investigated[5]).

E. laevigata. Coasts of California, particularly between P. Conception and San Diego. The blades of this alga vary from being simple and linear to pinnate structures with filiform segments. The alga grows both in shallow and in deeper waters[6]). An alginic acid raw material, the dry algae contains approximately 18 to 20 per cent[3]).

E. planifolia. The alga occurs in the same regions[6]).

Literature

1) Scagel, R. F., Prov. Dep. of Fish. Victoria, B. C. (1): 1947
2) Cutcheon, S. Mc. et al., J. Am. Pharm. Ass. 38, 196–200, 1949
3) Humm, H. J., ex Tressler, D. K. et Lemon, J. Mc. W., Marine Products of Commerce. Reinhold Publishing Corporation, New York, 1951
4) Pratt, R. et al., J. Am. Pharm. Ass. 40, 575–579, 1951
5) Mautner, H. G., Ec. Bot., 8 (2): 174, 1954
6) Chapman, V. J., Bot. Mar. 3 (2): 33, 1961; id. 3 (3/4): 101, 1961

General Literature

Chapman, V. J., 1950; Fritsch, F. E., 1952; Hoppe, H. A., 1958; Hoppe, H. A. and Schmid, O. J., 1962; Maass, H., 1959; see Index of Literature.

Eisenia

E. bicyclis (Kjellm.) Setch. Coasts of Japan. The type species[1]. Branched stipe, part of the primary lamina. Japanese name "Asame". The alga is used in Japan for the production of iodine. The content amounts to 0.3 per cent of the fresh raw material, ash content approximately 50 per cent. Laminaran, D-glucose, and fucose have been found[2].

The alga contains no mannitol[3]. The constituents of *Eisenia*, such as crude protein, fat, sugar, vitamins etc. have been reported by Black[4]. *E. bicyclis* yields maximum laminaran in August[5]. The alga is eaten, being used in soups or taken with soya-sauce. It is also used as a fertilizer material in agriculture.

E. arborea Aresch. Inhabitant of the coasts of California and Japan. On the coast of California this alga forms a "forest" under the *Macrocystis* populations together with *Laminaria farlowii* and *Pterygophora californica*[6].

E. Cokeri Howe. Peru.

Literature

1) Miyabe, K., J. of the Sapporo Agric. College, 1: 1–50, 1957
2) Smith, F. and Montgomery, R., The Chemistry of Plant Gums and Mucilages. Reinhold Publishing Corporation, New York, 1959
3) Handa, N. and Nisizawa, K., Nature 1961, 192, 1708
4) Black, W. A. P., The Algae ex Altschul, "Processed Plant Proteins Foodstuffs". Academic Press Inc., New York, 1958
5) Meeuse, B. J. D. ex Lewin R. A., Physiology and Biochemistry of Algae. Academic Press Inc., New York, 1962
6) Aleem, A. A., Second Int. Seaweed Symp., 1955

General Literature

Fritsch, F. E., 1952; Hoppe, H. A., 1958; Lewin, R. A., 1962; Segawa, S., 1957; see Index of Literature.

Fucus

The *Fucaceae* form highly extensive meadows exposed at low water. There are approximately 15 species, mostly in the Northern Hemisphere.

The thalli are mostly strap-shaped, with a centre rib, often more than 1 metre long, sometimes with floating bladders. They are affixed to rocks, stones, walls or wooden piers by means of hapteres or holdfasts. The Fucaceae are perennial. The conceptacles on inflated tips of the thallus. Species of *Fucus*, together with *Ascophyllum nodosum*, are sometimes termed as "rockweeds".

The quantitative composition of *Fucus* zones has been investigated[1]). The species and distribution of Fucaceae in the world has been inquired by POWELL, who also studied their taxonomy, nomenclature, and ecology. This author investigated the following types: *F. vesiculosus*, L., *F. ceranoides* L. emend., *F. distichus* L. emend. Powell, *F. serratus* L., *F. spiralis* L. emend. (or *F. sherardii* Stackhouse emend.), *F. virsoides*[2]).

NIZZAMUDDIN has written about the classification and distribution[3]). Variation in populations of Fucaceae, for example on the Côte Basque from 1894 to 1959, has been studied[113]), and populations on the White Sea coast have been reported on by KUSNETSOV[114]).

The problem of speciation in *Fucus* has been studied by various authors[4]) [5]). LEVRING has reported on the submicroscopical structure of eggs and spermatozoids of *Fucus* and related genera[7]). Other investigations include biological and ecological studies[6]), physiological studies[14]) [15]), studies on reproduction and growth[8]) [9]), on the influence of salinity on growth rate[10]), on the constitution and the behaviour of a fertilization inhibitor[112]), and on seasonal fluctuation in resistance to low temperature[115]) [116]).

Ecological experiments have been carried out on various *Fucus* species, e.g. *F. serratus* and *F. ceranoides*[11]), while culture experiments are reported by FOTT and others[12]) [13]).

For physiological and biochemical investigations of *F. ceranoides*, *F. evanescens*, *F. platycarpus*, *F. serratus*, *F. spiralis*, *F. vesiculosus*, see LEWIN (1962).

The cellulose of Fucaceae[17]) and fluctuations of ascorbinic acid have been investigated[16]).

The "rockweeds" have long been used in coastal countries for fertilizing purposes[18]) [19]).

In Norway investigations have been made on the use of *Fucus* species, particularly *F. vesiculosus* for fodder[20]) [21]).

F. vesiculosus. Known in Germany as "Blasentang", "Schweinetang" and "Höckertang", and as "Paddy Tang", "Bladder Wrack", "Kelp Ware", "Lady Wrack", "Black Tang" in England, as "Varech vésiculeux" (France), "Smelltang" (Norway), or "Bodelha" in Portugal (Pl. IV: 7).

On the coasts of the Northern Atlantic and Pacific Oceans, also Baltic. Often in abundant stands in the littoral zone. PRINTZ reported on *F. vesiculosus* from Norwegian coasts[22]), and LEVRING from Swedish coasts[23]), while HOFFMANN studied the Baltic areas between Flensburger Förde and the Brodtner Ufer in detail. Major *Fucus* colonies are found on the West Coast of the Isle of Fehmarn in the Baltic, and *Fucus vesiculosus* also grows on the Baltic coasts of Finland[24]). The distinction of two different types, is made there one is a small foliose surf form, while the other is a tufted cartilaginous deep-water form. In addition to the normal attached form one finds *Fucus* masses drifting in the bays at some metres depth. In their life rhythm these algae have a certain similarity to drifting *Sargassum* species. Specimens of between 2 to 5 metres length have been found (MICHANEK).

The life cycle of *F. vesiculosus* on the coasts of Murmansk has been studied by Russian scientists[25]).

F. vesiculosus occurs near Heligoland and on the Scottish, Irish[19]), French[27]), Spanish, and Portugese[26]) coasts. It is also recorded from the Canary Islands. There are large stands on the rocks of Vancouver Island and elsewhere on the Canadian Pacific coast[28]) [29]).

The many growth-forms have been often described in literature[30]. LEVRING mentions *F. vesiculosus* β *balticus*, and forms deviating to a greater or lesser extent can be derived from this forma *typica*, for example, *f. plicata*, *f. nana*, *f. angustifolia*, *f. subcostata*, *f. filiformis*.

Reports have been published on habitat-conditioned growth-forms on the Island of Rügen[31] [32].

F. vesiculosus is a perennial form, bushy, the most times 1 metre long, flat, often very branched, with entire margin, and clearly marked mid-rib; of leathery consistency, brown to olivegreen. Air-bladders are arranged in pairs on both sides of the mid-rib. The plant is fixed to the rocky substrate by means of hapteres.

The numerous subspecies and growth-forms, ranging from few millimetres to several metres in length shall be mentioned here. Often the alga will be thrown up in great masses on to shores and beaches after storms.

The biology of this species has been studied[33] [34], and its reproduction has been investigated[7] [35] [36] [37], also the structure and development of the hapteres[38], regeneration of tissue[39], reproduction, growth, and growth-substances[8] [40].

Culture experiments in media with salt content varied have been described by MUNDA[41]. The water and electrolyte exchange of *F. vesiculosus* grown in media of different salt content also has been investigated[42].

Growth and harvesting conditions have been studied[43], also different methods of drying the raw material[44] [45].

The constituents and seasonal fluctuations in chemical composition of the alga have been studied[46–50] [38], and about its metabolism has been written[51–53].

According to HOFFMANN this alga contains from 18 to 28 per cent of alginic acid. Baltic *Fucus* contains approximately 12 per cent alginic acid in the months of July and August[54]. Numerous other papers have been published on the alginic acid content of *F. vesiculosus*[55–60] [118].

Mannit is formed during hydrolysis[61].

The alga yields fucoidan upon hotwater extraction[60] [62–64]. SMITH-MONTGOMERY reported in detail on the constituents of the alga[55]. Fucosterol has been first isolated from *F. vesiculosus* and later on also found in numerous other brown algae (HEGNAUER).

The carbohydrates of the alga have been investigated[65–67]. According to BLACK it contains approximately 15 per cent of alginic acid, 9 per cent of mannitol, 9.2 per cent of fucose and 2.6 per cent of laminaran[68] [69]. The nitrogen-containing constituents[18] and the distribution of nitrogen in the thallus[71] have also been studied. The cause of seasonal fluctuations in N-content is mainly attributed to physiological factors of development.

The proteins and amino acids have been analyzed by various authors[70] [72a] [72b] [73] [74]. Carotene and carotinoid content[76] [77] and post mortem changes in carotinoid content have been reported[75]. The influence of seasonal fluctuations on ascorbinic-acid and niacinic-acid content has been analyzed[79] [80]. The maximum content of ascorbinic acid has been found to occur in the early summer months. 77 milligrammes vitamin C per 100 grammes fresh weight have been found[78].

F. vesiculosus contains antimicrobial substances[81].

Violaxanthin and lutein are recorded from fresh thalli. By drying and storing violaxanthin is entirely or partly converted into zeaxanthin (HEGNAUER).

YOUNG and co-workers have investigated the inorganic constituents[82]. VINOGRADOV has reported the iodine content. Changes in nitrate content have been found[83].

The alga contains an ethereal oil, probably consisting of terpenes. The hydrocarbon n-hentriacontan has been isolated. Analyses are available on the fatty acid content of thallus fat (HEGNAUER).

Fucus species have been studied in the last years by experts of the perfumery industry with a view to their utilization in the manufacture of scent. They could, perhaps, be used in the development of perfumeries such as Fern and Lavender.

F. vesiculosus has been used for some time as a medicinal drug, mainly on account of its iodine content, for example in slimming teas, and in the homoeopathic treatment of gland defects, goitre, Basedow's disease etc.[84–86].

The alga is used, particularly, in coastal regions, as a fertilizer. Gibberellin and gibberellin-like compounds have been shown to be present[87] [88].

F. ceranoides. Often occurs in brackish water. On the Irish coasts, for example near the estuaries of rivers[19]. The alga also lives in the Baltic.

The content of alginic acid in a dried raw material collected from the Bay of Danzig amounts to 21 to 29 per cent[89]. The iodine content of the alga has been investigated[90]. Water and electrolyte exchange have been studied in *F. ceranoides* grown in media of different salt content[42].

F. distichus (including *F. inflatus*). Coasts of Great Britain and Ireland, Norway, Iceland, Sweden. Distribution and ecology have been studied[91–93].

F. evanescens. Coasts of Japan (Hokkaido), Kurile Islands, Sakhalin, Kamchatka, coasts of the sea of Okhotsk, Bering Sea, Pacific Coast of North America (Alaska to Oregon). Very common; also found drifting. This is a highly variable species, with more than 25 forms distinguishable[94–96]. *F. evanescens* also grows on the coasts of Canada (St. Andrews). The inorganic constituents of Canadian material have been studied[49] [82], and organic constituents have been stated by SMITH-MONTGOMERY[55]. The vitamin C content has been reported as 24 milligrams per 100 grammes of fresh weight[78].

F. mytili. This species reaches a length of about 20 to 50 centimetres. It is to be found on the North Frisian "Wattenmeer" (mud banks, exposed at low-tide), on mollusc and shell banks. *F. mytili* Nienb. is not attached to fixed substrata, but is retained on mussels (Mitilidae) by byssus filaments.

F. serratus. Known as "Sägetang" in German, "Serrated Wrack", "Black Wrack", "Prickly Tang" (English) or "Hummertang" (Norwegian) (Pl. I: 1, 2; Pl. III: 5).

Northern Atlantic Ocean, particularly Norway, Iceland, Scotland, Ireland[19], France, Spain, Portugal, Canada, and Australia[99]. Growths of *F. serratus* are also recorded from various parts of the Baltic coasts. HOFFMANN has investigated the areas between Flensburger Förde and Brodtner Ufer.

F. serratus often grows in very dense colonies, forming "belts". These belts are exposed at low-tide.

The alga is bushy, mostly up to 1 metre long. Thallus flat, bifurcated, leathery, of an olive-green to black-green colour. The margin is strongly serrated. Male and female reproduction organs are contained in different plants.

F. serratus has hapteres. LEVRING makes a distinction between *f. typica* and *f. arctica*[23].

The biology of this alga has been studied[33] [34] [35].

Reports have been published on various drying methods of the raw material[44] [45]. Seasonal fluctuations in chemical composition[46] [100], the constituents of the alga[50] [60], including carbohydrates[101] [102] and polysaccharides[66] have been investigated.

The alga is regarded as a raw material for the production of alginic acid[55]. It contains about 20 per cent laminaran on a dry weight basis[103].

Reports have been published on L-fucose and D-mannitol production[104]. L-fucose is required for certain partial syntheses. It can be produced from *F. serratus* relatively easily.

The alga has been examined with regard to the β carotene content[77]. Vitamin C content has been found to be 48 milligrams per 100 grammes of fresh weight[78]. The highest content has been found in samples that had been collected in early summer[79]. The influence of seasonal fluctuations and of age of the alga on niacin content have also been analyzed[80].

F. serratus contains antimicrobial substances[81] [52].

Young and co-workers have investigated the inorganic constituents of the alga. Variations in nitrate content have been found[83]. Vinogradov has investigated the iodine content.

F. spiralis (= F. platycarpus). (Pl. III: 6). Above the *Fucus vesiculosus* belt, on stones and embankment structures on the North Sea Coasts; below the *Pelvetia* belt on the west and south coast of Ireland[19]. The distribution and occurence of *F. spiralis* on the Côte Basque[110], and the coasts of Spain (Bay of Biscay)[26] and Portugal have been reported. Whereas this species is relatively rare in northern waters, it grows abundantly in southern areas. Atlantic Coasts of the U.S.A.

The thalli are 10 to 30 centimetres high, mostly twisted, without air-bladders. The mucilaginous conceptacles contain male and female reproduction organs in the same plant. Several varieties of *F. spiralis* have been reported and described by Powell[111].

The polysaccharides have been analyzed[66], and other constituents[55] [60] [105] [107] and variations in chemical composition have been investigated[46] [108]. Fucoidan has been found[106], while a substance similar to Fucoidan has been isolated from the reproductive branches (Lewin, 1962). On hydrolysis mannitol is produced[61]. A number of vitamins, *inter alia* vitamin B_{12}, have been proved to be present[109].

F. virsoides. Mediterranean. Thalli of 10 to 20 centimetres, without floating bladders; conceptacles bisexual.

Literature

1) Baardseth, E., Third Int. Seaweed Symp., 1958
2) Powell, H. T., Third Int. Seaweed Symp., 1958
3) Nizamuddin, M., Bot. Mar. IV (3/4): 1962
4) Burrows, E. M. and Lodge, S. M., J. Mar. Biol. Ass. U. K. 30 (1): 161–176, 1951
5) Burrows, E. M. and Lodge, S. M., J. E. Lensley (ed.) "Species Studies in the British Flora" pp 83–85, 1955
6) Bauch, R., Wiss. Z. der Ernst Moritz Arndt-Univ., Greifswald, Math.-naturwiss. Reihe 5/6. Jahrg. V, 1955/56
7) Levring, T., Physiologia Plantarum 5: 528–539, 1952
8) Blackler, H., Second Int. Seaweed Symp., 1955
9) Burrows, E. M., Second Int. Seaweed Symp., 1955
10) Munda, I., Fourth Int. Seaweed Symp., 1961
11) Burrows, E. M., Fourth Int. Seaweed Symp., 1961
12) Fott, B., Algenkunde, Gustav Fischer Verlag, Jena, 1959
13) Burrows, E. M. and Lodge, S. M., Nature, 172; 1009–1010, 1953
14) Sosa-Bourdouil, C., Ann. Biol. ser. 3–34 (11/12): 501–511, 1958

15) Brown, F. A., et al., Biol. Bull. 115 (2): 345, 1958
16) Hagene, P., C. R. Acad. Sc. (Paris) 248 (4): 579–582, 1959
17) Schurz, J., Naturwissenschaften 40: 438, 1953
18) Smith, B. and Young, E. G., Biol. Chem. 205: 849–858, 1953
19) Brennan, A. T., Notes on some common Irish Seaweeds. Inst. f. Ind. Res. and Stand., Dublin, 1950
20) Høie, J. and Sannan, F., Third Int. Seaweed Symp., 1958
21) Jensen, A., Third Int. Seaweed Symp., 1958
22) Printz, H., Tang og Tare, J. Grundt Tanum Forlag, Oslo, 1953
23) Levring, T., Studien über die Algenvegetation von Blekinge, Südschweden, Diss. Lund, 1940
24) Hägren, E., Mem. Soc. Fauna et Flora Fenn. 32: 143, 1957
25) Tikhovskaja, Z. P., Murmansk Biol. Struts Triedy 2: 93–107, 1955
26) Cabrero-Gomez, F., Estudio de las Algas marinas españoles C. S. J. C., Madrid, 1951
27) Davy de Virville, Ad. and Denizot, M., C. R. Acad. Sc. (Paris) 247 (3): 323–325. 1958
28) Wort, D. J., Canad. J. Bot. 33: 323–340, 1955
29) Scagel, R. F., Prov. Dep. of Fish., Victoria, B. C., Nr. 1, 1947
30) Waern, M., The Rocky-Shore Algae in the Öregrund Archipelago. Uppsala, 1952
31) Müller-Stoll, W. R. and Künzenbach, R., Arch. Protistenk. 101 (3): 289–334, 1956
32) Müller-Stoll, W. R., Third Int. Seaweed Symp. 1958
33) Knight, M., Proc. Linnean. Soc. London, 159 (2), 87–90, 1947
34) Knight, M., and Parke, M., J. Mar. Biol. Ass. U. K. 29 (2): 439–541, 1950
35) Cock, H. H. and Elvidge, J. A., Proc. Roy. Soc. Ser. B. 138: 97–114, 1951
36) Esping, U., Arkiv Kemi 11 (2): 107–115, 1957
37) Esping, U., ibid. 11 (2): 117–127, 1957
38) Moss, B., Ann. Bot. (London), 14: 411–419, 1950
39) Moss, B., Fourth Int. Seaweed Symp., 1961
40) Moss, B., Third Int. Seaweed Symp., 1958
41) Munda, I., Bot. Mar. V. 1963
42) Munda, I., Bot. Mar.
43) Faure, L., Science et Pèche 1 (16): 1, 1954
44) Booth, E., J. Sci. Food Agric., 7. 11. 1956
45) Mitchell, T. J. and Potts, C. S., J. Sci. Food Agric., 402–412, 1955
46) Black, W. A. P., J. Soc. Chem. Ind. 68: 183–189, 1949
47) Moss, B., Ann. Bot. 12: 267–279, 1948
48) Moss, B., Ann. Bot. 14: 395–410, 1950
49) MacPherson, M. G. and Young, E. G., Canad. J. Bot. 30: 67–77, 1952
50) Mautner, H. G., Ec. Bot. 8 (2): 174, 1954
51) Chambers, J. A. and Burnett, H. J., Third Int. Seaweed Symp., 1958
52) Institute of Seaweed Research. Annual Report 1957
53) ibid. 1958
54) Chechelska, B. and Urbanski, T., C. A. 49, 20, 14231 i, 1955
55) Smith, F. and Montgomery, R., The Chemistry of Plant Gums and Mucilages. Reinhold Publishing Corporation, New York, 1959
56) Vincent, D. L., Canad. J. of Technol. 34: 220–226, 1956
57) Vincent, D. L.; ibid. 34: 220–226, 1956
58) Rose, R. C.; ibid. 29: 19–28, 1951
59) Vincent, D. L. et al., J. of Applied Chemistry 1955, 374–378
60) Black, W. A. P., Ann. Rep. Chem. Soc. 1953, Vol. 50
61) Stoll and Jucker, Heteroside ex W. Ruhland (ed), Handbuch der Pflanzenphysiologie, Springer Verlag, 1958
62) Yaphe, W. and Morgan, K., Nature, 183: 761–762, 1959
63) O'Neill, A. N., J. Am. Chem. Soc. 76: 5074, 1954
64) Black, W. A. P. et al., J. Sci. Food Agric., 1952
65) Lindberg, B., Acta Chem. Scand. 7: 1119–1122, 1953
66) Jensen, A., Norsk. inst. for tang- og tareforskning, Rep. 9, 1956
67) Bidwell, R. G. S., Proc. IX. Int. Bot. Congr. 2: 32, 1959
68) Black, W. A. P. et al., J. Sci. Food Agric. 754–764, 1955
69) Black, W. A. P.; ibid. 445–448, 1954
70) Black, W. A. P., The Algae ex Altschul, "Processed Plant Proteins Foodstuffs". Academic Press Inc., New York, 1958
71) Jacobi, G., Kieler Meeresforsch. 10: 37–57, 1954
72a) Coulson, C. B., J. Sci. Food Agric. 674–682, 1955
72b) Coulson, C. B., Chem. and Ind. 971–972, 1953
73) Smith, B. and Young, E. G., J. of Biol. Chem. 217 (2): 845–853, 1955
74) Ericson, L. E., Acta Chem. Scand. 6: 805–807, 1952
75) Liaaen, S. and Sorensen, N. A., Second Int. Seaweed Symp., 1955
76) Owen, E. C., J. Sci. Food Agric. 5 (9): 449–453, 1954
77) Haug, A. and Larsen, B., Norsk inst. f. tang- og tareforskning, Rep. 16, 1957
78) Humm, H. J., The Brown Algae ex Tressler, D. K. et Lemon, J. Mc. W., Marine Products of Commerce. Reinhold Publishing Corporation, New York, 1951
79) Jansen, A., Fourth Int. Seaweed Symp., 1961

80) Larsen, B., Norsk inst. f. tang- og tareforskning, Rep. 19, 1958
81) Roos, H., Kieler Meeresforsch., XIII (1), 1957
82) Young, E. G. and Langille, W. M., Publ. No. 4636 of the Nat. Res. Council, Halifax, 1958
83) Larsen, B. and Jensen, A., Norsk inst. f. tang- og tareforskning, Rep. 15, 1957
84) Weber, U. and Gerhard, H., D. Apoth. Ztg. 53: 1350 et 1373, 1938
85) Hartwig
86) Volk, D. Apoth. Ztg. 97: 421, 1957
87) Radley, M., Nature (London) 1961, 191, 685
88) Institute of Seaweed Research. Annual Report 1961
89) Trzesinski, P. and Czapke, K., Fourth Int. Seaweed Symp., 1961
90) Klemperer, H. G., Biochem. J. 67 (3): 381–390, 1957
91) Powell, H. T., J. Mar. Biol. Ass. U. K. 36 (2); 407–432, 1957
92) Powell, H. T.; ibid. 36 (3): 663–693, 1957
93) Bagenal, T. B. and Powell, H. T., Nature 182 (4638): 775–776, 1958
94) Tokida, J., The Marine Algae of Southern Saghalien. Fac. of Fish. Hokkaido Univ. 2 (1): 1954
95) Segawa, S., The Seaweeds of Japan. Hoikusha, Osaka, 1957
96) Yabu, H. and Imai, A., Bull. Jap. Soc. Phyc. 5 (2): 44–49, 1957
97) Powell, H. T. and Lewis, J. R., Nature 169, 4399, 508, 1952
98) Sundene, O., Blyttia 14 (2): 69–70, 1956
99) Wisley, Austral. J. of Marine and Freshwater Research II (I): 55, 1960
100) Baardseth, E. and Haug, A., Norsk inst. f. tang- og tareforskning, Rep. 2, 1953
101) Jensen, A.; ibid. Rep. 10, 1956
102) Jensen, A.; ibid. Rep. 10, 1956
103) Black, W. A. P. et al., J. of Applied Chemistry 1951
104) Trabert, Archiv d. Pharm. Bd. 293 (3): 278, 1960
105) Quillet, M., Bull. du Laboratoire Maritime de Dinard (45): 63–68, 1959
106) Quillet, M., Coll. Int. du Centre Nat. de la Recherche Scient., Dinard, 1960
107) Bouveny, H. and Lindberg, B., Acta Chem. Scand 9 (1): 168–169, 1955
108) Primo, C., Second Int. Seaweed Symp., 1955
109) Teeri, A. E. and Bieber, R. E., Science 127 (3313): 1500, 1958
110) Dupérier, R., Bull. Cent. Etud. et Rech. Sc. Biarritz 3 (1): 63–65. 1960
111) Powell, H. T., Brit. Phyc. Bull. 2 (1): 17, 1960
112) Branham, J. M. and Metz, C. B., Biol. Bull 119 (2): 305, 1960
113) Fischer-Piette, E. and Dupérier, R., Bull. Cent. Etud. et Rech. Sc. Biarritz 3 (1): 67–90, 1960
114) Kusnetsov, V. V., Leningrad Izdatel'stva Akad. Nauk SSSR Chapter 4, 1960 (R)
115) Parker, J., Biol. Bull. 119 (3): 474–478, 1960
116) Parker, J., and Philpott, D. E., Biol. Bull. 119 (2): 330–331, 1960

General Literature

Chapman, V. J., 1950; Fritsch, F. E., 1952; Hoffmann, C., 1938; Hoppe, H. A., 1958; Hoppe, H. A. and Schmid, O. J., 1962; Fott, B., 1959; Hegnauer, R., 1962; Kuckuck, P. 1953; Lewin, R. A., 1962; Maass, H., 1959; Pax, F., 1962; Schmid, O. J., 1962; Uphof, J. C. Th., 1959; Vinogradow, A. P., 1953; see Index of Literature.

Glossophora

This is a genus of which very little is known.

G. Kunthii. Coasts of New Zealand[1]), Chatham Islands, Chile, Peru, Juan Fernandez[2]). The alga also occurs in large quantities on the coasts of Peru. *G. Kunthii* has small ligulate outgrowths appearing in large numbers from the upper parts of the forked band-shaped thalli.

The content of alginic acid is approximately 25 per cent. This alga can be regarded as a useful raw material[3]).

Literature

1) Bergquist, P. L., Bot. Mar. 1 (3/4): 86, 1960
2) Levring, T., Lunds Univ. Årsskrift N. F. Avd. 2, 56 (10), 1960
3) Soto, M. C., Anales fac. farm. y bioquim. Univ. San Marcos 1: 627–636, 1950

General Literature

Fritsch, F. E., 1952; Hoppe, H. A. and Schmid, O. J., 1962; Maass, H., 1959; see Index of Literature.

Halidrys

H. siliquosa (L.) Lyngb. Known also as "Meereiche", "Schotentang", "Sea-Oak".

Coasts of the Northern Atlantic Ocean. The plant is 1 to 2 metres long, bushy; the blades are flattened, double edged, pinnate, approximately 5 millimetres wide, without mid rib. The floating bladders are stalked and pod-shaped; the conceptacles are also on the ends of branches. Cytological studies have been published[1] [2].

The alga contains alginic acid and can be regarded as a useful raw material[3]. The polysaccharides have been analyzed[4]. *H. siliquosa* has been found to have antibiotic activity[5–7].

Physiological and biochemical investigations have been reported by Lewin[8].

Literature

1) Moss, B. and Elliot, E., Ann. Bot. n. s. 21 (81) 143–151, 1957
2) Naylor, M., Ann. Bot. n. s. 22 (86): 205–217, 1958
3) Smith, F. and Montgomery, R., The Chemistry of Plant Gums and Mucilages. Reinhold Publishing Corporation, New York, 1959
4) Jensen, A., Norsk inst. for tang- og tareforskning, Rep. 9, 1956
5) Chesters, C. G. C. and Stott, J. A., Second Int. Seaweed Sympos. 1955
6) Stott, C. G. C., Inst. of Seaweed Research, Ann Rep. 1957
7) Stott, C. G. C.; id. 1958
8) Lewin, R. A., Physiology and Biochemistry of Algae. Academic Press Inc., New York, 1962

General Literature

Brennan, A. T., 1950; Chapman, V. J., 1950; Fritsch, F. E., 1952; Kuckuck, P., 1953; see Index of Literature.

Hedophyllum

H. sessile (Ag.) Setch. Pacific Coast of North America, Vancouver Island, particularly on the North and North-East Coast of Deer Island. The alga is to be found abundantly in populations[1], attached by numerous hapteres arising from the thickened basal margins of the blade. Older blades are often split several times.

The alga contains alginic acid, laminaran and mannitol in the usual quantities[2].

H. kuroshioense. Japan.

The development of cultivated *Hedophyllum* species has been studied.

Literature

1) Scagel, R. F., Prov. Dep. of Fish. Victoria, B. C. (1), 1947
2) Cutcheon, S. Mc. et al., J. Am. Pharm. Ass. 38. 196–200, 1949
3) Cole, K., Int. Oceanogr. Congr. Prep., 195–196, 1959

General Literature

FRITSCH, F. E., 1952; MAASS, H., 1959; SEGAWA, S., 1957; see Index of Literature.

Heterochordaria

The genus *Heterochordaria* is still rarely unknown.

H. abietina (Rupr.) Setch. & Gardn. Coasts of Northern Japan. Also abundant on other coasts of the Pacific Ocean, Kurile Islands, Sakhalin, Siberia, Kamchatka, Bering Sea, Aleutian Islands, Alaska to California. Grows on rocks in littoral zone. A very small alga, its appearance resembling to that of a fir-tree. Japanese name "Matsumo". It is used for human consumption after being salt-cured and prepared with soya sauce. *H.* is also salt-cured with mushrooms.

According to the STANDARD TABLES OF FOOD COMPOSITION, Japan, the alga contains 22.4 per cent of raw protein and 46.5 per cent of sugar[1]). Content of ethereal oil is 0.01 per cent[2]). Physiological and biochemical investigations have been reported by LEWIN (1962).

Literature

1) BLACK, W. A. P., The Algae ex Altschul, "Processed Plant Proteins Foodstuffs". Academic Press Inc., New York, 1958
2) HEGNAUER, R., Chemotaxonomie der Pflanzen, Birkhäuser Verlag, Basel, 1962

General Literature

CHAPMAN, V. J., 1950; FRITSCH, F. E., 1952; LEWIN, R. A., 1962; TOKIDA, J., 1954; SEGAWA, S., 1957; UPHOF, J. C. TH., 1959; see Index of Literature.

Himanthalia

H. elongata S.F. Gray (= *H. lorea* (L.) Lyngb.). Also known as "Riementang" and "Button Weed" (Pl. III: 5).

Coasts of the Northern Atlantic Ocean from Norway to Spain. Frequent on Irish coasts[1]), on the French coast near Dinard[2]), on the West Coast of Sweden[3]). Growing in less exposed areas in the shore zone. Perennial thallus consists of a dish-shaped form, 3 to 4 centimetres diameter, with a short elastic stipe affixed to rock. Out of the dish-shaped part there develop after the second or third year several strap-like sprouts, approximately 1 centimetre wide and up to 4 metres long, often subdivided several times, bearing male and female reproduction organs. A portion of the sprouts is sterile. Without floating bladders.

Reproduction, rate of growth, cytology[4]) [5]), and epiphytic growth have been investigated[14]).

H. elongata has been formerly used in the kelp industry. In France the alga has been used for the fertilization of artichoks.

The content of alginic acid is very high, amounting to approximately 35 to 38 per cent. Spanish *H. elongata* contains a considerable amount of alginic acid. The Spanish name is "correas". However, the species occurs rather seldon in these areas[6]).

The alga can be regarded as an alginic acid raw material[7]. Its chemical composition has been investigated by several workers[8–12]. Physiological and biochemical investigations have been reported by LEWIN[13].

Literature

1) BRENNAN, A. T., Inst. f. Ind. Res. and Stand., Dublin, 1950
2) LAMI, R., Bull. Lab. Marit. Dinard 44: 55, 1959
3) WENNBERG, T., Medd. f. Göteborgs Botaniska Trädgard XVIII, 1950
4) BLACKLER, H., Second Int. Seaweed Symp., 1955
5) MOSS, B., Phyc. Bull. 6: 31–32, 1958
6) CABRERO-GOMEZ, F., Estudio de las Algas marinas espanoles C. S. I. C., Madrid, 1951
7) SMITH, F. and MONTGOMERY, R., The Chemistry of Plant Gums and Mucilages. Reinhold Publishing Corporation, New York, 1959
8) MOSS, B., J. Mar. Biol. Ass. U. K. 31: 20–34, 1952
9) BAARDSETH, E. and HAUG, A., Norsk inst. f. tang- of tareforskning, Rep. 2, 1953
10) JONES, R. F., Biol. Bull. 110: 169–178, 1956
11) JONES, R. F., Biol. Bull. 112 (1): 81–91, 1957
12) JENSEN, A., Norsk inst. f. tang- of tareforskning, Rep. Nr. 9, 1956
13) LEWIN, R. A., Physiology and Biochemistry of Algae. Academic Press Inc., New York, 1962
14) v. D. ENDE, G. and v. OORSCHOT, R., Bot. Mar., 5 (4): . . ., 1963

General Literature

CHAPMAN, V. J., 1950; FOTT, B., 1959; FRITSCH, F. E., 1952; HOFFMANN, C., 1938; HOPPE, H. A. and SCHMID, O. J., 1962; KUCKUCK, P., 1953; MAASS, H., 1959; see Index of Literature.

Hizikia

H. fusiforme (Harvey) Okamura. Coasts of Japan and Southern Korea, and the offshore island Che Yu Do. The alga grows on rocks near the low-water level. It is collected during the first months of the year when it is small and tender, and dried in the sun. "Hijiki", as it is called in Japan, is edible and is boiled as a kind of vegetable in fresh water.

A number of results have been published on reproduction and growth of the alga[1] [2]. The composition has been reported by BLACK. According to "Standard Tables of Food Composition", Japan, the alga contains 6 to 7 per cent of raw protein, approximately 35 per cent of sugar. Provitamin A, vitamin B_1, B_2, and niacin have also been determined by BLACK[3].

Physiological and biochemical investigations have been reported by LEWIN[4].

Literature

1) SUTO, S., Bull. Jap. Soc. Sci. Fish. 17 (1): 9–12, 1951
2) SUTO, S., Bull. Jap. Soc. Sci. Fish. 17 (1): 13–41, 1951
3) BLACK, W. A. P., The Algae ex Altschul "Processed Plant Proteins Foodstuffs". Academic Press Inc., New York, 1958
4) LEWIN, R. A., Physiology and Biochemistry of Algae. Academic Press Inc., New York, 1962

General Literature

CHAPMAN, V. J., 1950; SEGAWA, S., 1957; SCHMID, O. J. and HOPPE, H. A., 1962; see Index of Literature.

Hormophysa

Very little is known about this genus.

H. truqueta. Coasts of Ceylon (Gulf of Mannar) and India (East Coast of Madras). It is containing 20 to 22 per cent alginic acid, and could, therefore, be used as a raw material for the production of alginates, if it would exist in sufficient quantity.

Literature

VALSAN, A. P., Current Science, India 24 (10): 343–345, 1955.

Hormosira

H. banksii. Coasts of Australia, Tasmania, and New Zealand in the mid littoral zone, especially in sheltered areas. CHAPMAN and TREVARTHEN[1]) have studied the ecology of this alga. A "loose lying" form has been reported by MOORE[2]), whereas the influence of various factors on *H. banksii* has been studied by BERGQUIST[3]) [4]) [5]) and CHAPMAN[6]) (Pl. X: 23, 24, 25).

BERGQUIST published a statistical report on the ecology and morphology of *Hormosira* in BOTANICA MARINA[7]). Physiological and biochemical investigations have been noted by LEWIN[8]).

Hormosira is a plant of characteristic habit. The branched thalli consist of numerous, globular hollow-bodies connected by short terete intersections.

The alga serves as a feeding material in New Zealand, particularly for sheep, for which purpose it is very often used in combination with *Sargassum* species.

Literature

1) CHAPMAN, V. J. and TREVARTHEN, C. B., Coll. Int. du Centre Nat. de la Recherche Scient. 81: 231–248, 1959
2) MOORE, L. B., Trans. Roy. Soc., New Zealand, 78 Part 1: 48–53, 1950
3) BERGQUIST, P. L., Physiol. Pl. 11 (4): 760–770, 1958
4) id; Nature 181 (4618): 1270, 1958
5) id.; Physiol. Pl. 12 (1): 30–36, 1959
6) CHAPMAN, V. J., Proc. IX. Int. Bot. Congr. 2: 66, 1959
7) BERGQUIST, P. L., Bot. Mar. 1 (1/2): 22, 1959
8) LEWIN, R. A., Physiology and Biochemistry of Algae. Academic Press Inc., New York, 1962

General Literature
CHAPMAN, V. J., 1950; FRITSCH, F. E., 1952; see Index of Literature.

Hydroclathrus

Widespread in tropical and subtropical seas.

H. clathratus (Bory ex C.A. Agardh) Howe (= *H. cancellatus*). Coasts of Egypt (Red Sea), India (East Coast of Madras), Indonesia (Borneo, Celebes, Java, Sumlawa, Timor, and other insular areas), New Guinea, Singapore, Philippines (Lingaya Gulf, Hundred Islands in Pangasinan Province), Japan, China, Australia, Hawaiian Islands.

The adult algae have a spongy appearance. They are loosely attached on the substratum. The young thalli are hollow, not perforated. Often they are shaken off by the action of wind and waves, so that material accumulates in the lagoons of coral reefs. The orginal yellow-green colour of the thalli turns dark brown upon drying.

H. clathratus contains approximately 16 per cent of alginic acid (Indian Province). It is eaten on the Hawaiian Islands, in China and on the Philippine Islands it is used as a fertilizer.

Literature

FRITSCH, F. E., 1952; SEGAWA, S., 1957; SULIT, J. I. and SAN JUAN, R. C., PHILIPP. J. Fish. 3 (1): 147–151. 1955; UPHOF, J. C. TH., 1959; VELASQUEZ, G. T., PHILIPP. Seaweeds Report, 1955; ZANEVELD, J. S., Ec. Bot. 13 (2): 89. 1959; see Index of Literature.

Kjellmaniella

K. gyrata (Kjellman) Miyabe (= *Laminaria gyrata*). Coasts of Japan, particularly in the colder regions of Kushiro up to the Northern Kurile Islands. Japanese name "Tororo-kombu".

The stipes of the alga are 3 to 4 centimetres long and approximately 4 millimetres thick, the blades are 1 to 4.5 metres long and about 7 to 12 centimetres wide.

The alga is rich in mucilaginous substances. Dry algae are prepared with hot water and consumed as a soup.

K. crassifolia. Coasts of Japan, from Shiokuhi to the Oshima Province, Sakhalin. Japanese name "Gagome". The fishermen harvesting the alga also call it "Nagakiri-kombu". It is sold on the local market, particularly in Hakodate.

Literature

FRITSCH, F. E., 1952; MIYABE, K., 1957; SEGAWA, S., 1957; TOKIDA, J., 1954; see Index of Literature.

Laminaria

Approximately 30 species of *Laminaria* are common on the coasts of the Northern Atlantic and Pacific Ocean, in the Arctic Ocean, on the coast of the White Sea[142]). They are also found on the coasts of Morocco and on the South African and the Patagonean coasts. *Laminaria* belongs to the group of "sublittoral" or "oarweeds" (Pl. V: 9; Pl. XV: 38).

The thallus of the *Laminaria* is frequently several metres long, consisting of stipe and blade. There is a sturdy holdfast (haptere), with which the plant is fixed to the rocky substratum. Most species are perennial and grow on the coasts of colder waters often in large groups, in depths of 3 to 15 metres. At the beginning of the vegetative period, in winter, a new thallus grows out of the stipe, the old blade dying off gradually, until it is finally shed from the alga. The shed blade material is thrown up on the shores by the surf in great masses. It is collected and processed into various algal products, in particular algal meal.

Particularly elaborate studies have been made of the *Laminaria* of the Scottish coasts and neighbouring islands, for instance, on the distribution of the Laminariaceae and their seasonal and cyclic changes[1–3]), periodicity[4]), and ecology[5–6]). Harvesting methods of *Laminaria* differ in the various areas of provenance. Drying of *Laminaria* has frequently been investigated. Thus, thorough investigations have been made on the through-circulation drying of *L. hyperborea*, *L. digitata* stipes and fronds and *L. saccharina*[7]). Methods of non-thermal drying[8]), of drying using a steam-heated drum-dryer[9]), and the preservation by ensiling[10]) have also been reported.

The constituents of *Laminaria* have been investigated by several authors.

In dried *Laminaria* fronds the mannitol fraction rises from 5 per cent in spring to 35 per cent in late summer.

Laminaran can be regarded as a reserve carbohydrate of the Brown Algae.

Laminaria contains 5 to 20 per cent of fucoidan.

Alginic acid is a constituent of the cell wall material. The content thereof in the individual species has been determined[11]) [12]). Research reports have been published on fucose [13]) [14]), amino acids[15–17]), volatile constituents[18]) [19]), vitamins[16]) [20]) [21]), nitrate content[22]), iodine content (Vinogradow), and antibiotic substances[23]).

The seasonal rhythm in chemical composition has been studied in the case of *L. hyperborea*, *L. digitata*, and *L. saccharina*[24–26]). The alginic acid content of the *Laminaria* has been found to be the highest in spring[27]). Also the content of inorganic substances shows seasonal fluctuations.

The changes in constituent composition with water depth (1 to 24 metres) have been investigated. Apparently the effects of light also play an essential rôle in these changes[28]).

Mautner reports comprehensively on the chemistry of Brown Algae, in particular *Laminaria*[29]); Black has published numerous papers on the subject. The researches of *Laminaria*, during the time of several years and producing important informations, are reported in the Annual Reports of the Institute of Seaweed Research. It has published a survey on carbohydrates, proteins, fats, vitamins, minerals, and trace elements etc. especially of *L. hyperborea*[31]).

Massoni (1960)[148]) has published on Breton species of *Laminaria* and their alginic acid content.

The most important species of *Laminaria*, industrially used as raw materials either fresh or dried, for the manufacture mainly of alginic acid and alginates, but also of algal meal for feeding purposes[15]) [30]), are *L. hyperborea* Gunn. Foslie (= *L. cloustoni* Edm.), *L. digitata* (L.) Lam. (= *L. flexicaulis* (Le Jolis)) and *L. saccharina* (L.) Lamour. The *Laminaria* meal used for fodder and fertilizers is mostly a mixture of the three species, and consists of the dried and ground blades or the whole plant. *L. hyperborea* has been found to prevail in Scottish material, *L. digitata* in Norwegian.

In former times *Laminaria* also has been used for the production of iodine, and iodine and potassium salts.

L. hyperborea (Gunn.) Foslie. Also known as "Tangle" (English), "Palmentang" (German), "Stokktare" (Norwegian). Inhabitant of the coasts of the North Sea of Norway [32]), down to France; Irish[33]) and Scottish coasts[34]), also on the Atlantic coast of the Iberian Peninsula[35]), Atlantic coast of Canada. Partly exposed at low tide (Pl. IV: 8; Pl. XV: 39, 40).

The leathery thallus grows 1 to 1,5 metres in length. Often much divided and hart shaped at its base. The long stipe is round. When cut at the cutting area a sticky jelly oozes from the stipe. The presence of mucilage anals in the stipe is one of the features distinguishing this species from *L. digitata*. The adhesive organs are cone-shaped with the hapteres, bent down at an angle from the stipe.

A distinction between *L. hyperborea* and *L. digitata* has been reported by JENSEN and HAUG: after a ten up to twelve hours extraction with a soda solution a dark brown extract is formed with *L. hyperborea* while a water-clear to yellowish extract is formed with *L. digitata*[41]).

Reports have been published on the ecology of the alga[34]) [36]), on growth and gradients[37]), on epiphytic growth[141]), and in detail, on perennial changes and the seasonal variations in chemical constitution[24]) [38]) [39]), considerable changes being found. The effect of ageing on growth and chemistry of this alga has been reported during the SECOND and THIRD INTERNATIONAL SEAWEED SYMPOSIA, 1955 and 1958.

The seasonal variations in the case of *L. hyperborea* almost exclusively refer to the branched part. The laminarian content of the entire plant is 0 per cent in June and approximately 18 per cent in September; the protein content was 12 per cent in January and 19 per cent in August; the mannitol content was 5 per cent in May and 18 per cent in July; the alginic acid content was 24 per cent in February and 14 per cent in September.

The content of mannitol varied between 18 and 36 per cent in plants from depths varying from $^3/_4$ to 6 m. In the case of Laminaran this ratio has been found to be reversed, declining under the same conditions of depth from 23 per cent to 14 per cent. The other constituents are hardly influenced by the depth of the habitat.

L. hyperborea is washed on to the beaches in large quantities, particularly on the coasts of Scotland and Ireland (Atlantic Coast), especially during the winter months. The collectors break off the fronds and discard them, heaping the stipes into piles. The drying process of the stipes sometimes takes several months[42]).

The carbohydrates of the alga have been investigated by several authors[43]), alginic acid, laminaran, mannitol, and fucose being found[13]) [44]) [45]). *L. hyperborea* is an important raw material in a l g i n i c a c i d and a l g i n a t e[13]), manufacture in many areas including the Canadian coasts[46]) [47]). *L. hyperborea* of the Scottish coasts has been thoroughly studied. It contains approximately 14 to 24 per cent of alginic acid. The extraction of alginic acid out of this species is more difficult than in the case of *L. digitata* and *L. saccharina*.

Researches on the isolation of fucoidan have been reported[48]). The laminaran from *L. hyperborea* fronds is insoluble in cold water, while the laminaran from *L. digitata* and other species is soluble. Laminaran and its derivates have been often investigated[49–52]) [140]). Laminaran is found mostly in the fronds of the algae. It is a reserve carbohydrate which the alga requires for fructification. It is present in greatest quantity in the period of slow growth from summer to late autumn (36 to 49 per cent), while in spring it is found only in very small quantity[53]). The chemistry of laminaran has been investigated[54]).

Laminitol, a cyclitol, has been shown to be present[55]) [56]). Reports appeared on nitrogenous constituents[57]), on proteins and amino acids[58–60]), on carotenes and carotinoids[61]), on the influences of season, habitat, and age on niacin content[62]), on inorganic constituents[63]), and on the distribution of iodine and other constituents in stipes of *L. hyperborea*[64]). Treated dried stipes of *L. hyperborea*, the so-called "Stipites Laminariae", formerly played a rôle in surgery and gynaecology. The stipes swell up in water to a size five times as large as their original circumference.

L. hyperborea has been collected on European coasts in former centuries and utilized for feeding purposes by the coastal population. Today the alga is an important constituent of algal feeding meals. On the coasts it is eaten by animals, particularly cattle[65]).

L. digitata (L.) Lam. (= *L. flexicaulis* (Le Jolis)). "Fingertang" (German), "Horsetail Kelp", "Sea Girdles", "Sea Wand", "Red Ware" (English), "Fingertare", "Silketare" (Norwegian). Coasts of the Northern Atlantic and Pacific Ocean. Norway[66–68]), Great Britain, Ireland[33]), France (Brittany)[69]) [70]), Spain (North and North-West Coast)[35]) [71]), Canada (Pacific Coast)[72]), Atlantic Coast of Canada and the U.S.A.[73]) from Maine to Nova Scotia. Often in abundant stands partially exposed at low tide.

The leathery thallus grows up to a length of three metres; it is strap-shaped, divided, narrowing gradually to the stipe. The stipe is 1 to 2 metres long, flattened distally. Upon cutting no mucilage escapes from the wound. The holdfast is flat. The alga is perennial. On the Irish coast two varieties are distinguished: *L. digitata* var. *typica*. The fronds are up to 1–5 m long, not deeply divided and yellowish brown, and *L. digitata* var. *stenophylla*. The plants are up to 3–6 m long, narrow, deeply divided, and dark in colour. The varieties have been described in detail by Brennan[33]).

L. stenophylla is mentioned in American literature. The alginic acid content has been estimated as approximately 40 per cent[74]). The identity of this form with *L. digitata* var. *stenophylla* of the Irish Atlantic Coast does not seem to be assured.

Cytological observations in *L. digitata* have been reported[75]). Measurements of abundance have been made for example, on the Norwegian coast. Experiments have been published concerning the most favourable frequency of harvesting[68]). Several experiments have been made on drying, particularly non-thermal drying of the alga[8]).

The chemical composition of *L. digitata* and its varieties in dependence of geographical and seasonal conditions have been investigated[24]) [32]) [68]) [71]) [76]) [77]). The annual seasonal variations in the case of *L. digitata* refer almost exclusively to the frond. Referring, however, to the entire plant, growing near the coast, the laminaran content is 2 per cent in April, 21 per cent in July; protein content is 12 per cent in March, 5 per cent in August; the mannitol content is 5 per cent in January and 21 per cent in June; the alginic acid content is 25 per cent in January and 19 per cent in July[78]). Perrin wrote about the material of French provenance[70]).

The carbohydrates of this species have been much investigated[79]). The content of alginic acid, laminaran, fucoidan, D-mannitol, D-glucose, and D-galactose has been analyzed[13]). The alginic acid content was found to amount to 15 to 40 per cent on a dry-weight basis. Scottish material contains approximately 19 to 27 per cent. The possibilities for extraction are very good. According to Humm the stipes contain about 33.3 per cent, the fronds approximately 32 per cent of algin[74]).

L. digitata is a most important raw material for alginic acid and alginates[74]) [80–83]) [70]) [68]), and numerous patents have been issued covering their production.

The laminaran content in autumn amounts to approximately 10 to 20 per cent on a dry-weight basis. Its culmination point lies in the autumn months, and its lov mark in spring. Laminaran has only been found in fronds. It is absent in the stipes and basal parts. Some work has been published on the constitution of the water-soluble laminaran from *L. digitata*[84] [50] [140]. DILLONS discussed laminaran and its compounds at the FOURTH INTERNATIONAL SEAWEED SYMPOSIUM 1961[85]. Studies on the production of laminaran has been reported[49] [52] in LEWIN (1962).

The antilipemic activity of laminaran sulphate has been investigated[86].

Investigations of fucoidan[84] and its anticoagulant effects have been made[146]. Fructose has been found to be present[87]. The proteins and amino acids[58] [60] and nutritional value have been investigated, the biological value being assessed as 75 per cent of the value of egg albumin[88]. Fronds and stipes of the alga have been analyzed for carotene and carotinoids[61]. The influence of season, habitat, and age on niacin content has been investigated[62]. Researches on iodine[89] and potassium[90] have been published. Iodine is present in the *Laminaria* chiefly in inorganic form[91]. Antimicrobial substances have been found[92] [93]. HEGNAUER (1962) has reported on the lipoids. Analyses have been made on fatty acids in thallus fat[140].

The alga is commonly used for fodder, mainly, as an additive for cattle fodder. In France it is also used for making glue (MAASS).

L. saccharina (L.) Lamour. "Zuckertang" (German), "Sugar Wrack", "Tangle", "Dulse", "Broadleaf Kelp" (English), "Sukkertare" (Norwegian). Coasts of the Northern Arctic Ocean, the Northern Atlantic, and Pacific Ocean. In large stands on the North-Sea coasts with other *Laminaria*, fully exposed at low tide. Particularly on the coasts of Norway[66], Great Britain (Scotland), Ireland[33], Spain[35], also on the Schleswig-Holstein Baltic Coast. On the coasts of North America (Vancouver Island) in large stands[94], Canada (Nova Scotia), North Atlantic Coast of the U.S.A.[73].

Various forms of *L. saccharina* also occur on the coasts of Sakhalin, the Sea of Okhotsk, and the Japan Sea.

L. saccharina prefers quiet waters. It mostly grows in depths of 3 to 15 metres. The thallus is strap-shaped, leathery, up to 4 metres long, and 30 centimetres wide, of a bright brown with a strongly undulated margin, narrowed at the base. The stipe is approximately 1 centimetre thick and almost 1 metre long. This perennial alga is fixed to the rocky substrate by means of its holdfast organs.

Ecology and growth of the alga have been studied[95] [96], and cytological observations made[75]. Chemical composition of *L. saccharina* and its variation with season have been studied[24] [97] [68]. The seasonal variations in *L. saccharina*, as in the other species, are almost exclusively limited to the frond part. The laminaran content referred to the entire plant, growing in shore is 1 per cent in March, 24 per cent in August; and the protein content likewise 14 per cent in March and 5 per cent in August; the mannitol content 9 per cent in February and 23 per cent in July, and the alginic acid content 21 per cent in January and 13 per cent in August[98].

This species is a very important source of alginic acid and alginates[81]. The alginic acid content usually amounts to 15 to 35 per cent on a dry-weight basis. Extraction possibilities are considered to be good.

The content of water insoluble laminaran is lower in *L. saccharina* than in *L. cloustoni*[31]. The laminaran is obtained from the fronds[49].

The proteins and amino acids have been investigated[60] [59]. In *L. saccharina* of the Baltic Coast it has been found that the protein content in late summer was approximately

4 per cent, in winter, on the other hand, 14 per cent[99]). Iodine mainly occurs in inorganic form in this alga[91]). Antimicrobial substances have been found[92]).

The alga is often used for fodder, for example in Iceland and on the Orkney Islands, where it is fed to sheep and horses. It is also used to make algal meals. Young stipes of *L. saccharina* were formerly sold by green-grocers in Scotland.

L. saccharina of the East-Asian coasts, Japanese name "Karafuto-kombu" or "Karafuto-tororo-kombu" has been described in detail by Tokida[100]). This species has a great variety of forms. A distinction is made between *f. linearis*, coasts from Sakhalin, and the Sea of Okhotsk, Siberian Coast of the Sea of Japan, *f. bullata* from sheltered localities of Hishitoma, Aniwa Bay, and Merei *f. membranacea*, on the West Coast, and Sakhalin on the east, on stones and shells in the interior of a lagoon, Aniwa Bay. *L. saccharina* play a major rôle in Japanese "kombu" production.

Japanese *Laminaria* species

A number of *L.* species of the Japanese coasts and other East Asian coasts locally are of great economic significance. Botanic classification of these species is not quite uniform. The most important are listed below:

L. angustata. Coasts of Japan, Hokkaido (Shiramuka, Kushiro to Muroran, Cape Esan), Hidaka[101]) [102]). The alga has been described by Miyabe[101]). Studies in its morphogenesis have been published[103]). Japanese name: "Misuishi-kombu". The alga is used in Western Honshû for the manufacture of shredded kombu, "Kizami-kombu" or "cut kombu", and exported to China in considerable quantities. Work has been published on the production of alginates from *L. angustata*[104]). The alga has a high content of iodine: fresh plants contain approximately 0.23 per cent. The extract contains antimicrobial substances[92]).

L. cichorioides. Coasts of Japan, Hokkaido (northern part of Shiribeshi Province north-eastward along the coasts of Teshio and Kitami Provinces, Kunashiri Island, Nemuro Bay, Oshoro Bay[101]) [102]) [105]). The alga has been described by Miyabe[101]). Japanese name: "Chijimi-kombu" or "Tororo-kombu". The alga is rich in mucilage.

L. cichorioides is harvested in very small quantities, as it is of a quality inferior to other species.

L. coriacea. Coasts of Japan (Kushiro)[101]). This alga has been described by Miyabe[101]). Japanese name: "Gaggara-kombu". The alga has a low content of sugar and salts, and is considered of inferior quality.

L. cuneifolia. Eastern part of the Siberian Sea, Bering Straits, northern coasts of Europe[101]).

L. dentigera. Kurile Islands, Sakhalin, Kamchatka, Bering Islands, Alaska. Japanese name: "Kumado-kombu". The alga has been mentioned by Tokida[100]).

L. diabolica Miyabe. Coasts of Japan, Hokkaido (Akkeshi, Kushiro Province, Nemuro Peninsular, Etorofu Island), Kurile Islands, Sakhalin[101]) [102]). This alga has been described by Miyabe[101]) and has also been studied by Yabu[106]).

Tokida distinguishes *f. genuina* and *f. longipes* from *L. diabolica*. The *f. longipes* also occurs on the Siberian coast of the Sea of Japan. It is larger than *f. genuina*[100]). Japanese name: "Oni-kombu". The alga is used for production of "Hanaori-kombu" and "Motozoroi-kombu", but the quality is not particularly good. Near Akkesti *L. diabolica* is used for the production of iodine.

L. fragilis. Coasts of Japan, Hokkaido (in the vicinity of Daiba, Hakodate, Muroran)[101]. This alga has been described by MIYABE[101]. Japanese name: "Yayan-kombu". *L. fragilis* has a high content of salt and iodine. It is exported in small quantities to China, because of its high salt content, it is used very little for kombu production in general.

L. japonica. Japan, Hokkaido (from Muroran to Kikonai), Korea, Coast of Siberia[101] [102]. This alga has been described by MIYABE[101], it grows up to 12 metres in length. Japanese name: "Makombu". The chemical composition of *L. japonica* has been investigated[107]. The alga contains, inter alia, glucose and fucose[13]. In the fresh alga 0.14 to 0.26 per cent iodine has been found[108]. Flavine and other compounds have been studied[109]. The changes of constituents during the salt-curing process before sun-drying have been investigated[110]. Reports have been published on the fermentation process[111]. *L. japonica* has the highest sugar content of all Japanese Laminarias. It is used for making sweets and for giving flavour to foods[101]. The best quality, "Shinori-kombu" is manufactured near the village of Osatsube. Complex utilization has been studied[112], and the various qualities were tested for their suitability as human food[113]. In Southern China *L. japonica* is used as a medical drug[114]; in Russia it is used as a source of iodine.

L. japonica is cultivated. Studies have been published on growth[115] [116], reproduction [117], harvesting[118], and fertilization[119]. The application of fertilizer (urea) in the cultivation of *Laminaria* has been studied[145].

L. longipedalis. Coasts of Japan, Hokkaido (Kushiro Province)[101]. This alga has been described by MIYABE[101]. Japanese name: "Enaga-kombu". Due to its low quality harvesting of this species is insignificant and amounts to no more than 50 to 60 tons annually.

L. longipes. Kurile Islands, Sakhalin, Kamchatka, and on the Aleutian Islands, Bering Sea. *L. longipes* f. *typica* has been described by TOKIDA[100].

L. longissima. Coasts of Japan (Nemuro and Kushiro Provinces), Kurile Islands[101]. This alga has been described by MIYABE[101]. Japanese name: "Naga-kombu", "Kimberi", "Gimberi". This species is exported in considerable quantities to China. It has a high salt content and is a raw material for alginic acid and alginates. The production of algin from *L. longissima* has been described[104]. It is also harvested for iodine production. Fresh raw material contains approximately 0.24 per cent.

L. ochotensis. Coasts of Hokkaido on the Sea of Japan (Ishikasi Province, Yagishiri Islands, Teuri Islands, Rishiri Islands, and Rebun Islands), on the Sea of Okhotsk, West Coast of Sakhalin, coasts of Korea and China[100] [101] [102]. The alga has been described by MIYABE and TOKIDA[101].

L. ochotensis is one of the most important raw materials for the production of "kombu" products. The plant grows up to 12 metres in length. The Teshio Province produces the best quality. Material from Rishiri and Rebun is not quite as valuable, but is produced in very large quantity. Japanese name of these qualities: "Rishiri-kombu" or "Hosome-kombu".

L. ochotensis is used like *L. japonica*. The Teshio quality in particular has a very high sugar content and is useful for flavouring food. It is a raw material for alginic acid and alginates. The production of algin from *L. ochotensis* has been described[104].

L. religiosa. Japan, Hokkaido (Matsumae, Shirbeshi Province, Bay of Oshoro), Honshû[101] [102] [105]. The alga has been described by MIYABE; it grows up to 10 metres in length[101]). Japanese name: "Hosome-kombu". Utilized for making "kombu" of very high quality. Also used for the production of iodine. The fresh alga contains 0.16 per cent.

L. Ruprechtii. Japan (Etorofu Island), Kurile Islands[101]). This alga has been described by MIYABE[101]). Japanese name: "Chishima-gohai-kombu". Nothing is known about utilization.

L. saccharina cf. p. 178.

L. yezoensis. Japan (Kushiro), Kurile Islands[101]). This alga has been described by MIYABE[101]). Japanese name: "Gobei-kombu". *L. yezoensis* is scarcely used for "kombu".

There have been numerous publications on *Laminaria* species on the East-Asian coasts. A survey of the propagation of *Laminaria* has been published in 1957[120]). Further studies have been published on Japanese *Laminaria* subsequently[143]). Growth, reproduction, and development have been described[121]) [122]) [123]). The algae have also been chemically analyzed. INAGAKI[124]) analyzed the sterenes and also found vitamin A in Japanese *Laminaria* in similar amounts to those quoted for cod liver oil. The volatile constituents have been investigated[18]) [19]), and the sapid components have been studied[125]).

Various qualities of edible *Laminaria* have been analyzed[126]). *Laminaria* as "Sea-cabbage" are also eaten by the Russians[127]). A number of Japanese *Laminaria* are utilized for the manufacture of alginic acid and alginate, others for the production of iodine. The literature mentions in particular *L. angustata, L. japonica, L. longissima*, and *L. ochotensis*.

Experiments on *Laminaria* cultivation in Japan were made as early as 300 years ago. Since then methods have considerably improved. YAMADA has reported on this topic in detail in 1957[102]). *L. japonica* is also cultivated on the Chinese coasts. The spores are collected and the gametophytes bred, whereafter the ensuing sporophytes are subsequently exposed in the ocean.

At the Russian coasts of the White Sea according to JEWTUCHENKO there are good conditions for the growth of *Laminaria*. In Russia they are utilized for the manufacture of alginates[147]).

Various *Laminaria* species

A number of further *L.* species have been investigated.

L. Agardhii. Atlantic Coast of U.S.A. and Canada (Nova Scotia). Growth regulating substances have been investigated[128]). A number of vitamins, inter alia, B_{12}, niacin, and pantothenic acid[129]) have been proved to occur in them. The properties of sodium alginate from *L. Agardhii* have been investigated[130]).

L. Andersonii. California. The alga mostly grows under the *Macrocystis*[131]). *L. Andersonii* is richer in nitrogen than the giant kelps. The iodine content has been estimated to be 0.6 per cent. The amount of alginic acid is 14 to 24 per cent[74]).

L. bullata. Pacific Coast of North America (Vancouver Island)[94]). The iodine content is reported to be 0.4 per cent.

L. cuneifolia. This species has been described in 1959 by WILCE[132]).

L. esculenta. The fresh alga contains 29 milligrams vitamin C per 100 grammes of fresh weight[20]).

L. groenlandica. This species has been described by WILCE[144]).

L. longicruris. Canada, particularly on the South-West Coast of Nova Scotia. The alga has stipes 5 metres in length, fronds 4 metres long, and 1.5 metres wide. Experiments with through-circulation drying have been made with this species[133]). The inorganic constituents have been studied[134]).

L. ochroleuca. Great Britain[135]) and France. Coast of Morocco (Mazagan, Mogador, Cape Blanc)[136]). Cytological observations have been made by NAYLOR[75]). The chemical composition of the alga has been investigated. Alginic acid content is approximately 36 to 37 per cent; D-mannitol content approximately 21 to 22 per cent, and laminaran content approximately 20 per cent[136]).

L. pallida. South Africa, Cape Peninsula, occurs in large quantities from Cape Ahulas to the West Coast, particularly near Port Nolloth. The alga prefers colder regions, it grows submerged and gets visible only at floodtide. Harvesting is very difficult, unless the algae are worked up on to the beach. The seasonal variations in the main constituents, such as alginic acid, mannitol etc. have been analyzed[13]) [138]).

L. Rodriguezii. Spain (Catalonia). This species is the only one of previously known Laminariaceae found on the Western Mediterranean Coast. It grows on muddy substrate at a depth of 150 metres. Its occurence is limited[139]).

Physiological and biochemical investigations with a number of *Laminaria* spp. have been reported in LEWIN (1962), inter alia, the following: *L. Agardhii, L. angustata, L. cloustoni, L. digitata, L. japonica,* and *L. saccharina.*

Literature

1) MARSHALL, S. M. et al., A Study of Certain British Seaweeds and their Utilization in the Preparation of Agar. His Majesty's Stat. Off., London, 1949
2) WALKER, F. T., Proc. Int. Bot. Congr., Paris, 8, Sect. 17, 138–139. 1954
3) WALKER, F. T. and RICHARDSON, W. D., J. Mar. Res. 15. 123–133. 1956
4) WALKER, F.T., Nature 177 (4522): 1246. 1956
5) WALKER, F. T., Trans. Proc. Bot. Soc. Edinburgh, 37. 182–199. 1958
6) WALKER, F. T., Acta Adriat. 8 (13): 1959
7) GARDNER, R. G. and MITCHELL, T. J., J. Sci. Food Agric. 4. 113–129. 1953; ibid. 4. 237–245. 1953; ibid. 4. 364–373. 1953
8) REID, K. C. and JACKSON, PH., J. Sci. Food Agric. 7. 4. 1956
9) BOOTH, E.; ibid. 7. 11. 1956
10) BLACK, W. A. P.; ibid. 14–23. 1955
11) BLACK, W. A. P., Ann. Rep. Chem. Soc. Vol. 50, 1953
12) JENSEN, A., Norsk inst. f. tang- og tareforskning, Rep. 9, 1956
13) SMITH, F. and MONTGOMERY, R., The Chemistry of Plant Gums and Mucilages. Reinhold Publishing Corp., New York, 1959
14) BLACK, W. A. P., J. Sci. Food Agric. 445–448. 1954
15) BLACK, W. A. P., The Algae ex Altschul, Processed Plant Proteins Foodstuffs. Academic Press Inc., New York, 1958
16) BLACK, W. A. P., Ann. Rep. Chem. Soc. Vol. 50, 1953
17) TAKAGI, M. and KURIYAMA, Bull. Fac. Fish. Hokkaido Imp. Univ. 10 (1): 72–76. 1959
18) KATAYAMA, T., Bull. Jap. Soc. Sci. Fish. 24 (5): 346–354. 1958 CA 53. 12. 11532c. 1959
19) KATAYAMA, T.; ibid. 24 (11): 925–932. 1959
20) HUMM, H. J., The Brown Algae ex Tressler, D. K. et Lemon, J. Mc. W., Marine Products of Commerce. Reinhold Publishing Corporation, New York, 1951

21) Larsen, B., Norsk inst. f. tang- og tareforskning, Rep. 19, 1958
22) Larsen, B. and Jensen, A.; ibid. Rep. 15, 1957
23) Stott, Ann. Rep. 1957
24) Baardseth, E. and Haug, A., Norsk inst. f. tang – og tareforskning, Rep. 2, 1953
25) Black, W. A. P., Nature, 161. 174. 1948
26) Haug, A. and Jensen, A., Norsk inst. f. tang – og tareforskning, Rep. 4, 1954
27) Black, W. A. P., J. Mar. Biol. Ass. UK 29. 45–72. 1950 CA 44. 8429. 1956
28) Black, W. A. P., J. Soc. Chem. Ind. 69. 161–165. 1950
29) Mautner, H. G., Ec. Bot. 8 (2): 174. 1954
30) Black, W. A. P., Chem. and Ind. 1640. 1955
31) Institut of Seaweed Research, Midlothian, 1956
32) Jensen, A. and Haug, A., Norsk inst. f. tang- og tareforskning, Rep. 14, 1956
33) Brennan, A. T., Inst. f. Ind. Res. and Stand., Dublin, 1950
34) Walker, F. T. and Richardson, W. D., J. Ecol. 43 (1): 26–38. 1955
35) Cabrero-Gomez, F., Estudio de las Algas marinas espanoles C. S. I. C., Madrid, 1951
36) Kain, J. M., Fourth Int. Seaweed Symp., 1961
37) Black, W. A. P. et al., Third Int. Seaweed Symp., 1958
38) Walker, F. T. and Richardson, W. D., J. Conseil perm. int. exploration mer 22 (3): 298–308. 1957
39) Black, W. A. P. et al., Econ. Proc. R. Dublin Soc. 4. 137. 1959
40) Black, W. A. P., J. Soc. Chem. Ind. 67. 165–168. 1948
41) Jensen, A. and Haug, A., Tidschr. Kjemi Bergves. Metallurgi 12. 138–139. 1952
42) Moore, L. B., New Zealand Science Rev. 9. 180–184. 1951
43) Lindberg, B. and Mc. Pherson, J., Acta Chem. Scand. 8. 1547–1550. 1954 CA 49. 10. 7063f. 1955
44) Black, W. A. P. et al., J. Sci. Food Agric. 754–764. 1955
45) Stoll and Jucker, Heteroside ex W. Ruhland, Handbuch der Pflanzenphysiologie, Springer-Verlag, 1958
46) Rose, R. C., Canad. J. of Technology 29. 19–28. 1951 CA 45. 4377. 1951
47) Vincent, D. L., Canad. J. of Technology 34. 220–226. 1956
48) Black, W. A. P. et al., J. Sci. Food Agric. 1952 WFA (5): 41. 1953
49) Black, W. A. P. et al., J. of appl. Chem., 1951 WFA (4): 41. 1952
50) Black, W. A. P., J. Sci. Food Agric. 4. 58–64. 1953
51) Black, W. A. P.; ibid. 5. 176–183. 1954
52) Friedländer, M. H. G. et al., Biochem. et Biophys. Acta 14. 136–144. 1954
53) Whistler, R. L. and Smart, Ch. L., Polysaccharide Chemistry, Academic Press Inc., New York, 1953
54) Connell, J. J. et al., J. Chem. Soc. 3494–3500. 1950 CA 45. 16. 7023c. 1951
55) Lindberg, B. et McPherson, J., Acta Chem. Scand. 8. 1875–1876. 1954
56) Lindberg, B., Second Int. Seaweed Symp., 1955
57) Scott, R., Ann. Rep. 1957
58) Coulson, C. B., Chem. and Ind. 997–998, 1953
59) Coulson, C. B.; ibid. 971–972, 1953
60) Coulson, C. B., J. Sci. Food Agric. (6): 674–682. 1955
61) Owen, E. C., J. Sci. Food Agric. 5 (9): 449–453. 1954
62) Larsen, B., Third Int. Seaweed Symp., 1958
63) Dunbar, J. C. and Mitchell, T. J., J. Sci. Food Agric. 4 (4): 183. 1953
64) Larsen, B. and Haug, A., Bot. Mar. II (3/4): 250. 1961
65) Høie, J. and Sannan, F., Third Int. Seaweed Symp., 1958
66) Printz, H., Tang og Tare, Johann Grundt Tanum Forlag, Oslo, 1953
67) Printz, H., Utilization of Seaweeds in Norway
68) Haug, A., Teknisk Ukeblad, Oslo, 1953 WFA (4): 39. 1854
69) Richard, R., Chimie et Industrie 65 (6:) 793–800. 1951
70) Perrin, M.; ibid. 74 (2). 1955
71) Primo, C., Second Int. Seaweed Symp., 1955
72) Wort, D. J., Canad. J. Bot. 33. 323–340. 1955
73) Sanford, F. B., Fishery Leaflet 469. 1958 US Dep. of the Interior, Washington, D. C.
74) Humm, H. J., The Brown Algae ex Tressler, D. K. et Lemon, J. Mc. W. Marine Products of Commerce. Reinhold Publishing Corporation, New York, 1951
75) Naylor, M., Second Int. Seaweed Symp., 1955
76) Butler, M. R., Fish. Res. Board Can. Ms. Repts. Exptl. Stat. 47, 1930
77) Haug, A. and Jensen, A., Second Int. Seaweed Symp., 1955
78) Black, W. A. P., J. Soc. Chem. Ind. 67. 169–172. 1948
79) Jensen, A., Norsk inst. f. tang- og tareforskning Rep. 10, 1956
80) Vincent, D. L., Canad. J. of Technology 34. 220–226. 1956
81) Bashford, L. A. et al., J. Soc. Chem. Ind. 69 (11): 337. 1950
82) Vincent, D. L. et al., J. of appl. Chemistry 374–378. 1955
83) Chanda, S. K. et al., J. Chem. Soc. (London) 1833. 1952
84) Percival, E. G. and Ross, A. G., J. chem. Soc. 717, 1950

85) Dillon, Th., Fourth Int. Seaweed Symp., 1961
86) Mookerjea, S. and Hawkins, W. W., Can. J. Biochem. Physiol. 36. 261–268. 1958
87) Quillet, M., C. R. Acad. Sci. 238. 926–928. 1954. CA 48. 12. 7128b
88) Larsen, B. and Hawkins, W. W., J. Sci. Food Agric. (12): 523. 1961
89) Shaw, T., Proc. Roy. Soc. 946B. 109–117. 1960
90) Bronardel, J. and Rinck, E., Bull. Inst. Oceanogr. (967). 1950
91) Roche, J. and Lafon, M., C. R. Acad. Sci. 229–481–482. 1949
92) Roos, H., Kieler Meeresforsch. 13 (1): 1957
93) Chesters, C. G. C. and Stott, J. A., Second Int. Seaweed Symp., 1955
94) Scagel, R. F., Prov. Dep. of Fish, Victoria, B. C. (1). 1947
95) Burrows, E. M., Proc. IX. Int. Bot. Congr. 2A. 5–6. 1959
96) Parke, M., J. Mar. Biol. Ass. UK 27. 651–709. 1948
97) Black, W. A. P., J. Mar. Biol. Ass. UK 33 (1): 49–59. 1954
98) Black, W. A. P., J. Soc. Chem. Ind. 67. 172–176. 1948
99) Ericson, L. E., Acta Chem. Scand. 6. 805–807. 1952
100) Tokida, J., Fac. Fish. Hokkaido Univ. 2 (1). 1954
101) Miyabe, K., J. of the Sapporo Agric. College 1. 1–50. 1957
102) Yamada, Y., Coll. Int. du Centre Nat. de la Recherche Scient., Dinard, 1957
103) Nishibayashi, T. and Inoh, S., Biol. J. Okayama Univ. 2 (4): 147–158. 1956
104) Suzuki, Bull. Fac. Fish. Hokkaido Univ. 1953. WFA (6): 43. 1954
105) Tokida, J. and Masaki, T., Bull. Fac. Fish. Hokkaido Univ. 10 (3): 173–195. 1959
106) Yabu, H., Bull. Jap. Soc. Phyc. 6 (2): 57–60. 1958
107) Kizevetter, J. V., Bull. Far East Branch Acad. Sci. USSR. 19. 21–30. 1936. CA 31. 12. 44603. 1937
108) Morita, M. and Sato, H., J. Chem. Soc. Japan Pure Chem. Sect. 75 (4): 416–418. 1954
109) Tsujimura, M. et al., Nat. Sci. Rept. Ochanomizu Univ. 4 (1): 100–104. 1953. CA 51. 3. 2142. 1957
110) Ji, M. H. et al., Scientia (Peking) 21. 666–667. 1958
111) Gotovets, V. H. and Kizevetter, J. V., Bull. Far East Branch Acad. Sci USSR 22. 54–61. 1937
112) Ji, M. H. et al., Chem. World 6. 273–276. 1959
113) Murata, K. et al., Bull. Fac. Fish Hokkaido Univ. 9 (4): 299–305. 1959
114) Chiu, B. T., Second Int. Seaweed Symp. 1955
115) Tseng, C. K. et al., Acta Bot Sinica 6 (2): 103–130. 1957
116) Tseng, C. K. et al., Scientia (Peking) 15. 475–477. 1957
117) Tseng, C. K. et al.; ibid. 4. 129–130. 1959
118) Inst. of Oceanography, Academia Sinica; ibid. 21. 739. 1959
119) Wu, C. Y. et al., Scientia (Peking) 13. 432–433. 1959
120) Hasegawa, Y. and Fukuhara, E., Hokosuishi-Geppo 13. 434–444. 1957
121) Suto, S., Bull. Jap. Soc. Sci. Fish 18 (1): 1–5. 1952
122) Tokida, J. and Jabu, H., Abstr. 9th Pac. Sci. Congr. 53. 1957
123) Nishibayashi, T. and Inoh, S., Bot. Mag. Tokyo 71 (835): 32–36. 1958
124) Inagaki, Ch. K. J., Science of Foods 1. 194–195. 1946/47. CA 46. 11497. 1952
125) Morita, M., Nippon Kagaku 2. 76. 692–694. 1955. CA 51. 22. 18151c. 1957
126) Oishi, K. et al., Bull. Fac. Fish. Hokkaido Univ. 9 (4): 283–298. 1959
127) Barbaianow, K. A., Rybnoe Khoziaistvo 25 (5): 57–61. 1959. WFA (3): 41. 1961
128) Williams, L. G., Science 11o. 169. 1949
129) Teeri, A. E. and Bieber, R. E.; ibid. 127 (1): 500. 1958
130) Vincent, D. L. et al., J. of appl. Chemistry 374–378. 1955
131) Aleem, A. A., Second Int. Seaweed Symp. 1955
132) Wilce, R. T., Bot. Not. 112 (2): 158–174. 1959
133) Merrit, J. H., J. Sci. Food Agric. 11 (Oct./Nov.). 1960
134) Young, E. G. and Langille, W. M., Publ. 4636 of the Nat. Res. Council, Halifax, 1958
135) Parke, M., Nature 162 (4112): 295–296. 1948
136) Gero, J. B. de, Bull. de l'Inst. des Pêches Mar. du Maroc 4: 1960
137) Shuttleworth, R. G., South Afric. Ind. Chem. June: 115–117. 1951
138) Holdt, M. M. v. et al., J. Sci. Food Agric. 6. 193–196. 1955
139) Bas, C., Third Int. Seaweed Symp., 1958
140) Hegnauer, R., Chemotaxonomie der Pflanzen, Birkhäuser Verlag, Stuttgart, 1962
141) Marshall, W., Brit. Phyc. Bull. 2 (1): 18–19. 1960
142) Kuznetsov, V. V., Leningrad Iz datel'stva Akad. Nauk SSSR Chapter 4, 1960
143) Hasegawa, J., Month. J. Hokkaido Fish. Exp. Stat. 17 (2): 20–22. 1960
144) Wilce, R. T., Bot. Not. 113 (2): 203–209. 1960
145) Hasegawa, J. et al., Month. J. Hokkaido Fish. Exp. Stat. 17 (12): 11–24. 1960
146) Garcia Pineda, M. D. and Montequi, D., Bol. del Inst. Espanol de Oceanogr. Nr. 72. 1955. WFA (6): 41. 1956

147) SCHULZEN, H., Melliand 37 (9): 1087–1093. 1956
148) MASSONI, R., Coll. Int. du Centre National de la Recherche Scient., Dinard, 1960

General Literature
Annual Reports 1955–1962; CHAPMAN, V. J., 1950; FOTT, B., 1959; FRITSCH, F. E., 1952; HEGNAUER, R., 1962; HOFFMANN, C., 1938; HOPPE, H. A., 1958; HOPPE, H. A. and SCHMID, O. J., 1962; HUMM, H. J., 1951; KUCKUCK, P., 1953; LEWIN, R. A., 1962; MAASS, H., 1959; PAX, F., 1962; SCHMID, O. J. and HOPPE, H. A., 1962; SEGAWA, S., 1957; VINOGRADOV, A. P., 1953; see Index of Literature.

Lessonia

Four species, in Southern Hemisphere, often forming entire submarine forests.

L. flavicans Bory (= *L. fusca*). Southern Chile, Argentina (Patagonia), Falkland Islands, Southern Georgia, Kerguelen, Heard Island[1]).

L. nigrescens Bory. Peru and Chile, Falkland Islands, often in very large populations[1]).

The plants have a tree-like appearance. The stipes often get 10 centimetres thick and up to 4 metres long. They are subdivided at many points and end in large lancet-shaped fronds.

On the Chilean Coast *Lessonia* species are known as "algina" or "chascon". They are used as a source of alginic acid and alginates.

L. variegata. New Zealand[2]).

Literature

1) LEVRING, T., Contributions to the Marine Algal Flora of Chile, Lunds Univ. Årsskrift N. F. Avd. 2. 56 (10). 1960
2) BERGQUIST, P. L., Bot. Mar. I (3/4): 86. 1960

General Literature
CHAPMAN, V. J., 1950; FOTT, B., 1959; FRITSCH, F. E., 1952; HOPPE, H. A., 1958; HOPPE, H. A. and SCHMID, O. J., 1962; see Index of Literature.

Macrocystis

As shown by WOMERSLEY[39]) the genus consists of three species: *M. pyrifera* (L.) C. Ag., *M. angustifolia* Bory, and *M. integrifolia* Bory. It can be accepted that earlier studies made with material named *M. pyrifera* not necessarely refer to this species in its recent sense but one of the two other species instead of it. *Macrocystis* is known under several English names: "Giant Kelp", "Long Bladder Kelp", "Kelp Flag", "Sea-Ivy", "Brown Kelp". Known as "Birnentang" in German (Pl. V: 10; Pl. IX: 22).

The kelp beds along Southern California and North-Western Mexico have been investigated by DAWSON and his co-workers in 1960[1]), and ALEEM (1956) undertook underwater studies for the ascertainment of quantity. Extensive beds of 200 to 500 metres length have been found in depths of between 7 and 25 metres. In the centre of the beds closely congested specimens have been found with holdfasts of up to 2 metres in diameter, while towards the perifery individuals became more scattered. The author valued the *Macrocystis* populations in these areas to be between 60 and 100 tons per hectare, with an annual growth rate of 100 tons per hectare[2]).

As early as in 1915 estimates of the US Department of Agriculture have been made on *Macrocystis* beds on the Pacific Coast of the U.S.A. These calculations amounted to an annual harvesting potential of 60 million tons[3]).

The alga also occurs on the coasts of South America: Peru, Chile (from Chiloe to the Magalhaes Straits)[4]). Argentina (Patagonia), Australia (Tasmania)[5]), New Zealand. It also lives at Kerguelen[6]) and on the coasts of South Africa. The beds are widespread and isolated, often two to three miles long. The algae grow in widespread and scattered stands, and they are smaller than those of the American Pacific Coast. Harvesting is regarded to be rather difficult. Considerable populations are found out near the Isle of Tristan da Cunha[7]).

M. pyrifera is regarded to be the largest marine alga. But indications of several hundred metres length, however, can be considered as an exaggeration. Average dimensions of 50 to 70 metres have been found, with a fresh weight of 120 to 140 kilos. The algae mostly grow in depths of between 6 to 40 metres. They are anchored on the rocky substratum with strong holdfasts approximately 1 metre in diameter. From the basis the alga develops numerous, sometimes up to 100, stipes. The stipe, only about 1 centimetre thick, in the upper part on the one side has lancet-like thallus lobes, 1 to 1.5 metres in length, and on the base a pear-shaped floating bladder. The thallus floats on the water surface.

M. pyrifera is a perennial and reaches an age of 8 to 10 years. Growth is remarkable; about 30 centimetres per day in young plants.

The alga is the most important raw material for alginic acid and alginates, as well as for other brown alga products. It is used as a feeding material and fertilizer[8–11]). Extensive fields, often exclusively *Macrocystis*, permit mechanical harvesting methods, therefore giving special significance to this alga as a raw material.

Studies on the growth of the alga have been made by Haxo and Neushul on the Californian Coast[12]). Quantitative measurements of important parameters of *Macrocystis* populations have been made on the coasts of California and Baja California, using modern methods, by North[13]). North also studied the influence of water clarity upon the growth of *Macrocystis*[14]) and on the life span of the fronds, and reported on experimental transplanation of *M. pyrifera*[15]) [16]). Development studies in culture have also been made[17]). Attachment in fine sediments and its biological and geological significance have been studied[18]). Clendenning has reported on photosynthesis and growth in *Macrocystis* to the Fourth International Seaweed Symposium[19]). This author also made determinations of fresh weight, solids, ash, and equilibrium moisture, and published them in Botanica Marina[20]). Kelp harvesting on the Californian Coast has been studied[21]). The algae are harvested there with special ships equipped with underwater mowing equipment. In this way approximately 700 tons of the large algae can be harvested per day. Such special harvesting ships have been used by Hercules Powder Company, Los Angeles. Due to the fast growth of the alga under certain circumstances a good harvest can be collected 3 to 4 times a year.

Physiological and biochemical investigations of *Macrocystis pyrifera* have been described by Lewin (1962). There are numerous publications on the constituents of the alga. The alga contains alginic acid in amounts varying between 14 and 21 per cent[22]) [9]). Various reports and patents give the details of alginic acid and alginate production[23]) [24]). The composition of alginic acid made from *Macrocystis pyrifera* has been described by WHISTLER and KIRBY (1959)[25]), who found 20 to 40 per cent L-guluronic acid. Reports have also been published on alkaline degradation of the alginates[26]). A further constituent is L-fucose[27]).

The inorganic constituents have been investigated, and numerous trace elements have been shown to be present. The value of *Macrocystis pyrifera* for human consumption, due to its content of trace elements, has been reported at the SECOND INTERNATIONAL SEAWEED SYMPOSIUM[29]). The extract shows antibiotic activity[30]). The aqueous extract has toxic effects upon mice[31]). Also because of the content of vitamins *Macrocystis* is used on the Pacific Coast, for example in New Zealand, for human and animal foods. Kelp meal is utilized as an animal feed supplement and also as a supplement in the breeding of calves and in poultry farming. The content of riboflavin is considerable. Other constituents are auxins and growth-regulating substances[32]). A significant kelp industry for the production of potassium salts and fertilizers – potassium content approximately 10 to 22 per cent – existed in former years on the American Pacific Coast. Also iodine and iodine salts have been produced from *Macrocystis* and *Nereocystis*. During the First World War the kelp industry produced acetone and other solvents, such as ethyl acetate and other ethyl compounds, as well as a carbon product for decolouring purposes known as "kelpchar". This industry had a considerable scope. More than 6 million tons of potassium chloride and approximately 20,000 tons of iodine salt were produced every year[11]). Production was later on discontinued, because cheaper raw material sources were available, and because labour was too expensive. This does not imply that such a kelp industry could not be established on other coasts where raw materials are available in sufficient quantity and where there will be also cheap labour. While the kelp industry is no longer operated in the U.S.A., where the algae have been mostly used for alginic acid and alginate production, the production of alcohol, acetone, and organic solvents is continued in the U.S.S.R. (FOTT).

M. pyrifera grows in deep water (20–120 or even 200 feet) of a fairly wide temperature range (from close to zero in the Subantarctic in winter to about 18° C in California in summer. This species occurs circum-subantarctic (most subantarctic islands), in South America (west coast to Peru and on the east coast up to about. lat. 50° S), at the west coast of North America from Baja California to the Monterey peninsula, Australia (Tasmania) and New Zealand (mainly east and south coast of S. island).

M. angustifolia occurs in shallow water (extreme low water to about 20 feet) in Australia (South Australia, Victoria, Tasmania) and South Africa (Cape). Temperature range 8–16° C.

M. integrifolia grows from water down to about 20 feet, usually as a relatively narrow belt along the coast. Temperature range 5–19° C. It occurs along the West coast of S. America from Valparaiso to Peru and in N. America from the Monterey peninsula to Alaska. Rather extended populations are reported from Vancouver[33]).

Literature

1) Dawson, E. X. et al., Pac. Nat. 1 (14): 1960
2) Aleem, A. A., Second Int. Seaweed Symp., 1955
3) Whistler, R. L. and Smart, Ch. L., Polysaccharide Chemistry. Academic Press Inc., New York, 1953
4) Levring, T., Contributions to the marine Algal Flora of Chile. Lunds Univ. Årsskrift N. F. Avd. 2. 56 (10). 1960
5) Cribb, A. B., Austral. J. Marine and Freshwater Res. 5 (1): 1–34. 1954
6) Sannié, C., C. r. Acad. Sci. Paris, 232. 2040–2041. 1951
7) Shuttleworth, R. G., South Afric. Ind. Chem. June 1951, 115–117
8) Tseng, C. K., Science 101 (2633): 597–602. 1945
9) Steiner, A. B. and McNeely, W. H., Ind. and Eng. Chem. 43. 2073. 1951
10) Stoloff, L. and Lee, Ch. F., Fishery Leaflet 173. 1946. US Dep. of the Interior, Fish and Wildlife Service, Washington
11) Sanford, F. B., Fishery Leaflet 469, 1958: ibid.
12) Haxo, F. T. and Neushul, M., Third Int. Seaweed Symp., 1958
13) North, W. J.; ibid.
14) North, W. J., Int. Oceanogr. Congr. Prep. 347–349. 1959
15) North, W. J., Nature (London) 190. 1214. 1961
16) North, W. J., Fourth Int. Seaweed Symp., 1961
17) Cole, K., Int. Oceanogr. Congr. Prep. 195–196. 1959
18) Thompson, W. C.; ibid. 586–587. 1959
19) Clendenning, K. A., Fourth Int. Seaweed Symp., 1961
20) Clendenning, K. A., Bot. Mar. (3/4): 1962
21) Scofield, W. L., Calif. Fish & Game 45 (3): 135–157. 1959
22) Nelson, W. L. and Cretscher, L. H., J. Am. Chem. Soc. 51. 1914–1932. 1929
23) Rose, R. C., Canad. J. Technol. 29. 19–28. 1954
24) Algin Corp. A. P. 2. 18. 551 (ex Maass)
25) Whistler, R. L. and Kirby, K. W., Hoppe-Seyler's Z. Physiol. Chem. 314 (1): 46–48. 1959
26) Whistler, R. L. and BeMiller, J. N., J. Amer. Chem. Soc. 82 (2): 456–459. 1960
27) Smith, F. and Montgomery, R., The Chemistry of Plant Gums and Mucilages. Reinhold Publishing Corporation, New York, 1959
28) Young, E. G. and Langille, W. M., Publ. 4636 of the Nat. Res. Council, Halifax, Canada, 1958
29) Seifert, G. L. and Wood, H. C., Second Int. Seaweed Symp., 1955
30) Roos, H., Kieler Meeresforsch. XIII (1): 1957
31) Stoloff, L., Ec. Bot. 16 (2): 86–94. 1962
32) Mautner, H. G., Ec. Bot. 8 (2): 174. 1954
33) Scagel, R. F., Prov. Dep. of Fish. Victoria, B. C. (1). 1947
34) Walker, F. T., Ann. Bot. (London) 16. 23–26. 1952
35)
36) Wort, D. J., Canad. J. Bot. 33. 323–340. 1955
37) Vincent, D. L.; ibid. 34. 220–226. 1956
38) Rose, R. C., Canad. J. of Technol. 29. 19–28. 1951. CA 45. 4377. 1951
39) Womersley, H. B. S., Univ. Calif. publ. Botany 27, 1954

General Literature

Chapman, V. J., 1950; Fott, B., 1959; Fritsch, F. E., 1959; Hoffmann, C., 1938; Hoppe, H. A., 1958; Hoppe, H. A. and Schmid, O. J., 1962; Humm, H. J., 1951; Kelco Co., 1960; Lewin, R. A., 1962; Maass, H., 1959; Mc. Neely, W. H. Algin ex Whistler, R. L. and BeMiller, J. N., Industrial Gums, 1959; Uphof, J. C. Th., 1959; Vinogradov, A. P., 1953; see Index of Literature.

Mesogloia

Thallus soft and gelatinous, a few centimetres thick, simple or branched, intertwined.

M. crassa. Japan. Japanese name "Futo-mozuku". The alga is eaten in the fresh state as a vegetable.

M. decipiens. Japan. Japanese name "Mozuku". The alga is collected in the months of April and May and salt-cured. Prepared with soya sauce and vinegar, it is eaten.

Literature

Chapman, V. J., 1952; Fritsch, F. E., 1952 Hoffmann, C., 1938; Uphof, J. C. Th., 1959; see Index of Literature.

Nemacystus

N. decipiens. Coasts of Japan. This species is perhaps identical with *Mesogloea decipiens* (quod vide).

The constituents, according to the Standard Table of Food Composition in Japan, are 2.7 per cent raw protein, 1.5 per cent fat, and 2.3 per cent sugar. Vitamin B_1 and B_2 have been shown to be present. The ash content is very high, some 93.7 per cent[1]).

Physiological and biochemical investigations have been reported in Lewin[2]).

Literature

1) Black, W. A. P., The Algae ex Altschul. Processed Plant Proteins Foodstuffs. Academic Press Inc., New York, 1958

2) Lewin, R. A., Physiology and Biochemistry of Algae. Academic Press Inc., New York, 1962

General Literature

Fritsch, F. E., 1952; Segawa, S., 1957; see Index of Literature.

Nereocystis

N. luetkeana Post. et. Rupr. Pacific Coast of North America, from Alaska to California (Los Angeles). Widespread and abundant stands, particularly on the Coast of British Columbia (Vancouver Island in the vicinity of Malcolm Island) and Alaska[1]) [2]) [3]) (Pl. IX, 21).

This is an annual alga. The stipe reaches 25 metres in length and the terminates in a large, gas-filled floating bladder from which arise several five metre long thallus lobes.

The literature records individual specimens of *Nereocystis* of a total length of 100 metres, with floating bladders two to three metres in length and 15 centimetres in diameter, and blades 15 metres in length.

Hoffmann reports on one specimen 40 metres long, blade area 85 square metres, and fresh weight 55 kg.

Nereocystis is known by many popular names, such as "Bull Kelp", "Black Kelp", "Bladder Kelp", "Ribbon Kelp", "Sea Otters", "Cabbage", "Sea Whip", "Sea Onion".

Before the First World War *Nereocystis* was utilized in the North-American kelp industry together with other large brown algae of the Pacific Coast, particularly *Macrocystis* and *Alaria*, for the production of potash salts and fertilizers.

The organic[4]) and inorganic[5]) constituents of this alga have been analyzed and the seasonal variations in chemical composition studied[2]).

Nereocystis luetkeana is a valuable raw material for alginic acid and alginates. Production has been described[6]) [7]). The content of alginic acid reaches about 13 to 15 per cent in the stipes and approximately 14 to 20 per cent in the blades. The metabolism of iodine has been investigated[8]).

Development studies have been made with the aid of cultural experiments[9]). Physiological and biochemical investigations have been reported in LEWIN[10]).

The alga is eaten by Red Indians. Stipes and parts of the floating bladders are also salt-cured or candied, being sold as a delicacy under the name of "Seatron". Kelp meal is also used as an animal feed supplement[11]).

Literature

1) SCAGEL, R. F., Prov. Dep. of Fish. Victoria, B. C., (1). 1947
2) WORT, D. J., Canad. J. Bot. 33. 323–340. 1955
3) MAUTNER, H. G., Ec. Bot. 8 (2): 174. 1954
4) SAGAR, W. et al., Am. J. Pharm. 118. 156–163. 1946. CA 40. 6123. 1946
5) YOUNG, E. G. and LANGILLE, W. M., Publ. 4636 of the Nat. Res. Council, Halifax, Canada, 1958
6) VINCENT, D. L., Canad. J. Technol. 34. 220–226. 1956
7) ROSE, R. C., Canad. J. Technol. 29. 19–28. 1951. CA 45. 4377. 1951
8) TONG, W. and CHAIKOFF, J. L., J. Biol. Chem. 215. 473–485. 1955. CA 49. 20. 14116e. 1955
9) COLE, K., Int. Oceanogr. Congr. Prep. 195–196. 1959
10) LEWIN, R. A., Physiology and Biochemistry of Algae. Academic Press Inc., New York, 1962
11) STOLOFF, L., Ec. Bot. 16 (2): 86–94. 1962

General Literature

CHAPMAN, V. J., 1950; FOTT, B., 1959; FRITSCH, F. E., 1952; HOFFMANN, C., 1938; HOPPE, H. A., 1958; HOPPE, H. A. and SCHMID, O. J., 1962; HUMM, H. J., 1951; MAASS, H. 1959; SCHMID, O. J. and HOPPE, H. A., 1962; UPHOF, J. C. TH., 1959; see Index of Literature.

Padina

Some 12 species are recorded mostly from warmer regions.

P. pavonia. Frequently on Mediterranean coasts, but also in the other warm oceans. The thallus is flat, fan-shaped, conical, 5 to 12 centimetres long. Patches of calcium carbonate are often visible. The alga contains 2.2 per cent D-mannuronic acid, 1.8 per cent L-guluronic acid[1]).

P. australis. Red Sea, Ceylon (Gulf of Mannar), India (East Coast of Madras), Indonesia (Sumatra, Java, Celebes, and other islands), the Chinese Sea, and the Philippines[2]).

An alginic acid content of approximately 15 per cent has been found[3]). The sterenes have been investigated[4]). *P. australis* is often eaten as a salad or used for the production of a jelly-like sweetmeat. The alga is also used in the Philippines as a fertilizer.

P. japonica. Japan[5]). The morphogenesis of this alga has been studied[7]).

P. arborescens, *P. minor*, and *P. crassa* are also recorded from Japan.

Physiological and biochemical investigations of *P. arborescens* have been reported in LEWIN[6]).

P. commersonii and *P. tetrastromatica*, from Ceylon, are utilized in a manner similar to *P. australis*[2]).

Literature

1) SMITH, F. and MONTGOMERY, R., The Chemistry of Plant Gums and Mucilages. Reinhold Publishing Corporation, New York, 1959
2) ZANEVELD, J. S., Ec. Bot. 13 (2): 89. 1959
3) VALSAN, A. P., Current Science 24 (10): 343–345. 1955
4) GOVINDACHARI, T. R. and RAJAGOPALAN, S., First Int. Seaweed Symp., 1952

5) NISHIBAYASHI, T. and INOH, S., Bot. Mag. Tokyo, 852. 261–268. 1959
6) LEWIN, R. A., Physiology and Biochemistry of Algae. Academic Press Inc., New York, 1962
7) KUMAGAE, N. and INOH, S., Kromosomo 46/47. 1521–1530. 1960

General Literature

FOTT, B. 1959; FRITSCH, F. E., 1952; SEGAWA, S., 1957; see Index of Literature.

Pelagophycus

P. porra. Southern California. Known also as "Elk Kelp".

The alga occurs in depths of 20 to 30 metres, often together with *Macrocystis pyrifera*. The stipes reach 6 to 10 metres, 45 metres in the largest specimens. The weight of one plant is between 15 and 70 lbs. A large air-bladder develops at the end of the stipe which then bifurcates into two horizontal branches, each bearing four to six tender blades about 6 metres in length, on thin stalks[1]).

The alga is harvested chiefly with other brown algae. The potash content on an average is higher than, for example, that of *Macrocystis* or *Nereocystis* and amounts to approximately 20 per cent on a dry-weight basis. The alga is used as a fertilizer material together with other kelps.

It serves as a raw material for alginic acids or alginates. The alginic acid content is approximately 16 per cent[2]).

Physiological and biochemical investigations have been reported in LEWIN[3]).

Literature

1) ALEEM, A. A., Second Int. Seaweed Symp., 1955
2) HUMM, H. J., The Brown Algae ex Tressler, D. K. et Lemon, J. Mc. W., Marine Products of Commerce. Reinhold Publishing Corp., New York, 1951
3) LEWIN, R. A., Physiology and Biochemistry of Algae. Academic Press Inc., New York, 1962

General Literature

CHAPMAN, V. J., 1950; FRITSCH, F. E., 1952; HOFFMANN, C., 1938; HOPPE, H. A., 1958; HOPPE, H. A. and SCHMID, O. J., 1962; MAASS, H., 1959; see Index of Literature.

Pelvetia

P. canaliculata (L.) Dec. et Thur. Known as "Channel Wrack", "Cow Tang". Northern Atlantic, often in abundant populations in the marginal zones along the high-tide marks.

The thallus of this alga is much divided, 10 to 15 centimetres high, bushy, attached by hapteres. There is no centre rib. There is a characteristic channel-shaped pit. The colour is olive-brown in winter and somewhat lighter in summer. The receptacles are yellowish. The channel-shaped pit holds water, which prevents the alga from drying out when exposed at low tide.

The populations on various European coasts have been investigated, for example in Ireland[1]) or Portugal[2]).

What are known as salt-marsh varieties of *P. canaliculata* have been described. A distinction is made between var. *libera*, var. *coralloides*, and var. *radicans*. Anatomy, cytology, development of the reproductive structures, fertilization, and embryology have been studied[3]).

Booth has described drying methods by means of steam-heated drum-dryers[4]).

Chemical composition[11]) and seasonal variations[5]) [6]) have been investigated. Carbohydrates[7]) and component sugars have been analyzed[8]). The alga contains alginic acid[9]) [10]), laminaran[12]) and on hydrolysis yields mannite and volemite[13]) [14]) [15]). The share of fucoidin is relatively high, often above 20 per cent of the dried alga[9])[16]). *P. canaliculata* contains L-fucose[17]). The seasonal variation in combined L-fucose has been studied, and a maximum content of 13 per cent has been found[18]). Protein and amino acids have been investigated[19–21]). The alga contains a peptide (0.73 per cent of the dry weight)[22]), ascorbinic acid[23]), and vitamin B_{12}[26]), and has antibiotic activity[27]). Investigations on the carotene and carotinoid content have been made[24]) [25]).

P. canaliculata is used in various areas, for example on the Orkney Islands, in a boiled state as an additional feed material for cows. In Ireland it is used as a pig feed[1]). The alga is used as a fertilizer in many coastal areas, where it is abundantly found.

P. canaliculata var. *libera* contains 6.3 per cent fat, which probably helps to prevent drying out in the case of exposed marsh-plants[15]).

P. fastigiata. Central California. This species resembles *P. canaliculata*, but is larger.

It contains fastigiatin, an acidic tripeptide[2]), and antimicrobial substances[28]), and a toxin of a, as yet, unknown structure[29]) [30]).

P. siliquosa. Coasts of East Asia. The geographical distribution has been shown[31]).

P. wrightii. Coasts of Japan (Northern Honshû, Hokkaido), Kurile Islands, Sakhalin, Korea, Siberian coast of the Japan Sea. Growing on rocks in the littoral zone.

A distinction is made between *f. typica*, *f. Babingtonii*, and *f. japonica*, which usually grow together with *Fucus evanescens*[32–34]).

Physiological and biochemical investigations of *P. canaliculata*, *P. fastigiata*, and *P. wrightii* have been reported in Lewin (1962).

Literature

1) Brennan, A. T., Inst. f. Ind. Res. and Stand., Dublin, 1950
2) Fischer-Piette, E., Coll. Int. du Centre Nat. de la Recherche Scient., Dinard, 1959
3) Subrahmanyan, R., J. Indian Bot. Soc. 36 (1): 12–34. 1957; ibid. 36 (3): 373–395. 1957
4) Booth, E., J. Sci. Food Agric. 7. Nov., 1956
5) Black, W. A. P., J. Soc. Chem. Ind. 68. 183–189. 1949
6) Baardseth, E. and Haug, A., Norsk inst. f. tang- of tareforskning, Rep. 2. 1953
7) Lindberg, B. et al., Acta Chem. Scand. 8. 817–820. 1954. CA 49. 11. 7501b, 1955
8) Jensen, A., Norsk inst. f. tang- og tareforskning, Rep. 9. 1956
9) Black, W. A. P., Ann. Rep. Chem. Soc. 1953, Vol. 50
10) Smith, F. and Montgomery, R., The Chemistry of Plant Gums and Mucilages. Reinhold Publishing Corporation, New York, 1959
11) Quillet, M., Bull. du Laboratoire Maritime de Dinard 45. 63–68. 1959
12) Black, W. A. P. et al., J. of appl. Chem. 1951. WFA (4): 41. 1952
13) Lindberg, B., Second Int. Seaweed Symp., 1955
14) Stoll and Jucker, Heteroside ex W. Ruhland. Handbuch der Pflanzenphysiologie, Springer-Verlag, 1958
15) Mautner, H. G., Ec. Bot. 8 (2): 174. 1954
16) Black, W. A. P. et al., J. Sci. Food Agric. 1952. WFA (5): 41. 1953
17) Black, W. A. P. er al., J. Sci. Food Agric. 4. 85–91. 1953
18) Black, W. A. P.; ibid. 5. 445–448. 1954
19) Coulson, C. B., Chem. and Ind. 1953, 997–998
20) Coulson, C. B., J. Sci. Food Agric. 6. 674–682. 1955
21) Coulson, C. B., Chem. and Ind. 1953, 971–972

22) YENN, E. W., The Plant Proteins and Peptides and their Localization in Cells and Tissues ex W. Ruhland. Handbuch der Pflanzenphysiologie, Springer-Verlag, 1958
23) HAGENE, P., C. R. Acad. Sc. Paris, 248 (4): 579–582. 1959
24) HAUG, A. and LARSEN, B., Norsk inst. f. tang- og tareforskning, Rep. 16. 1957
25) OWEN, E. C., J. Sc. Food Agric. 5 (9): 449–453. 1954
26) ERICSON, L. E., Chem. and Ind. 829–830. 1952
27) CHESTERS, C. G. C. and STOTT, J. A., Second Int. Seaweed Symp., 1955
28) ROOS, H., Kieler Meeresforsch. XIII (1). 1957
29) HABEKOST, R. C. et al., J. Washington Acad. Sci. 45. 101–103. 1955
30) HEGNAUER, R., Chemotaxonomie der Pflanzen, Birkhäuser Verlag, Stuttgart, 1962
31) TSENG, C. K. and CHANG, C. F., Oceanol. et Limnol. Sinica 1 (2): 209–217. 1958
32) TOKIDA, J., Fac. of Fish. Hokkaido Univ. 2 (1): 1954
33) SEGAWA, S., The Seaweeds of Japan. Hoikusha, Osaka, Japan. 1957
34) YABU, H. and IMAI, A., Bull. Jap. Soc. Phyc. 5 (2): 44–49. 1957

General Literature

CHAPMAN, V. J., 1950; FOTT, B., 1959; FRITSCH, F. E., 1952; HOFFMANN, C., 1938; HOPPE, H. A., 1958; LEWIN, R. A., 1962; MAASS, H., 1959; see Index of Literature

Petalonia

P. Fascia (= *Phyllitis Fascia* Ktz. = *Ilea Fascia* (Müller) Fries). East Asian coastal regions, particularly Formosa, Japan, China, Sakhalin, Kamchatka. Coasts of Malaya and Pacific Coast of North America, as well as in other parts of the world.

Thallus of greenish or olive-brown colour, approximately 30 centimetres long, narrows basally into a short flattened stalk.

TOKIDA makes a distinction between *f. typica* and *f. caespitosa.*

Sun-dried algae are often eaten in East Asia, particularly by the rural population of the Japanese districts of Awa and Sagami. Japanese name: "Seiyô-hahanori".

General Literature

CHAPMAN, V. J., 1950; FRITSCH, F. E., 1952; HOFFMANN, C., 1938; SEGAWA, S., 1957; TOKIDA, J., 1954; UPHOF, J. C. TH., 1959; see Index of Literature.

Phyllogigas

The Antarctic genus *Phyllogigas* is identical with the genus *Himantothallus*. The various species have been combined by SKOTTSBERG and NEUSHUL to form the *Ph. grandifolius* (A. et E.S. Gepp) Skottsb.

South Georgia, South Orkney Islands, South Shetland Islands, Palma Peninsular, Queen Mary Coast, Wilkes Land, Victoria Land.

Ph. grandifolius is the largest Antarctic kelp. The alga grows in a depth of 20 to 30 metres. It has been described in detail by SKOTTSBERG and NEUSHUL[1]) [2]).

Phyllogigas and *Himantothallus* have been studied by ZINOVA[3]). There is at present no information on industrial utilization of the Antarctic kelp. An investigation of this would appear to be very interesting and useful.

Literature

1) Skottsberg, C. and Neushul, M., Bot. Mar. II (1/2): 164. 1960
2) Neushul, M., Bot. Mar. V (1): 19. 1963
3) Zinova, A. D., Bot. Zhurn. 44 (3): 372–379. 1959

General Literature
Fritsch, F. E., 1952; see Index of Literature.

Postelsia

P. palmaeformis. Pacific coasts of North America, particularly in the surf zone. Also known as "Sea Palm", "Meerespalme". Resembles in appearance to the fronds of a palm, the stalks are 40 to 60 centimetres long, bearing numerous narrow drooping blades. At low tide the alga protrudes from the water. The "sea-palm" forests have a very peculiar appearance. *P. palmaeformis* has been shown to have antibacterial activity[1] [2].

Literature

1) Mautner, H. G., Ec. Bot. 8 (2): 174. 1954
2) Pratt, R. et al., J. Am. Pharm. Ass. 40. 575–579. 1951

General Literature
Fott, B., 1959; Fritsch, F. E., 1952; see Index of Literature.

Pterygophora

P. californica. Coast of California. The woody stipe of this alga is round at its base and flat in its upper part. The sporophylls reach 1.5 metres in length, and the terminal blade is often as long.

The alga grows together with other algae, for example *Eisenia arborea* and *Laminaria farlowii*, under *Macrocystis pyrifera*[1].

Literature

1) Aleem, A. A., Second Int. Seaweed Symp., 1955

General Literature
Fritsch, F. E., 1952; see Index of Literature.

Saccorhiza

S. bulbosa de la Pylaie (= *S. polyschides* (Lightf.) Batt). Atlantic coasts of Europe and North Africa, sporadically also in the western Mediterranean region, Norway[1], Great Britain, Ireland (Kerry and Aran Islands)[2], Spain and Portugal[3], Canada (South-West Coast of Nova Scotia)[4]. The alga grows in sandy or rocky places, below the low-water mark; the blade is digitate, sometimes to 2 metres in length. The stipe is flattened, in the lower part of a spiral shape, approximately 60 centimetres long. The hapteres are covered by a bell-shaped outgrowth up to 40 centimetres in diameter. The bell survives the winter, while the rest of the plant perishes.

The constituents, particularly the inorganic, have been investigated[4].

The seasonal variation in chemical composition[5]) and the sugaı content have been analysed[6]).

The alga formerly was used in the kelp industry, and today it can be regarded as a raw material for alginic acid and alginates, provided that it can be harvested in sufficient quantity.

Literature

1) Wennberg, T., Meddel. f. Göteborgs Bot. Trädg. XVIII, 1950
2) Brennan, A. T., Inst. f. Ind. Res. and Stand., Dublin, 1950
3) Cabrero-Gomez, F., Estudio de las Algas marinas espanoles C. S. I. C., Madrid, 1951
4) Young, E. G. and Langille, W. M., Publ. 4636 of the Nat. Res. Council, Halifax, Canada, 1958
5) Black, W. A. P., J. Soc. Chem. Ind. 67. 172–176. 1948
6) Jensen, A., Norsk inst. f. tang- og tareforsk. Rep. 9, 1956

General Literature

Chapman, V. J., 1950; Fritsch, F. E., 1952; Hoppe, H. A., 1958; Maass, H., 1959; see Index of Literature.

Sargassum

Approximately 250 species are recorded, mostly on rocky coasts of the temporate and warmer seas, particularly in the Southern Hemisphere.

The thallus of *S.* species – also known as "Beerentang", "Gulf Weed" or "Sargassum Weed" – is mostly richly branched, with flat, stalked, oblong blades, and air-bladders similarly stalked located between the branches. Short ramuli carrying the reproductive organs arise from the blade stalks (Pl. VI: 11).

The alginic acid of some *S.* species has been investigated. A floating *S.* species contains 6.8 to 9 per cent of alginic acid, while sedentary species 13.3 to 17.2 per cent, or, if investigated by other methods, 22.2 to 33.5 per cent[1]). Reference should be made to further investigations on individual species of different provenance. Analyses of nitrogen and potash have been investigated[2]). *S.* contains volatile constituents[3]), lipids[4]). The sterenes, particularly of Indian material, have been investigated[5]), and the constitution of sargasterol studied[6]). *S.* species contain catechinlike tanning substances[51]).

Physiological and biochemical investigations of numerous *S.* species, inter alia *S. confusum*, *S. enerve*, *S. linifolium*, *S. ringgoldianum*, *S. thunbergii*, and *S. tortile* have been reported in Lewin (1962).

S. aquifolium. Occurs on the coasts of the Red Sea and of the Indian Ocean, Indonesia, Australia, and other coastal areas of the Pacific Ocean.

The alga is eaten locally[7]).

S. bacciferum. Cf. *S. natans.*

S. confusum. China, Korea, Japan (Bay of Oshoro, and neighbouring coasts), Kurile Islands, Sakhalin, Siberian coast of the Sea of Japan[8]) [9]).

S. confusum f. *validum* occurs on the coasts of Korea, Japan (Honshû, Hokkaido) and Sakhalin[9]).

S. confusum contains a blood-sugar reducing phenol sargalin, which disintegrates into arginine and anthelmintic phenols on acid hydrolysis. 0.19 per cent ethereal oil has been found to be present[46]).

S. echinocarpum. Coasts of the Pacific Ocean. In Hawaii known as "Limu Kala", either eaten fresh or in numerous preparations, particularly with fish and meat dishes.

S. enerve. Coasts of the Pacific Ocean and of the adjoining seas, particularly of Japan, Korea (southern off-shore island of Che Ju Do), the Philippines (Lingayen Gulf, Hundred Islands).

Raw material for alginic acid and alginates[10] [11]). Used in Japan for the preparation of soups or a soyabean sauce called "Hondawara".

S. filipendula. Coasts of Texas, Yucatán, and Cuba[12] [13]).

S. fluitans. Coasts of Texas, Cuba, Panama Canal zone[12] [13]).

S. fulvellum. Korea (southern off-shore island Che Ju Do), coasts of Japan.

S. fusiforme. China and Japan, Chushan district[14]). In Japan the alga is used as a raw material for alginic acid and alginates. Seasonal variations of the chemical constituents have been investigated[14]).

In China the alga is used in pharmacy. Young algae are eaten as vegetables or used for the preparation of teas, after drying[15]).

In East Asia *S. fusiforme* is used as a fertilizer together with other algal species.

S. granuliferum. Coasts of Malaya, Singapore, Indonesia, New Guinea, Australia, Chinese Sea, Japan. The alga is eaten on Amboina[7]).

S. hemiphyllum. Coasts of the Pacific Ocean, Japan, the Philippines (Lingayen Gulf, Hundred Islands). Raw material for alginic acid and alginates[11]). This alga is used as a fertilizer particularly in China.

S. Horneri. Coasts of the Pacific Ocean and of the adjoining seas, Formosa, China, Korea, Japan (Honshû, Hokkaido, particularly Bay of Oshoro)[8]), Kurile Islands, Sakhalin[9]).

Raw material for alginic acid and alginates. The alga supplies a good quality of sodium alginate[16]).

In many parts of China this alga is used as a fertilizer.

S. Horneri has also been found in drifting seaweeds in the Sea of Japan and has been investigated[17]).

S. hystrix var. *spinulosa*. Grenada (St. Georges Harbour) floating[12]). Brazil (Urubé, Fortaleza, Ceará) washed up on the beaches[12]). Cuba[13]).

S. kjellmannianum. Japan, particularly Bay of Oshoro and the neighbouring coastal areas[8]), Southern Kurile Islands, and Sakhalin[9]).

S. lendigerum. Coast of Cuba[13]).

S. linifolium. Mediterranean region, Indian Ocean. Raw material for alginic acid and alginates[18]).

S. longifolium. South Africa.

S. micracanthum. Coasts of Japan. Raw material for the production of alginic acid and alginates. The alga supplies a good quality of sodium alginate[16]).

S. Miyabei. Coasts of Japan, particularly Bay of Oshoro and neighbouring coastal areas[8]), Southern Kurile Islands, and Sakhalin[9]).

S. muticum. Coasts of Japan. This alga has also been found in the North-East Pacific[19]).

S. myriocystum. Coasts of India, particularly East Coast of Madras. The alginic acid content is reported to be 22 per cent.

S. natans (= *S. bacciferum*). Coasts of Texas, Mexico, and Cuba, drifting in the Gulf stream[12] [13] [20]), often washed up on the beaches in large quantities. Sometimes washed-up specimens of this species are found on the coasts of Great Britain. In South America the alga is used as a medicinal preparation (Pl. XI: 27).

S. natans. Coasts of Puerto Rico. Investigated by MARTINEZ NADAL et al. This alga has antibiotical activity. The substance has been isolated and is known as sarganin[51–56]). Low molecurlar weight carbohydrates have been analysed[57]).

S. nigrifolium. Japan, the Philippines (Lingayen Gulf, Hundred Islands). Raw material for alginic acid and alginates[11]).

S. phlagjophyllum. Coasts of India, particularly East Coast of Madras. The alginic acid content is reported to be approximately 27 to 28 per cent.

S. platicarpum. Coast of Cuba[13]).

S. polyceratium. Coasts of Cuba, Panama Canal zone[12]) [13]), Puerto Rico[47]). This alga has been shown to have antibiotic activity[47]).

S. polycystum. Coasts of the Indian Ocean, Malaya, Singapore, Indonesia, New Guinea, Australia, the Philippines. The alga is eaten in many places and in a great variety of different manners of preparations[7]).

S. ringgoldianum. Coasts of Japan. Reports have been published on the growth rings[48]). Observations and experiments have been made on the behaviour of fishes towards floating objects[49]).

The oils and sterols have been investigated[21]). Fucosterol and sargasterene were found[22]). The tannin contained in the alga has been studied[23]).

S. serratifolium. Coasts of the Indian Ocean, China, Japan, Sakhalin [9]), the Philippines (Lingayen Gulf, Hundred Islands). Raw material for alginic acid and alginates[11]).

S. siliquastrum. Coasts of the Pacific Ocean, particularly of southern China[15]), and Japan, the Philippines (Lingayen Gulf, Hundred Islands). Raw material for alginic acid and alginates[11]). Young algae are eaten in Japan.

S. siliquosum. Coasts of Malaya, Singapore, Indonesia, New Guinea, Australia, the Philippines, Chinese Sea. The alga is used on the Philippines as a salad or vegetable[7]).

S. spinuligerum. Coast of New Zealand[24]).

S. tennerimum. Coasts of India, particularly East Coast of Madras, Cape Comorin. The alginic acid content has been indicated as approximately 29 to 30 per cent. Another estimate is 34.6 per cent alginic acid and 9.4 per cent mannitol[7]). *S. tennerimum* is used together with *S. wrightii*, particularly in the State of Madras, as a fertilizer for coco-palm plantations.

S. thunbergii. Coasts of South China[15]), Korea, Japan (particularly Bay of Oshoro and neighbouring coastal areas), Kurile Islands, Sakhalin. Many different varieties and forms are known. The *S.* species of Chushan district and their seasonal variations in chemical constitution have been investigated in detail[14]). The polysaccharide has been investigated[15]). In China *S. thunbergii* is described as a medicinal drug[15]), but it is also used as a pig feed and fertilizer.

S. vulgare. Coasts of the Pacific and Atlantic oceans, West Coast of Africa (Mauretania), Cameroons[26]). Coasts of Cuba[13]), Venezuela, and Columbia[12]). The alga is eaten on the Philippines.

S. wrightii. Coasts of India, particularly East Coast of Madras. The alginic acid content is reported to be approximately 25 per cent. Another estimate runs up to 29.8 per cent of alginic acid and 7.3 per cent of mannitol[7]. *S. wrightii* is used together with *S. tennerimum*, particularly in the State of Madras, for fertilizing coco-palm trees.

S. species are fairly widespread on the coasts of many countries and often are washed up on the beaches in large quantities.

Egypt: The *S.* species and varieties of the coasts of the Red Sea are utilized as a feed material in Egypt.

Argentina: A number of species are known to occur on the long Argentinian coasts.

Brazil: Species occurs on the North Brazilian Coast, near Recife[27].

Ceylon: *S.* species have been investigated in view of their alginic acid content in the Gulf of Mannar[28].

China: *S.* species exist in large quantities on the South Chinese coasts, used as feeding materials and fertilizers.

Gold Coast: Local *S.* species have been described[26] [29].

India: A number of *S.* species are known to occur on the Indian coast and have been both investigated and utilized[30] [31] [33] [34]. A report was submitted by PILLAI to the THIRD INTERNATIONAL SEAWEED SYMPOSIUM on the alginic acid content of the Indian *Sargassum* species and its extraction on a cottage industry basis[32] [51]. The mannitol of this alga has also been investigated[33].

Indonesia: According to WEBER VAN BOSSE, 58 different *S.* species are known to occur on the Indonesian coasts.

Japan: Numerous *S.* species are known to occur on the Japanese coasts. Apart from those species mentioned above, SEGAWA has listed the following species: *S. crispifolium*, *S. filicinum*, *S. giganteifolium*, *S. nipponicum*, *S. patens*, *S. piluliferum*, *S. racemosum*, *S. saganianum*, *S. sandei*, *S. tenuifolium*, *S. tortile*, *S. tosaense*, and *S. Yendoi*.

YAMADA (1955) has written on the distribution of *Sargassum* on the coasts of Japan and its neighbouring regions, in a paper submitted to the SECOND INTERNATIONAL SEAWEED SYMPOSIUM. He mentioned approximately 70 species which he subdivided into two groups:

(a) Southern types and non-Southern types.

(a) Of Indo-Pacific origin being distributed in the region in question, in Formosa, Loochoo Islands, southern parts of Kyushi and Sikoku Ku-Peninsular, Hachijo Island.

(b) 30 species on the Japanese coasts, and in its adjacent waters, especially *S. Horneri*, *S. thunbergii*, *S. serratifolium*, *S. tortile*, *S. enerve*, *S. kjellmannianum*, *S. micracanthum*.

S. Horneri and *S. thunbergii* are most common and are also recorded at Hong Kong or Formosa in the south and in the southern Kurile Islands and Sakhalin in the north.

The following species are recorded as occuring only in the southern parts of the region: *S. piluliferum*, *S. patens*, *S. pinnatifidum*, *S. kashiwajimanum*, *S. tosaense*, *S. setaceum*, *S. nipponicum*, *S. hemiphyllum*, *S. saganianum*, *S. tenuifolium*, *S. giganteifolium*, *S. kushimotense*, *S. assimile*, *S. crispifolium*.

In the central parts of Japan *S. Yendoi*, *S. ringgoldianum*, *S. filicinum*, *S. nigrifolium*, and *S. racemosum* have been found.

Tsushima current, near the southern end of Kyushu:

S. microceratium, *S. confusum*, *S. Turneri*, *S. Miyabei*[35].

The Japanese name for *Sargassum* is "Ginbaso". In Japan the *S.* species are utilized industrially for the production of alginic acid and alginates. Reports have been published on production[44] [45].

New Zealand: *S.* species, mostly rockweed-forms, are used together with *Hormosira* and other algae as a feeding material for sheep.

Peru: *S.* species are recorded from the Peruvian Coast.

Philippines: Several *Sargassum* species are recorded from the Philippines[36]. They are often eaten by the indigenous population, and are used in large quantities as fertilizer[7].) The production of alginic acid and alginates from the Philippine *S.* species has been investigated[11]. The content is indicated to be between 13 to 17 per cent.

Singapore: Several not quite identifiable *S.* species are used for feeding purposes, particularly for pigs and ducks.

U. S. A.: A considerable quantity of *Sargassum* is known to be found at the Florida coast, which supplies a good quality of alginate; the material can be utilized industrially[37].

The *Sargassum* in the Sargasso Sea

The Sargasso Sea is a large area of more than two million square miles of the Atlantic Ocean between the Azores and the West-Indian island, in latitude between 20° and 35° Northern. Masses of drifting algae are contained in the Sargasso Sea, as well as in other parts of the world's oceans, for example in the Caribbean Sea and in the Gulf of Mexico. These masses accumulate due to the action of currents to form more or less sizeable "islands" or "meadows"[30]. The quantity of floating algae in the Sargasso Sea has been valued to be between 4 and 11 million tons fresh weight[39]. The vegetation of the Sargasso Sea has been described as follows by WINGE in 1923:

> "The vegetation of the Sargasso Sea consists of three common species of *Sargassum*, spread over the whole area, they live and propagate vegetatively year after year, and have a distinct optimum period of growth in late summer. Besides, five less common species are found, but generally only in the western regions, the Gulf Stream territory, they are supposed to be identical with existing coastal forms. Of these five species, the lifetime of which very likely does not extend over more than a few months as a rule.
>
> It is uncertain whether the three inhabitants proper of the Sargasso Sea, belong to existing coastal forms, no individuals of these species have been found that give evidence of coastal origin."

Other authors have assumed that the algae of the Sargasso Sea originally descended from the species growing on the coasts of the Caribbean Sea, from where they have been torn loose and carried into the open ocean.

FRITSCH comments as follows:

"The view that this vegetation consists in the main of an accumulation of detached plants, has found little support in recent years. Attaching disks have never been found on the plants of the Sargasso Sea, and investigation on the spot has shown that there is continuous growth of the younger parts, as the older die away."

It appears that some of the floating *S.* species cannot be identified with those occurring on the coast, while in others this evolutionary descent is still clearly apparent. In literature *S. natans* and *S. hystrix* var. *fluitans* are mentioned as the most important floating types. Drifting *S.* species also have been found in the Northern Atlantic.

The floating weed masses in the Gulf of Mexico are estimated as approximately 90,000 tons. DAVIS says that "it is indicated that the seaweed in the Gulf is the result of fresh weed floating in from the Atlantic"[41]).

Harvesting is considered to be possible under certain circumstances. The algae have been investigated. Approximately 13 to 23 per cent of algin, less than 1 per cent mannitol and approximately 4 per cent of fucoidan have been found. Laminaran has not been found. DAVIS has reported on the possibilities for commercial exploitation[41]). Other authors, too, have referred to this interesting possibility[42]).

A report has been published on the effect of *Sargassum* distribution on its epibiotic and antibacterial activity[50]). In particular, the occurence of sterile *S.* species *S. natans* (L.) J. Meyen and *S. fluitans* Børgesen in the northern and southern Sargasso Sea eddies have been investigated. The plants outside the eddying zone were less healthy, the number of epibiotic plants, animals, and bacteria was larger than in the eddies of the Sargasso Sea. The *S.* species apparently thrive better in the eddies[50]).

Floating seaweeds, particularly *Sargassum* have also been found and studied in other regions, for example in the Tsuyazaki region and in the Iki-Tusima regions[43]).

Literature

1) DAVIS, F. W., Science 111 (2876): 150. 1950
2) CHIDAMBARAM, K. and UNNY, M. M., First Int. Seaweed Symp., 1952
3) KATAYAMA, T., Bull. Jap. Soc. Sci. Fish 21 (6): 425–428. 1955. CA 50. 18. 13184g. 1956
4) KANEDA, T. and ISHII, S.; ibid. 15 (11): 608–610. 1950
5) GOVINDACHARI, T. R. and RAJAGOPALAN, S., First Int. Seaweed Symp., 1952
6) TSUDA, K. et al., J. Amer. Chem. Soc. 80 (4): 921–925. 1959
7) ZANEVELD, J. S., Ec. Bot. 13 (2): 89. 1959
8) TOKIDA, J. and MASAKI, T., Fac. of Fish. Hokkaido Univ. 10 (3): 173–195. 1959
9) TOKIDA, J., Fac. of Fish. Hokkaido Univ. 2 (1): 1954
10) SULIT, J. I. et al., 4th Meeting, Sect. II, 165–170, 1952. CA 48. 9585. 1954
11) SULIT, J. I. and SAN JUAN, R. C., Philipp J. of Fish. 3 (1): 147–151. 1955. WFA (6): 1956
12) TAYLOR, W. R., Bot. Ser. Field Museum of Nat. Hist. 20 (4): 1941
13) DIAZ-PIFERRER, N. and LÓPEZ, H., Inst. Cubano de Invest. Tecnol. (6): 1959
14) ZHOU, Z. R., Nutrit. Acta 3 (3): 170–172. 1958
15) CHIU, B. T., Second Int. Seaweed Symp., 1955
16) WHISTLER, R. L., and SMARTH, CH. L., Polysaccharide Chemistry. Academic Press Inc., New York, 1953
17) KIMURA, K. et al., Bull. Tohoku Reg. Fish. Res. Lab. 12. 28–45. 1958
18) SMITH, F. and MONTGOMERY, R., The Chemistry of Plant Gums and Mucilages. Reinhold Publishing Corporation, New York, 1959
19) SCAGEL, R. F., Fish. Res. Papers, Wash. Dept. Fish. 1 (4): 49–58. 1956
20) SOLONI-TOURAL, F. G., Centro de Invest. Pesqueras (4): 1954
21) KANEDA, T., Bull. Jap. Soc. Sci. Fish. 17 (8/9): 20–24. 1952. CA 48. 22. 13828d. 1954
22) TSUDA, K. et al., Pharm. Bull. Tokyo 5 (1): 85–86. 1957. CA 2043a. 1958
23) OGINO, CH. and TAKI, Y., J. Tokyo Univ. Fish. 43 (1): 1–5, 1957. CA 52. 21. 18668a. 1958
24) BERGQUIST, P. L., Bot. Mar. I (3/4): 86. 1960
25) JANECKE, H. and KEHR, W., Pharmaz. Ztg. 105 (25): 684. 1960
26) LAWSON, G. W., J. West Afric. Science Ass. 1 (2): 1955
27) GONÇALVES DA COSTA LIMA, E., Algas de nosso litoral Universidade de Recife, 1952
28) VALSAN, A. P., Current Science, Bangalore, India 24 (10): 343–345. 1955
29) LAWSON, G. W., J. Ecol. 44 (1): 1954
30) PILLAI, K. S. and VARIER, N. S., First Int. Seaweed Symp., 1952
31) PILLAI, K. S. and VARIER, N. S., J. Proc. Inst. Chemists, India 24. 205–212. 1952. CA 48. 4. 19613. 1954
32) PILLAI, K. S., Third Int. Seaweed Symp., 1958
33) VARIER, N. S. and PILLAI, K. S., Bull. Central Res. Inst. Travancore 11 (1): 37–62. 1952. CA 47. 1871. 1953
34) PILLAI, K. S.; ibid. Ser. A 3. 73–81. 1954. CA 8526f. 1955
35) YAMADA, Y., Second. Int. Seaweed Symp., 1955

36) VELASQUEZ, G. T., Philippine Seaweeds Report, 1955
37) WILLIAMS, R. H., Univ. of Miami, Marine Laboratory, 1949
38) SCHOTT, G., Geographie des Atlantischen Ozeans, Hamburg, 1912
39) BLACK, W. A. P., The Algae ex Altschul. Processed Plant Proteins Foodstuffs. Academic Press Inc., New York, 1958
40) WINGE, O., Rep. on the Danish Oceanographical Expeditions, 1908–1910, 1923
41) DAVIS, F. W., Univ. of Miami, Gulf and Caribbean Fish. Inst., Coral Gables, Florida, March 1950
42) MAUTNER, H. G., Ec. Bot. 8 (2): 174. 1954
43) SEGAWA, S. et al., Sc. Bull. Fac. Agr. Kyushu Univ. 17 (1): 83–89. 1959; ibid. 17 (3): 291–297. 1959; ibid. 17 (3): 299–305. 1959
44) TOMIYASU, Y., J. Agr. Chem. Soc. Japan 23. 19–22. 1949. CA 44. 3441. 1950
45) HAYASHI, K., J. Soc. Trop. Agr. Taihoku Imp. Univ. 13. 321–323. 1941. CA 42. 5425. 1948
46) HEGNAUER, R., Chemotaxonomie der Pflanzen, Birkhäuser Verlag, Stuttgart, 1962
47) ALMODOVAR. L. R., Bot. Mar. VI (1/2): 143–146. 1964
48) YOSHIDA, T., Bull. Jap. Soc. Sc. Fish. 26 (7): 673–678. 1960
49) HIROSAKI, Y., J. Fac. Sc. Hokkaido Univ. Zool. 14 (3): 320–326. 1960
50) CONOVER, J. T. and SIEBURTH, J. MC. N., Bot. Mar. VI (1/2): 147–157. 1964
51) MARTINEZ NADAL, N. G., Res. Dep. Univ. of Puerto Rico, Inf. Bull. 1958–59
52) ibid. 1961–62
53) ibid. 1962–63
54) MARTINEZ NADAL, N. G., J. Amer. Pharm. Ass. Sci. Ed. 50. 356. 1961
55) MARTINEZ NADAL, N. G. et al., Antimicr. Agents and Chemother. 1963, 68–72. Amer. Soc. f. Microbiology, 1964
56) MARTINEZ NADAL, N. G. et al., Fourth Interscience Conf. on Antimicrobial Agents and Chemotherapy, Oct. 1964
57) MARTINEZ NADAL, N. G., et al., J. of Pharm. Sci. 52 (5). 1963

General Literature

CHAPMAN, V. J., 1950; FOTT, B., 1959; FRITSCH, F. E., 1952; HOFFMANN, C., 1938; HOPPE, H. A., 1958; HOPPE, H. A. and SCHMID, O. J., 1962; LEWIN, R. A., 1962; MAASS, H., 1959; SCHMID, O. J. and HOPPE, H. A., 1962; SEGAWA, S., 1957; UPHOF, J. C. TH., 1959; see Index of Literature.

Sphacelaria

Approximately 25 species are known. In all seas. The thallus from a few millimetres to several centimetres, mostly much branched with large apical cells terminating the branches. The branches arise from below the apical cell.

Sp. bipinnata (Kütz.) Sauv. The alga contains alginic acid[1]), but can have little practical significance as a raw material.

Sp. arctica. The amino acids of this alga have been investigated[2]).

Physiological and biochemical investigations have been reported in LEWIN[3]).

Literature

1) SMITH, F. and MONTGOMERY, R., The Chemistry of Plant Gums and Mucilages. Reinhold Publishing Corporation, New York, 1959
2) MAUTNER, H. G., Ec. Bot. 8 (2): 174. 1954
3) LEWIN, R. A., Physiology and Biochemistry of Algae. Academic Press Inc., New York, 1962

General Literature

FOTT, B., 1959; FRITSCH, F. E., 1952; see Index of Literature.

Splachnidium

Sp. rugosum. Southern Hemisphere, particularly Cape province, Indonesia, Australia. The alga has a gelatinous, hollow, branched thallus, affixed to the substratum by a basal disc. It is common in the Singapore region and is used for food.

General Literature: FRITSCH, F. E., 1952; see Index of Literature.

Taonia

T. atomaria (= *Dictyota atomaria* Grev.). The wedge-shaped segments have irregularly toothed margins. The fronds reach a length of 30 centimetres. They arise singly or in tufts from the rhizoids. The mature thallus is six-layered. The alga contains alginic acid[1]).

Literature

1) SMITH, F. and MONTGOMERY, R., The Chemistry of Plant Gums and Mucilages. Reinhold Publishing Corp., New York, 1959

Turbinaria

T. species are recorded from the coasts of South Africa, India, Ceylon, Malaya, Indonesia, South China, and the Philippines. They also have been found on the South American Pacific Coast (Chile) (Pl. VI: 11).

The algae are of a tough consistency. The lateral blades are membranaceous, and sometimes there are air-bladders on the membranaceous lobes.

T. conoides J. Ag. India (East Coast of Madras, Malabar), Ceylon (Gulf of Mannar), Malaya, Indonesia, New Guinea, southern China, Australia, the Philippines.

The alga contains alginic acid[1]), according to personal information, approximately 23 per cent. The sterols have been investigated[2]).

T. dentata. India (Madras). The alga contains approximately 20 to 22 per cent of alginic acid according to personal information.

T. ornata. Ceylon, Indonesia, Malaya, southern China[3]), Japan, the Philippines. The alga is eaten locally in many areas.

The distribution of the alga has been studied by YAMADA[4]).

T. turbinata. Cuba[5]) [6]) and Columbia[7]).

The *T.* species can be considered to be a source of alginic acid and alginates. Material from the Philippines (Lingayen Gulf, Hundred Islands, Pangasinan province) has been investigated[8–10]). It has been found to contain approximately 30 per cent of alginic acid.

T. species are used in India and the Philippines as fertilizers particularly for coco-nut plantations.

Literature

1) Valsan, A. P., Current Science 24 (10): 343–345, 1955
2) Govindachari, T. R. and Rajagopalan, S., First Int. Seaweed Symp., 1952
3) Chiu, B. T., Second Int. Seaweed Symp., 1955
4) Yamada, Y., Bull. Jap. Soc. Phyc. 5 (1): 24. 1957
5) Soloni-Toural, F. G., Centro de Inves. Pesqueras (4). 1954
6) Diaz-Piferrer, M. and López, H., Inst. Cubano de Invest. Tecnol. 66. 1959
7) Taylor, W. R., Bot. Ser. Field Mus. of Nat. History 20 (4). 1941
8) Sulit, J. I. et al., Indo-Pacific Fish. Council Proc. 4th Meeting, Sect. II. 165–170. 1952. CA 48. 9585. 1954
9) Velasquez, G. T., Philippine Seaweeds Report, 1955
10) Sulit, J. I. and San Juan, R. C., Philip. J. of Fish. 3 (1): 147–151. 1955

General Literature

Chapman, V. J., 1950; Fritsch, F. E., 1952; Hoppe, H. A. and Schmid, O. J., 1962; Maass, H., 1959; Segawa, S., 1957; Uphof, J. C. Th., 1959; Zaneveld, J. S., 1959; see Index of Literature.

Undaria

U. pinnatifida (Harv.) Suringar. Japan, West Coast of Hokkaido, particularly near Otaru. Coasts of China and Korea. The alga grows on rocks, in depths of about 20 to 40 feet. It has a flat stipe and a midrib with undulating wing-like pinnate blades. It is between 50 to 100 centimetres long.

U. pinnatifida is used in Japan for making a product known as "Wakame". The alga is dragged ashore by long hooks. Harvesting is between February and June; the material is often washed in fresh water, the midrib cut out, and the thalli put up on lines for drying. The semi-dried product is kneaded and then dried completely. "Wakame" is sold in bundles, or cut up as "Kizami-wakame", or roasted and candied with sugar; it is often also sold in tins as a finished product known as "Ito-wakame".

The quantity of harvested raw material is considerably high. Before the war only in Japan approximately 45,000 tons of raw material yielded 9,000 tons of finished product. Also today "Wakame" products play a major rôle on the East Asian markets.

Today *U. pinnatifida* is often cultivated[1]. Ecology, reproduction, culture, and the constituents of the alga have been the subject of several studies[2–7] [17].

The alginic acid of *U. pinnatifida* has been investigated (Maass). According to the Standard Table of Food Composition, Japan, the alga contains about 15 per cent of raw protein, approximately 57 per cent of sugar, provitamin A, vitamin B_1, B_2, niacin, and vitamin C[8]. Xylose and methylpentose[9], and 0.03 per cent of ethereal oils have been found. The extract has an antibacterial activity[10]. Physiological and biochemical investigations have been reported in Lewin (1962).

U. distans Miyabe et Okamura. Coasts of the Sea of Japan at Hokkaido, on the Eastern Coast of Muroran, and on the Coast of Honshû. This species is a little larger than *U. pinnatifida*. Its Japanese name is "Nambu-wakame". It is also described in the literature as *U. pinnatifida* (Harvey) Suringar f. *distans* Miyabe et Okamura[11].

The alga can also be used as a raw material for "Wakame". Miyabe reported details of this species and its comparison with *U. pinnatifida*[12].

U. peterseniana (Kjellm.) Okamura. Japanese coasts.

U. undarioides (Yendo) Okamura. This alga has been described in detail [13] [14], and morphogenetical studies were carried out with it[16]). Investigations on correlations between water temperature and geographical distribution of some economic seaweeds, for example *U. pinnatifida*, have been made by Japanese researchers[15]).

Literature

1) YAMADA, Y., Coll. Int. du Centre Nat. de la Recherche Scient., Dinard, 1957
2) SAITO, Y., Bull. Jap. Soc. Sci. Fish. 22 (4): 229–234. 1956
3) SAITO, Y.; ibid. 22 (4): 235–239. 1956
4) UJÜE, Y., Bull. Jap. Soc. Phyc. 5 (3): 80–83. 1957
5) ARASAKI, S., Aquiculture 5: 60–65. 1958; ibid. 6: 27–33. 1958
6) KUROGI, M. and AKIYAMA, K., Bull. Tôhoku Reg. Fish. Res. Lab. 10: 95–117. 1957
7) SAITO, Y., Aquiculture 5 (3): 16–20. 1958. Bull. Jap. Soc. Sci. Fish. 24 (6/7): 484–486. 1958. Bull. Jap. Soc. Phyc. 7 (2): 58–63. 1959
8) BLACK, W. A. P., The Algae ex Altschul. Processed Plant Proteins Foodstuffs. Academic Press Inc., New York, 1958
9) YOUNG, E. G., Carbohydrate Accumulation by lower Plants (Rhodophyta) ex W. Ruhland. Handbuch der Pflanzenphysiologie, Springer-Verlag, 1958
10) ROOS, H., Kieler Meeresforsch. XIII (1): 1957
11) TOKIDA, J. et MASAKI, T., Fac. of Fish. Hokkaido Univ. 10 (3): 173–195. 1959
12) MIYABE, K., J. of the Sapporo Agric. College I. 1–50. 1957
13) SEGI, T. and KIDA, W., Rep. Fac. of Fish. Univ. of Mie 2 (3): 517–526. 1957
14) SEGI, T. and KIDA, W.; ibid. 3 (1): 236–246. 1958
15) ENDO, T. and MATSUDAIRA, Y., Bull. Jap. Soc. Sc. Fish. 26 (9): 871–876. 1960
16) NISHIBAYASHI, T. and INOH, S., Biol. J. Okayama Univ. 6 (1/2): 27–36. 1960
17) INOH, S. and NISHIBAYASHI, T., Kromosomo 44/45. 1498–1499. 1960

General Literature

CHAPMAN, V. J., 1950; FRITSCH, F. E., 1952; HEGNAUER, R., 1962; HOFFMANN, C., 1938; LEWIN, R. A., 1962; MAASS, H., 1959; SEGAWA, S., 1957; SCHMID, O. J. and HOPPE, H. A., 1959; UPHOF, J. C. TH., 1959; see Index of Literature.

Zonaria

This alga resembles to *Taonia* in some respects. The branched, usually erect fronds show transverse zones of hairs.

Z. latissima. India (East Coast of Madras). The content of alginic acid is estimated to be approximately 16 per cent. However, the alga, is relatively small, and it can hardly be expected to play a major rôle as a source of industrial raw material.

General Literature: FRITSCH, F. E., 1952; see Index of Literature.

Rhodophyceae

Acanthopeltis

A species occurring along the Japanese and Korean coasts.

A. japonica Okamura Schottmüllera. Thallus erect, 5 to 12 centimetres high, with lateral ramifications, petiolate below. This alga consists of circular, thick, dentate, undulating small discs, closely packed together.

A. japonica, in Japanese called "Dorakusa" or "Toriashi", is used mixed with other agarophytes for agar production[1–5]. STOLOFF (1962) refers to the polysacolloid as agar[11]. The yield of agar, a solid gelatinous substance, is approximately 5%[6] [7]. Isolation and purification have been described[10]. This alga is also used for food[8]. Cholesterol has been identified as a constituent[9]. For physiological and biochemical investigations see LEWIN (1962).

Literature

1) Hypothekenbank von Japan. Die Agarherstellung in Japan, Tokyo, Nr. 17, 1950
2) ARAKI, CH., Seaweed Polysaccharides ex Wolfrom, M. L., Carbohydrate Chemistry of Substances of Biological Interest. Pergamon Press Ltd., London, 1959
3) STOLOFF, L. and SILVA, P., Ec. Bot. 11 (4): 327–330, 157
4) SELBY, H. H. and SELBY, T. A., Agar ex Whistler, R. L. and BeMiller, J. N., Industrial Gums. Academic Press Inc., New York, 1959
5) SCHULZE, W., D. Apoth. Ztg. 98 (15): 340, 1958
6) HUMM, H. J., The Red Algae of Economic Importance ex Tressler, D. and Lemon, J. Mc. W., Marine Products of Commerce. Reinhold Publishing Corporation, New York, 1951
7) SMITH, F. and MONTGOMERY, R., The Chemistry of Plant Gums and Mucilages. Reinhold Publishing Corporation, New York, 1959
8) CHAPMAN, V. J., Seaweeds and their Uses. Methuen & Co. Ltd., London, 1950
9) TSUDA, K. et al., Chem. & Pharm. Bull 6 (1): 101–104, 1958
10) MURAKAMI, S., Sc. Rep. Saitama Univ. 3 B (3): 251–254, 1960
11) STOLOFF, L., Ec. Bot. 16 (2): 86–94. 1962

General Literature

HOFFMANN, C., 1938; HOPPE, H. A., 1958; HOPPE, H. A. and SCHMID, O. J., 1962; KYLIN, H., 1956; LEWIN, R. A., 1962; SEGAWA, S., 1957; UPHOFF, J. C. TH., 1959; see Index of Literature.

Acanthophora

Some 7 species, very variable and difficult to separate. In warmer seas.

A. spicifera (Vahl.) Börgesen. Coasts of the Indian Ocean, particularly around the Seychelles, Pakistan, Malaya, Indonesia, Philippines, and Marianas, also in West Indian waters. The type species. Thallus 2 to 15 centimetres high, terete, erect, ramified, with shoots bearing short cartilaginous spines arranged in spirals. This alga is eaten in Java, on the Philippines, and the Kangean Islands[1–4]. Material of *A. spicifera* collected near Puerto Rico showed antibiotic activity[5].

In addition to *A. spicifera*, *A. deliles*, and *A. dendroides* have been found on the Pakistani coast[6]. *A. orientalis* is known to occur along the Japanese coasts.

Literature

1) Uphof, J. C. Th., Dictionary of Economic Plants. H. R. Engelmann (J. Cramer), Weinheim, 1959
2) Zaneveld, J. S., Ec. Bot. 13 (2): 89, 1959
3) Chapman, V. J., Seaweeds and their Uses. Methuen & Co. Ltd., London, 1950
4) Eisses, J., Indonesian J. of Nat. Sc. (1, 2, 3): 1953
5) Burkholder, P. R. et al., Bot. Mar. II (1/2): 149, 1960
6) Amand, P. L., Marine Algae from Karachi. The University of the Panjab, Lahore, 1943

General Literature

Fritsch, F. E., 1952 & 156; Kylin, H., 1956; Segawa, S., 1957; see Index of Literature.

Aeodes

Some 6 species, in warmer seas.

A. nitidissima J. Agardh. Coasts of New Zealand, a characteristic species. Thallus erect, 25 to 75 centimetres high, flat, entire or irregularly lobed, or thin and thread-like.

A. orbitosa. Coasts of South Africa, very frequent. This alga contains carrageenan[1–3]). Stoloff (1962) refers to the polysacolloid as carrageenan[5]). In South Africa a product extracted from this alga is used for the clarification of beer.

A. lanceolata. On Japanese coasts. Botanical investigations have been reported[4]).

Literature

1) Stoloff, L. and Silva, P., Ec. Bot. 11 (4): 327–330, 1957
2) Smith, F. and Montgomery, R., The Chemistry of Plant Gums and Mucilages. Reinhold Publishing Corporation, New York, 1959
3) Stoloff, L., Carrageenan ex Whistler, R. L. and BeMiller, J. N., Industrial Gums. Academic Press Inc., New York, 1959
4) Tazawa, N., Bull. Jap. Soc. Phyc. 8 (3): 112–117. 1960
5) Stoloff, L., Ec. Bot. 16 (2): 86–94. 1962

General Literature

Fritsch, F. E., 1952 & 1956; Kylin, H., 1956; see Index of Literature.

Agardhiella

3 species, mostly coasts of North America, in warmer regions.

A. tenera (J. Ag.) Schmitz. East coast of North America, West Indies, especially Cuba, Gulf of Mexico, Pacific coast of America, from Vancouver Island to Chile[1]). The type species. Thallus erect, 15 to 30 centimetres high, terete, ramified, often flattened. The alga contains an agaroid[2–4]) resp. carrageenan[5]). Stoloff (1962) refers to the polysacolloid as gelan[9]). It is used for commercial purposes on the east coast of North America[6]).

Certain species, such as *A. robusta,* occur also in Pakistan[7]), and in Manila Bay, where they are processed for "gulaman", a gelatinous product in widespread use[8]).

Literature

1) Levring, T., Contributions to the marine Algal Flora of Chile. Lunds Univ. Årsskrift N. F. Avd. 2. Bd. 56, Nr. 10, 1960
2) Humm, J., The Red Algae of Economic Importance ex Tressler, B. and Lemon, J. Mc. W.,

Marine Products of Commerce. Reinhold Publishing Corporation, New York, 1951
3) Stoloff, L. and Silva, P., Ec. Bot. 11 (4): 327–330, 1957
4) Selby, H. H. and Selby, T. A., Agar ex Whistler, R. L. and BeMiller, J. N., Industrial Gums. Academic Press Inc., New York, 1959
5) Smith, F. and Montgomery, R., The Chemistry of Plant Gums and Mucilages. Reinhold Publishing Corporation, New York, 1959
6) Humm, J., Science 100 (2593): 209–212, 1944
7) Anand, P. L., Marine Algae from Karachi. The University of the Panjab, Lahore, 1943
8) Zaneveld, J. S., Ec. Bot. 13 (2): 89, 1959

General Literature

Diaz-Piferrer, M. and López, H., Inst. Cubano de Investi Tecnol. (6), 1959; Fritsch, F. E. 1952 & 1956; Kylin, H., 1956; Taylor, W. R., Bot Ser. Field Museum of Natural History 20 (4): 1941; see Index of Literature.

Ahnfeltia

Some 5 species, in both cold and warm regions.

A. plicata (Huds.) Fries. Widespread. In the Northern Atlantic, coasts of Norway[1]), Finland[2]), and Eire[28]), rarely, however, Baltic coast of Schleswig-Holstein and Sweden[3]). Largest quantities are found on the borders of the White Sea, on the coasts of Sakhalin, of Korea, the Japanese Islands, and Formosa, along the Atlantic coast of Canada, U.S.A., and Mexico, on the Californian coast, South Chile, the Falkland Islands, Graham Land, Crozet, Kerguelen[4]), and on the Pakistani coast[5]).

The type species. Thallus erect, 5 to 15 centimetres high, terete, bifurcated or irregularly branched, of dense, smallcelled structure and of granular consistency.

A. plicata is the most important raw material for Russian agar, which was sold in Europe before the Second World War under the name of "Sakhalin agar"[5–10]). In Japan this alga was of minor economic significance, being mainly utilized in a mixture with other red algae[11]) [12]). "Itani agar" is made from *A. plicata* today in Japan[13]). Views differ whether *A. plicata* should be considered to be an agarophyte or an agaroidophyte. The Russian material yields an agar-like substance, while the North American, like the Irish variant is said to yield an agaroid[15]). Stoloff (1962) refers to the polysacolloid as agar[30]). This alga and the production of *Ahnfeltia*-phycocolloid has been investigated by several authors[16–20]). In a series of experiments an ash content of 12,3% was found, which is very low, compared with *Enteromorpha intestinalis* which contains 45,4% of ash[21]). The phycocolloid contains agarose[22]). For physiological and biochemical investigations see Lewin (1962). New methods for the production of agar from *A. plicata* have been investigated[25]) [29]).

A. species thriving along the European coasts have been investigated[23]).

A. gigartinoides. Southern California and Mexico, can be regarded as an agaroid raw material[15]).

A. durvillaei. West coast of South America, especially of Chile[24]).

A. setacea. Is also to be found along the Chilean coast (Costa del Corral)[24]).

A. concinna. This alga is eaten by the native population of Hawaii[3]) [26]). In Japan this species also yields a funoran, but of an inferior quality[27]).

A. paradoxa. Coasts of Japan.

Literature

1) Printz, H., Seaweeds, their properties and general uses, especially in Norway, 1952
2) Mathiesen, Memoranda Soc. Fauna et Flora Fenn. 34: 78–84, 1959
3) Hoffmann, C., Kieler Meeresforsch. 9 (1): 5–14, 1952
4) Levring, T., Lunds Univ. Årsskrift N. F. Avd. 2, 56 (10): 1960
5) Anand, P. L., Marine Algae from Karachi. The University of the Panjab, Lahore, 1943
6) Humm, H. J. and Wolf, A., Duke University Press, (3): 1946
7) Stoloff, L. and Silva, P., Ec. Bot. 11 (4): 327–330, 1957
8) Fott, B., Algenkunde. Gustav Fischer Verlag, Jena, 1959
9) Morel, Th., Medeling van het Centraal Inst. voor Technische Onderzoekingen te Djakarta, Oct. 1949
10) Smith, F. and Montgomery, R., The Chemistry of Plant Gums and Mucilages. Reinhold Publishing Corporation, New York, 1959
11) Selby, H. H. and Selby, T. A., Agar ex Whistler, R. L. and BeMiller, J. N., Industrial Gums. Academic Press Inc., New York, 1959
12) Tokida, J. and Masaki, T., Fac. Fish. Hokkaido University 10 (3): 173–195, 1959
13) Kojima, Y et al., J. of the Shimonoseki College of Fisheries
    9 (1): 43–52, 1960
    9 (3): 317–322, 1960
    9 (3): 323–328, 1960
14) Stoloff, L. and Lee, Ch. F., Fishery Leaflet 173, 1946, US. Dep. of the Interior, Fish and Wildlife Service
15) Selby, H. H. and Selby, T. A., Agar ex Whistler, R. L. and BeMiller, J. N., Industrial Gums. Academic Press Inc., New York, 1959
16) Gregory, B. D., Proc. Linnean. Soc. (London) p. 63, 1934
17) Gryuner, V. and Gorshkov, J., J. Appl. Chem. (USSR), 10, 2054–2063, 1937
18) Kizevetter, J. V., Bull. Pacific Sci. Inst. Fish. & Oceanogr. 13, 1–133, 1937
19) Pleskatsevich, U. P., Trudy Archangel Vosoroslevogo Nauch Issledovatel Belogo Mor. 234–254, 1938
20) Kizevetter, J. V., J. Appl. Chem. (USSR), 14, 250–255, 1941
21) MacPherson, M. G. and Young, E. G., Canad. J. of Bot. 30. 67–77. 1952
22) Yaphe, W., Canad. J. of Bot. 37, 751–757, 1959
23) Ross, A. G., J. Sci. Food Agric. (7): 333–335, 1953
24) Levring, T., Lunds Univ. Årsskrift N. F. Avd. 2, 56 (10), 1960
25) Tagawa, S. et al., J. Shimonoseki Coll. Fish. 10 (1): 35–41, 1960
26) Uphof, J. C. Th., Dictionary of Economic Plants. H. R. Engelmann (J. Cramer), Weinheim, 1959
27) Chapman, V. J., Seaweeds and their Uses. Methuen & Co. Ltd., London, 1950
28) Brennan, A. T., Notes on some common Irish Seaweeds. Inst. f. Ind. Res. and Standards, Dublin, 1950
29) Kojima, Y. et al., 10 (1): 43–56, 1960 cf. Tagawa, Kojima et Kono
30) Stoloff, L., Ec. Bot. 16 (2): 86–94. 1962

General Literature

Dangeard, P., Coll. Int. du Centre Nat. de la Recherche Scient., Dinard, 1960; Fritsch, F. E., 1952; Hoppe, H. A., 1958; Hoppe, H. A. and Schmid, O. J., 1962; Humm, H. J., 1951; Kylin, H., 1956; Lewin, R. A., 1962; Levring, T., 1940; Pax, F., 1962; Segawa, S., 1957; Tokida, J., 1954; Tseng, C. K., 1, Science 101 (2633): 597–602, 1945; see Index of Literature.

Alsidium

3 species, Mediterranean Sea and Atlantic Ocean.

A. corallinum Agardh. Mediterranean. The type species. Thallus erect, 5 to 15 centimetres high, fleshy, cartilaginous, ramified, terete.

A. helminthochorton. Mediterranean, especially near Corsica and Sicily. This species yields "helminthochorton", "Corsican Moss", a vermifugant. This material, as sold, often contains a mixture of various species of algae and is uneaven in effect. The alga contains a floridoside[1]), an alkaloid-like substance, the active principle of which is not known, though it is presumably amino acid in nature[2]).

Literature

1) Augier, J., Rev. gen. bot. 60, 257–283, 1953
2) Hegnauer, R., Chemotaxonomie der Pflanzen, Birkhäuser Verlag, Basel, Stuttgart, 1962

General Literature
Chapman, V. J., 1950; Fott, B., 1959; Hoffmann, C., 1938; Hoppe, H. A., 1958; Kylin, H., 1956; Pax, F., 1962; Uphof, J. C. Th., 1959; see Index of Literature.

Amansia

Several species, warmer seas.

A. multifida Lamour. In warmer parts of the Atlantic coast of America. The type species. Thallus erect, 5 to 25 centimetres high, flat, strap-shaped, branched, membranaceous. Material from Puerto Rico has been found to possess antibiotic activity[1]).

A. glomerata, *A. japonica*, and *A. misui* occur in Japan.

Literature

1) Burkholder, P. R. et al., Bot. Mar. II (1/2): 149, 1960

General Literature
Fritsch, F. E., 1952 & 1956; Kylin, H., 1956; Segawa, S., 1957; see Index of Literature.

Amphiroa

Numerous species, in temperate and warm seas.

A. fragilissima (L.) Lamour. West Indies. The type species. Thallus erect, 2 to 10 centimetres high, branched.

A. capensis, South Africa. This species contains a peptide[1]).

A. cretacea, Japan, Sakhalin, and Pacific coasts of North America[2]).

Numerous other species of *A.* occur on Japanese coasts.

For physiological and biochemical investigations see Lewin (1962).

General Literature
Fritsch, F. E., 1952 & 1956; Kylin, H., 1956; Lewin, R. A., 1962; Segawa, S., 1957; Tokida, J., 1954; see Index of Literature.

Asparagopsis

4 species, in warm waters, mainly in the southern hemisphere.

A. taxiformis (Delile) Collins and Harvey. The type species. Thallus erect, 5 to 10 centimetres high, terete, very branched.

A. sandfordiana. Coasts of the Pacific Islands and of Japan. This species is called "limu kohu", "koko" or "red limu" on Hawaii, and is eaten.

A. hamifera, coasts of Japan.

General Literature
Chapman, V. J., 1950; Fritsch, F. E., 1952; Kylin, H., 1956; Segawa, S., 1957; Uphof, J. C. Th., 1959; see Index of Literature.

Ballia

Some 6 species, temperate and colder seas.

B. callitricha (Ag.) Montagne. On the coasts of the Antarctic Ocean, where it is found in abundance. The type species. Thallus erect, 2 to 20 centimetres high, fine and thread-like, much branched.

B. arctica. Occurring on the coast of the Baltic. The amino acids of this species have been investigated[1]).

Literature

1) Ericson, L. E., Acta Chem. Scand. 6: 805–807, 1952

General Literature
Fritsch, F. E., 1952; Kylin, H., 1956; see Index of Literature.

Bangia

Some 8 species, difficult to distinguish, distributed widely.

B. fusco-purpurea (Dillw.) Lyngb. Widespread, particularly in the North Atlantic, southward to the Mediterranean. This alga occurs at the upper tidal limit, not infrequently forming a coating on stones and wood. The type species. Thallus erect, filamentous, unbranched, sometimes irregularly indented, dark-red or brown-violet in colour. DL-galactose has been isolated from this alga[1]).

For physiological and biochemical investigations of *B. fusco-purpurea* and *B. pumila*, see Lewin (1962).

Literature

1) Black, W. A. P. and Cornhill, W. J., Chem. and Ind. 1954, 514–516

General Literature
Fritsch, F. E., 1952; Hegnauer, R., 1962; Kuckuck, P., 1953; Kylin, H. 1956; Levring, T., 1940; Lewin, R. A., 1962; Segawa, S., 1957; see Index of Literature.

Batrachospermum

Approximately 50 species, of world wide distribution, mostly in fresh water.

B. moniliforme Roth. This alga is found in the Gulf of Bothnia. The type species. Thallus erect, approximately 10 to 20 centimetres high, much branched, of gelatinous consistency, violet or bluegreen coloured. The polysaccharide from this alga has been investigated[1]) [2]). For physiological and biochemical investigations see Lewin (1962).

Literature

1) Janecke, H. and Kehr, W., Pharmaz. Ztg. 105 (25): 684, 1960
2) Augier, J., Rev. gen. bot. 60: 257–283, 1953. CA 49, 438b

General Literature
Fritsch, F. E., 1952; Kylin, H., 1956; Lewin, R. A., 1962; see Index of Literature.

Bostrychia

Approximately 25 species, temperate and warmer seas, often also in rivers and brackish water, in the tropics, on the roots of mangroves.

B. scorpioides (Gmel.) Mont. Atlantic coasts of France and Great Britain, in salt-marches. Characteristic species. Thallus terete or flat, mostly creeping, moss-like. Dulcitol[1]) [2]), sorbitol[3]) [5]), and fluorescent substances[3]) have been isolated. Physiological and biochemical investigations have been reported in Lewin (1962).

B. radicans. Ocurring on the coasts of Burma, Indonesia (Java), Pacific Islands, Columbia, Ecuador, West Africa. This alga is eaten in Burma[4]).

Detailed investigations have been reported of *Bostrychia* and the *Bostrychietum* and its characteristics[6]).

Literature

1) Schachat, R. E. and Glicksman, M., Some Lesser Known Seaweed Extracts ex Industrial Gums. Academic Press Inc., New York, 1959
2) Young, E. G., Carbohydrate accumulation by lower plants (Rhodophyta) ex Rhuhland, Handbuch d. Pflanzenphysiologie, Springer Verlag, 1958
3) h'Eocha, C. O. et al., Third Int. Seaweed Symp., 1958
4) Zaneveld, J. S., Ec. Bot. 13 (2): 89, 1959
5) Black, W. A. P., Ann. Rep. Chem. Soc. 50, 1953
6) Post, E., Bot. Mar. V (1): 9, 1963

General Literature
Fritsch, F. E., 1952; Kylin, H., 1956; Lewin, R. A., 1962; Segawa, S., 1957; see Index of Literature.

Callithamnion

Approximately 50 species, in various seas.

C. corymbosum (Smith) Lyngb. Northern regions of the Atlantic, southward to the Mediterranean Sea. The type species. Thallus 1 to 10 centimetres high, branching lateral or dichotomous, tufted, rose-pink, often growing on top of other algae or on stones.

Phycoerythrin has been isolated from a number of *C.* species[1]). Physiological and biochemical investigations of *C. tetragonum* and *C. tetricum* have been reported in Lewin (1962).

Literature

1) Coulson, C. B., Chem. and Ind. 1953, pp 997–998

General Literature
Fritsch, F. E., 1952; Kuckuck, P., 1953; Kylin, H., 1956; Levring, T., 1940; Lewin, R. A., 1962; Segawa, S., 1957; see Index of Literature.

Callophyllis

Approximately 50 species, mostly in temperate and warmer seas. Species of *C.* are very diverse in habit.

C. variegata (Bory) Kütz. The type species. Thallus erect, 5 to 45 centimetres high, flat or leaf-shaped, dichotomously or sometimes pinnately branched.

C. megalocarpa. Investigations showed an antibiotic activity[1]).

Some species of *C.* occur along the Japanese coasts.

Physiological and biochemical investigations of *C. laciniata* and *C. marginifructa* have been reported in LEWIN (1962).

Literature

1) PRATT, R. et al., J. Am. Pharm. Ass. 40: 575–579, 1951

General Literature

FRITSCH, F. E., 1952; KYLIN, H., 1956; LEWIN, R. A., 1962; SEGAWA, S., 1957; see Index of Literature.

Caloglossa

Some 6 species, in salt, fresh and brackish waters of warmer regions, particularly in the tropics.

C. Leprieurii (Mont.) J. Ag. In all warmer seas, particularly along the coasts of Ceylon, Burma, Indonesia, the Pacific Islands, Columbia, Ecuador, and the Atlantic Ocean. The type species. Thallus erect, 3 to 5 centimetres high, leaf-like or strong-shaped, branching dichotomous. This alga is eaten in Burma.

C. adnata. Coasts of India, Burma, and Indonesia (Borneo). This species is also used for food.

C. ogasawaraensis. Coasts of Japan.

Literature

FRITSCH, F. E., 1952; KYLIN, H., 1956; SEGAWA, S., 1957; ZANEVELD, J. S., Ec. Bot. 13 (2): 89, 1959; see Index of Literature.

Campylaephora

Few species are known.

C. hypnaeoides (J. Ag.) Okamura (= *Ceramium hypnaeoides)*. Coasts of Japan, China, Korea, Sakhalin, Siberia, and the Kurile Islands. Thallus erect, 10 to 15 centimetres high, terete, much branched, the uppermost curved and pincer-like. In Japan this alga is used as agar or agaroid raw material[1]) [2]). It yields a soft agar and contains 3,6-anhydro-L-galactose[3]). Physiological and biochemical investigations are reported in LEWIN (1962). The Japanese call it "Ego" or "Yego nori", and it is also eaten in Japan[4]).

C. crassa. Occurs in various forms along the coasts of Japan and Sakhalin[5]).

Literature

1) STOLOFF, L. and LEE, CH. F., Fishery Leaflet 173, 1946, US Dep. of the Interior, Fish and Wildlife Service
2) ARAKI, CH., Seaweed Polysaccharides ex Wolfrom, M. L., Carbohydrate Chemistry of Substances of Biological Interest. Pergamon Press. Ltd., London, 1959
3) SMITH, F. and MONTGOMERY, R., The Chemistry of Plant Gums and Mucilages. Reinhold Publishing Corporation, New York, 1959
4) HOFFMANN, C., Kieler Meeresforsch. III (1), 1938
5) TOKIDA, J., Fac. of Fish. Hokkaido University 2 (1), 1954

General Literature

CHAPMAN, V. J., 1950; KYLIN, H., 1956; LEWIN, R. A., 1962; SEGAWA, S., 1957; see Index of Literature.

Catenella

5 species, warmer and temperate sea coasts.

C. opuntia (Good. et Woodw.) Grev. Around the high-water mark, very often in salt-marches or mangrove swamps. The type species. Thallus 3 to 5 centimetres high, creeping or erect, terete or flattened, branched, articulated.

C. impudica. Coasts of Burma, Indonesia (Java), Pacific Islands, Northern Australia, and on the coasts of Central and South America. This species is collected in large quantities on the coast of Tenasserim. Mixed with other algae, such as species of *Bostrychia*, it is sold for food on Burmese markets.

C. nipae. Burma, Indonesia, Eastern Australia. This species is also eaten. It is sold, for example, on the markets of Rangoon.

Literature

CHAPMAN, V. J., 1950; FRITSCH, F. E., 1952; KYLIN, H., 1956; SEGAWA, S., 1957; UPHOF, J. C. TH., 1959; ZANEVELD, J. S., Ec. Bot. 13 (2): 89, 1959; see Index of Literature.

Ceramium

Approximately 60 species, widespread.

C. rubrum (Huds.) Agardh. "Gemeiner Horntang". Occurring in the North Atlantic including Great Britain[23]), coast of Japan (Gulf of Oshoro)[1]) (Pl. XIII: 32). The type species. Thallus erect, 5–15 centimetres high, terete, branching mostly dichotomous, branches curved like pincers, of brown or brown-red colour. The alga has been investigated chemically[2]) [3]), with regard to amino acids[8]), antimicrobial substances[9]) [10]), sugar residues, of which galactose, glucose, and xylose have been identified[5]) [7]), and pigments[11]) [12]).

C. rubrum has no major economic importance, but in Japan it is used mixed with other algae for the production of agaroid[4]) [5]) [6]), and the pigment can be used as a dye-stuff for textiles and cosmetics[28]).

C. boydenii. Japanese coasts. Less important for agar or agaroid production[20]), but it is used for commercial purposes in mixture with other red algae[6]) [4]). The japanese term is "Igusu"[13]). The alga yields a soft gelatinous substance[5]). There is a report on the chemical properties of this alga[24]).

C. byssoideum. Material of this species from the coast of Puerto Rico showed antibiotic activity[14]).

C. codii. Coasts of Great Britain[15]).

C. crassum. Japanese coasts. Not important as an agaroid raw material, but it is used commercially in mixture with other red algae[4]). The alga yields a soft gelatinous substance[5]).

C. diaphanum. Material of this species collected in the Kieler Förde contained an antimicrobial substance[9]).

C. hypnaeoides cf. *Campylaephora hypnaeoides.* Used commercially, together with other red algae, for the production of agar or agaroid, "ego" in Japanese[16]) [17]). There are reports on the isolation and purification of agar.

C. manorensis and *C. miniatum.* Coast of Pakistan[18]).

C. shuttleworthianum. This alga has been investigated[27]).

C. tenuicorme. This species contains niacin[8]).

In addition to *C. rubrum* the following species are represented along the Roumanian coast of the Black Sea:

C. circinatum, C. corticulatum, C. diaphanum, C. elegans, and *C. fastigiatum*[19]).

Several authors investigated species occurring on the Pacific Coast of North America[20]) and the coasts of Texas, Yukatan, and Columbia[21]).

Various *C.* species are eaten in East Asia[22]).

Physiological and biochemical investigations of *C. boydenii, C. ciliatum, C. deslongchampsii, C. diaphanum, C. rubrum,* and *C. tenuissimum* are reported in LEWIN (1962).

Investigations of the genus *Ceramium* from British coasts have been published[26]).

Literature

1) TOKIDA, J. and MASAKI, T., Fac. of Fish. Hokkaido University 10 (3): 173–195, 1959
2) ROSS, A. G., J. Sci. Food Agric. 7: 333–335, 1953
3) YOUNG, E. G. and LANGILLE, W. M., Publ. Nr. 4636 Nat. Res. Council Halifax, 1958
4) SELBY, H. H. and SELBY, T. A., Agar ex Whistler, R. L. and BeMiller, J. N., Industrial Gums. Academic Press Inc., New York, 1959
5) SMITH, F. and MONTGOMERY, R., The Chemistry of Plant Gums and Mucilages. Reinhold Publishing Corporation, New York, 1959
6) ARAKI, CH., Seaweed Polysaccharides ex Wolfrom, M. L., Pergamon Press Ltd., London, 1959
7) BLACK, W. A. P. and Cornhill, W. J., Chem. and Ind. 1954, 514–516
8) ERICSON, L. E. and CARLSON, B., Kungl. Svenska Vetenskapsakademien 6 (49), 1953
9) ROOS, H., Kieler Meeresforsch. XIII (1), 1957
10) CHESTERS, C. G. C. and STOTT, J. A., Second Int. Seaweed Symp., 1955
11) COULSON, C. B., Chem. and Ind. 1953, 997–998
12) H'EOCHA, C. O., ex Comparative Biochemistry of Photoreactive Systems. Academic Press Inc., New York, 1960
13) Bericht der Hypothekenbank von Japan, Tokyo, 1950, Nr. 17
14) BURKHOLDER, P. R. et al., Bot. Mar. II (1/2), 149, 1960
15) DIXON, P. S., Ann. & Mag. Nat. Hist., ser. 13, 1 (1): 14–16, 1958
16) STOLOFF, L. and SILVA, P., Ec. Bot. 11 (4): 327–330, 1957
17) SCHULZE, W., D. Apoth. Ztg. 98 (15): 340, 1958
18) ANAND, P. L., Marine Algae from Karachi. The University of the Panjab, Lahore, 1943
19) CELAN, M. and SERBANESCU, G., Stat. Zool. Mar., Agigea, 1959, 531–562
20) DAWSON, E. Y., Farlowia, 4, 1950
21) TAYLOR, W. R., Bot. Ser. Field Museum of Natural History 20 (4): 1941
22) HOFFMANN, C., Kieler Meeresforschungen 3 (1): 1938
23) MARSHALL, S. M. and ORR, A. P., Natural Plant Hydrocolloids, 1954, p. 101
24) TAMI, K. and WATANABE, T., Bull. Tohoku Reg. Fish. Res. Lab. 16: 145–149, 1960

25) MURAKAMI, S., Sc. Rep. Saitama Univ. 3 B (3): 251–254, 1960
26) DIXON, P. S., J. Mar. Biol. Ass. U. K. 39 (2): 331–347, 1960
27) id.; id. 39 (2): 375–390, 1960
28) STOLOFF, L., Ec. Bot. 16 (2): 86–94, 1962

General Literature

FRITSCH, F. E., 1952; HOPPE, H. A., 1958; HUMM, H. J., 1951; KYLIN, H., 1956; LEVRING, T., 1940; LEWIN, R. A., 1962; PAX, F., 1962; SEGAWA, S., 1957; UPHOF, J. C. TH., 1959; see Index of Literature.

Chondria

Approximately 25 species in warmer seas.

Ch. tenuissima (Good.) et Woodw.) Agardh. Ocurring along the coasts of Europe, North America, Pakistan[1]), and elsewhere at the coasts of warmer regions. The type species. Thallus erect, 5 to 25 centimetres high, terete, though sometimes flattened, branching lateral, cartilaginous or fleshy in consistency.

The chemical components of various *Ch.* species have been investigated. Floridoside has been identified[2]).

Ch. armata. Coasts of South East Asia, Malayan Archipelago, Polynesia, Formosa, and of the southern Japanese islands. The Japanese term of this alga is "Domoi" or "Hanayanagi". The plant reaches a height of approximately 5 centimetres, is 2 to 5 centimetres wide, and in fresh condition has a red-violet colour, but is dark-violet when dried. *Ch. armata* is of importance as medical drug; it has been known as an anthelmintic since earlier times. The effective anthelmintic agent is domoic acid, whose constitutional formula has been ascertained. L-citrulline and D-aspartic acid, as well as a number of free amino acids have been isolated from this species[3–7]) [14]).

Ch. dasyphylla. Coasts of Europe (near low-water level in the North Atlantic), Mediterranean, Indian Ocean, and Japan[1]) [8]). This species is intricately branched and of reddish-violet colour.

Ch. atropurpurea. Coast of Columbia[9]).

Ch. caerulescens. The composition[10]) and fluorescent substances in particular[11]) have been investigated in this species.

Ch. cornuata. Shores of Pakistan[1]).

Ch. littoralis. Puerto Rico. A compound with antibiotic activity, termed chonalgine, has been isolated from this species[16]) [17]).

Ch. crassicaulis, Ch. expansa, Ch. intricata, Ch. lancifolia, Ch. ryukyuensis on Japanese coasts (SEGAWA), have been investigated to some extent[13]).

Physiological and biochemical investigations of *Ch. armata* and *Ch. crassicaulis* are reported in LEWIN (1962). A new amino acid chondrine has been isolated from *Ch. crassicaulis*[15]).

Literature

1) ANAND, P. L., Marine Algae from Karachi. The University of the Panjab, Lahore, 1943
2) AUGIER, J., Rev. gen. bot. 60, 257–283, 1953
3) TAKEMOTO, T. and DAIGO, K., Chem. and Pharm. Bull. 6 (5): 578–580, 1958
4) TAKEMOTO, T. and DAIGO, K., Bull. Jap. Soc. Phyc. 7 (3): 91–93, 1959
5) DAIGO, K., Pharm. Fac. Univ. of Osaka 79 (3), 1959
6) TAKEMOTO, T. and DAIGO, K., Archiv der Pharmazie 293/65 (6): 627, 1960
7) TAKEMOTO, T., Jap. J. of Pharm. and Chem. XXXI, Nr. 8
8) TOKIDA, J., Fac. of Fish. Hokkaido University 2 (1), 1954

9) TAYLOR, W. R., Bot. Ser. Field Museum of Natural History 20 (4): 1941
10) AUGIER, J., Bull. Centre Etud. et Rech. Sc., Biarritz 1 (3): 395–396, 1957
11) H'EOCHA, C. O. et al., Third Int. Seaweed Symp. 1958
12) BURKHOLDER, P. R. et al., Bot. Mar. II (1/2): 149, 1960
13) YABU, H. and KAWAMURA, Mem. Fac. Fish. Hokkaido Univ. 7 (1/2): 61–71, 1959
14) HEGNAUER, R., Chemotaxonomie der Pflanzen, Birkhäuser Verlag, Basel, Stuttgart, 1962
15) KURIYAMA, M. et al., Bull. Jap. Soc. Sc. Fish. 26 (6): 627, 1960
16) MARTINEZ NADAL, N. G., Inf. Bull. 1962–1963. Res. Dep. Univ. of Puerto Rico
17) MARTINEZ NADAL, N. G. et al., Antimicr. Agents and Chemother. 1963, 68–72. Amer. Soc. f. Microbiology, 1964

General Literature

CHAPMAN, V. J., 1950; FRITSCH, F. E., 1952; KYLIN, H., 1956; LEWIN, R. A., 1962; SEGAWA, S., 1957; see Index of Literature.

Chondrus

According to KYLIN there is only one species of *Chondrus* in the northern parts of the Atlantic Ocean, but according to other authors there are quite a number of taxa, whose status, however, is up to now unconfirmed.

Ch. crispus (L.) Stackh. Atlantic coasts of Europe and Northern America, and Morocco. Main occurrence and most important harvesting areas: Coasts of Iceland, Norway[1–3]), Great Britain (Scotland, Hebrides, Orkney Islands), Ireland[4]), France (Brittany), Spain, Portugal. *Ch. crispus* occurs in industrially exploitable quantities in numerous places on the Atlantic coast of North America, from New Jersey to Newfoundland, particularly in Massachusetts (Scituate) and Maine (Portland, Rockland), where the North American carrageenan industry is chiefly situated[5]) [6]). The Irish-moss industry is the oldest seaweed industry in the U.S.A. Since 1944–1945 it has developed in a continously rising rate to its modern significance. Widespread on the coast of Nova Scotia (Yarmouth), very scarce in the Bay of Fundy, but abundant in Fundy Approaches, Northumberland Strait, Prince Edward Island, and parts of the Atlantic coast[7–15]). This alga is also recorded from New Zealand (Pl. III: 5; Pl. XVI: 41, 42).

Ch. crispus mostly grows on rocky habitats, from low water to approximately 6 metres. Thallus perennial, erect, 5–15 centimetres high, branching mostly dichotomous, flat and fan-like, or curled, variable in form, of a leathery or cartilaginous consistency. The colour varies between greenish yellow, bright violet, and brownish purple.

Ch. crispus often is found in a mixture with *Gigartina stellata* (cf. 238). The two species are mostly harvested and processed jointly and are known jointly in the trade as Carrageen or Irish Moss. Investigations have been made in 1936 of the seasonal variations in *Ch. crispus* by BUTLER[16]). A study has been published of the growth of this alga in submerged populations[17]) as well as of investigations of the physiology of germination[96]). In the North Atlantic area the alga is usually harvested between its third and fifth years, from April to September. The most favourable harvesting time for each locality has to be ascertained by phycocolloid measurement.

The fresh alga contains ca. 80% of water, the dried alga ca. 50 to 60% of phycocolloid.

Investigations concerning the best drying process, particularly by circulation drying, have been carried out mainly in Canada[18]) [19]). Artificial drying is preferable to open air drying because the influence of weather on the duration of the operation results in naturally dried products of very uneven quality.

Two qualities have been marketed for a number of decades, "unbleached" and "bleached". Bleaching is effected by washing and by exposure to sunlight. Bleaching is often done artificially with sulphur dioxide[20]).

Ch. crispus has been used for a long time in the areas of origin, for example, on the West coast of Ireland, in Great Britain and France either as a medical drug or as food, and nowadaysis utilized in various technical fields, for which it is being exported. In Iceland the alga is used for fodder, especially for sheep, cattle, and horses. Carrageen is boiled in Norway, Scotland, and Ireland, mixed with bran, and fed to domestic animals[7]).

Originally, at the beginning of the last century, the alga was imported into the U.S.A. from Europe. In 1835, however, abundant occurences have been discovered on rocky coasts from Massachusetts to Newfoundland[21]). The alga and the extract produced from it, known as „vegetable jelly", found widespread and varied application[22]). In 1866 a comprehensive work has been published in the U.S.A. on "Irish Moss"[23]), and in modern literature there is no lack of works showing investigations on this species. Thus Stoloff (1949) has produced a comprehensive compilation entitled "Irish Moss, from an Art to an Industry"[24]) in which he discusses the development of raw materials, harvesting, drying and bleaching, extraction, and utilization. Further publications deal with the same subject[25]) [26]).

The alga served as a food and has also been used as a thickening agent for sweets, puddings etc. The nutritive value of carrageen has been the subject of various investigations[27]) [28]).

The official and trade names of the alga, if used as a drug, are "Carragheen", "Carrageen", "Fucus crispus", "Lichen irlandicus", "Alga perlada" (Latin), "Irländisches Moos", "Felsenmoos", "Knorpeltang", "Perlmoos" (German), "Irish Moss", "Dorset Weed" (English), "Mousse d'Irlande" (French), "Gelatintang" (Norway). In pharmacy carrageen is used as a mucilage for diarrhoea and as a component of numerous health teas, for example, of those used against colds. The alga is used as a direct agent in cosmetics[29]). Carrageen powder is also utilized by the soap industry. The alga is used as a clarifying agent for wine, beer, and honey, as a finishing substance in textile industry and paper industry (Hoppe, Schulzen)[30]). In decorating it has long been known as "Painters' Moss". The leather industry uses it for smoke-curing of leather.

A number of publications in recent decades deal with the direct use of carrageen, for example, in regard of its peculiar reactions towards certain dyes used in therapeutics[31]) and its use as copper finings[32]), as stabilizing material[33]), and in the canning of meat[34]). Patents refer to the use of carrageen in the food industry[35]) and in pharmacy (as an anticoagulant)[36]) [97]). *Ch. crispus* together with various *Gigartina* species, particularly *G. stellata*, is today the most important raw material for the production of Irish Moss Extract or Carrageenan (cf. 319)[8]) [14]) [37–46]). Stoloff (1962) refers to the polysacolloid as a carrageenan[100]).

In the early stage of the industry in Great Britain carrageenan has been described as "British agar"[46]) [47]). As it has been accentuated, *Chondrus crispus* belongs to the red algae, whose direct use has been known for many years in European countries. The constituent substances of the alga attracted the interest of research workers at a very early time. In

1876 descriptions of levulinic acid and carrageen sugar have been published[48]. In 1887 the first account of the Irish Moss industry has been published[49] [50], and of Irish Moss gelatin[51], too. In 1900 SEBOR wrote on the carbohydrates of carrageen "moss"[52]. After 1920 publications have appeared, on the composition[53–57], on hydrolysis of carrageen mucilage[58], on stability of the phycocolloid in the dried alga[59], and on the chemical composition and the polysaccharide complex[60–63]. Further reports have been published during and after the Second World War on the development of the Irish Moss industry in Canada and in the U.S.A.[65–67]. Chemical investigations have also been published[68] [147] [30] [69–76] [95].

For the production of carrageenan, its constituent substances, properties, and utilization cf. p. 319. Chemical investigation of *Ch. crispus* has included amino acids[76] [77], peptides, and proteins[78] [79], carotene[81], inorganic constituents[82], and vitamins[80] (of which several, including vitamin B_{12}, have been found).

Chondrus crispus varieties. *Ch. crispus* shows numerous forms, depending on origin and habitat, which have often been described as varieties in the literature. Thus LAMOUROUX (1813) has described[35], DAWSON-TURNER (1819)[10], BATTERS (1902)[9] and THOMAS (1938)[22] wrote about varieties. NEWTON describes[6] namely var. *filiformis*, var. *stellata*, var. *aequalis*, var. *dilatata*, var. *typica*, and var. *nana*.

Two varieties of *Ch. crispus* are distinguished on the coast of the Iberian Peninsula, namely var. *angustifrons*, and var. *latifrons*.

Ch. armatus. Coasts of Japan, especially in the Bay of Oshoro[83–84], Northern Honshû, Hokkaido, Sakhalin, particularly on the Western coast[85]. The alga is called "Toge-tsunomata" in Japan and used industrially for the production of slime[85]. The gelatine type product is used in food. *Ch. armatus* and other Japanese *Ch.* varieties are also used for the production of funoran[7], under the name of "saini".

Ch. canaliculatus. Coasts of Peru and Chile, particularly near Arica, Cobija, and Valparaiso[86]. The alga is used for carrageen production and can be regarded as a carrageenan raw material[39].

Ch. elatus. Coasts of Japan (SEGAWA). Japanese name, "Naga-tsunomata". This alga contains L-arabinose, D-galactose, floridose[71]. There is a publication[98] on cooking dried *Ch. elatus*.

Ch. ocellatus. Coasts of Japan. This alga, "tsunomata", is used for sizing cloth, calcimines, and plaster for house walks, often together with other Japanese *Ch.* varieties. The constituent substances have been investigated[71] [87]. Carrageenan raw material[14] [39] [44] [71] [80–90]. The polysaccharides correspond to kappa-carrageenan, but no evidence has been obtained concerning the position of the sulfate groups[95]. STOLOFF (1962) describes the polysacolloid as a carrageenan[100]. The most important component of this polysacolloid is described as carrabiose, a sugar similar to agarobiose[94].

L-citrulline[91] and a taurin compound[92] have been isolated from *Ch. ocellatus*.

Various forms of *Ch. ocellatus* can be distinguished, for example, f. *crispus*, particularly in the Bay of Oshoro[83]. SEGAWA recognizes f. *canaliculatus*, f. *giganteus*, f. *nipponicus*, f. *typicus*.

Ch. pinnulatus. Highly variable; f. *typicus* at the coasts of Japan, particularly in the Bay of Oshoro[83], Northern Honshû, Hokkaido, Kurile Islands, Sakhalin, Korea, and Siberian coast of the Japanese Sea. TOKIDA also describes f. *flabellatus* f. nov., f. *longicornis* f. nov.,

f. *ciliatus* subf. nov., f. *cervicornis* f. nov. The species is exploited industrially in Japan for the production of algal slime[85]).

Ch. platynus. Coasts of Japan. The alga is used for home consumption.

Ch. yendoi. Northern coast of Japan[83]). Reports have been published of changes in chemical composition of the discoloured dry product[99]).

Various *Ch.* species are also reported from the coasts of the Philippines[93]).

One up to now undefined red alga from Indonesia has been offered commercially as "Indonesian Irish Moss".

Physiological and biochemical investigations of *Ch. crispus*, *Ch. giganteus*, and *Ch. ocellatus* are reported in LEWIN (1962).

Literature

1) PRINTZ, H., Utilization of Seaweeds in Norway
2) PRINTZ, H., Seaweeds, their properties and general uses, espicially in Norway, 1952
3) PRINTZ, H., Tang og Tare. Johann Grundt Tanum Forlag, Oslo, 1953
4) BRENNAN, A. T., Notes on some common Irish Seaweeds. Inst. f. Ind. Res. and Stand., Dublin, 1950
5) MAC FARLANE, C. J., Nova Scotia Res. Found., Canada, 1956
6) SANFORD, F. B., Fishery Leaflet 469, 1958. US Dep. of the Interior, Washington, D. C.
7) HUMM, H. J., The Red Algae of Economic Importance ex Tressler, D. K. and Lemon, J. Mc. W., Marine Products of Commerce. Reinhold Publishing Corporation, New York, 1951
8) STOLOFF, L. and LEE, CH. F., Fishery Leaflet 173, 1946. US Dep. of the Interior, Washington DC.
9) GODSTON, J., Food Ind. 51. 731. 1949
10) WILLIAMS, R. H., University of Miami, Marine Laboratory, 1949
11) FOSTER, W. S., Dept. Sea and Shore Fish., Augusta, Maine, Gen. Bull. (3) 1953 and (5) 1955
12) RHYDWEN, Trade News (Dep. of Fish., Ottawa) 7 (2), 1954
13) YOUNG, E. G. et al., J. Fish. Res. Bd. Canada, 16 (1): 1959
14) STOLOFF, L., Carrageenan ex Industrial Gums. Academic Press Inc., New York, 1959
15) MAC FARLANE, C. J., Third Int. Seaweed Symp., 1958
16) BUTLER, M. R., Biochem. J. 30. 1338–1344, 1936
17) TAYLOR, A. R. A., Proc. IX. Int. Bot. Congr. 2, 394, 1959
18) MERRIT, J. H. et al., J. Sci. Food. Agric 1961, 12, 718
19) MERRIT, J. H. et al., Fourth Int. Seaweed Symp., 1961
20) LAWALL, C. H. and HARRISON, J. W. E., J. Amer. Pharm. Ass., Sci. Ed., 21. 1146–1153, 1929
21) The Marine Laboratory, University of Miami, Sea Secrets, July 20, 1954
22) STOLOFF, L. S., Natural Plant Hydrocolloids, 1954, p. 92
23) BATES, H. G., Rept. U. S. Comm. Agr. 1866, 423–430
24) STOLOFF, L. S., Ec. Bot. 3 (4): 428, 1949
25) MAC LEAN, D. B. et al., Nova Scotia Res. Foundation, Canada, 1954
26) RHYDWEN, Trade News (Dep. of Fish., Ottawa) 7 (2), 1954
27) NILSON, H. W. and SCHALLER, J. W., Food Res. 6: 461–469. 1941
28) LARSEN, B. and HAWKINS, W. W., J. Sci. Fd. Agric. 1961, 12, 523
29) SMITH, P. J., Am. Perfum. Ess. Oil. Rev. 66: 55, 1955 CA 50. 4. 29151, 1956
30) SCHULZEN, H., Melliand Textilberichte 41 (5): 596–600, (8): 992–995, 1960
31) JUSTIN-MUELLER, E., J. pharm. Chim. 30. 154–161, 1924
32) RANKIN, C., J. Inst. Brev. 35. 287–291, 1929
33) SELTZER, W., US Pat. Off. 2, 11, 594; 1935
34) REEDMAN, E. J. and BUCKBY, L., Can. J. Res. D, 21: 348–357, 1943
35) FRIEDEN, A. et al., US Pat. Off. 2, 427, 594; 1947
36) ELSNER, H., US. Pat. Off. 2, 284, 566; 1942
37) TSENG, C. K., Science 101 (2633): 597–602, 1945
38) WHISTLER, R. L. and SMART, CH. L., Polysaccharide Chemistry. Academic Press Inc., New York, 1953
39) STOLOFF, L. and SILVA, P., Ec. Bot. 11 (4): 327–330, 1957
40) STOLOFF, L., 1954
41) MAC LEAN, D. B., Chem. in Canada 6, 39–42, 1954
42) NEWTON, L. et al., Coll. Int. du Centre Nat. de la Recherche Scient., Dinard, 1957
43) YOUNG, E. G., Coll. Int. du Centre Nat. de la Recherche Scient., Dinard, 1960
44) SCHMITT, M., Coll. Int. du Centre Nat. de la Recherche Scient., Dinard, 1960

45) DANGEARD, P., Coll. Int. du Centre Nat. de la Recherche Scient., Dinard, 1960
46) MARSHALL, S. M. et al., A Study of Certain British Seaweeds and their Utilization in the Preparation of Agar. His Majesty's Stationary Office, London, 1949
47) Institute of Seaweed Research, Inveresk, March 1956
48) BENTE, F., Ber. Deut. Chem. Ges. 9, 1157–1158, 1876
49) CLARK, A. H., Fish and Fish. Ind. of the United States 2 (3): 219–220. 1887
50) WILCOX, W. A., idem. Sect. 2, Geogr. Rev. 219–220, 1887
51) PAINTER, E., Proc. Am. Pharm. Ass. 35. 578–582. 1887
52) SEBOR, Z., Oesterr. Chem. Ztg. 3, 441–444, 1900
53) HAAS, P., Pharm. J. 106, 485, 1921
54) HAAS, P., Biochem. J. 15, 469–475, 1921
55) HAAS, P. and HILL, T. C., Ann. Appl. Biol. 7, 352–362, 1921
56) RUSSELL-WELLS, B., Biochem. J. 16, 578–586, 1922. CA 17, 798, 1923
57) HARWOOD, F. C., J. Chem. Soc. 123, 2254–2258, 1923
58) HAAS, P. et al., Biochem. J. 23, 425–429, 1929
59) YOUNG, E. G. and GORING, D. A. I., Third Int. Seaweed Symp. 1958
60) BUTLER, M. R., Fish. Res. Board Can. Rept. Nr. 47, 1930
61) BUTLER, M. R., Biochem. J. 28, 759–796, 1934
62) BUTLER, M. R., Biochem. J. 29, 1025–1027, 1934
63) BUTLER, M. R., Biol. Bull. 73, 143–146, 1937
64)
65) COGSWELL, J. F., Am. Mag. 5, 42–43, 1942
66) NEEDLER, A., Atl. Biol. Sta. Circ. 9, 3, 1944. Fish. Res. Bd. of Canada
67) NEEDLER, A., Atl. Biol. Sta. Circ. 10, 1947. Fish. Res. Bd. of Canada
68) ROSS, A. G., J. Sci. Food Agric. Nr. 7. 333–335, 1953
69) BAKER, G. L., Food Eng. 26 (6): 245, 1954
70) PERCIVAL, E., Chem. and Ind. (London) 1954, 1487. CA 49, 22, 15748d, 1955
71) SMITH, F. and MONTGOMERY, R., The Chemistry of Plant Gums and Mucilages. Reinhold Publishing Corporation, New York, 1959
72) FOSTER and STACY, Polysaccharides from Algae and Lichens ex W. Ruhland (ed.), Handbuch der Pflanzenphysiologie, Springer-Verlag, 1958
73) HIRST, E. L. and JONES, I. K. N., Gums and Mucilages of Plants ex W. Ruhland (ed.), Handbuch der Pflanzenphysiologie, Springer-Verlag, 1958
74) YOUNG, E. G., Carbohydrate Accumulation by lower Plants (Rhodophyta) ex W. Ruhland (ed.), Handbuch der Pflanzenphysiologie, Springer-Verlag, Berlin, 1958
75) JANECKE, H. and KEHR, W., Pharmaz. Ztg. 105 (25): 684, 1960
76) BLACK, W. A. P., The Algae ex Altschul "Processed Plant Proteins Foodstuffs". Academic Press Inc., New York, 1958
77) SMITH, D. B. and YOUNG, E. G., J. of Biol. Chem. 217 (2): 845–853, 1955
78) YOUNG, E. G. and SMITH, D. G., J. of Biol. Chem. 233 (2): 406–410, 1958
79) YOUNG, E. G. and SMITH, D. G., Third Int. Seaweed Symp. 1958
80) TEERI, A. E. and BIEBER, R. E., Science 127, 1500, 1958 Nr. 3313
81) HAUG, A. and LARSEN, B., Second Int. Seaweed Symp., 1955
82) YOUNG, E. G. and LANGILLE, W. M., Publ. Nr. 4636 of the Nat. Research Council, Halifax, Canada, 1958
83) TOKIDA, J. and MASAKI, T., Fac. of Fish. Hokkaido University 10 (3): 173–195, 1959
84) SEGAWA, S., The Seaweeds of Japan. Hoikusha, Osaka, Japan. 1957
85) TOKIDA, J., Fac. of Fish. Hokkaido University 2 (1): 1954
86) LEVRING, T., Contributions to the marine Algal flora of Chile. Lunds Univ. Årsskrift N. F. Avd. 2. Bd. 56 (10), 1960
87) ARAKI, C. and HIRASE, S., Bull. Chem. Soc. Japan 29, 7770–755, 1956. CA 51, 11, 8015d, 1957
88) O'NEILL, A. N., J. Am. Chem. Soc. 77, 2837, 1955
89) ARAKI, CH., Seaweed Polysaccharides ex Wolfrom, M. L., Carbohydrate Chemistry of Substances of Biological Interest. Pergamon Press Ltd., London, 1959
90) MORGAN, K. and O'NEILL, A. N., Can. J. Chem. 37, 1201–1209, 1959
91) KURIYAMA, M. et al., Bull. Fac. Fish. Hokkaido Univ. 10 (3): 277–283, 1959
92) KURIYAMA, M., Nature, 1961, 192, 969
93) VELASQUEZ, G. T., Philippine Seaweeds Report, 1955
94) HEGNAUER, R., Chemotaxonomie der Pflanzen, Birkhäuser Verlag, Stuttgart, 1962
95) LEWIN, R. A., Physiology and Biochemistry of Algae. Academic Press Inc., New York, 1962
96) WEBER, W., Z. Bot. 48 (3): 251–257, 1960
97) GARCIA PINEDA, M. D. and MONTEQUI, D., Bol. del Institu'o Espanol de Oceanografia Nr. 72, Nov., 1955. WFA Nov./Dec. 1956
98) HISAUCHI, K., Bull. Jap. Soc. Phyc. 8 (2): 71–73, 1960
99) KOMAKI, S. and MATSUMURA, M., Month. J. Hokkaido Fish. Exper. Stat. 17 (3): 41–45, 1960
100) STOLOFF, L., Ec. Bot. 16 (2): 86–94, 1962

General Literature

CHAPMAN, V. J., 1950; FOTT, B., 1959; FRITSCH, F. E., 1952; HOFFMANN, C., 1938; HOPPE, H. A., 1958; HOPPE, H. A. and SCHMID, O. J., 1962; KUCKUCK, P., 1953; KYLIN, H., 1956; LEWIN, R. A., 1962; PAX, F., 1962; SCHMID, O. J., 1962; SCHULZEN, H., 1962; UPHOF, J. C. TH., 1959; see Index of Literature.

Corallina

Approximately 25 species in various seas.

C. officinalis L. "Coral Moss". Abundant and widespread in the northern hemisphere, especially in the northern regions of the Atlantic Ocean and in the Mediterranean. The type species. Thallus erect, 3 to 12 centimetres high, articulated, terete or flattened, branched, strongly calcified, brittle, often forming meadows. The alga has been investigated chemically[1]), in addition of its mineral constituents. It contains glucose[2]), glyceritol glycoside[3]), and antimicrobial substances[4]).

C. mediterranea. Especially Mediterranean.

C. squamata. Coasts of Japan. This alga contains a peptide[5]).

C. pilulifera, *C. sessilis*, and other East Asiatic species have been described[6]) [7]).

The mineral components of the Corallinaceae have been investigated repeatedly in recent years[8]) [9]). Almost all Corallinaceae are strongly calcified (with calcium and magnesium carbonates). In the Pacific Ocean these algae contribute to the formation of coral reefs. Fossil corallinaceae are said to have already existed in the Palaeozoic age. They have, moreover, been ascertained with complete certainty for the Cretaceous period[10]). Comparisons have been made concerning the calcareous algal flora of recent and fossil reefs[11]).

Corallina species have been used as a vermifuge towards the end of the eighteenth century[12]).

For physiological and biochemical investigations of *C. officinalis* and *C. squamata*, see LEWIN (1962).

Literature

1) ROSS, A. G., J. Sci. Food Agric. (7): 333–335, 1953
2) BLACK, W. A. P. and CORNHILL, W. J., Chem. and Ind. 1954, pp 514–516
3) WICKBERG, B., Acta Chem. Scand. 12 (6): 1183–1186, 1958
4) ROOS, H., Kieler Meeresforschungen, Bd. XIII (1), 1957
5) YEMM, E. W., The Plant Proteins and Peptides and their Localization in Cells and Tissues ex W. Ruhland, Handbuch der Pflanzenphysiologie, Springer Verlag, 1958
6) TOKIDA, J., The Marine Algae of Southern Saghaline. Fac. of Fish. Hokkaido University, 2 (1): 1954
7) SEGAWA, S., Rec. Oceanogr. Works Japan spec. no. 3: 221–223, 1959
8) WALTER-LEY, L. and STRAUSS, R., Coll. Int. du Centre Nat. de la Recherche Scient., Dinard, 1960
9) FURUYA, K., Bot. Mag. Tokyo (867): 355–359, 1960
10) FOTT, B., Algenkunde, Gustav Fischer Verlag, Jena, 1959
11) JOHNSON, J. H., Pac. Sci. Ass. 9th Congr. Abstr. 54–55, 1957
12) CHAPMAN, V. J., Seaweeds and their Uses. Methuen & Co., Ltd., London, 1950

General Literature

FRITSCH, F. E., 1952; KUCKUCK, P., 1953; KYLIN, H., 1956; LEWIN, R. A., 1962; SEGAWA, S., 1957; see Index of Literature.

Corallopsis

6 species in warmer seas.

C. salicornia (Ag.) Greville (= *Sphaerococcus salicornia)* Agardh. Pacific Ocean. The type species. Thallus erect, 10 to 30 centimetres high, terete, incised, articulated, frequently branched, surface prickly. In their anatomic structure similar to *Gracilaria* species. Raw material for agar[1] [2] [3].

C. salicornia var. *minor*. Sonder. Malaya, Indonesia (particularly Sumatra, Bali, Sumbawa, and other islands), islands in the Pacific. This species contains agar, and on Bali it is eaten as a vegetable[4]. Its Bali name is "bulung buku". Stoloff (1962) refers to the polysacolloid as agar[5].

The various species of *Corallopsis* for the most part grow in lower littoral or sublittoral regions. Around Sumbawa *C. salicornia* var. *minor* is found even at depths of 15 to 30 metres.

C. opuntia. Occurs on the coral reefs off Ceylon and off the Japanese coasts.

Literature

1) Stoloff, L. and Silva, P., Ec. Bot. 11 (4): 327–330, 1957
2) Selby, H. H. and Selby, T. A., Agar ex Whistler, R. L. and BeMiller, J. N., Industrial Gums. Academic Press Inc., New York, 1959
3) Smith, F. and Nontgomery, R., The Chemistry of Plant Gums and Mucilages. Reinhold Publishing Corporation, New York, 1959
4) Eisses, J., Indonesian J. of Nat. Sc. (1, 2, 3), 1953
5) Stoloff, L., Ec. Bot. 16 (2): 86–94. 1962

General Literature

Chapman, V. J., 1950; Fritsch, F. E., 1952; Kylin, H., 1956; Segawa, S., 1957; Zaneveld, J. S., Ec. Bot. 13 (2): 89, 1959; see Index of Literature.

Cryptonemia

Some 15 species, warmer seas.

C. lomation (Bertol.) J. Ag. Mediterranean; the type species. Thallus erect, 5 to 30 centimetres high, stalked in the lower part, in the upper flattened, leaf-like.

C. decumbens. Pacific islands. The Polynesians utilize this species for food.

C. schmitziana. Japanese coasts.

Literature

Fritsch, F. E., 1952; Kylin, H., 1956; Segawa, S., 1957; Uphof, J. C. Th., 1959; see Index of Literature.

Cystoclonium

4 species, North Atlantic, shores of Brazil, Falkland Islands, and Kamchatka.

C. purpurascens (Huds.) Kuetz. Northern Atlantic, Baltic. The type species. Thallus erect, 10 to 30 centimetres high, terete, much branched, cartilaginous. Material of this species collected in the Kieler Förde contained antimicrobial substances[1].

Physiological and biochemical investigations are reported in LEWIN (1962). The extraction of the alga produces a soft gel[2]).

C. armatum. Pacific coasts. This species is eaten in Japan.

Literature

1) Roos, H., Kieler Meeresforsch. XIII (1), 1957
2) MARSHALL, S. M. et al., A Study of Certain British Seaweeds and their Utilization in the Preparation of Agar. His Majesty's Stationary Office, London, 1949

General Literature

FRITSCH, F. E., 1952; KUCKUCK, P. 1953; KYLIN, H. 1956; LEWIN, R. A., 1962; UPHOF, J. C. TH., 1959; see Index of Literature.

Dasya

Approximately 40 species, in temperate and warmer seas.

D. pedicellata Ag. European coasts of the Atlantic, of North and Central America, especially in the warmer regions, Mediterranean. The type species. Thallus erect, 2 to 60 centimetres high, irregularly branched. Material of this species collected off Puerto Rico showed antibiotic activity[1]). This alga is also recorded from the coasts of the Netherlands[2]) and Sweden, where the specimens were strikingly large[3]).

D. sessilis. Japanese coasts.

Literature

1) BURKHOLDER, P. R., et al., Bot. Mar. II (1/2): 149, 1960
2) HARTOG, C. DEN, Fourth Int. Seaweed Symp., 1961
3) MICHANEK, G., Fourth Int. Seaweed Symp. 1961

General Literature
FRITSCH, F. E., 1952; KYLIN, H., 1956; SEGAWA, S., 1957; see Index of Literature.

Delesseria

Approximately 15 species, European coasts of the Atlantic, Baltic, North American Westcoast, Falkland Islands, Tierra del Fuego, New Zealand.

D. sanguinea (L.) Lamouroux. "Blutroter Seeampfer", "Dock-leaved Delesseria". European Atlantic and Baltic coasts, in the tidal zone, but also in deeper regions. The type species. Thallus erect, 10 to 30 centimetres high, flat, leaf-like, lobed, with midrib and lateral nerves. In the lower part very often stalked. This alga is strikingly carmine-red. It occurs in various shapes: The specimens found in the Baltic per instance have more slender 'leaves' than those from the North Sea (Pl. XIII: 33).

In spring this species contains a blood anticoagulant. This substance is similar to heparin in its chemical and physiological properties[1]) [2]) [6]). Material from the Baltic has been found to contain antimicrobial substances[3]). Extraction gives a viscous extract[7]).

D. elata. The constituents have also been investigated[4]) [5]).

D. violacea. Coasts of Japan.

Physiological and biochemical investigations of *D. decipiens* and *D. sanguinea* are reported in LEWIN (1962).

Literature

1) ELSNER, H. et al., Naunym-Schmiedebergs Archiv für experimentelle Pathologie und Pharmakologie 190 (5): 510, 1938
2) FOTT, B., Algenkunde, Gustav Fischer Verlag, Jena, 1959
3) ROOS, H., Kieler Meeresforsch. XIII (1), 1957
4) SMITH, F. and MONTGOMERY, R., The Chemistry of Plant Gums and Mucilages Reinhold Publishing Corporation, New York, 1959
5) GERVAIS, Rev. Gen. Bot. 66 (785): 395–401, 1959
6) GARCIA PINEDA, M. D. and MONTEQUI, D., Bol. del Inst. Espanol de Oceanografia Nr. 72, 1955
7) MARSHALL, S. M. et al., A Study of Certain British Seaweeds and their Utilization in the Preparation of Agar. His Majesty's Stationary Office, London, 1949

General Literature
CHAPMAN, V. J., 1950; FRITSCH, F. E., 1952; KUCKUCK, P., 1953; KYLIN, H., 1956; LEVRING, T., 1940; LEWIN, R. A., 1962; PAX, F., 1962; see Index of Literature.

Digenea

1 species. Mediterranean, warmer regions of the Atlantic and Indian Oceans, Japan, Philippines[1]), and Australia. Large quantities are also recorded from the coasts of Florida, the Gulf of Mexico, and the West Indian area[2–5]).

D. simplex (Wulf.) Agardh. Thallus erect, 10 to 20 centimetres high, terete, branching lateral of bifurcate, consistency cartilaginous. Shoots consist of thick, sturdy long shoots, and slender, thin short shoots. This alga may be considered as an agar or agaroid raw material[6–8]). Because of the low agar content, however, (10 to 15% of the dried material) it seems doubtful, whether *D. simplex* will gain any commercial importance, although it occurs in large quantities[2]). In East Asia this species serves for food[9]). In Japanese it is called "Makuri-Nori".

D. simplex is of significance because of its anthelmintic character, especially for ascarids[10]) [26]). It contains the vermifugal components kainic acid and alpha-allokainic acid[11–13]) [27]). The ethereal oils and other components of this species have been investigated[10]) [14–25]) [26]). There is evidently a rather high percentage of iodine, although reports relating to the height differ widely.

Physiological and biochemical investigations are reported by LEWIN (1962).

Literature

1) VELASQUEZ, G. T., Philippine Seaweeds Report, 1955
2) HUMM, H. J., Agar from *Digenea simplex*. Florida State University Studies. Contributions to Science
3) WILLIAMS, R. H., University of Miami, Marine Laboratory, 1949
4) TAYLOR, W. R., Bot. Ser., Field Museum of Natural History 20 (4), 1941
5) DIAZ-PIFERRER, M. and LÓPEZ, H., Inst. Cubano de Invest. Tecnol. (6), 1959
6) SELBY, H. H. and SELBY, T. A., Agar ex Whistler, R. L. and BeMiller, J. N., Industrial Gums. Academic Press Inc., New York, 1959
7) STOLOFF, L. and SILVA, P., Ec. Bot. 11 (4): 327–333, 1957
8) SMITH, F. and Montgomery, R., The Chemistry of Plant Gums and Mucilages. Reinhold Publishing Corporation, New York, 1959
9) UPHOF, J. C. TH., Dictionary of Economic Plants., H. E. Engelmann (J. Cramer), Weinheim, 1959

10) Kawaguchi et al., Bull. Jap. Soc. Sc. Fish. 19 (4): 481–486, 1953. CA 48031/1955
11) Takemoto, T., Jap. J. of Pharmacy and Chemistry XXXI, (8)
12) Takemoto, T., Shiraishi Pharm. Comp., Osaka, Japan, 1958
13) Nitta, J. et al., Nature, 181 (4611): 761–762, 1958
14) Katayama, T., Bull. Jap. Soc. Sc. Fish. 24 (3): 205–208, 1958
15) Miyamoto, M., J. Pharm. Soc. Japan XXXVIII 77 (6): 568–570, 1957; idem XXXIX–XLII (6): 571–588, 1957
16) Honjo, M., J. Pharm. Soc. Japan XLIII 77 (6): 589–592, 1957; idem XLIV 77 (6): 593–597, 1957; idem XLV 77 (6): 598–603, 1957
17) Ueyanagi, J. et al., J. Pharm. Soc. Japan XLVI/XLIX 77 (6): 603–621, 1957
18) Sanno, Y., J. Pharm. Soc. Japan L 77 (6): 622–625, 1957; idem LI 77 (6): 626–629, 1957
19) Osugi, K., J. Pharm. Soc. Japan LII 77 (6): 630–634, 1957
20) Uchibayashi, J. Pharm. Soc. Japan LIII 78 (8): 845–849, 1958
21) Sugawa, J. Pharm. Soc. Japan LIV–LVIII 78 (8): 849–872, 1958
22) Honjo, M., J. Pharm. Soc. Japan LIX–LXI 78 (8): 888–907, 1958
23) Sanno, Y., J. Pharm. Soc. Japan LXII–LXIV 78 (10): 1113–1127, 1958
24) Osugi, K., J. Pharm. Soc. Japan LXV–LXXVI 78 (12): 1313–1374, 1958
25) Kashiwada et al., Bull. Jap. Soc. Sc. Fish. 25 (2): 100–102, 1959
26) Hegnauer, R., Chemotaxonomie der Pflanzen, Birkhäuser Verlag, Stuttgart, 1962
27) Sanno, Y., J. Pharm. Soc. Japan LXXVIII 80 (5): 603–607, 1960

General Literature

Chapman, V. J., 1950; Fritsch, F. E., 1952; Hoppe, H. A., 1958; Kylin, H., 1956; Lewin, R. A., 1962; Pax, F., 1962; Segawa, S.; see Index of Literature.

Dilsea

3 species, Northern Atlantic and Pacific.

D. edulis Stackhouse. European coasts, including Ireland, in deep water[15]. The type species. Thallus erect, 10 to 30 centimetres high, leaf-like, stalked, undivided or irregularly lobed or slotted. The constituent substances of this species have been investigated by a number of authors[1–8]. They include water-soluble polysaccharides[9] and floridean starch[10–12]. The physical properties and structure of floridean starch have been investigated[17]. The extracts do not gelate like agar or carrageenan[13], although they are viscous[16]. This alga is eaten in some coastal regions, e.g. on the Irish and North American coasts[14]. Physiological and biochemical investigations are reported by Lewin (1962).

Literature

1) Janecke, H. and Kehr, W., Pharmaz. Ztg. 105 (25): 684, 1960
2) Smith, F. and Montgomery, R., The Chemistry of Plant Gums and Mucilages. Reinhold Publishing Corporation, New York, 1959
3) Foster and Stacy, Polysaccharides from Algae and Lichens ex W. Ruhland, Handbuch der Pflanzenphysiologie, Springer Verlag, 1958
4) Young, E. G., Coll. Int. du Centre Nat. de la Recherche Scient., Dinard, 1960
5) Ross, A. G., J. Sci. Food Agric. (7): 333–335, 1953
6) Barry, V. C. and McCormick, J. E., J. Chem. Soc. (London) 1957. 2777–2783
7) Black, W. A. P., Ann. Rep. Chem. Soc. 1953, Vol. 50
8) Black, W. A. P., The Algae ex Altschul, Processed Plant Proteins Foodstuffs. Academic Press Inc., New York, 1958
9) Stoloff, L. and Silva, P., Ec. Bot. 11 (4): 327–330, 1957
10) Peat, S. et al., J. Chem. Soc. 1959, 3223
11) Peat, S. et al., J. Chem. Soc. 1959, 3341
12) Barry, V. C. et al., J. Chem. Soc. (London) 1949, 1468–1470
13) Whistler, R. L. and Smart, Ch. L., Polysaccharide Chemistry. Academic Press Inc., New York, 1953
14) Hirst, E. L. and Jones, J. K. N., Gums and Mucilages of Plants ex W. Ruhland (ed.), Handbuch der Pflanzenphysiologie, Springer Verlag, 1958
15) Brannan, A. T., Notes on some common

Irish Seaweeds. Inst. f. Ind. Res. and Stand., Dublin, 1950
16) Marshall, S. M. and Orr, A. P., Natural Plant Hydrocolloids, 1954, p. 101
17) Greenwood, C. T. and Thomsen, J., J. Chem. Soc. 1961, 1534

General Literature

Fritsch, F. E., 1952; Hegnauer, R., 1962; Kylin, H., 1956; Lewin, R. A., 1962; Uphof, J. C. Th., 1959; see Index of Literature.

Dumontia

2 species, Northern coasts of the Atlantic and Pacific, including Canada, Japan, and Korea.

D. incrassata (Müller) Lamouroux. Atlantic and Pacific Oceans, widespread. The type species. Thallus erect, 10 to 40 centimetres high, terete, hollow, irregularly branched or unbranched, reddish- to violet-brown, mucilaginous. The constituent substances of this species have been investigated by various authors[1–2]. They include soluble polysaccharides[3] [4]. For physiological and biochemical investigations see Lewin (1962). A soluble mucilage similar to the *Dilsea edulis* mucilage has been investigated; it is a viscous extract[6]. This species is to be found in large quantities near Tomei (Japan), where it is harvested and eaten or used as a substitute for *Gloiopeltis*[5].

D. simplex. Northern Japan, Korea, Sakhalin, and the Kurile Islands[5].

Literature

1) Young, E. G., Coll. Int. du Centre Nat. de la Recherche Scient., Dinard, 1960
2) Smith, F. and Montgomery, R., The Chemistry of Plant Gums and Mucilages. Reinhold Publishing Corporation, New York, 1959
3) Stoloff, L. and Silva, P., Ec. Bot. 11 (4): 327–330, 1957
4) Marshall, S. M. et al., A Study of Certain British Seaweeds and their Utilization in the Preparation of Agar. His Majesty's Stationary Office, London, 1949
5) Tokida, J., The Marine Algae of Southern Saghalien. Fac. Fish. Hokkaido Univ. 2 (1), 1954

General Literature

Fritsch, F. E., 1952; Hegnauer, R., 1962; Kuckuck, P., 1953; Kylin, H., 1956; Lewin, R. A., 1962; Segawa, S., 1957; see Index of Literature.

Endocladia

3 species, coasts of the Atlantic Ocean and Pacific.

E. muricata J. Agardh. Coasts of Brazil and the U.S.A. (California). The type species. Thallus erect, 3 to 6 centimetres high, terete, branched, with tuberculate surface. The species has been chemically investigated[1]. According to some authors it contains agar[2–5], according to others an agaroid[6] [7] or carrageenan with some kappa-carrageenan, like *Chondrus crispus*[8]. Stoloff (1962) refers to the polysacolloid as carrageenan[9].

Literature

1) Ross, A. G., J. Sci. Food Agric. (7): 333–335, 1953
2) Smith, F. and Montgomery, R., The Chemistry of Plant Gums and Mucilages. Reinhold Publishing Corporation, New York, 1959
3) Humm, H. J. and Wolf, F. A., Duke University Press (3): 1946
4) Stoloff, L. and Lee, C. F., Fishery Leaflet 173, 1946. US. Dep. of the Interior, Fish and Wildlife Service
5) Stoloff, L. and Silva, P., Ec. Bot. 11 (4), 327–330, 1957
6) Selby, H. H., Fishery Leaflet 118, 1948, US Dep. of the Interior, Fish and Wildlife Service
7) Selby, H. H. and Selby, T. A., Agar ex Whistler, R. L. and BeMiller, J. N., Industrial Gums. Academic Press Inc., New York, 1959
8) Yaphe, W., Can. J. Bot. 37: 751–757, 1959
9) Stoloff, L., Ec. Bot. 16 (2): 86–94. 1962

General Literature

Chapman, V. J., 1950; Fritsch, F. E., 1952; Hoppe, H. A., 1958; Kylin, H., 1956; Pax, F., 1962; see Index of Literature.

Eucheuma

Approximately 20 species, widespread in warmer seas.

E. muricatum (Gmel.) Web. van B. = *(E. spinosum = E. denticulatum = Fucus muricatum)*. Malaya, Indonesia (Cheribon/Western Java, Madura, Sunda Islands, particularly Makassar, Ceram) are the main areas of origin, the species also occurs on the Chinese and Japanese coasts, the Philippines, the Northern coast of Australia, and at various places of the Indian Ocean. The type species. Thallus erect, 5 to 30 centimetres high, stem rounded or flat, branching radial or complanate, often abundantly covered with simple or branched papillae. The alga grows in various forms on the coral reefs along the coasts and between the numerous islands, rarely deeper than 2 metres. It is harvested by hand by the natives in small boats, washed in sea water and dried in the sun. The algae are bleached completely by this process. Harvesting season in Indonesia begins in September and ends in May, the main harvest takes place between November and February.

E. muricatum, often described as *E. spinosum*, is traded via Singapore as "Singapore Weed".

The alga has been used in Japan for a long time as an additional agar raw material in mixture with other species of algae. However, it is also processed in other countries where phycocolloids are produced.

The phycocolloid from *Eucheuma* is described as eucheuman in more recent literature; it was formerly included among the carrageenans[1–7]). In Indonesia the alga is used as a food[2]) [8]) [9]). It is known there under the names of "agar-agar", "crude agar", or "ruwe agar", but also as "agar agar pulu" or "agar agar seru laut". In Makassar it is also known as "agar agar kasar" and in Ceram as "agar agar geser". *E. muricatum* is also the raw material for "Makassar agar".

The name of "agar agar" has been taken over by the Japanese for the well-known Japanese trade product (cf. p. 289).

The alga is used in Eastern Asia in the production of textiles, and in China it is used in the paper production. In China it is called "chilin-t' sai".

The chemical composition of the phycocolloid has been investigated, particularly with regard to the identification of the component sugars and glucuronic acid[10–12]).

E. acanthocladum. Gulf of Mexico, Cuba[13–14]).

E. amakusaensis. Coasts of China, Formosa, Japan. A less important agaroid raw material of the soft type, which in Japan is exploited economically in mixture with other raw materials[3–4]).

E. cottonii. Occurs on the coast of Zanzibar. The alga contains a phycocolloid similar to carrageenan, which, however, is better regarded as a variety of eucheuman[4]) [7]), because it corresponds chemically very nearly to the eucheuman derived from *E. muricatum*[19]).

The alga is described as a "coarse stranded quality". Sometimes *E. cottonii* is also numbered under *E. striata*, *E. edule*, or *E. spinosum*. Further names are *E. cervicorne* or *E. vermiculare*.

That is the reason that a clear classification of the various *E.* varieties seems to be difficult and somewhat confusing. In trade the *E. cottonii* of Zanzibar, like *E. serra*, is described as "Zanzibar Weed".

E. crustaeforme. Coasts of China, Formosa, Japan. A less important agaroid raw material of the soft type, which is exploited commercially in Japan in mixture with other raw materials[3]) [4]).

E. edule. Coasts of Indonesia, New Caledonia, Celebes, Makassar, islands in the Pacific Ocean. The alga is described as var. *major.* It contains a phycocolloid similar to carrageenan. On Java it is used in preparations of jellies[2]) [7]) [3]) [4]) [5]).

E. gelatinae. This alga is also referred to as *E. gelatinosa.* Coasts of Bourbon, India, Japan, China, Formosa, Philippines, Indonesia, New Caledonia, Australia. A less important agaroid raw material of the soft type, exploited commercially in mixture with other species of algae. The native Philippine population produces an agar substance from this alga. It is also used for the production of jelling materials and as a creaming agent in the manufacture of natural rubber.

The phycocolloid is used in mixture with various drugs as a remedy against coughs. In China and Indonesia the alga is used for food; in Japan it is used in the colour, paint, and textile industries[2]) [3]) [4]) [5]) [9]).

E. gelidium. Coasts of Cuba[14]).

E. horridum. (= *Gigartina horrida)* Malaya, Indonesia, Singapore, Indian Ocean (Mauritius, Bourbon). This alga is used with other species in the manufacture of jellies and of agar[2]) [3]) [5]).

E. isiforme. Coasts of China and Japan, Florida between Miami and Key West, Yucatan, Cuba (Playá Habana-Baracoa)[13]) [15]) [16]) [14]) [4]).

The alga contains an agaroid or a phycocolloid similar to carrageenan[3]) [4]).

Like *E. muricatum* this alga is often referred to as *E. spinosum* or *E. denticulatum.* It is used in a considerable extent in Japan and China for the production of agar[1]).

The establishment of *E. isiforme* off the coast of Florida is as yet not confirmed.

E. papillosum. Also referred to as *Kallymenia dentata.* Occurs in the coasts of China and Formosa, Japan, Hawaii, on submerged reefs. The alga is thrown on the shores in large quantities by gales, mainly in August and September, when it is collected and dried for the use as a condiment or as an ingredient in soya-bean sauce. It is also collected by divers. In Japan it is known under the name of "Tosaka nori"[3]).

E. papillosum is also reported from the coasts of the Red Sea and the Somaliland[9]).

E. serra. Coasts of Zanzibar and Tanganyika. *E. serra* is also reported from Japan

(SEGAWA, 1957) and Indonesia[2]) [5]). The alga is collected in large quantities between June and December, especially on the coasts of Zanzibar. The trade designation for it is "fine stranded quality"[17]). This commodity consists mainly of mixtures of various species and is described as "Zanzibar Weed". In a mixture with other algae it is used as an additional raw material for manufacturing various phycocolloids or for the making of an eucheuman, differing from the product of *E. muricatum* or "Singapore Weed".

E. speciosa. West coast of Australia, New Zealand. A very gelatinous seaweed used for jellies, especially in Australia[9]). It is an agaroid raw material[3]), but its commercial exploitation appears to be insignificant. *Eucheuma* varieties, according to other information, are used in Australia for the production of Australian agar.

E. striata. Coasts of Zanzibar. The alga is collected when thrown up on the shores and dried and exported.

In the last years, however, only various forms of *E. cottonii* and *E. serra* of doubtful identity have been traded from Zanzibar.

E. species are also known to thrive at the coasts of the Indian Ocean. Reports in literature mention 2 species[18]).

According to the view of BOERGESEN (1950) the species *E. jugatum* and *E. nodulosum* mentioned in the literature are, like *E. horridum* and *E. serra*, forms of a polymorphous species.

The authors have had several samples of *Eucheuma* from Indonesia and from the Island of Zanzibar, which could not yet be clearly identified, but appear to be various forms of *E. muricatum* (= *E. spinosum)* from Indonesia and *E. serra* from Zanzibar, respectively. The phycocolloids from these algae very often differ in their physical properties. The investigations are not yet concluded.

STOLOFF (1962) describes the polysacolloids from *E. cottonii*, *E. edule*, and *E. muricatum* as carrageenans[20]).

Literature

1) STOLOFF, L. and LEE, CH. F., Fishery Leaflet 173, 1946. US Dep. of the Interior
2) EISSES, J., Indonesian J. of Nat. Sc. Nr. 1–3, 1953
3) WHISTLER, R. L. and BEMILLER, J. N., Industrial Gums. Academic Press Inc., New York, 1959
4) SMITH, F. and MONTGOMERY, R., The Chemistry of Plant Gums and Mucilages. Reinhold Publishing Corporation, New York, 1959
5) ZANEVELD, J. S., Ec. Bot. 13 (2): 89, 1959
6) YOUNG, E. G., Coll. Int. du Centre Nat. de la Recherche Scient., Dinard, 1960
7) STOLOFF, L. and SILVA, P., Ec. Bot. 11 (4): 327–330, 1957
8) HUMM, H. J. and WOLF, F. A., Duke University Press, Nr. 3, 1946
9) UPHOF, J. C. TH., Dictionary of Economic Plants. H. R. Engelmann (J. Cramer), Weinheim, 1959
10) NAKAMURA, T., Bull. Jap. Soc. Sci. Fish. 20, 501–505, 1954; idem 21, 1053–1056, 1956; idem 21, 1057–1061, 1956; idem 23, 647–651, 1958; idem 23, 652–655, 1958; idem 24, 285–288, 1958
11) ARAKI, CH., Seaweed Polysaccharides ex Wolfrom, M. L., Pergamon Press Ltd., London, 1959
12) JANECKE, H. and KEHR, W., Pharmaz. Ztg. 105 (25): 684, 1960
13) TAYLOR, W. R., Bot. Ser. Field Museum of Natural History 20 (4): 1941
14) DIAZ-PIFERRER, M. and LÓPEZ, H., Inst. Cubano de Invest. Tecnol. (6): 1959
15) HUMM, H. J., Science 100 (2593): 209–212, 1944
16) SOLONI-TOURAL, F. G., Centre de Invest. Pesqu. (4): 1954
17) ANDERSON, G. W., First Int. Seaweed Symp., 1952
18) DICKENSON, C. J., Kew Bull. Roy. Bot. Gardens 4, 535–539, 1954

19) O'Colla ex Lewin, R. A., Physiology and Biochemistry of Algae. Academic Press Inc., New York, 1962
20) Stoloff, L., Ec. Bot. 16 (2): 86–94, 1962

General Literature
Chapman, V. J., 1950; Fott, B., 1959; Fritsch, F. E., 1952; Hoffmann, C., 1938; Hoppe, H. A., 1958; Hoppe, H. A. and Schmid, O. J., 1962; Kylin, H., 1956; Segawa, S., 1957; see Index of Literature.

Falkenbergia

4 species in different areas.

F. rufolanosa (Harvey) Schmitz. Coasts of Australia and Japan. This species has also been recorded from European coasts[1] [2]. The type species. Thallus globular, diameter approximately 2 centimetres. The filaments making up the globes are thin, branched and for the most part intricately entangled.

F. hildenbrandii. Coast of Puerto Rico. Material collected there revealed an antibiotic activity[3].

Literature

1) Dizerbo, A. H., Fourth Int. Seaweed Symp., 1961
2) Conway, E., Nature 186 (4724): 566–567. 1960
3) Burkholder, P. R. et al., Bot. Mar. II (1/2): 149, 1960

General Literature

Fritsch, F. E., 1952; Kylin, H., 1956; Segawa, S., 1957; see Index of Literature.

Furcellaria

One species, northern parts of the Atlantic and in the Baltic.

F. fastigiata (Huds.) Lamouroux (= *Fucus fastigiatus*). This alga particularly occurs in the Kattegat, but is also found in other coastal areas, for example, on the Baltic, Swedish, Norwegian, and Irish coasts, and also on the Canadian Atlantic coast (Nova Scotia).

Thallus erect, 10 to 25 centimetres high, terete, branched of a cartilaginous consistency.

Furcellaria fastigiata has a considerable resemblance to *Polyides rotundus*. Comparative anatomical investigations were made early on it[1]. The typification of *P. rotundus* and *F. fastigiata* has been published[2].

Furcellaria grows in the Kattegat in large quantities in depths of about 6 to 8 metres, and it separates from the substratum to float in the sea, often accumulating in layers several metres in thickness, due to currents. The raw material is mostly collected in the colder months. It is the basis of the Danish phycocolloid industry[3–7].

In the Baltic (for example Boknis Eck) there have been abundant occurrences of the alga in 4 to 7 metres depth apparently only since 1953 (cf. Hoffmann). Attempts have been made to produce what is known as "Ostsee-Agar" (Baltic agar) from *F. fastigiata*[8], however, the amount of this alga on the German Baltic coast appeared to be insufficient for industrial utilization.

Quantity and quality investigations for this purpose have also been made on the Baltic coast[9–10] [35] [36].

Several forms of *F. fastigiata* occur also in the Swedish western and southern coast areas[11]), and on the Norwegian[12]), and Irish coasts, often together with *Chondrus crispus* and *Gigartina stellata*.

Because *F. fastigiata* does not bleach when dried in the sun, it is often called "black carrageen" in Ireland[13]). The quantities occurring on the Canadian coast (Gulf of St. Lawrence, Northumberland Strait) should be sufficient for industrial exploitation[14]) [15]).

This alga has also been found on the Pakistani coasts[16]). Growth and development of *F. fastigiata* have been studied[17]) [18]) [32–34]). Whereas the Danish industry called the phycocolloid produced from *F. fastigiata* after the Second World War "Danish agar", it is now described as furcellaran. Meanwhile it was also classified as gelan[19]) [37]), carrageenan[20]), or agaroid[21]).

Chemical investigations have been undertaken, in particular on the polysaccharides [22–24]) [15]). Carrageenan with some kappa-carrageenan, corresponding to that from *Chondrus crispus* have been found[25]), also galactose, glucose, xylose[26]), and a glycoside[27–28]).

The alga contains D-mannitol and floridoside (3-floridoside-alpha-mannoside)[29–30]). Antibiotic substances have been established[31]).

Physiological and biochemical investigations are reported in Lewin (1962).

Literature

1) Denys, G., Mitt. Bot. Staatsinstitut Hamburg, 27, 1–31, 1909
2) Drew, K. M., J. Linnean Soc. London, Botany 55 (363): 744–752, 1958
3) Lund, S. and Bjerre-Petersen, E., First Int. Seaweed Symp., 1952
4) Schachat, R. E. and Glicksman, M. ex Whistler, R. L. and BeMiller, J. N., Industrial Gums. Academic Press Inc., New York, 1959
5) Christiansen, E. ex ibid, Industrial Gums. Academic Press Inc., New York, 1959
6) Schachat, R. E. and Glicksman, M., Ec. Bot. 13 (4): 365–370, 1959
7) Lund, S. and Bjerre-Petersen, E., Fourth Int. Seaweed Symp., 1961
8) Förster, Pharmaz. Zentralhalle 89, 409, 1950
9) Baltic Belt Seas Committee 1957. Cons. Int. Explor. Mer. Rapp. et Proc. Verb. 146 (2): 12–13, 1959
10) Czapke, K., Fourth Int. Seaweed Symp., 1961
11) Levring, T., Studien über die Algenvegetation von Blekinge, Südschweden. Diss., Lund, 1940
12) Printz, H., Seaweeds, their properties and general uses, expecially in Norway. WFA Nr. 1, 33, 1952
13) Brennan, A. T., Notes on some common Irish Seaweeds. Inst. f. Ind. Res. and Stand., Dublin, 1950
14) Mac Farlane, C. I., Third Int. Seaweed Symp., 1958
15) Painter, T. J., Canad. J. of Chem. 38, 112, 1960
16) Anand, P. L., Marine Algae from Karachi. The University of the Panjab, Lahore, 1943
17) Austin, A. P., Phyc. Bull. 5, 16–17, 1957
18) Austin, A. P., Ann. Bot. London 95 (2966): 310, 1960
19) Stoloff, L. and Silva, P., Ec. Bot. 11 (4): 327–330, 1957
20) Young, E. G., Carbohydrate accumulation by lower plants (Rhodophyta) ex W. Ruhland, Handbuch der Pflanzenphysiologie, Springer-Verlag, 1958
21) Selby, H. H. and Selby, T. A. ex Whistler, R. L. and BeMiller, J. N., Industrial Gums. Academic Press Inc., New York, 1959
22) Ross, A. G., J. Sci. Food Agric. (7): 333–335, 1953
23) Smith, F. and Montgomery, R., The Chemistry of Plant Gums and Mucilages. Reinhold Publishing Corporation, New York, 1959
24) Painter, T. J., Chem. and Ind. (47): 1488, 1959
25) Yaphe, W., Can. J. Bot. 37, 751–757, 1959
26) Black, W. A. P. and Cornhill, W. J., Chem. and Ind. 514–516, 1954
27) Lindberg, B., Acta Chem. Scand. 8, 869, 1954
28) Lindberg, B., Acta Chem. Scand. 9, 1093–1096, 1955
29) Lindberg, B., Second Int. Seaweed Symp., 1955
30) Stoll and Jucker, Heteroside ex W. Ruhland, Handbuch der Pflanzenphysiologie, Springer-Verlag, 1958
31) Chesters, C. G. C. and Stott, J. A., Second Int. Seaweed Symp., 1955

32) Austin, A. P., Hydrobiologia 15 (3): 193–207, 1960
33) Austin, A. P., Ann. Bot. 24 (94): 257–274, 1960
34) Austin, A. P., Ann. Bot. 24 (95): 296–310. 1960
35) Czapke, K., Prace Morsk. Inst. Ryback. Gdyni 11/B: 95–111, 1960
36) Kireeva, M., Annal. Biol. 15: 65–66. 1960
37) Stoloff, L., Ec. Bot. 16 (2): 86–94. 1962

General Literature

Hegnauer, R., 1962; Hoffmann, C., 1938; Hoppe, H. A., 1958; Hoppe, H. A. and Schmid, O. J., 1962; Kylin, H., 1956; Lewin, R. A., 1962; Schmid, O. J., 1962; see Index of Literature.

Gelidiella

11 species, widespread in warmer seas.

G. acerosa (Forsk.) Feldm. et Hamel (= *Gelidium rigidum*). Warm seas, e.g. off Indonesia, Japan, Cuba. The type species. Thallus 1–5 centimetres high, terete, sometimes flattened, creeping, with erect branches.

Raw material for agar, especially in Japan, together with other agarophytes[1] [2] [5]. *G. acerosa* yields a type of agar that forms hard jellies[3].

In Indonesia the species is harvested by the population at low water and is used for food[4].

Literature

1) Stoloff, L. and Silva, P., Ec. Bot. 11 (4): 327–330, 1957
2) Selby, H. H. and Selby, T. A., Agar ex Whistler, R. L. and BeMiller, J. N., Industrial Gums. Academic Press Inc., New York, 1959
3) Smith, F. and Montgomery, R., The Chemistry of Plant Gums and Mucilages. Reinhold Publishing Corporation, New York, 1959
4) Zaneveld, J. S., Ec. Bot. 13 (2): 89, 1959

General Literature
Diaz-Piferrer, M. and López, H., Inst. Cubano de Invest. Tecnol. (6), 1959; Levring, T., 1960; Fritsch, F. E., 1952; Kylin, H., 1956; Segawa, S., 1957; see Index of Literature.

Gelidiopsis

6 species, boarders of warmer seas. Off the shores of Formosa, Japan[1], and Guinea[2].

G. variabilis (J. Ag.) Schmitz. Coasts of the Indian Ocean. The type species. Thallus erect, 3 to 10 centimetres high, terete, irregularly branched, of cartilaginous, sometimes even hornlike consistency.

G. rigida. Malaya, Indonesia, Philippines. This species is collected as raw material for agar production in these areas. Javanese algae yielded an agar content of 24,4%[3]. *G. rigida* is also used as food[4]. The Malayan name is "intip-intip"[5].

Literature

1) Yamada, Y., Bull. Jap. Soc. Phyc. 5 (2): 60–63, 1957
2) Lawson, G. W., J. Ecol. 44 (1), 1954
3) Zaneveld, J. S., Ec. Bot. 13 (2): 89, 1959
4) Eisses, J., Indonesian J. of Nat. Sc. No. 1,2,3, 1953
5) Morel, Th., Medeling van het Centraal Inst. voor Technische Onderzoekingen te Djakarta, Oct. 1949

General Literature
CHAPMAN, V. J., 1950; KYLIN, H., 1956; SEGAWA, S., 1957; UPHOF, J. C. TH., 1959; see Index of Literature.

Gelidium

About 40 species, many insufficiently investigated, in warmer regions, in littoral habitats and in deeper waters. There is a great variety of forms and shapes in this genus, and individual species are often very difficult to distinguish clearly.

G. corneum (Huds.) Lamouroux. Often designated *G. corneum* var. *sesquipedale*; European habitats: coasts of Spain (Galicia, Asturia, Santander)[1], Portugal, Italy (Straits of Messina), Great Britain[2]. Africa: coast of Morocco (Tangier to Agadir; with main collecting regions at Mazagan and Safi). South Africa. Asia: coasts of Japan and China. North America: coasts of California; isolated specimens on coast of Florida, coast of Cuba[3]. The type species. Thallus terete or flattened to a two-edged blade without midrib, often pinnately, sometimes irregulaly branched up to 30 centimetres high. Colour: bright to dark red. This alga grows in depths of up to 30 metres. Investigations have been made on shedding, floating and fixing of the spores[4].

Considered to be an important agar raw material, particularly for commercial production in Spain, Portugal, and Morocco[5–11] [80].

The agar content of the *Gelidium* species has been studied[12], as well as that of the polysaccharide[13]. Cholesterol has been found in it[14]. The *Gelidium* varieties supply the standard agar according to TSENG[15].

A number of *G.* species in addition *G. corneum* are important agar raw materials.

G. amansii. Coasts of Japan, particularly on the west coast of Hokkaido, Korea[16], China. This species is also found on various coasts of the Indian Ocean (Madagascar, Mauritius), as well as on the coasts of Malaya, the Philippines, Moluccas, California, and Mexico[17]. In Japan this raw material is known as "Tengusa", "Tokoro-tengusa", "Makusa" or "Kinukusa". A number of forms are recorded from Japan. The f. *typicum* is very common and widely distributed in the Pacific and the Sea of Japan. Other forms are: f. *elatum*, f. *elegans*, f. *latioris*, f. *teretius-culum* (SEGAWA, 1957)[18] [19].

SEGI has reported on the 'type' specimens[20]. The life history and development of this alga has been the subject of numerous research efforts[21–24].

In harvesting, which happens in the months of April to September, the algae are raked out of shallow water; whereas in deeper water, particularly on Japanese coasts, this work is also carried out by divers. Algae from the deeper regions are preferred for agar production. The plants are bleached by sun and fresh water and afterwards dried. Reports have been published on harvesting[79], on growth, and development of the alga, including growths on artificial stone-bed substrata[75] [76], and on the correlation between water temperature and geographical distribution[77].

This constitutes the most important agar raw material of Japanese production (hard type). The phycocolloid content is approximately 25 to 30 per cent of the dry weight basis. [5] [8] [9] [10] [25–29] [80]. Polysaccharides have been studied[30–31] [73]. The agar contains D-galactose, L-galactose, 3.6-anhydro-galactose[10]. Investigations have also been made on the agarose[32].

The alga is eaten in East and South East Asia in various preparations[17) 33)].

G. arborescens. Recorded for the U.S.A. (Southern California), Japan, China. Grows in deeper water. Agar raw material of small economic significance, mostly utilized in mixture with other species[8–10) 26) 34) 80)].

G. attenuatum. An agar raw material[8–10) 80)]. Used in Japan and China for the preparation of home-made agar "Isingglass"[33)].

G. arbuscula. Canary Islands (Teneriffe) produces agar of good quality[8–10) 80)].

G. australe (= *G. pulchrum*). Recorded from the coasts of California. Often described as an agar raw material [10)], but the phycocolloid, resembles much more to carrageenan from *Chondrus crispus*[5) 25)]. The phycocolloid content amounts to about 32 to 37 per cent (Chapman, 1950).

G. capillaceum. Coasts of Portugal, and Mediterranean coasts, e.g. Sardinia. In combination with *G. corneum* this alga serves as a raw material for agar in Italy.

G. cartilagineum. Canary Islands, South Africa (False Bay), U.S.A. (California, particularly near San Diego) Mexico, China. In California *G. cartilagineum* grows on rock substrata in fast moving water (Pl. XII: 28).

An important agar raw material of good quality, which is used commercially, particularly in the U.S.A. This alga supplies the greatest deal of the raw materials for American Agar. The phycocolloid content is approximately 40 to 45 per cent[5) 8–10) 25–27) 34–36) 80)].

The highest agar content in *G. cartilagineum* has been shown to reach 70 per cent[29)]. The content apparently is subject to seasonal fluctuations, and is about by 30 per cent higher in July than in January. The Mexican material contains approximately 10 to 35 per cent of the dry weight of agar[9)]. Work has been done on the structure of agar from *G. cartilagineum*[37)], and the alga has been investigated chemically[38)] and its inorganic content analysed[39)]. Agarose has been shown to be present[40–42)], and a report has appeared on the floridoside[43)].

G. caulacanthum. Coasts of New Zealand (Bay of Islands to Bluff)[44)]. An agar raw material for commercial utilization when mixed with other species[9) 26)]. The alga is small and therefore of lesser commercial importance[45)].

G. coulteri. Coasts of the U.S.A. (Southern California), Mexico. This alga supplies agar[9) 40) 80)]. It has been investigated chemically[38)], and the presence of agarose established[41)].

G. crinale. Temperate and warmer coastal areas: Japan, Gulf of Guinea[46–48)]. An agar raw material (hard type)[8) 10) 28)], commercially utilized in mixture with other types[9)]. The product contains D-galactose, L-galactose, 3.6-anhydro-L-galactose[10)].

G. crispum. Coasts of Peru and Central Chile. This is a very small alga. Specimens from Chile (Valparaiso) were only a few centimetres long[48)]. This alga, therefore, has no chance to be used as a commercial raw material.

G. densum. Southern California, used with *G. cartilagineum* as an agar raw material.

G. divaricatum. China, Japan, and other areas of the Pacific Ocean. An agar raw material (hard type). Commercially exploited in mixture with other species[10) 9)]. This alga is eaten in China; in Japan it is used for the preparation of home-made agar "Isingglass"[33)].

G. elegans. Coasts of Japan. Like *G. amansii* this species is described as "Tengusa" and serves as an agar raw material too[49)].

G. filicinum. Peru and Chile (Concepcion and southern Chile)[48)].

G. grubbae. China, Shantung Province. This new species has been described by Fan[50)].

G. japonicum. Japan and Korea[16]. The Japanese name is "Onikusa". This algae is an important agar raw material of the hard type for Japanese commercial production, being utilized in mixture with other species[8–10] [26] [29] [49] [80]. It is also used in Japan and China for the preparation of home-made agar "Isingglass"[35].

G. latifolium. Great Britain and Ireland, Mediterranean and Black Sea, Malaya, Indonesia (Java, Timor), Japan, China, Hawaiian Islands, Pacific Coast of the U.S.A. An agar raw material[8–10] [29] [51] [52] [80]. Material from the west coast of Java contains approximately 23 per cent, and that from Bandoeng approximately 35 per cent[17] [53]. A report recently appeared on this species in Britain[54]. Home made agar "Isingglass" is made from this species in Japan and China[33].

G. liatulum. Japan. A most important agar raw material of the hard type, utilized commercially in mixture with other species[9] [10].

G. lingulatum. Chile (from Pichidangui to Coquimbo, south of San Antonio, Chiloé)[48] [55]. An agar raw material utilized commercially. The agar content reaches a height of 18 to 23 per cent.

G. linoides. Japan, Korea, China. An agaroid raw material of the soft type, of small significance, but commercially utilized in mixture with other species[9] [10] [16] [26]. Also used in Japan and China for the preparation of home-made agar "Isingglass"[33].

G. micropterum. India (East coast of the State of Madras). An agaroid raw material[9]. This alga is reputed to contain 43 per cent phycocolloids.

G. nudifrons. Coast of the U.S.A. (California), Japan; from deeper water. An agar raw material, utilized commercially in mixture with other species[8–10] [26] [27] [34] [80]. It is also used in Japan for the preparation of home-made agar "Isingglass"[33].

G. pacificum (= G. polycladum). Japan, Korea, Formosa[16]. An important agar raw material (hard type), found in quantity on Japanese coasts; utilized commercially in mixture with other species[8–10] [26] [29] [80]. Also used in Japan and China for the preparation of home-made agar "Isingglass"[33]. The phycocolloid of this species contains D-galactose[10].

G. planiusculum. Japan. Japanese name "Hirakusa". An agar raw material (hard type) of little commercial significance, utilized in mixture with other species[9] [10].

G. pristoides. South Africa (Strandfontain). An agar or agaroid raw material utilized commercially[5] [8–10] [26] [40]. This species has been investigated chemically[35] [56]; STOLOFF (1962) describes the phycocolloid as agar[80]; it is similar to carrageenan[25]. Agarose has been proved[41], and the alga also contains floridoside[57].

G. pseudointricatum. Coasts of Chile (Chiloé), Juan Fernandez. This species, which is apparently common in this area, is a true *Gelidium*[48].

G. pulchellum. Great Britain and Ireland, in extensive beds[54]. An agar raw material [8–10] [51] [80]. The alga occurs in Ireland at various places and is collected during the summer months.

G. pulvinatum. An agar raw material[8–10] [80].

G. pusillum. Coasts of Japan (f. *foliaceum*) (Japanese name "Hai-ten-gusa"), Juan Fernandez[48]. Cameroon[59]. Coast of Pakistan (Karachi)[60] and in most warm and temperate waters[48]. This alga also occurs on the Columbian coast, on rocks near the tide line[58]. An agar raw material (hard type), commercially utilized in mixture with other species [8–10] [80].

G. pyramidale. Pacific coast of North America. This alga can be used as an agar raw material, if it occurs in sufficient quantities.

G. rigens. Coasts of Japan, where it is eaten[33]).

G. serrulatum. Coasts of Venezuela. An agaroid raw material[9]).

G. sesquipedale. Cf. *G. corneum* var. *sesquipedale*.

G. spinulosum. Morocco. An agar raw material with a phycocolloid content of approximately 33 per cent, utilized commercially in mixture with other species[9]).

G. subcostatum. Japan, Korea, Formosa. Japanese name "Hirakusa". An agar raw material (hard type)[8]) [10]) [26]) [49]), but according to other authorities this alga, which is utilized commercially in mixture with other species, is an agaroid raw material[9]). STOLOFF (1962) describes the polysacolloid as agar[80]). The phycocolloid contains L-galactose[10]). The Korean material supplies an agar of inferior quality[16]). In Japan this alga is used for the preparation of home-made agar, "Isingglass"[33]). Isolation and purification of agar from it has been described.

G. subfastigiatum. Coasts of Japan, west coast of Hokkaido. An agar raw material (hard type), utilized commercially in mixture with other species[9])[10])[19]).

G. tenue. Coasts of Japan. Japanese name "Hirakusa". An agar raw material (hard type)[10]), but according to other authorities it is agaroid raw material, of small importance, utilized commercially in mixture with other species[9]).

G. tsengii. Hong Kong, a new species described by FAN[50]).

G. vagum. Japan, particularly on the west coast of Hokkaido, Honshû, Sakhalin. An agar raw material (hard type), utilized commercially in mixture with other species[9]) [10]) [19]). Limited quantities are produced, only for home made "tocoroten"[18]). The chemical properties of the alga have been investigated[73]).

A number of other species of *Gelidium* occur in various coastal areas.

Ceylon. *Gelidium* occurs on the south and west coast of the island, but *Gracilaria* is the most important agar raw material of Ceylon[61]).

Chile. Species of *Gelidium* also supply agar in Chile.

Great Britain. The structure and development of the thallus of the British *Gelidium* species has been investigated [62]).

India. Native species are marketed in India.

Indonesia. Species of *Gelidium* are containing 40 per cent agar, and reputed to be better than *G. amansii* and *G. japonicum* in quality, are sold in Indonesia.

Italy. In Italy certain species of *Gelidium* are used in agar production[63]).

Japan. Papers have been published on the species of the genus *Gelidium* from Japan and its vicinity[64]) [65]).

Korea. *Gelidium* species are found among the raw material of Korean agar[66]).

Philippine Islands. Various species of *Gelidium* occur in these islands, but apparently have not yet been further investigated[67]).

U.S.A. Various *Gelidium* species grow on rock substrata in fairly shallow waters on the Californian coast[68]) [78]).

Eight species of *Gelidium* have been described from the Pacific coast of the U.S.A. and the Hawaiian Islands; six are new, including *G. undulatum*, which resembles to *Gelidium crispum* from Peru and Chile[69]).

Morphology, ecology, taxonomy, and classification of the Gelidiacae have been studied [70–72].

Physiological and biochemical investigations have been reported in LEWIN (1962) on: *G. amansii*, *G. cartilagineum*, *G. corneum*, *G. japonicum*, *G. latifolium*, *G. pacificum*, and *G. subcostatum*.

Literature

1) CABRERO-GOMEZ, F., Estudio de las Algas marinas espanoles C. S. I. C., Madrid, 1951
2) DIXON, P. S., Phyc. Bull. 6: 47–48. 1958
3) DIAZ-PIFERRER, M. and LÓPEZ, H., Inst. Cubano de Invest. Tecnol. (6), 1959
4) SUTO, S., Bull. Jap. Soc. Sci. Fish. 15 (11): 671–673, 1950
5) STOLOFF, L. and LEE, F., Fishery Leaflet 173, 1946. US Dep. of the Interior, Fish and Wildlife Service
6) VEIGUINHA, A. S., Ind. agr. alim. 1951, 261–264
7) MUROLO, M. and MUROLO, V., Brevet d'Invention Nr. 1.017. 156 Rep. Française 1950/1952
8) STOLOFF, L. and SILVA, P., Ec. Bot. 11 (4): 327–330, 1957
9) SELBY, H. H. and SELBY, T. A., Agar ex Industrial Gums. Academic Press Inc., New York, 1959
10) SMITH, F. and MONTGOMERY, R., The Chemistry of Plant Gums and Mucilages. Reinhold Publishing Corporation, New York, 1959
11) DANGEARD, P., Coll. Int. du Centre National de la Recherche Scient., Dinard, 1960
12) AOKI, K., Bull. Jap. Soc. Sci. Fish. 8, 223–226, 1940
13) JANECKE, H. and KEHR, W., Pharmaz. Ztg. 105 (25): 684, 1960
14) TSUDA, K. et al., Chem. & Pharm. Bull. 6 (1): 101–104, 1958
15) TSENG, C. K., Science 101 (2633): 597–602, 1945
16) SCHULZE, W., D. Apoth. Ztg. 98 (15): 340, 1958
17) ZANEVELD, J. S., Ec. Bot. 13 (2): 89, 1959
18) TOKIDA, J., Fac. of Fish., Hokkaido Univ., 2 (1); 1954
19) TOKIDA, J. and MASAKI, T.; ibid. 10 (3): 173–195, 1959
20) SEGI, T., REPT. Fac. of Fish., Univ. of Mie 3 (2): 251–256, 1959
21) UEDA, S. and KATADA, M., Bull. Jap. Soc. Sci. Fish. 15 (7): 354–358, 1949
22) KATADA, M.; ibid. 15 (7): 359–362, 1949
23) KATADA, M. et al.; ibid. 19 (4): 471–473, 1953
24) KATADA, M. and MATSUI, T., J. Shimonoseki Coll. Fish. 3: 229–233, 1954
25) HUMM, H. J. and WOLF, F. A., Duke Univ. Press (3): 1946
26) HUMM, H. J., The Red Algae of Economic Importance ex Tressler, D. K. et Lemon, J. Mc. W., Marine Products of Commerce. Reinhold Publishing Corporation, New York, 1951
27) BAKER, G. L., US Pat. Off. 2. 669. 519, 1952–1954
28) WHISTLER, R. L. and SMART, CH. L., Polysaccharide Chemistry. Academic Press Inc., New York, 1953
29) ARAKI, CH., Seaweed Polysaccharides ex Wolfrom, M. L., Carbohydrate Chemistry of Substances of Biological Interest. Pergamon Press Ltd., London, 1959
30) MIYABE, S. and HAYASI, H., J. Soc. Trop. Agr., Taihoku Imp. Univ. 11, 200–203, 1939
31) FOSTER and STACY, Polysaccharides from Algae and Lichens ex W. Ruhland, Handbuch der Pflanzenphysiologie, Springer-Verlag, 1958
32) ARAKI, C. and HIRASE, S., Bull. Chem. Soc. Jap. 33: 291–295. 1960
33) UPHOF, J. C. TH., Dictionary of Economic Plants. H. R. Engelmann (J. Cramer), Weinheim, 1959
34) SANFORD, F. B., Fishery Leaflet 469, 1958. US Dep. of the Interior, Washington D. C.
35) SELBY, H. H., Fishery Leaflet 118, 1948. US Dep. of the Interior, Washington D. C.
36) MCINNES, A. A., Second Int. Seaweed Symp., 1955
37) O'NEILL, A. N. and STEWART, D. K. R., CANAD. J. Chem. 34: 1700–1703, 1956
38) ROSS, A. G., J. Sci. Food Agric. (7): 333–335, 1953
39) YOUNG, E. G. and LANGILLE, W. M., Publ. 4636 of the Nat. Res. Council, Halifax, Canada, 1958
40) YAPHE, W., Can. J. Microbiol. 3. 987–993. 1957
41) YAPHE, W., Can. J. Bot. 37. 751–757. 1959
42) YAPHE, W., Analytical Chemistry 32. 1327. 1960
43) LINDBERG, B., Second Int. Seaweed Symp., 1955
44) BERGQUIST, P. L., Bot. Mar. I (3/4): 86. 1960
45) MOORE, L. B., New Zealand J. of Science and Technology 25 (5): 183–209. 1944 (Sec. B)
46) LAWSON, G. W., J. Ecol. 44 (1): 1954

47) LEVRING, T., Contributions to the marine algal Flora of Chile. Lunds Univ. Årsskrift N. F. Avd. 2. 56 (10), 1960
48) VALERA, M. DE, Inst. for Ind. Res. and Stand., Dublin, 1958
49) Hypothekenbank von Japan. Die Agar-Herstellung in Japan, Tokyo (17): 1950
50) FAN, K.-CH., Bot. Mar. II (3/4): 247. 1961
51) BRENNAN, A. T., Inst. for Ind. Res. and Stand., Dublin, 1950
52) DILLON, T., First Int. Seaweed Symp., 1952
53) EISSES, J., Indonesian J. of Nat. Sc. (1, 2, 3): 1953
54) DIXON, P. S., Ann. Bot. n. s. 23 (91): 397–407, 1959
55) HARING, G., Informe sobre Agar-Agar. Compania Industrial, Vina del Mar, Chile, 1953
56) BLACK, W. A. P., The Algae ex Altschul. Processed Plant Proteins Foodstuffs. Academic Press Inc., New York, 1958
57) STOLL and JUCKER, Heteroside ex W. Ruhland, Handbuch der Pflanzenphysiologie, Springer-Verlag, 1958
58) TAYLOR, W. R., Bot. Ser. Field Museum of Natural History 20 (4): 1941
59) LAWSON, G. W., J. West Afric. Science 1 (2): 1955
60) ANAND, P. L., Marine Algae from Karachi. The University of the Panjab, Lahore, 1943
61) HOLSINGER, E. C. T., First Int. Seaweed Symp., 1952
62) DIXON, P. S., Ann. Bot. n. s. 22 (87): 353–368. 1958
63) BUZZONETTI, G. C., Chimica (Mailand) 12. 157–161. 1956. CA 17253e. 1956
64) SEGI, T., Rept. Fac. of Fish. Univ. of Mie 2 (1): 124–137. 1955
65) SEGI, T.; ibid. 2 (3): 456–462. 1957
66) Foreign Commerce Weekly (Washington). *Gelidium* processed in the Republic of Korea (5): 1954
67) VELASQUEZ, G. T., Philippine Seaweeds Report, 1955
68) WILLIAMS, R. H., The Commercial Seaweed Industry. Univ. of Miami, Marine Laboratory, 1949
69) LOONIS, N. H., All. Hancock Found. Occ. Pap. (24): 1–10. 1960. Rev. Biol. Mar., Valparaiso, 1960
70) DIXON, P. S., Phyc. Bull. 6: 32–33. 1958
71) DIXON, P. S., Bot. Mar. III (1): 1. 1961
72) FAN, K.-CH., Univ. Calif. Publ. Bot. 32 (5): 315–368. 1961
73) TAMI, K. and WATANABE, T., Bull. Tohoku Reg. Fish. Res. Lab. 16: 145–149. 1960
74) MURAKAMI, S., Sc. Rep. Saitama Univ. 3 B (3): 251–254. 1960
75) YAMASAKI, H. and OSUGA, H., Bull. Jap. Soc. Sc. Fish. 26 (1): 9–12. 1960
76) YAMASAKI, H.; ibid. 26 (2): 116–122. 1960
77) ENDO, T. and MATSUDAIRA, Y., Bull. Jap. Soc. Sc. Fish. 26 (9): 871–876. 1960
78) DAWSON, E. Y., Pac. Nat. 1 (19/20): 31–52. 1960
79) SUTO, S., Aquiculture Sci. Sc. 8. 1–53. 1954
80) STOLOFF, L., Ec. Bot. 16 (2): 86–94. 1962

General Literature

CHAPMANN, V. J., 1950; FOTT, B., 1959; FRITSCH, F. E., 1952; HOFFMANN, C., 1938; HOPPE, H. A., 1958; HOPPE, H. A. and SCHMID, O. J., 1962; KYLIN, H., 1956; LEWIN, R. A., 1962; PAX, F., 1962; SEGAWA, S., 1957; UPHOF, J. C. TH., 1959; see Index of Literature.

Gigartina

Approximately 90 species, widespread, a genus including several distinct forms regarded as subgenera by KYLIN.

G. pistillata (Gmel.) Stackh. Atlantic coasts of Europe; Morocco, South Africa[1]), Canada, Malaya[2]). *G. pistillata* is the type species. Thallus erect, 1–90 centimetres high, terete of flattened and thickly foliose, more or less branched or lobed. The alga contains a carrageenan[3–6]) of low kappa-carrageenan content[7]). The inorganic constituents have also been investigated[8]).

G. pistillata (= *Mastocarpus Klenzianus*). It is eaten and used as agar raw material[2]).

G. stellata (= *G. mamillosa*). The most important species of the genus are commercially utilized. Atlantic coasts of Norway, Great Britain, Ireland, France, Spain, Portugal, Morocco, less widespread on the coasts of the U.S.A., but abundant on the Nova Scotia coast (Bay of Fundy, Digby Neck)[4]) [9–15]). According to SEGAWA *Gigartina stellata* also occurs on Japanese coasts. Several *G.* species, particularly *G. stellata*, are often utilized together with *Chondrus crispus* as a carrageenan raw material on an industrial scale. The gatherers usually make no distinction between these algae.

According to the literature, *G. stellata* occurring on British coasts includes several varieties (var. *acuta* Good and Woodw.) and growth forms (var. *prolifera* Turn. and var. *incurvata* Turn.)[16–17].

Commercial utilization of this alga has been investigated by the authors of this volume. The phycocolloid obtained was originally described as "British agar" but is more correctly described as a carrageenan.

G. stellata is an important carrageenan[3] [6] [18–25], or according to some authors, agaroid raw material[19] [26]. The content of kappa-carrageenan is comparable to that of *Chondrus crispus*[7]. The polysaccharides[5] [27–30], amino acids[33], and other constituents have been studied[31–32]. It has been found to contain floridoside and also D-mannitol, which previously has not been found in red algae[34].

G. stellata has been investigated with regard to the effect of different ions on gel strength of extracts[19], carotene, carotenoid and vitamin C content[35] [37] [56], antibiotics[38], anticoagulent effects[57], inorganic constituents[8], and variation in chemical composition[39].

A number of other species of *Gigartina* are reported to be used in some regard as phycocolloid raw material and to be of some commercial interest.

G. acicularis. Portugal, Morocco (Tangier to Agadir), Gold Coast[40], Cameroons[41], Canada[8].

The alga contains a carrageenan[3] [4] with low kappa-carrageenan value[7]. The phycocolloid is clearly distinct from the usual carrageenan. *G. acicularis* is of commercial significance in many areas[42]. Its inorganic constituents have been investigated[8].

Portuguese commercial raw material of this species is often mixed with *G. pistillata*.

G. alveata. Coasts of New Zealand[43].

G. angulata. Coasts of New Zealand, where the plant is used for the clarification of beer[17].

G. asperifolia. California. The plant contains carrageenan[3] [5] and was formerly described as an agar or agaroid raw material respectively[44] [26] [4]. It requires special treatment for use[45].

G. california. California. This species has been mentioned by Fritsch. Nothing has been found out about its commercial significance.

G. canaliculata. California. This alga contains carrageenan[3] [5] and was formerly described as an agar or agaroid raw material respectively[44] [26] [4]. It requires special treatment prior to use[45] [17].

G. chamissoi. Peru and Chile (Bay of Pacasmayo, Chiloé)[46], in quantity. This alga contains carrageenan and is used for the production of Chilean agar.

G. chauvinii. Occurs at the Pacific coast of America, from California to Chile, particularly at Concepcion and in the Cape Horn region[46] [58].

G. cristata. Coasts of California. This alga contains carrageenan with a low value of kappa-carrageenan[7]. The alga has been studied by several authors[31].

G. clavifera. New Zealand, where it is used for the clarification of beer[17].

G. corymbifera. California. An agaroid raw material[26].

G. decipiens. New Zealand. The alga contains carrageenan[3] [4] [5], it is used as a substitute for *Chondrus crispus*, particularly in the brewery industry, as a clarifying agent[17][47].

G. glomerata. Peru and Chile, particularly at Chiloé[46].

G. horrida. Pacific Ocean. Used in Malaya as an agar raw material[2]. According to some authorities it is not certain, whether this alga is really a *Gigartina* species. Sometimes it has been placed in *Eucheuma*[48].

G. intermedia. Bay of Oshoro, Japan[49].

G. leptorhynchos. This alga has been studied chemically: Taurines have been isolated from it[50].

G. lessonii. Coasts of Peru and Chile, particularly at Arica and Juan Fernandez[46]

G. ochotensis. Coasts of Japan, particularly in the Bay of Oshoro[49], northern Honshû, Hokkaido, Kurile Islands, Sakhalin, Korea, coasts of the Okhotsk Sea, Kamchatka[51]. In Japan used industrially for the production of an "alga mucilage"[51].

G. pacifica. Coasts of Japan, particularly coast of Oshoro[49], northern Honshû, Hokkaido, Kurile Islands, Sakhalin, coasts of Sea of Okhotsk, and the Bering Islands[51]. Utilized industrially in Japan for the production of an "alga mucilage"[51].

G. papillata. This species is reported to contain a remarkably high Vitamin C content[17].

G. radula. Coasts of Australia and South Africa (Kommetje). The alga contains carrageenan[3] [4] [5] with a low kappa-carrageenan value[7]. It has been analysed chemically[31]. Commercial raw material from South Africa consists partly of the very similar *G. spatula* (Levring).

G. serrata. California. The alga contains carrageenan[5] and was formerly described as an agar raw material[44] requiring special treatment for use[45] [17].

G. skottsbergii. Recorded from the coasts of Chile (Fuegia, St. Magdalena, Puerto Bueno, Valparaiso), Falkland Islands, Patagonia[46]. The phycocolloid has been analysed[52].

G. stiriata. South Africa (Kommetje). The alga contains carrageenan with a share of kappa-carrageenan content corresponding to that of *Chondrus crispus*[7]. This alga has also been investigated by other authors[31].

G. teedii. Japan, also Portugal together with *G. stellata* to which it closely resembles. Used as a food product in Japan[2] [17]. A variety of this species has been described[53].

G. tenella. Japan, Bay of Oshoro[49]. The alga contains galactose[5].

G. tuberculosa. Peru, Chile (Southern Patagonia, Fuegia), New Zealand, Auckland Islands[46].

Gigartina species are not clearly identified till now, also recorded from South Africa[54], from the south and west coast of Ceylon[55], the Singapore area, and New Zealand.

Physiological and biochemical investigations on *G. acicularis*, *G. leptorhynchos*, *G. papillata*, *G. pistillata*, *G. radula*, and *G. stellata* are reported in Lewin (1962).

Stoloff (1962) describes the polysacolloids of *G. acicularis*, *G. asperifolia*, *G. canaliculata*, *G. chamissoi*, *G. chauvinii*, *G. cristata*, *G. decipiens*, *G. leptorhynchos*, *G. pistillata*. *G. radula*, *G. skottsbergii*, *G. stellata* (= *mamillosa*), *G. stiriata* as carrageenans[59].

Literature

1) Isaac, W. E. and Simons, S. M., J. S. Afric. Bot. 20. 117–124. 1954

2) Uphof, J. C. Th., Dictionary of Economic Plants. H. R. Engelmann (J. Cramer), Weinheim, 1959

3) Stoloff, L. and Silva, P., Ec. Bot. 11 (4): 327–330. 1957

4) Stoloff, L., Carrageenan ex Whistler, R. L. and BeMiller, J. N., Industrial Gums. Academic Press Inc., New York, 1959

5) SMITH, F. and MONTGOMERY, R., The Chemistry of Plant Gums and Mucilages. Reinhold Publishing Corporation, New York, 1959
6) DANGEARD, P., Coll. Int. du Centre Nat. de la Recherche Scient., Dinard, 1960
7) YAPHE, W., Can. J. Bot. 37. 751–757. 1959
8) YOUNG, E. G. and LANGILLE, W. M., Publ. 4636 of the Nat. Res. Council, Halifax, Canada, 1958
9) PRINTZ, H., Utilization of Seaweeds in Norway
10) PRINTZ, H., Seaweeds, their properties and general uses, especially in Norway, 1952
11) PRINTZ, H., Tang og Tare. Johann Grundt Tanum Forlag, Oslo, 1953
12) BRENNAN, A. T., Notes on some common Irish Seaweeds. Inst. f. Ind. Res. and Stand., Dublin, 1950
13) CABRERO-GOMEZ, F., Estudio de las Algas marinas espanoles C. S. I. C., Madrid, 1951
14) MAC FARLANE, C. I., Third Int. Seaweed Symp., 1958
15) MAC FARLANE, C. I. Second Int. Seaweed Symp., 1955
16) MARSHALL, S. M. et al., A Study of Certain British Seaweeds and their Utilization in the Preparation of Agar. His Majesty's Stationary Office, London, 1949
17) CHAPMANN, V. J., Seaweeds and their Uses. Methuen & Co., Ltd., London, 1950
18) TSENG, C. K., Science 101 (2633): 597–602. 1945
19) MARSHALL, S. M. and ORR, A. P., Natural Plant Hydrocolloids, 1954, 101
20) O'NEILL, A. N., J. Am. Chem. Soc. 77. 2837. 1955
21) YAPHE, W. and BAXTER, B., Applied Microbiology 3 (6): 380–383. 1955
22) NEWTON, L. et al., Coll. Int. du Centre Nat. de la Recherche Scient., Dinard, 1957
23) YOUNG, E. G., Carbohydrate accumulation by lower plants (Rhodophyta) ex W. Ruhland, Handbuch der Pflanzenphysiologie, Springer-Verlag, 1958
24) YOUNG, E. G., Coll. Int. du Centre Nat. de la Recherche Scient., Dinard, 1960
25) SCHMITT. M.; ibid.
26) SELBY, H. H. and SELBY, T. A., Agar ex Whistler, R. L. and BeMiller, J. N., Industrial Gums. Academic Press Inc., New York, 1959
27) DEWAR, E. T. and PERCIVAL, E. G., Nature 156. 633–634. 1945
28) BLACK, W. A. P. and CORNHILL, W. J., Chem. and Ind. 1954, 514–516
29) HIRST, E. L. and JONES, I. K. N., Gums and Mucilages of Plants ex W. Ruhland, Handbuch der Pflanzenphysiologie, Springer-Verlag, 1958
30) JANECKE, H. and KEHR, W., Pharmaz. Ztg. 105 (25): 684. 1960
31) ROSS, A. G., J. Sci. Food Agric. (7): 333–335. 1953
32) BLACK, W. A. P., The Algae ex Altschul. Processed Plant Proteins Foodstuffs. Academic Press Inc., New York, 1958
33) COULSON, C. B., Chem. and Ind., 1953, 971–972
34) LINDBERG, B., Second Int. Seaweed Symp., 1955
35) HUMM, H. J., The Red Algae of Economic Importance ex Tressler, D. K. and Lemon, J. Mc. W., Marine Products of Commerce. Reinhold Publishing Corporation, New York, 1951
36) OWEN, E. C., J. Sci. Food Agric. 5 (9): 449–453. 1954
37) HAUG, A. and LARSEN, B., Second Int. Seaweed Symp., 1955
38) CHESTERS, C. G. C. and STOTT, J. A., Second Int. Seaweed Symp. 1955
39) BAARDSETH, E. and HAUG, A., Norsk inst. f. tang- og tareforskning, Rep. 2, 1953
40) LAWSON, G. W., First Int. Seaweed Symp., 1952
41) LAWSON, G. W., J. West Afric. Science 1 (2). 1955
42) SCHACHAT, R. E. and GLICKSMAN, M., Some Lesser Known Seaweed Extracts ex Whistler, R. L. and BeMiller, J. N., Industrial Gums. Academic Press Inc., New York, 1959
43) BERGQUIST, P. L., Bot. Mar. I (3/4): 86. 1960
44) HUMM, H. J. and WOLF, F. A., Duke Univ. Press, (3). 1946
45) STOLOFF, L. and LEE, CH. F., Fishery Leaflet 173. 1946. US Dep. of the Interior, Washington
46) LEVRING, T., Contributions to the marine algal Flora of Chile. Lunds Univ. Årsskrift N. F. Avd. 2. 56 (10). 1960
47) Bull. Imp. Inst. (London) 41. 163–165. 1943
48) ZANEVELD, J. S., Ec. Bot. 13 (2): 89. 1959
49) TOKIDA, J. and MASAKI, T., Fac. Fish. Hokkaido Univ. 10 (3): 173–195. 1959
50) WICKBERG, B., Third Int. Seaweed Symp. 1958
51) TOKIDA, J., Fac. Fish. Hokkaido Univ. 2 (1). 1954
52) LACOSTE and DIAZ, E. N., Rev. Invests. Agr. 11 (2): 163–178. 1957
53) RODRIGUES and MESQUITA, J. E., Bol. Soc. Bot. n. s. 32. 91–94. 1958
54) HEWITT, F. E., Univ. Calif. Publ. Bot. 32 (3): 195–233. 1960
55) HOLSINGER, E. C. T., First Int. Seaweed Symp. 1952
56) JENSEN, A., Acta Chem. Scand. 14 (9): 2051. 1960
57) GARCIA PINEDA, M. D. and MONTEQUI, D., Bol. del Inst. Espanol de Oceanogr. (72). 1955

58) FELDMANN, J., Rev. Alg. n. s. 5 (2): 151–152. 1960
59) STOLOFF, L., Ec. Bot. 16 (2): 86–94. 1962

General Literature
FOTT, B., 1959; FRITSCH, F. E., 1952; HOFFMANN, C., 1938; HOPPE, H. A., 1958; HOPPE, H. A. and SCHMID, O. J., 1962; Institute of Seaweed Research, Inveresk, 1956; KYLIN, H., 1956; LEWIN, R. A., 1962; PAX, F., 1962; SEGAWA, S., 1957; see Index of Literature.

Gloiopeltis

Some seven, in part insufficiently distinct species in warmer areas of East Asia, growing on rocky substrata.

G. tenax (Turn.) J. Ag. Coasts of Japan, particularly near Osaka; in China, at Chekiang and Amoi[1]). Coast of California[2]). The type species. Thallus erect, 1–4 centimetres high, terete, irregularly branched, very loose inside and hollow, swelling readily into mucilage. Japanese name "Ma-funori"; raw material of "Funori" (Japanese) or "Hailo" (Chinese), described as Funoran[2]) or Funorin[3]) [4]) in modern scientific literature. Some authors regard the product from *Gloiopeltis* as carrageenan[5]) [6]) or carrageenan-like[7]). The raw material is harvested throughout the year. The plants are cut from rocks by means of long-handled hooks, or are carried up by divers from deeper zones. The annual production of fresh algae reaches approximately 5,000 tons in Japan and from 500 to 1.000 tons in China[2]).

G. furcata. Coasts of Japan, Kurile Islands, Sakhalin, Korea, Siberia on the Sea of Japan, China, Pacific coast of North America[8]). This species is abundant in East Asian waters. Japanese name: "Funori" or "Fukoro-funori". The constituents are similar to those of *G. tenax*[9]). A raw material for Funoran, particularly in southern China. The product is used for sizing black silk and is made by a great part of the population during the summer months. Material of inferior quality is used for paper making[10]).

A number of publications appeared on the funoran from *G. furcata*[11–14]). STOLOFF (1962) describes the polysacolloids of *G. tenax* and *G. furcata* as carrageenans[27]). Results of physiological and biochemical investigations are reported by LEWIN (1962).

G. coliformis and *G. intricata*. Japan, Bay of Oshoro. These two species often have been regarded as particular forms of *G. furcata* f. *coliformis*[2]) [15]) [16]). They are recorded from the coast of California, where they have been investigated[6]).

G. complanata. Coasts of Japan. Japanese name "Hana-funori". The alga is also regarded as a funoran raw material[2]).

The mucilage of *Gloiopeltis* species has been the subject of a number of investigations [17–20]).

Reports have been published on the *Gloiopeltis* culture[21–24]). Detailed reports were submitted in 1957 at the Colloquium in Dinard. The "méthode des rochers posés" has been known for some time and is still practised now. However, other methods have also been developed. The growers arrange ropes and nets in littoral areas where *Gloiopeltis* grows. The spores of the alga then colonize the nets and after about two weeks the nets are laid out in quiet bays, to allow the algae to grow just as in *Porphyra* culture[25]). Algal diseases, such as the "black dots" disease of *G. tenax*, have been reported[26]).

Literature

1) Tseng, C. K., Lignan Sci. J. 12: 43–63, 1933
2) Schachat, R. E. and Glicksman, M., Some Lesser Known Seaweed Extracts ex Whistler, R. L. and BeMiller. J. N., Industrial Gums. Academic Press Inc., New York, 1959
3) Young, E. G., Carbohydrate accumulation by lower Plants ex W. Ruhland, Handbuch der Pflanzenphysiologie, Springer-Verlag, 1958
4) Young, E. G., Coll. Int. du Centre Nat. de la Recherche. Scient., Dinard, 1960
5) Stoloff, L. and Silva, P., Ec. Bot. 11 (4): 327–330, 1957
6) Smith, F. and Montgomery, R., The Chemistry of Plant Gums and Mucilages. Reinhold Publishing Corporation, New York, 1959
7) Dangeard, P., Coll. Int. du Centre Nat. de la Recherche. Scient., Dinard, 1960
8) Tokida, J., Fac. of Fish. Hokkaido University, 2 (1): 1954
9) Araki, C. et al., Third Int. Seaweed Symp., 1958
10) Chiu, B. T., Second Int. Seaweed Symp., 1955
11) Araki, C., Seaweed Polysaccharides ex Wolfrom, M. L., Carbohydrate Chemistry of Substances of Biological Interest. Pergamon Press Ltd., London, 1959
12) Arasaki, S., Bull. Jap. Soc. Sci. Fish. 14 (4): 207–210, 1949
13) Mita, K., Bull. Jap. Soc. Sci. Fish. 22 (9): 558–560, 1957
14) Janecke, H. and Kehr, W., Pharmaz. Ztg. 105 (25): 648, 1960
15) Tokida, J. and Masaki, T., Fac. of Fish. Hokkaido University 10 (3): 173–195, 1959
16) Humm, H. J., The Red Algae of Economic Importance ex Tressler, D. K. and Lemon, J. Mc. W., Marine Products of Commerce. Reinhold Publishing Corporation
17) Katsuura, K., J. Chem. Soc. Jap. Ind. Chem. Sect. 56: 903–905, 1953. CA 48. 22. 142631. 1954
18) Katsuura, K., J. Chem. Soc. Jap. Ind. Chem. Sect. 57: 765, 1954
19) Hirase, S. et al., Bull. Chem. Soc. Japan 31: 428–431, 1958. CA 53, 7, 6361a, 1959
20) Hirase, S. et al., Bull. Chem. Soc. Japan 29: 985–987, 1956. CA 52. 11. 8972g, 1958
21) Suto, S., Bull. Jap. Soc. Sci. Fish. 14 (4): 184–188, 1949
22) Suto, S., Bull. Jap. Soc. Sci. Fish. 15 (5): 225–228, 1949
23) Matsui, T., J. Shimonoseki Coll. Fish. 6 (3): 353–358, 1957
24) Matsui, T., J. Shimonoseki Coll. Fish. 8 (1): 113–120. 1959
25) Yamada, Y., Coll. Int. du Centre Nat. de la Recherche, Scient., Dinard, 1957
26) Matsui, T., J. Ag. I. Jap. Bot. 33 (2): 38–40. 1958
27) Stoloff, L., Ec. Bot. 16 (2): 86–94, 1962

General Literature

Chapman, V. J., 1950; Fott, B., 1959; Fritsch, F. E., 1952; Hoffmann, C., 1938; Hoppe, H. A. and Schmid, O. J., 1962; Kylin, H., 1956; Lewin, R. A., 1962; Segawa, S., 1957; see Index of Literature.

Goniotrichum

2 species, in various seas.

G. elegans (Chauv.) Le Jolis. Thallus erect, filamentous, branched, adhering to the substratum by a widened basal cell. *G. elegans* requires vitamin B_{12}[1]). Physiological and biochemical investigations of this species have been reported by Lewin (1962).

Literature

1) Fries, L., Nature, 183 (4660): 558–559, 1959

General Literature: Fritsch, F. E., 1952; Kylin, H., 1956; Lewin, R. A., 1962; see Index of Literature.

Gracilaria

Approximately 100 species, in various oceans.

G. confervoides (L.) Grev. (= *G. verrucosa* (Hudson) Papenfuss). Found at low water and below, a characteristic inhabitant of sandy bottom in the Mediterranean and other seas. (Pl. XII: 30).

Europe: South-West coast of Norway (LEVRING), southern Trondheim, depths 1–6 metres[1–3]. North Sea, North Frisian mud shoals (KUCKUCK), Mediterranean coasts[4–7]. This species also occurs on the Irish coasts, but not in industrially utilizable quantities[8].

Africa: South Africa ("Sea string" – abundant in the Saldanha-Langebaab-Lagoon and at the southern part of the West Coast)[9]. Individuals grow to a length of up to two metres. Habitat below low water. Much of the weed is cast ashore after the winter storms [10]. *Gracilaria* also inhabits the Egyptian Mediterranean coast.

Asia: Widely distributed on the Japanese coast. Japanese name: "Ogo" resp. "Ogo-nori"[11–13]. Coasts of the Kurile Islands, Sakhalin, Korea, Formosa, China ("Chinese Moss"), Siberian coast of the Sea of Japan, the Philippines[14]. Coasts of Thailand and Indonesia ("Gulaman Dagat"), Ceylon ("Ceylon Moss"), India, East Coast of South India (in the brackish water lakes of Orissa)[10], coasts of Pakistan[15], coasts of the Black Sea[10].

Australasia: South-East coast of Australia, New Zealand (Manukau, Tauranga, Porirua)[16].

North and Central America: Coasts of California, Washington, British Columbia. South Atlantic Coast of the Beaufort Region, Gulf of Mexico, Yucatan, North Carolina, Cuba.

In these areas *G. confervoides* grows together with other *G.* species. Individuals from the Atlantic coast are smaller than the African plants, being mostly 20–90 centimetres long.

Gracilaria on the Pacific Coast (British Columbia), on the other hand, often grows to a length of 3 to 4 metres.

On the South Atlantic coast two phases are distinguished: The one growing on the rocks, is fertile, but only grows in small quantities, the other is sterile and drifts in large masses in the bay. These algae are "fished" in the summer months and are used for industrial processing[17–24] [10].

South America: Coasts of Columbia[17], Ecuador, St. Helena[10], Chile (Costa del Talcahuana, Chiloé, Ancud), Argentina, and Uruguay. This is a species of many varieties[25].

G. confervoides is the type species of this genus. Thallus erect, of various lengths (a few centimetres to three or four metres long), very thin filaments, terete or flattened, branching dichotomous or lateral; colour dark red when fresh. Commercial qualities are for the most part bleached to some extent. The growth form of this alga varies with the location.

As a raw material *G. confervoides* is the most important *Gracilaria*. Investigations have been made on growth, development, ecology, and culture experiments[11] [26–29], and harvesting and treatment of raw materials are described[30–32].

Gracilaria supplies an agar or an agaroid raw material, depending on the origin and time of harvesting. The amount of phycocolloid content apparently depends on the time of harvesting, too, and is subject to strong fluctuations[21] [22] [33–42]. In a number of production areas *Gracilaria* is utilized together with other algal raw materials. In Japan, for example, it is processed with *Gelidium* and other species.

The *Gracilaria* phycocolloid was formerly regarded as of inferior value, but "it has become utilizable as the principle material by some special methods in Japan as well as in California" (TOKIDA, 1954[11]). The South African plant is regarded as a good agar raw material, whereas other varieties yield an agaroid material[10]. Quality differences in products from *Gracilaria* of various provenance have been reported in 1946 by STOLOFF and LEE. Distinctions have been made between the raw materials of the Californian and those of the Atlantic Coasts. The product from Atlantic Coast *Gracilaria* is similar to that from South African material[34].

A number of works have been published on the production of agar from *G. confervoides*[43–48].

The phycocolloid from many species of *Gracilaria* has also been described as Gracilaria Gum. It is differs in many respects from agar and has other fields of application. SCHACHAT and GLICKSMAN (1959), writing on the subject, state that "with the growth of knowledge of the chemical structure and physical properties it might be that new uses will be found for these potentially valuable Gracilaria extracts"[42] [49–51].

Other investigations on *G. confervoides* include: Dependence of composition on season of the year[52]; effects of environmental factors (water current and light intensity)[53]; chemistry of the raw material[54–56]; particularly of the polysaccharide[57] [58], (D-galactose, L-galactose[41], [59]), agarose[69], floridoside[61]) being identified as constituents); sterols[62].

G. confervoides is eaten in Asian countries as a jelly in various dishes[33] [63] [42], it is also a useful fertilizer, for example, for cocoa plantations[64].

Gracilarias other than the type species, *G. confervoides*, commercial significance to varying degrees:

G. arcuta. Indonesia, Makassar. This species, according to previous investigations, contains no agar material[65].

G. armata. Tropical and subtropical Atlantic, West Coast of U.S.A., from Cape Sahle to Tampa Bay, Mediterranean. An agar or an agaroid raw material[10] [18] [38] [41].

G. blodgettii. Coasts of Japan, U.S.A. (Florida, eastern coast of North Carolina), Cuba [24], Indonesia (Java, Surabaya, Lambok)[59]. An agaroid raw material (soft type)[10] [18] [20] [38] [41]. Indonesian material is mainly exported to Japan[65].

G. canaliculata. Philippines. This alga supplies an agar raw material corresponding to that of USP[66].

G. caudata. U.S.A., West Coast of Cape Sahle to Tampa Bay, Cuba[24]. Used as an agar raw material[10] [18] [38] [41].

G. cervicornis. U.S.A., (Florida, between Miami and Key West), Cuba[24]. Used as an agaroid raw material[10] [18].

G. chorda. Coasts of Japan. An agaroid raw material (soft type), used commercially mixed with other species[10] [41].

G. compressa. Coasts of Japan, India (east coast of State of Madras), Cuba[24]. An agaroid raw material (soft type), used commercially in mixture with other algal species. The phycocolloid content of Indian material of this species amounts to approximately 17 per cent[10] [41]. This alga is eaten in Japan[63].

G. cornea. U.S.A., West coast of Florida, West Indies (Barbados), Northern Brazil (Recife)[41]. An agar or an agaroid raw material[10] [18] [20] [21] [38] [67] [68]. This alga is used as a jelly in the West Indies like carrageenan in Europe[22].

G. corneum. A species is listed under this designation in literature for the coast of South Africa. The Japanese are also said to produce an agar of good quality from this material[42].

G. coronipifolia. Coasts of the Pacific Ocean. Used on Hawaiian Islands in numerous food preparations[42] [63] [69] [70].

G. corticata. Recorded for India, East Coast of the Madras State, coasts of Pakistan (Manora). Often occurring as drift weed. An agaroid raw material, content approximately 17 per cent[10] [15].

G. crassa. Coasts of Ceylon, India (East Coast of State of Madras), Thailand, Vietnam, Indonesia (South Coast of Java, Celebes), the Philippines. An agaroid raw material, content about 30 per cent. Eaten in Vietnam and the Philippines[10] [42] [64].

G. cylindrica. Coasts of Cuba. This alga has been investigated[24].

G. damaecornis. Coasts of Cuba. This alga has been studied[24].

G. dentata. Coasts of India and Pakistan[15].

G. divaricata. Coasts of Cuba. This alga has been investigated[24].

G. dura. Coasts of Morocco. An agaroid raw material, commercially utilized in mixture with other algal species[10].

G. erecta. The taxonomy of this alga has been described[26].

G. eucheumoides. Coasts of Indian and Pacific Oceans, particularly Japan, Vietnam, Malaya, Indonesia (Java, Celebes, Lambok, Sumbawa, Amboina), and the Philippines. Used for cooking in Vietnam and the Philippines[42] [64].

G. ferox. Coasts of Cuba[24] and Cameroon[71]. This alga has been investigated.

G. floridiana. U.S.A., (Florida). This alga supplies a viscous extract and can be regarded as an agar or agaroid raw material[21] [42].

G. foliifera. India (East Coast of the Madras State), U.S.A., Coasts of North Carolina and Florida (West Coast, Beaufort)[12] [20] [22] [34], Cuba[24], Great Britain. An agar or agaroid raw material, of increasing commercial significance in the U.S.A. Considerable differences have been found in this species concerning the quality of the phycocolloid, depending on location of collecting point and harvesting time[22]. Content of Indian material approximately 12 per cent. This alga has been chemically investigated[54]. The sterols have been investigated.

G. gigas. Coasts of Japan. An agaroid raw material (soft type), commercially utilized in mixture with other algal species[10] [41].

G. henriquesiana. Coast of the Gulf of Guinea. Of frequent occurence, particularly in deeper regions. In view of the other raw materials available the quantity of this species is not sufficient to be of commercial importance[73] [74]. An agar or agaroid raw material[38] [41] [76]. The phycocolloid has been described in detail; it has a melting point of 86 to 87° C and a gelation point of 46 to 47° C for 1 and 2 per cent solutions[10] [42].

G. lemanaeformis. Coast of Chile (Coquimbo to Concepcion), in considerable quantities. Industrial exploitation of this raw material appears possible.

G. lichenoides. Coasts of Japan, China, Australia, Malaya, Indonesia (South Sumatra, Madura, South Celebes, Java). The material referred to it is f. *lichenoides* and f. *lemania* (Soerjodinoto, 1958) or f. *coronipifolia* (Zaneveld)[77] [64]. Pacific Islands, coasts of the

Indian Ocean, India (East coast of the Madras State, Cape Comorin, Muttam Varkilai – sometimes free floating[78]), Ceylon [79–81]).

An agar or agaroid raw material, often mixed with other species and used commercially. Phycocolloid content of Indian material approximately 19 per cent[21] [38] [82] [10] [41] [42]).

Production of agar has been reported from Indian *G. lichenoides*[83]). Material from Ceylon varies in colour from green, straw-yellow to light purple. Work has been published on methods of collection, costs of production and bleaching, and on chemical composition[78] [81] [84]). Ceylonese *Gracilaria* was exported in a dried and bleached state as "Ceylon Moss" as early as 150 years ago to Europe and afterwards also to the U.S.A. It was used as a condiment. It was also sold on the Ceylonese and Indian markets. The main areas of production are Kalpitiya, Trincomalee and Mannar. In Kalpitiya harvesting takes place from January to May, in Trincomalee and Mannar from July to October[80]). *Gracilaria* is also used as a fertilizer on the island of Ceylon[85]).

Indonesian material is much exported to China and Japan.

The alga is used in East Asian areas for numerous cullinary preparations and delicacies, and as a pharmaceutical substance[42] [64] [65]).

G. lichenoides was already known to Chinese medicine in the 8th Century A.D.[70]).

G. mamillaris. Coasts of Cuba. This alga has been investigated[24]).

G. multipartita (= *G. lacinulata* = *G. foliifera*). East coast of North America (North California, Beaufort), in significant quantities, coasts of South India. An agar or an agaroid raw material. The quality of U.S. material corresponds roughly to that of *G. confervoides*, while the Indian plant produces a softer gel[18] [21] [38] [41]).

G. panamensis. Galapagos Islands. An agaroid raw material[10]).

G. peruana. Coasts of Peru and Chile; in depths of 3 to 14 metres. It is doubtful whether this is a true species or whether it is only a less branched variety of *G. confervoides* (LEVRING)[25]).

G. pygmea. Coasts of Pakistan[15]).

G. salicornia. Coasts of the Philippines[14]).

G. subsecundata. Coasts of the Philippines[14]).

G. taenioides. Ceylon, Malaya, Indonesia (Bangka), Korea, Japan, Australia. An agar or an agaroid raw material[38] [41] [42] [65] [82]). This alga is eaten in Indonesia, and in Singapore it is used to make agar[64]).

A number of *G.* species not exactly classified and studied up to now, are recorded from various coastal areas.

Australia: The genus *Gracilaria* has been investigated on the coasts of Australia[86]).

Chile: A Chilean *G.* species is said to yield in 5 to 6 per cent agar.

Indonesia: A *G.* species is collected near Buton, Rahamuna, and Kendari, and supplied to Java and Hong Kong via Makassar[65]).

Italy: Italian *G.* species are used for making agar[42] [87]).

Japan: The *G.* species on the Japanese coasts and in adjoining waters have been investigated and described[88]). In addition to the previously established species, SEGAWA describes *G. incurvata* and *G. textorii*.

Philippines: Various *G.* species have been reported by VELASQUEZ[14]).

Physiological and biochemical investigations of *G. acicularis*, *G. confervoides*, and *G. verrucosa* have been reported in LEWIN (1962).

STOLOFF (1962) describes the polysacolloids from *G. confervoides*, *G. armata*, *G. blodgettii*, *G. caudata*, *G. cornea*, *G. henriquesiana*, *G. lichenoides*, *G. multipartita*, *G. taenioides* as agar[89]).

Literature

1) WENNBERG, T., Medd. f. Göteborgs Botaniska Trädgard XVIII, 1950
2) STOKKE, K., Second Int. Seaweed Symp. 1955
3) STOKKE, K., Nytt. Mag. Bot. 5: 101–111. 1957
4) MINIO, M., Istituto di Studi Adriatici Venezia (1). 1949
5) POLLI, S.; ibid. (4). 1951
6) MUROLO, M. and MUROLO, V., Brevet d'Invention 1.017. 156 Rep. Française 1950–1952
7) MINIO, M. and SPADA, N., Instituto di Studi Adriatici Venezia (5). 1952
8) BRENNAN, A. T., Inst. f. Ind. Res. and Stand., Dublin, 1950
9) ISAAC, Wm. E., Second Int. Seaweed Symp. 1955
10) SELBY, H. H. and SELBY, T. A., Agar ex Whistler, R. L. and BeMiller, J. N., Industrial Gums. Academic Press Inc., New York, 1959
11) TOKIDA, J., Fac. of Fish. Hokkaido Univ. 2 (1): 1954
12) SASAKI, S., Bull. Hokkaido Reg. Fish. Res. Lab. 16: 56–58. 1957
13) TOKIDA, J. and MASAKI, T., Fac. of Fish. Hokkaido Univ. 10 (3): 173–195. 1959
14) VELASQUEZ, G. T., Philippine Seaweeds Report 1955
15) ANAND, P. L., Marine Algae from Karachi. The University of the Panjab, Lahore, 1943
16) MOORE, L. B., New Zealand J. of Science and Technology 25 (5) Sec. B.: 183–209. 1944
17) TAYLOR, W. R., Bot. Ser. Field Museum of Natural History 20 (4). 1941
18) HUMM, H. J., Science 100 (2593): 209–212. 1944
19) Duke University Press (3). 1946
20) CAUSEY, B. et al.; ibid. (3). 1946
21) STOLOFF, L. and LEE, CH. F., Fishery Leaflet 173, 1946. US Dep. of the Interior, Fish and Wildlife Service, Washington DC
22) HUMM, H. J., The Red Algae of Economic Importance ex Tressler, D. K. and Lemon, J. Mc. W., Marine Products of Commerce. Reinhold Publishing Corporation, New York, 1951
23) SOLONI TOURAL, F. G., Centro de Invest. Pesqu. (4). 1954
24) DIAZ-PIFERRER, M. and LOPEZ, H., Instituto Cubano de Invest. Tecnol. (6). 1959
25) LEVRING, T., Contributions to the marine algal Flora of Chile. Lunds Univ. Årsskrift N. F. Avd. 2. 56 (10). 1960
26) JONES, W. E., Phyc. Bull. 5. 16. 1957
27) SAWADA, T., Sc. Bull. Fac. Agr. Kyushu Univ. 16 (3): 387–396. 1958
28) JONES, W. E., J. Mar. Biol. Ass. UK 38 (1): 47–56. 1959
29) TSENG, C. K. et al., Kexue Tongbao (Scientia) (6): 202–203. 1959
30) WOOD, E. J. F., J. Council Sci. and Ind. Res., Australia 14. 315–316. 1941
31) USHIYAMA, S. and SATO, S., Jap. Pat. 4037. 1953. CA 48. 15. 8989c, 1954
32) SULIT, J. J. et al., Indo-Pacific Fish. Council Proc. 6. 280–283. 1955. CA 53. 12. 11715f. 1959
33) DELOACH, W. S. et al., Duke University Press (3). 1946
34) SANFORD, F. B., Fishery Leaflet 469. 1958. US Dep. of the Interior, Washington DC
35) Hypothekenbank von Japan. Die Agar-Agar-Herstellung in Japan. Tokyo (17). 1950
36) BAKER, G. L., US Pat. 2. 669. 519–1952/54
37) WHISTLER, R. L. and SMART, CH. L., Polysaccharide Chemistry. Academic Press Inc., New York, 1953
38) STOLOFF, L. and SILVA, P., Ec. Bot. 11 (4): 327–330. 1957
39) YAPHE, W., Can. J. Microbiol. 3. 987–993. 1957
40) ARAKI, CH., Seaweed Polysaccharides ex Wolfrom, M. L., Carbohydrate Chemistry of Substances of Biological Interest. Pergamon Press Ltd., London, 1959
41) SMITH, F. and MONTGOMERY, R., The Chemistry of Plant Gums and Mucilages. Reinhold Publishing Corporation, New York, 1959
42) SCHACHAT, R. E. and GLICKSMAN, M., Some Lesser Known Seaweed Extracts ex Whistler, R. L. and BeMiller, J. N., Industrial Gums. Academic Press Inc., New York, 1959
43) FUNAKI, K. and KOJIMA, Y., Bull. Jap. Soc. Sci. Fish. 16 (9): 401–404. 1951
44) KOJIMA, Y. and FUNAKI, K.; ibid. 16 (9): 405–410. 1951
45) KOJIMA, Y. and FUNAKI, K.; ibid. 16 (9): 411–418. 1951
46) KOJIMA, Y. and FUNAKI, K.; ibid. 16 (9): 419–422. 1951
47) KOJIMA, Y. and FUNAKI, K.; ibid. 18 (6): 245–248. 1952
48) AKIYA, S. and TANIMURA, A., J. Pharm. Soc. Japan 74: 125–128. 1954. CA 48. 10. 6076e. 1954

49) Stoloff, L., Fishery Leaflet 307. 1948. US Dep. of the Interior, Fish and Wildlife Service, Washington DC
50) Selby, H.; ibid. 118. 1948
51) Nilson, H. W. and Bender, M., Suppl. Comm. Fish. Rev. 1950, 263; ibid.
52) Kojima, Y. and Kusakabe, J., Bull. Jap. Soc. Sci. Fish. 17. (11): 345–348. 1952. CA 48. 2. 818h. 1954
53) Jones, W. E., J. Mar. Biol. Ass. UK 38 (1): 153–167. 1959
54) Ross, A. G., J. Sci. Food Agric. (7): 333–335. 1953
55) Young, E. G. and Langille, W. M., Publ. 4636 of the Nat. Res. Council, Halifax, 1958
56) Black, W. A. P., The Algae ex Altschul. Processed Plant Proteins Foodstuffs. Academic Press Inc., New York, 1958
57) Clingman, A. L. et al., J. Chem. Soc. (London) 1957, 197–203. CA 51. 14. 10385b. 1957
58) Janecke, H. and Kehr, W., Pharmaz. Ztg. 105 (25): 684. 1960
59) Black, W. A. P. and Cornhill, W. J., Chem. and Ind., 1954, 514–516
60) Yaphe, W., Can. J. Bot. 37. 751–757. 1959
61) Stoll and Jucker, Heteroside ex W. Ruhland, Handbuch der Pflanzenphysiologie, Springer Verlag 1958
62) Govindachari, T. R. and Rajagopalan, S., First Int. Seaweed Symp., 1952
63) Uphof, J. C. Th., Dictionary of Economic Plants. H. R. Engelmann (J. Cramer), Weinheim, 1959
64) Zaneveld, J. S., Ec. Bot. 13 (2): 89. 1959
65) Eisses, J., Indonesian J. of Nat. Sci. (1, 2, 3). 1953
66) Foreign Commerce Weekly, Washington, Results of Research in Philippine Agar (8). 1952
67) Humm, H. J. and Williams, L. G., Amer. J. Bot. 35 (5): 287–292. 1948
68) Goncalves da Costa Lima, E., Algas de nosso litoral. Universidade do Recife, 1952
69) Hoffmann, C., Die praktische Bedeutung der Meeresalgen. Kieler Meeresforsch. III (1): 1938
70) Chapman, V. J., Seaweeds and their Uses. Methuen & Co., Ltd., London, 1950
71) Lawson, G. W., J. West Afric. Science Ass. 1 (2): 1955
72) Humm, H. J. and Wolf, F. A., Duke University Press (3). 1946
73) Lawson, G. W., First Int. Seaweed Symp., 1952
74) Lawson, G. W., J. Ecol. 44 (1). 1954
75) Ohmi, H., Bull. Jap. Soc. Phyc. 6 (1): 4–7. 1958
76) Lawson, G. W., Amer. J. Bot. 41. 212–214. 1954. CA 48. 13. 7718c. 1954
77) Zaneveld, J. S., Chron. Nat. 105 (1) 1949
78) Pillai, K. S. and Varier, N. S., First Int. Seaweed Symp., 1952
79) Holsinger, E. C. T., First Int. Seaweed Symp., 1952
80) Durairatnam, M. and Medcof, J. C., The Ceylon Trade J. 19. 1–6. 1954
81) Durairatnam, M., Ceylon Dep. Fish. Pogr. Rep. Biol. and Technol (2): 54. 1956
82) Morel, Th., Med. van het Centraal Inst. v. Technische Onderzoekingen, Djakarta, Oct. 1949
83) Karanakar, P. D. et al., Indian Vet. J. 24. 274–282. 1948
84) Pillai, K. S., J. Sci. and Ind. Res. India 14B (9): 473. 1955
85) Chidambaram, K. and Unny, M. M., First Int. Seaweed Symp., 1952
86) May, V., Australia Commonwealth Council, Sci. Ind. Res. Bull. 235. 1–64. 1948
87) Buzzonetti, G. C., Chimica (Mailand) 12. 157–161. 1956. CA 17253e. 1956
88) Ohmi, H., Mem. Fac. Fish. Hokkaido Univ. 6 (1): 1–66. 1958
89) Stoloff, L., Ec. Bot. 16 (2): 86–94. 1962

General Literature

Chapman, V. J., 1950; Fott, B., 1959; Fritsch, F. E., 1952; Hoffmann, C., 1938; Hoppe, H. A., 1958; Hoppe, H. A. and Schmid, O. J., 1962; Kuckuck, P., 1953; Kylin, H., 1956; Lewin, R. A., 1962; Pax, F., 1962; Segawa, S., 1957; Tseng, C. K., 1945; Uphof, J. C. Th., 1959; see Index of Literature.

Gracilariopsis

Several species on the coasts of the Atlantic and Pacific Ocean.

G. sjöstedtii (Kylin) Dawson. There are no connecting filaments between the gonimoblasts and cystocarps.

Various species on the Japanese coasts have been investigated[1]).

G. vermiculophylla Ohmi. Japan (Hokkaido); considered as an agaroid raw material[2]).

Literature

1) Ohmi, H., Mem. Fac. Fish. Hokkaido Univ. 6 (1): 1–66, 1958
2) Sasaki, S., Bull. Hokkaido Reg. Fish. Res. Lab. 16: 56–58, 1957

General Literature

Kylin, H., 1956; see Index of Literature.

Grateloupia

Approximately 40 highly variable species, recorded from warmer seas.

G. filicina (Wulf.) Ag. Coasts of the Mediterranean, the warmer regions of the Atlantic Ocean, India, Indonesia, the Philippines, Japan, Hawaiian Islands, and Cuba[1]. The type species. Thallus erect, 5 to 75 centimetres high, flat and leaf like, tough, often ramified. The alga is eaten on the Batan Islands and Hawaiian Islands.

G. species have been studied by various authors. They can be regarded as a carrageenan raw material[2] [3]. In Japan *G.* species serve as an additional raw material for funoran production.

The botanically and chemically studied species are, inter alia: *G. litophila*, India[4], *G. carnosa*, *G. elliptica*, *G. imbricata*, *G. ramosissima*, *G. divaricata*, *G. prolongata*, *G. turuturu*, *G. Okamurai*, *G. livida*, Japan[5–7] [10], *G. cuneifolia*, Venezuela[8], *G. cutleria*, *G. doryophora*, *G. schizophylla*, Peru, Chile[9].

Physiological and biochemical investigations of *G. elliptica* have been reported in Lewin (1962).

Literature

1) Diaz-Piferrer, M. and López, H., Inst. Cubano de Invest. Tecnol. (6): 1959
2) Dawson, E. Y., Pac. Nat. 1 (1): 3–5, 1958
3) Dangeard, P., Coll. Int. du Centre Nat. de la Recherche Scient., Dinard, 1960
4) Govindachari, T. R. and Rajagopalan, S., First Int. Seaweed Symp., 1952
5) Tokida, J., Fac. of Fish. Hokkaido University. Vol. 2, (1), 1954
6) Tokida, J. and Masaki, T., Fac. of Fish. Hokkaido University. Vol. 10 (3): 173–195, 1959
7) Yabu, H., Fac. Fish. Hokkaido University. Vol. 8 (4): 278–289, 1958
8) Taylor, W. R., Bot. Ser. Field Museum of Natural History. Vol. 20 (4): 1941
9) Levring, T., Contributions to the marine Algal Flora of Chile. Lunds Univ. Årsskrift N. F. Avd. 2, Bd. 56 (10), 1960
10) Tazawa, N., Bull. Jap. Soc. Phyc. 8 (3): 112–117, 1960

General Literature

Chapman, V. J., 1950; Fritsch, F. E., 1952; Hoffmann, C., 1938; Kylin, H., 1956; Lewin, R. A., 1962; Segawa, S., 1957; Zaneveld, J. S., Ec. Bot. 13 (2), 89, 1959; see Index of Literature.

Griffithsia

Approximately 30 species are recorded from various oceans.

G. corallina (Lightf.) Ag. European coasts of the Atlantic Ocean. The type species. Thallus erect, 2 to 15 centimetres high, branching dichotomous or lateral.

G. flosculosa. This alga has been investigated chemically.

G. globifera. Material collected at Puerto Rico showed antibiotic activity[1]).

G. rhizophora. First recorded in India, this alga also occurs on the coasts of Ceylon, Borneo, Hainan, and Hawaiian Islands[2]).

G. japonica and *G. subcylindrica* occur on the coasts of Japan.

For physiological and biochemical investigations of *G. flosculosa* and *G. setacea* see LEWIN (1962).

Literature

1) BURKHOLDER, P. R. et al., Bot. Mar. II (1/2): 149, 1960

2) THIVY, F. and IYENGAR, E. R. R., Bot. Mar.

General Literature

FRITSCH, F. E., 1952; KYLIN, H., 1956; LEWIN, R. A., 1962; SEGAWA, S., 1957; see Index of Literature.

Gymnogongrus

Approximately 40 species are recorded from warmer and temperate oceans.

G. griffithsiae (Thurn.) Martius. Coasts of Northern Atlantic Ocean, Sakhalin, and the Sea of Japan, Latin America, widespread. The type species. Thallus erect, 10 to 15 centimetres high, terete, somewhat compressed at apices dichotomously branched, of a rather horny consistency[1]) [2]). Utilized as phycocolloid raw material in East Asia[2]).

G. norvegicus[3]) [4]) and *G. patens* have been studied botanically and chemically. Both species were found to be carrageenan raw materials[5–8]). STOLOFF (1962) describes the polysacolloids as a carrageenan material[16]). *G. norvegicus* also occurs in the Mediterranean area[15]).

G. flabelliformis. Japan, Bay of Oshoro[9]) [10]). This alga is eaten in Japan[11]).

G. pinnulata. Coasts of Japan and China. The alga is eaten in Japan; it is also taken for a paste which is used as a hair dressing[11]). It also serves as a funoran raw material.

G. disciplinalis and *G. furcellatus* occur on the coasts of Peru, Chile, and the Falkland Islands[12]). The possibilities of commercial and industrial utilization are being investigated by the authors of this volume.

G. javanicus. Indonesia, Hawaii, coasts of tropical ocean areas. This alga is eaten on the Hawaiian Islands, being known as "dry limu"[13]).

G. nigricans. British Cameroons[14]).

G. species. Coasts of India. This alga is regarded as an agaroid raw material.

Literature

1) GREGORY, B. D., Proc. Linnean. Soc. (London) p. 63, 1934

2) TOKIDA, J., Fac. of Fish. Hokkaido University. Vol. 2 (1), 1954

3) SCHOTTER, G., C. R. Acad. Sci., Paris 251 (11): 1174–1175, 1960

4) 251 (16): 1647–1649, 1960

5) SMITH, F. and MONTGOMERY, R., The Chemistry of Plant Gums and Mucilages. Reinhold Publishing Corporation, New York, 1959

6) SELBY, H. H. and SELBY, T. A., Agar ex Whistler, R. L. and BeMiller, J. N., Industrial Gums. Academic Press Inc., New York, 1959

7) STOLOFF, L. and SILVA, P., Ec. Bot. 11 (4): 327–330, 1957

8) DANGEARD, P., Coll. Int. du Centre Nat. de la Recherche Scient., Dinard, 1960

9) TOKIDA, J. and MASAKI, T., Fac. of Fish. Hokkaido University. Vol. 10 (3): 173–195, 1959

10) TOKIDA J. and MASAKI, T., Fac. of Fish. Hokkaido University. Vol. 10 (2): 87–96, 1959

11) Uphof, J. C. Th., Dictionary of Economic Plants. H. R. Engelmann (J. Cramer), Weinheim, 1959
12) Levring, T., Contributions to the Marine Algal Flora of Chile. Lunds Univ. Årsskrift N. F. Avd. 2, Bd. 56 (10), 1960
13) Zaneveld, J. S., Ec. Bot. 13 (2): 89, 1959
14) Lawson, G. W., West Afric. Science Assoc. 1 (2): 1955
15) Schotter, G., C. R. Acad. Sc. (Paris) 251 (11) 1174–1175. 1960; ibid. 251 (16) 1647–1649. 1960
16) Stoloff, L., Ec. Bot. 16 (2): 86–94. 1962

General Literature

Chapman, V. J., 1950; Fritsch, F. E., 1952; Hoffmann, C., 1938; Kylin, H., 1956; Segawa, S., 1957; see Index of Literature.

Halosaccion

Approximately 10 species are recorded from the colder oceans.

H. hydrophora (Post. et Rupr.) Kütz. Coasts of the Northern Pacific. The type species. Thallus erect, 5 to 15 centimetres high, mostly terete, branched or unbranched.

H. glandiforme. Extracts show antibiotic activity[1]). This alga is used in Kamchatka for human consumption[2]).

H. ramentaceum. Coasts of northern East Asia[2]) and North America. This species has been studied by several authors[3]) [4]). Physiological and biochemical investigations have been reported in Lewin (1962).

H. saccatum. Coasts of Japan.

Literature

1) Pratt, R. et al., J. Am. Pharm. Ass. 40, 575–579, 1951
2) Tokida, J., Fac. of Fish. Hokkaido University 2 (1), 1954
3) Furuya, K., Bot. Mag. Tokyo (867): 355–359, 1960
4) Young, E. and Langille, W. M., Publ. No. 4636 of the Nat. Research Council, Halifax, Canada, 1958

General Literature

Fritsch, F. E., 1952; Kylin, H., 1956; Lewin, R. A., 1962; Segawa, S., 1957; see Index of Literature.

Halymenia

Approximately 25 species are recorded from warmer seas.

H. floresia (Clem.) Agardh. Mediterranean and warmer parts of the Atlantic. The type species. Thallus erect, 5 to 35 centimetres high, terete, compressed or flat, branched, of a jelly-like consistency.

H. durvilliae var. *formosa.* Coasts of India, Indonesia, Pacific Islands, China, Japan. The alga is eaten on the Philippines and Hawaiian Islands.

Various *H.* species are recorded from Japan.

General Literature

Fritsch, F. E., 1952; Kylin, H., 1956; Segawa, S., 1957; Zaneveld, J. S., Ec. Bot. 13 (2): 89, 1959; see Index of Literature.

Hypnea

Approximately 25 species widespread in warmer seas.

H. musciformis (Wulfen) Lamouroux. In the warmer parts of the Atlantic Ocean and the Mediterranean, particularly on the East Coast of the U.S.A., from Florida to Nantucket, Caribbean, Gulf of Mexico, Cuba[1–5]), Brazil, between Sao Francisco River and Natal. The largest stands recorded exist near Recife, on reefs[5]) [6]), Columbia[2]), Morocco, between Tangier and Agadier[7]), Mauritania[8]), Gulf of Guinea often mixed with *Gigartina acicularis*[9]), Cameroon[10]), relatively large quantities in the region of Luanda, between the coast and the offshore islands[7]). Coasts of Pakistan. The weed is harvested mainly near Manora and Karachi by fishermen using nets[11]). India, Coast of Madras. The alga from these areas has a phycocolloid content of approximately 24 per cent[7]), Malaya, Indonesia, particularly in the district of Madura, coasts of Makassar, Bali, Timor, the Philippines, China, Japan[12–16]). The type species. Thallus erect, 5 to 30 centimetres high, terete, much ramified, with short thorny-shaped branches.

The main harvesting season on the coast of North Carolina takes places between May and July, on the coast of Florida between November and February, while in Brazil it can be harvested throughout the year. Collection, however, is intentionally limited to the dry season during the months of October to March, because the alga must be dried immediately after harvesting in order to avoid loss of substance by fresh water. The crop is usually loosened by long hooks from the rock substrata, but floating masses of algae are also collected[5]).

The economic significance of *Hypnea* has risen substantially in recent years. Utilization began in the U.S.A. in the years 1945–46. Detailed reports have been published on the production and the properties of the product[19]) [20]).

Species of *H.* occur mostly in tropical waters. Four of five species are found in the Caribbean, but on the warmer Atlantic coasts of the U.S.A. The genus appears to be represented only by *H. musciformis*[21–23]).

Reports on the phycocolloid contained in *H.* species differ. According to recent reports it is variously regarded as hypnean[5]) [24]), a gelan[25]) or carrageenan[26]) with a kappa-carrageenan value higher than that of *Chondrus crispus*[27]). Other studies on the polysaccharide of *Hypnea*[28–30]) and descriptions of methods for hypnean production[21]) [31]) [32]) have been published.

Because of its anthelmintic properties this plant is used in Indonesia as a vermifuge. On the Philippines it is eaten, while in coastal areas of Brazil it is used as a fertilizer for young cocoa trees[5]).

H. cenomyce. Coasts of Indonesia, particularly of Borneo, Timor, the Sulu Archipelago; the Philippines, and Australia. The alga is used for food by the local population in Indonesia[5]) [12]) [14]) [17]).

H. cervicornis. Coasts of Mexico (Yucatan), Columbia[2]), Cuba[4]), Gulf of Guinea[9]) [33]), Indonesia, north coast of East Java, Makassar, Bali, Tawitawi Islands, the Philippines[14]). It is a widespread inhabitant of many warmer ocean regions. The phycocolloid has been investigated[5]) [28]) and is described as a gelan[25]). In Indonesia the alga is used to produce a jelly, which is eaten with palm sugar. Phycocolloid content of Indonesian material is very low[5]) [14]) [21]) [12]). The alga is also eaten in China[17]).

H. cornuta. Coasts of Cuba[4]).

H. divaricata. Coasts of Indonesia (Sumatra, Timor), the Philippines, New Guinea, and in many warmer areas. The alga is eaten on Amboina, while on Timor it is made into agar for local use[5]) [14]).

H. japonica. Inhabitant of the coasts of the southern Japanese islands. The alga has been studied chemically. A cholesterol compound has been isolated[34]) [38]).

Further Japanese species are: *H. charoides*, *H. saidana*, and *H. variabilis*.

H. nidifica. Hawaii Islands. This species is used both as a fodder and a food, also medically as a stomach treatment[5]) [17]).

H. spicifera. South Africa. The morphology of this alga has been studied[39]), its phycocolloid investigated[28]), and described as a gelan[25]) or as an agar raw material[21]). The alga contains a polysaccharide similar to kappa-carrageenan[36]). Various authors have proved the presence of carrabiose, which is related to agarobiose[38]). The alga occurs in large quantities and is of a bright olive-green colour. In some areas the *Hypnea* beds almost exclusively consist of this species, and harvesting is essentially facilitated thereby[5]). *H. spicifera* is often used for food.

H. spinella. This alga is found on the coast of Columbia[2]).

H. valentiae. Coasts of Pakistan, in the Bay of Manora, often together with *H. musciformis*[11]), and on the Indian and Indonesian coasts. The commercial significance of this species, for example, in Indonesia, seems not to be very great. The sterols have been investigated[5]) [37]).

Physiological and biochemical investigations of *H. japonica*, *H. musciformis*, and *H. spicifera* have been reported in Lewin (1962).

Stoloff (1962) has described the polysacolloid of *H. musciformis*, *H. cervicornis*, and *H. spicifera* as a gelan[40]).

Literature

1) Humm, H. J., Science 100 (2593): 209–212, 1944
2) Taylor, W. R., Bot. Ser. Field Museum of Natural History 20 (4): Nov. 1941
3) Selby, H. H. and Selby, T. A. Agar ex Whistler, R. L. and BeMiller, J. N., Industrial Gums. Academic Press Inc., New York, 1959
4) Diaz-Piferrer, M. and López, H., Inst. Cubano de Invest. Tecnol. (6), 1959
5) Schachat, R. E. and Glicksman, M. Some Lesser Known Seaweed Extracts ex Whistler, R. L. and BeMiller J. N., Industrial Gums. Academic Press Inc., New York, 1959
6) Goncalves da Costa Lima, E., Algas de nosso litoral. Universidade do Recife, 1952
7) Hoppe, H. A., Persönliche Mitteilungen, 1963
8) Seoane-Camba
9) Lawson, G. W., J. Ec. 44 (1); 1954
10) Lawson, G. W., J. West Afric. Science Ass. 1 (2), 1955
11) Anand, P. L., Marine Algae from Karachi. The University of the Panjab, Lahore, 1943
12) Eisses, J., Indonesian J. of Nat. Science (1, 2, 3), 1953
13) Zaneveld, J. S., Chron. Nat. 105 (1); 1949
14) Zaneveld, J. S., Ec. Bot. 13 (2): 89, 1929
15) Velasquez, G. T., Philippine Seaweeds Report, 1955
16) Foreign Commerce Weekly, Washington, (8), 1952
17) Uphof, J. C. Th., Dictionary of Economic Plants. H. R. Engelmann (J. Cramer), Weinheim, 1959
18) Humm, H. J. and Williams, L. G., Am. J. Bot. 35 (5): 287–292, 1948
19) Humm, H. J., The Red Algae of Economic Importance ex Tressler, D. K. and Lemon, J. Mc. W., Reinhold Publishing Corporation, New York, 1951

20) Sanford, F. B., Fishery Leaflet 469, 1958. US Dep. of the Interior, Washington, D. C.
21) De Loach, W. et al., Duke University Press (3): 31–42, 1946
22) Utilization of Seaweeds from the South Atlantic and Gulf Coasts for Agar and its Decomposition by Bacteria. Duke University Press (3), 1946
23) Williams, R. H., University of Miami, Marine Laboratory, 1949
24) Schachat, R. E. and Glicksman, M., Ec. Bot. 13 (4): 365–370, 1959
25) Stoloff, L. and Silva, P., Ec. Bot. 11 (4): 327–330, 1957
26) Young, E. G., Carbohydrate accumulation by lower plants (Rhodophyta) ex W. Ruhland, Handbuch der Pflanzenphysiologie, Springer Verlag, 1958
27) Yaphe, W., Can. J. Bot. 37, 751–757, 1959
28) Smith, F. and Montgomery, R., The Chemistry of Plant Gums and Mucilages. Reinhold Publishing Corporation, New York, 1959
29) Dangeard, P., Coll. Int. du Centre Nat. de la Recherche Scient., Dinard, 1960
30) Young, E. G., idem
31) Humm, H. J., US Pat. 2, 446, 091, 1948
32) Baker, G. L., US Pat. 2, 669, 519, 1952–1954
33) Lawson, G. W., First Int. Seaweed Symp., 1952
34) Tsuda, K. et al., Chem. & Pharm. Bull. 7 (6): 747–749, 1959
35) Humm, H. J. and Wolf, F. A., Duke University Press (3), 1946
36) Clingman, A. L. and Nunn, J. R., J. Chem. Soc. (London) 493, 1959
37) Govindachari, T. R. and Rajagopalan, S., First Int. Seaweed Symp., 1952
38) Hegnauer, R., Chemotaxonomie der Pflanzen, Birkhäuser Verlag, Stuttgart, 1962
39) Hewitt, F. E., Univ. Calif. Publ. Bot. 32 (3): 195–233, 1960
40) Stoloff, L., Ec. Bot. 16 (2): 86–94, 1962

General Literature
Hoppe, H. A., 1958; Hoppe, H. A. and Schmid, O. J., 1962; Kylin, H. 1956; Lewin, R. A., 1962; Segawa, S., 1957; see Index of Literature.

Iridea

Approximately 25 species recorded from temperate and colder seas.

I. cordata (Turn.) Bory (= *Iridophycus cordata)*. West Coast of North America. The type species. Thallus erect, 5 to 125 centimetres high, flat or blade-like, undivided or irregularly lobed, rarely bifurcated or ramified. Remarkable features of living *Iridea* are its brilliant appearence and splendid red colour.

I. flaccida = *I. laminaroides*. Central California to Oregon, mostly in the intertidal zone, where it frequently forms colonies, but often in greater depths[1]). This species also occurs on the South American Pacific coast (Chile)[2]) and the Japanese coasts. The alga is utilized commercially in North America and supplies iridophycan[1]). The water-soluble polysaccharide is a galactan[3]). The properties of iridophycan are similar to those of carrageenan[4–6]), or according to one report, to a carrageenan with a low content of kappa-carrageenan[7]). Stoloff (1962) has described the polysacolloid as a carrageenan[39]).

Other sulfated galactans have been found[8]). The alga also contains dulcitol[17–20]), sorbitol, and a floridoside[21]). There is no cellulose[3]). Further investigations of the constituents of *I. flaccida* and other species have been made[9–17]).

Nomenclatur of the various species of *Iridea* (= *Iridophycus)* is not uniform, and often different designations are allotted to the very same species, for example, *I. flaccida* and *I. laminaroides*[1]).

Harvesting of *I.* species is relatively easy on account of their size. In a low-tide period of two to three hours one man, for example, on the Californian coast, can collect 200 to 300 lbs of fresh alga[1]). In the last years this alga has got economic significance in California.

In Japan species of *I.* are used as the basic raw material for funoran production[22]) [23]).

In Chile they are used for the production of agar, but because of the special properties of iridophycan, this does not appear to be expedient.

I. Boryana. Coast of Chile, Kerguelen (Desolation Island), and Crozet Island[2]).

I. caespitipes. St. Magdalena, Fuegia, and the Falkland Islands[2]).

I. capensis. Coasts of South Africa. The alga contains a substance very similar to carrageenan[4]) [6]), which, however, is also held to be iridophycan; according to STOLOFF it is a carrageenan[30]).

I. ciliata. Coasts of Central and South Chile[2]).

I. convuluta. Coasts of South Africa. The morphology of this alga has been investigated[29]).

I. cornucopiae (= *I. laminaroides* var. *cornucopias*). Northern coasts of Honshû, coasts of Hokkaido, the Kurile Islands, Sakhalin, Korea, and the Siberian Coast of the Sea of Japan. Commercially this species is one of the most important algae of Northern Japan for the production of "Ginnansô", an algal mucilage product[24]). This alga has been investigated chemically[16]). Other works have dealt with ecological and other problems[25–28]).

I. crispata. Occurs on the coast of Central and South Chile[2]).

I. dichotoma. Grows on the Falkland Islands and on the South-Chilean coasts[2]).

I. edulis. Coasts of Scotland, where it has been eaten[23]). Like *Rhodymenia palmata* this plant is referred to as "dulse".

I. membranacea. Occurs near Valparaiso[2]).

I. microcca. Has been found near San Carlos and Chiloé (Chile)[2]).

I. obovata. Coasts of Southern South America (Patagonia), Falkland Islands, Grahamland, Crozet[2]). This alga has been studied from chemical und technological points of view, but the investigations are not yet concluded.

I. pulchra. Pacific coasts. A raw material for iridophycan and dulcitol[1]).

I. splendens. Coasts of California and Oregon. The alga contains a galactan with similar properties to carrageenan[3]). It is regarded as a raw material for iridophycan production[1]).

I. subdichotomum. Coasts of Hokkaido, Kurile Islands, and Sakhalin. This alga is used for the production of "Ginnansô"[24]).

I. undulosa. Central and South Chile, Falkland Islands.

Results of physiological and biochemical investigations of *I.* species have been reported in LEWIN (1962).

Literature

1) SCHACHAT, R. E. and GLICKSMAN, M., Some Lesser Known Seaweed Extracts ex Industrial Gums. Academic Press Inc., New York, 1959
2) LEVRING, T., Contributions to the marine algal flora of Chile. Lunds Univ. Årsskrift N. F. Avd. 2 Bd. 56 (10), 1960
3) WHISTLER, R. L. and SMART, CH. L., Polysaccharide Chemistry. Academic Press Inc., New York, 1953
4) STOLOFF, L. and SILVA, P., Ec. Bot. 11 (4): 327–330, 1957
5) YOUNG, E. G., Carbohydrate accumulation by lower plants (Rhodophyta) ex W. Ruhland, Handbuch der Pflanzenphysiologie, Springer Verlag, 1958
6) SMITH, F. and MONTGOMERY, R., The Chemistry of Plant Gums and Mucilages. Reinhold Publishing Corporation, New York, 1959
7) YAPHE, W., Can. J. Bot. 37. 751–757. 1959
8) FOSTER and STACY, Polysaccharides from algae and lichens ex W. Ruhland, Handbuch der Pflanzenphysiologie, Springer Verlag, 1958
9) HASSID, W. Z., J. Am. Chem. Soc. 55, 4163–4167. 1933
10) HASSID, W. Z.; idem 57, 2046–2050, 1935
11) HASSID, W. Z., Plant Physiol. 11, 461–463, 1936
12) PUTMAN, E. W. and HASSID, W. Z., J. Am. Chem. Soc. 76. 2221. 1954

13) BEAN, R. C. and HASSID, W. Z., J. Biol. Chem. 212. 411–425. 1955. CA 49, 7. 4809c, 1955
14) BEAN, R. C. and HASSID, W. Z., J. Biol. Chem. 218. 425. 1956
15) TAKAGI, M. and KURIYAMA, Bull. Fac. Fish. Hokkaido Imp. Univ. 10 (1): 72–76, 1959
16) SMITH, F. and MONTGOMERY, R., The Chemistry of Plant Gums and Mucilages. Reinhold Publishing Corporation, New York, 1959
17) BLACK, W. A. P., Ann. Rep. Chem. Soc. 50. 1953
18) HASSID, W. Z., Plant Physiol. 8. 480–482. 1933
19) LINDBERG, B., Second Int. Seaweed Symp., 1955
20) YOUNG, E. G., Carbohydrate accumulation by lower plants (Rhodophyta) ex W. Ruhland, Handbuch der Pflanzenphysiologie, Springer Verlag, 1958
21) STOLL and JUCKER, Heteroside; idem
22) HUMM, H. J., The Red Algae of Economic Importance ex Tressler, D. K. et Lemon, J. Mc. W., Marine Products of Commerce. Reinhold Publishing Corporation, New York, 1951
23) HOFFMANN, C., Die praktische Bedeutung der Meeresalgen. Kieler Meeresforsch. III (1), 1938
24) TOKIDA, J., Fac. of Fish., Hokkaido Univ. 2 (1), 1954
25) HASEGAWA, Y. and FUKUHARA, E., Bull. Hokkaido Reg. Fish. Res. Lab. 9. 67–74. 1953
26) FUKUHARA, E., Bull. Hokkaido Reg. Fish. Res. Lab. 16. 70–73, 1957
27) FUKUHARA, E., Bull. Hokkaido Reg. Fish. Res. Lab. 17. 137–154. 1958
28) MIKAMI, H., Bull. Jap. Soc. Phyc. 6 (2): 80–83. 1958
29) HEWITT, F. E., Univ. Calif. Pub. Bot. 32 (3): 195–233. 1960
30) STOLOFF, L., Ec. Bot. 16 (2): 86–94. 1962

General Literature

CHAPMAN, V. J., 1950; FRITSCH, F. E., 1952; HOPPE, H. A., 1958; HOPPE, H. A. and SCHMID, O. J., 1962; KYLIN, H., 1956; LEWIN, R. A., 1962; see Index of Literature.

Laurencia

Approximately 80 species are recorded, mostly from warmer oceans. The species are highly variable and it is often very difficult to distinguish them clearly.

L. obtusa (Huds) Lamoroux. Atlantic coasts from Britain to the Canary Islands, and probably in all warmer ocean areas, for example, Pakistan[1]), Indonesia[2]). The type species. Thallus erect, 2 to 25 centimetres high, terete or flattened, ramified, pulpy or cartilaginous.

The alga collected near Puerto Rico are proved to have antibiotic activity[3]). *L. obtusa* is eaten in a number of Indonesian regions[2]) [4]).

L. species have been investigated on various occasions. They contain floridoside[5]), fluorescent[6]) and antibiotic[7]) substances. *Laurencia* admixed with *Gracilaria* tends to cause partial hydrolysis of the agar during extraction[8]) [9]).

Species of *L.* also occur on the coasts of Japan[10]), Queensland, Cuba, Mexiko, and Columbia[11]).

L. botryoides. China, Malaya, Japan, Australia, Tasmania. This species is probably eaten locally[2]).

L. caespitosa. This alga has been studied under the electron microscope[12]).

L. filiformis. Occurs on the coasts of Pakistan near Manora[1]).

L. glandulifera. Coasts of Japan, Kurile Islands, Sakhalin[13]). The ethereal oils of the alga have been studied[14]).

L. hypnoides. Coast of Pakistan[1]).

L. papillosa. Pacific coasts, the Philippines, Sandwich Islands, Hawaiian Islands; Atlantic coasts, Brazil to Florida. The alga is often eaten locally[2]) [4]).

L. perforata and *L. pinnatifida* are also eaten on the Hawaiian Islands[2]). The latter species also occurs on the coasts of Pakistan[1]), the Atlantic Ocean, and the North Sea[15]) [16]).

In Scotland and Ireland it is used as a spice under the name of "pepper dulse"[17] [19]), and in Iceland as a kind of chewing tobacco. The thallus is cartilaginous, ramified, violet or yellow-red, dark. The species has been investigated[18]).

In the literature, are mentioned *L. grevilleana*, *L. hamata*, *L. intermedia*, *L. okamurai*, *L. pinnata*, *L. undulata*, Japan *L. nipponica*, China, Japan, Kurile Islands, Sakhalin[13]), *L. platyclada*, Pakistan[1]), *L. virgata*, Pakistan[1]), Hawaiian Islands[2]), *L. wrightii*, Pacific Islands[4]).

For physiological and biochemical investigations of *L. pinnatifida* see LEWIN (1962). Extraction of the material gives a weak gel[20]).

Literature

1) ANAND, P. L., Marine Algae from Karachi. The University of the Panjab, Lahore, 1943
2) ZANEVELD, J. S., Ec. Bot. 13 (2): 89, 1959
3) BURKHOLDER, P. R. et al., Bot. Mar. II (1/2): 149, 1960
4) UPHOF, J. C. TH., Dictionary of Economic Plants. H. R. Engelmann (J. Cramer), Weinheim, 1959
5) AUGIER, J., Rev. gen. bot. 60: 257–283, 1953
6) H'EOCHA, C. O. et al., Third Int. Seaweed Symp., 1958
7) CHESTERS, C. G. C. and STOTT, J. A., Second Int. Seaweed Symp., 1955
8) DELOACH, WILLIAM S. et al., Duke Univ. Press, No. 3, 1946
9) CHAPMAN, V. J., Seaweeds and their Uses. Methuen & Co. Ltd., London, 1950
10) YABU, H. and KAWAMURA, Mem. Fac. Fish. Hokkaido Univ. 7 (1/2): 61–71, 1959
11) TAYLOR, W. R., Bot. Ser. Field Museum of Natural History 20 (4): 1941
12) MYERS et al., Ann. Bot. n. s. 23 (90): 257–260, 1959
13) TOKIDA, J., The Marine Algae of Southern Saghalien. Fac. of Fish. Hokkaido Univ. 2 (1), 1954
14) OBATA, et al., J. Agr. Chem. Soc. Japan 27: 331–334, 1953
15) DE VALERA, M., Third Int. Seaweed Symposium, 1958
16) KUCKUCK, P., Der Strandwanderer. J. F. Lehmanns Verlag, München, 1953
17) HOFFMANN, C., Die praktische Bedeutung der Meeresalgen. Kieler Meeresf. III (1): 1938
18) DIZERBO, A. H., Coll. Int. du Centre Nat. de la Recherche Scient., Dinard, 1957
19) BRENNAN, A. T., Notes on some common Irish Seaweeds. Inst. f. Ind. Res. and Stand., Dublin, 1950
20) MARSHALL, S. M. and ORR, A. P., Natural Plant Hydrocolloids, 1954

General Literature

CHAPMAN, V. J., 1950; FRITSCH, F. E., 1952; KYLIN, H., 1956; LEWIN, R. A., 1962; SEGAWA, S., 1957; see Index of Literature.

Liagora

Approximately 60 species at the coasts of warmer seas, for the most part insufficiently studied.

L. viscida (Forsk.) Agardh. Mediterranean, Red Sea. The type species. Thallus erect, 5 to 20 centimetres high, terete or flattened, branched, often calcified. Physiological and biochemical investigations have been reported in LEWIN (1962).

L. canariensis. The polysaccharides of this alga have been studied[1]).

L. farinosa var. *cheyneana*. Warmer seas. The alga occurs on the coasts of the Philippines and New Guinea. It is eaten on the Philippines[2]).

L. decussata. Hawaiian Islands. The alga is used for human consumption[2]).

Various *L.* species occur on the Japanese coasts.

Literature

1) YENN, E. W., The Plant Proteins and Peptides ex W. Ruhland. Handbuch der Pflanzenphysiologie, Springer Verlag, 1958
2) ZANEVELD, J. S., Ec. Bot. 13 (2): 89, 1959

General Literature
FRITSCH, F. E., 1952; KYLIN, H., 1956; LEWIN, R. A., 1962; SEGAWA, S., 1957; see Index of Literature.

Lithophyllum

Approximately 100 species in temperate and warmer seas.

L. incrustans Philippi. Mediterranean and adjoining parts of the Atlantic Ocean. The type species. Thallus encrusted and calcified, mostly with a coral-like branching. The growth of this alga has been studied[1]). The plant contains a peptide[2]). Physiological and biochemical investigations have been reported in LEWIN (1962).

Literature

1) HUVÉ, P., Ph. Comm. Int. Explor. Médit. Rapp. et Proc. Verb. n. s. 15 (2): 65–79, 1960
2) YENN, E. W., The Plant Proteins and Peptides ex W. Ruhland, Handbuch der Pflanzenphysiologie, Springer Verlag, 1958

General Literature

FRITSCH, F. E., 1952; KYLIN, H., 1956; LEWIN, R. A., 1962; SEGAWA, S., 1957; see Index of Literature.

Lithothamnion

More than 100 species were recorded from various parts of the world.

L. calcareum (Pallas) Aresch. European Atlantic coasts and Mediterranean. The type species. Thallus encrusted, entirely calcified, often with coral-like appendices that break off easily. The use of this calcaricus alga as a neutralizer and as a filter element to remove undesirable components from water, such as poisonous elements, radioactive substances etc. has been investigated[1]). Its chemical composition and its potentiality as a soil improver have been investigated[8]). This species is used on the Atlantic coast of Ireland as a fertilizer to improve acid oils[9]).

L. crassum. This calcareous alga occurs particularly in the Mediterranean, covering solid substrata with its flat thalli[2]). In France, Great Britain, and Ireland *L.* species are used as fertilizers: there was a regular "maerl" industry[4]) [3]).

Numerous *L.* species participate in the building up of the coral rocks and reefs in the Indian Ocean and the Pacific[2]) [5]).

L. polymorphum and *L. sonderi.* Occur on the coasts of the North Sea. They form hard, thin crusts of red-violet or pink colour.

L. papillosum. Often occurs together with *L. tortuosum* in the Mediterranean, particularly Spain, Balearic Islands, Algeria, France, Sardinia, Adriatic, and Aegean[6]).

Electron microscopic investigations have been made of the *Lithothamnion* species *(L. calcareum).*

Physiological and biochemical investigations of the species have been reported in LEWIN (1962).

Literature

1) Neveu, M., Coll. Int. du Centre Nat. de la Recherche Scient., Dinard, 1960
2) Fott, B., Algenkunde. Gustav Fischer Verlag, Jena, 1959
3) Hoffmann, C., Die praktische Bedeutung der Meeresalgen. Kieler Meeresf. III (1), 1938
4) Chapman, V. J., Seaweeds and their Uses. Methuen & Co. Ltd., London, 1950
5) Zaneveld, J. S., Blumea, suppl. 4: 206–219, 1958
6) Huvé, H., Bot. Mar. IV (3/4): 219, 1962
7) Lanlignel, A. and Schmitt, M., Coll. Int. du Centre Nat. de la Recherche Scient., Dinard, 1960
8) López-Benito, M., Inv. Pesq. (23): 53–70, 1963
9) Brennan, A. T., Notes on some common Irish Seaweeds. Inst. f. Ind. Res. and Stand., Dublin, 1950

General Literature
Fritsch, F. E., 1952; Kuckuck, P., 1953; Kylin, H., 1956; Lewin, R. A., 1962; Segawa, 1957; see Index of Litrature.

Lophocladia

3 species are recorded.

L. trichoclados (Ag.) Schmitz. Coastal areas of the West Indies. The type species. Thallus erect, 5 to 10 centimetres high, terete, branched. Material collected near Puerto Rico revealed antibiotic activity[1]).

Literature

1) Burkholder, P. R. et al., Bot. Mar. II (1/2): 149, 1960

General Literature
Fritsch, F. E., 1952; Kylin, H., 1956; see Index of Literature.

Mastocarpus

Mastocarpus is regarded as a synonym for *Gigartina*[1]).

M. Klenzianus Kütz. Indonesia. The alga is eaten locally[2]).

Literature

1) Kylin, H., Die Gattungen der Rhodophyceen. C. W. K. Gleerups Förlag, Lund, 1956
2) Eisses, J., Indonesian J. of Nat. Sc., No. 1, 2, 3, 1953

Mastophora

3 species, mainly Pacific area.

M. rosea (Ag.) Setch. Coasts of Japan and China. The type species. Thallus leafy, flat, laterally lobed, more or less prostrate, shoots frequently calcified.

M. zonalis (= *Zonaria zonalis*). Coasts of Puerto Rico. The alga has been shown to have antibiotic activity[1]).

Literature

1) Burkholder, P. R. et al., Bot. Mar. II (1/2): 149, 1960

General Literature

Kylin, H., 1956; Segawa, S., 1957; see Index of Literature.

Membranoptera

10 species, in the Northern Atlantic.

M. alata (Huds.) Stackh. Atlantic coasts of Europe and North America. The type species. Thallus erect, 2 to 12 centimetres high, strapshaped, or with very narrow blades, more or less branched, of a bright red colour.

Several species of *M.* have been investigated chemically[1]). For physiological and biochemical investigations see Lewin (1962).

Literature

1) Ross, G., J. Sci. Food Agric. (7): 333–335, 1953

General Literature

Fritsch, F. E., 1952; Kylin, H., 1956; Lewin, R. A., 1962; see Index of Literature.

Murrayella

2 species, recorded from the West Indies.

M. periclados (Ag.) Schmitz. West Indies, on the roots of mango trees, the type species. Thallus erect or stoloniferous, approximately 1 centimetre high, branched.

Material collected near Puerto Rico has been proved to have antibiotic activity[1]).

Literature

1) Burkholder, P. R. et al., Bot. Mar. II (1/2): 149, 1960

General Literature

Fritsch, F. E., 1952; Kylin, H., 1956; see Index of Literature.

Nemalion

Approximately 10 species, in various regions, including European and East and West Coasts of North America.

N. lubricum Duby. Atlantic coasts of Europe and Mediterranean. The type species. Thallus erect, 10 to 20 centimetres high, terete, branched, of a jelly-like consistency.

N. multifidum. Coasts of Europe, North America, and Japan. The polysaccharide of this species has been investigated[1]). The alga is eaten in Japan as "Tsukomo-nori". The thallus is filamentous, jelly-like, dark to purple coloured. Physiological and biochemical investigations are reported in Lewin (1962). A water-soluble mucilage has been isolated.

N. vermiculare. Coasts of Japan. This mucilaginous alga is known in Japan as "Umi-sômen".

N. achminthoides. Coasts of Argentina[2]).

Literature

1) JANECKE, H. and KEHR, W., Pharmaz. Ztg. 105 (25): 684, 1960
2) PUJALS, C., Darwiniana 12 (1): 75–80, 1960

General Literature
CHAPMAN, V. J., 1950; FRITSCH, F. E., 1952; HOFFMANN, C., 1938; KUCKUCK, P., 1953; KYLIN, H., 1956; LEWIN, R. A., 1962; SEGAWA, S., 1957; see Index of Literature.

Neodilsea

This genus has been described by TOKIDA in 1943.

N. yendoana Tokida. Northern part of Honshû, Hokkaido, Kurile Islands, Sakhalin, widespread. Characteristic species. Some significant distinctions followed when being compared with European *Dilsea edulis*.

The alga is used in Japan as a raw material for the production of "Ginnansô". The product generally is made from *Iridea* and *Rhodoglossum* species.

Physiological and biochemical investigations are reported in LEWIN (1962).

Literature

KYLIN, H., 1956; LEWIN, R. A., 1962; SEGAWA, S., 1957; TOKIDA, J., 1954; see Index of Literature.

Odonthalia

Approximately 10 species, in colder seas.

O. dentata (L.) Lyngb. Northern parts of the Atlantic Ocean, Northern Arctic Sea, northern coasts of Eastern Asia and Canada. The type species. Thallus erect, 10 to 25 centimetres high, flattened, more rarely terete, branched, often with a prominent mid-rib, of tough consistency.

O. corymbifera. Coasts of Hokkaido, Kurile Islands, Sakhalin, Korea, and Siberia, coasts of the Sea of Japan. The dried alga contains almost 6% bromine, it is the most valuable raw material of the Japanese coasts for the production of bromides[2]).

Additional Eastern Asiatic species are *O. aleutica* and *O. camtschatica*.

Various *O.* species have been studied for their constituents (floridean starch)[1]).

Physiological and biochemical investigations have been reported in LEWIN (1962).

Literature

1) MEEUSE, B. J. D. and KREGER, D. R., Biochim. et Biophys. Acta 13: 593–595. 1954
2) TOKIDA, J., Fac. of Fish. Hokkaido Univ. 2 (1): 1954

General Literature
FRITSCH, F. E., 1952; KYLIN, H., 1956; LEWIN, R. A., 1962; SEGAWA, S., 1957; see Index oı Literature.

Pachymenia

Approximately 10 species, in warmer seas, particularly on the Australian and New Zealand coasts.

P. carnosa J. Ag. Coasts of the Cape Territory, Japan. The type species. Thallus erect, 15 to 60 centimetres high, flat, undivided or lobed, thickish. The alga resembles to *Grateloupia*[1]).

P. himantophora. New Zealand. The suitability of this alga as a fertilizer has been investigated[2]) [3]).

Literature

1) TAZAWA, N., Bull. Jap. Soc. Phyc. 8 (3): 112–117, 1960
2) FRANCKI, R. J. B., Third Int. Seaweed Symp., 1958
3) FRANCKI, R. J. B., Plant and Soil XII (4). 1960 (I and II)

General Literature

FRITSCH, F. E., 1952; KYLIN, H., 1956; see Index of Literature.

Phycodrys

6 species, North Atlantic and Pacific Oceans, also Tierra del Fuego and the Falkland Islands.

Ph. sinuosa (Good. et Woodw.) Kütz. Northern part of the Atlantic, deeply living on firm substrata or other algae. The type species. Thallus erect, 10 to 20 centimetres high, leafy, lobed, with centre rib, pink-red in colour.

Ph. rubens. Baltic Sea (Kieler Förde), Irish coasts[2]). The alga contains antimicrobial substances[1]).

Ph. fimbriata. Japanese coasts.

Several *Ph.* species have been chemically investigated. Physiological and biochemical investigations of *Ph. rubens* and *Ph. sinuosa* have been reported in LEWIN (1962).

Literature

1) ROOS, H., Kieler Meeresforsch. XIII (1): 1957
2) BRENNAN, A., Inst. f. Industrial Res. and Stand., Dublin, 1950

General Literature

FRITSCH, F. E., 1952; KUCKUCK, P., 1953; KYLIN, H., 1956; LEVRING, T., 1940; LEWIN R. A., 1962; TOKIDA, J., 1954; SEGAWA, S. 1957; see Index of Literature.

Phyllophora

Approximately 15 species, on coasts of temperate and colder seas.

Ph. rubens (L.) Greville. European coasts of the Atlantic Ocean to Morocco; in deep water on rocky substrata. Abundant in the north-western part of the Black Sea and on the Turkish Black Sea coast. The type species. Thallus 5 to 15 centimetres high, erect, petiolate below flattened in the upper part, leaf-like, undivided or bifurcated. The blade is of a vivid red colour.

Ph. nervosa. This alga occurs together with *Ph. rubens* in the Black Sea. It is often regarded as a variety of *Ph. rubens.*

The so – called "phyllophora fields" have been studied by Russian authors. The effects of organic substances on photosynthesis and respiration have been investigated[30–33]).

Both species supply phyllophoran, an agaroid, and form the basis of the Soviet-Russian agaroid industry[1–7]).

The raw material, the agaroid, and its production have been investigated and described by numerous Russian authors[8–21]). STOLOFF (1962) has described the polysacolloid as an agar[34]).

Ph. nervosa has considerable iodine content (0.27 to 0.45 per cent of the dried alga) and is utilized for industrial production of iodine[22–24]).

The antilipemic and anticoagulant properties of *Ph. nervosa* occurring on the Turkish coasts have been studied. One substance isolated from the alga contained, indeed, antilipemic but no anticoagulant activity[29]).

Ph. brodiaei. North Sea and Baltic coasts. This is an atypical species. It is dark red, older parts are of a leathery consistency. It contains antimicrobial substances[25]).

Ph. clevelandii. Coasts of Baja – California and Mexico. An agaroid raw material[7]).

Ph. membranifolia. Coasts of Canada[26]), European Atlantic coasts and Baltic. The Baltic alga contains antimicrobial substances[25]). That species from Great Britain has likewise been investigated: extraction gives a weak gel[27]).

Ph. epiphylla. Coasts of Great Britain and Ireland. Extraction gives a weak gel[27]) [28]).

Physiological and biochemical investigations of *Phyllophora* species have been reported by LEWIN (1962).

Literature

1) HUMM, H. J., The Red Algae of Economic Importance ex Tressler, D. K. and Lemon, J. Mc. W., Marine Products of Commerce. Reinhold Publishing Corporation, New York, 1951
2) STOLOFF, L. and SILVA, P., Ec. Bot. 11 (4): 327–330, 1957
3) STOLOFF, L. and LEE, CH. F., Fishery Leaflet 173, 1946. US Dep. of the Interior, Fish and Wildlife Service
4) HUMM, H. J. and Wolf, F. A., Duke University Press, (3), 1946
5) SMITH, F. and MONTGOMERY, R., The Chemimistry of Plant Gums and Mucilages. Reinhold Publishing Corporation, New York, 1959
6) LEBEDEV, S. J., Third Int. Seaweed Symp., 1958
7) SELBY, H. H. and SELBY, T. A., Agar ex Industrial Gums. Academic Press Inc., New York, 1959
8) ELIN, V. et al., Byull. Nauck. Issl. Khim-Farm. Inst. 140–141, 1931
9) ELIN, V., Vrach Delo 15, 186–190, 1932
10) KORENZVIT, A., Khim Farm Prom. 4, 36–37, 1934
11) OPOTZKII, V. F. and BORTNIK, L. A., Ukrain Khim. Zhur. 10, 331–335, 1935
12) GRYUNER, V. and VERONYAN, L., Colloid J. (USSR) 5. 851–863, 1939
13) LEBEDEV, S.J. and YARTSEWA, K. A., Priroda (2): 96–97, 1956
14) idem; Doklady Akad. Nauck (USSR) 109. 160–163. 1956
15) STAVROV, S. N. and MOROZOV, H. A., Ukrain. Khem. Zhur. 23, 721–727, 1957
16) KRASNOVYKII, E. A. et al., Biofizika 1. 328–333. 1956
17) KABANOVA, Y. G., Trudy Inst. Okeanol. Akad. Nauk USSR 30, 156–165, 1959
18) PSHENIN, L. N., Mikrobiologya 28 (6): 927–932. 1959
19) LITVINENKO, L. G., Ukrain. Bot. Zhur. 16 (6): 49–54. 1959
20) MOROZOV, A. A. and STAVROV, S. N., Kolloid Zhur. 20. 194–197. 1958. CA 52, 20, 16840a, Oct. 1958
21) GRYUNER, V., Coll. Int. du Centre Nat. de la Recherche Scient., Dinard, 1960
22) HOFFMANN, C., Die praktische Bedeutung der

Meeresalgen. Kieler Meeresforschungen III (1), 1938
23) CABRERO-GOMEZ, F., Estudio de las Algas marinas espanoles. C. S. I. C., Madrid, 1951
24) VINOGRADOV, A. R., The Elementary Chemical Composition of Marine Organisms. Sears Foundation for Marine Research, Yale University, New Haven, 1953
25) ROOS, H., Kieler Meeresforschungen XIII (1): 1957
26) YOUNG, E. G. and LANGILLE, W. M., Publ. No. 4636 of the Nat. Res. Council, Halifax, Canada, 1958
27) MARSHALL, S. M. and ORR, A. P., Natural Plant Hydrocolloids, 1954, 101
28) BRENNAN, A. T., Inst. f. Ind. Res. and Stand., Dublin, 1950
29) GÜVEN, K. C. and ATKIN, E., Eczacilik Bulteni Cild 4, Ayi 12, 1962
30) LITVINENKO, L. G., Ukrain. Bot. Zhurn 17 (1): 12–18. 1960
31) POGREBNYAK, I. I., Nauck. Ezhegod. Odessk. Univ. Biol. Fak. 2: 9–10. 1960
32) VINOGRADOV, K. A.; ibid. 2: 140–141. 1960;
33) ibid., 2: 141–142. 1960
34) STOLOFF, L., Ec. Bot. 16 (2): 86–94. 1962

General Literature

CHAPMAN, V. J., 1950; FOTT, B., 1959; FRITSCH, F. E., 1952; HOPPE, H. A., 1958; HOPPE, H. A. and SCHMID, O. J., 1962; KUCKUCK, P., 1953; KYLIN, H., 1956; LEVRING, T., 1940; LEWIN, R. A., 1962; PAX, F., 1962; UPHOF, J. C. TH., 1959; see Index of Literature.

Plocamium

Approximately 50 species are recorded from various oceans.

P. coccineum (Huds.) Lyngbye. Atlantic coasts of Europe. German name "Kammtang". The type species. Thallus erect, flattened, richly branched, rather cartilaginous, of red colour. In deep water on stone substrata or on other algae.

P. pacificum. The phycoerythrin of the alga has been investigated[1]).

P. leptophyllum, *P. oviforme*, *P. recurvatum*, *P. telfairiae* occur on Japanese coasts.

For physiological and biochemical investigations of *P. coccineum* and *P. pacificum* see LEWIN (1962).

Literature

1) H'EOCHA, C. O., Archives of Biochem. and Biophys. 73 (1): 207–219, 1958

General Literature

FRITSCH, F. E., 1952; KUCKUCK, P., 1953; KYLIN, H., 1956; LEWIN, R. A., 1962; PAX, F., 1962; SEGAWA, S., 1957; see Index of Literature.

Plumaria

Several species recorded from the North Atlantic.

P. elegans (Bonnem.) Schmitz. Atlantic coasts. German name "Zierliche Seefeder". The type species. Thallus erect, 4 to 10 centimetres high, flattened, branched, branches feathery, reddish brown to nearly black. This alga has been investigated chemically. Physiological and biochemical investigations have been reported in LEWIN (1962).

P. plumosa. The alga is washed ashore on the beaches of New Foundland in large quantities and is used for fertilizing.

Literature

FRITSCH, F. E., 1952; KUCKUCK, P., 1953; KYLIN, H., 1956; LEVRING, T., 1940; LEWIN, R. A., 1962; SMITH, F. and MONTGOMERY, R., 1959; see Index of Literature.

Polyides

A genus inhabiting the northern part of the Atlantic Ocean.

P. rotundus (Gmel.) Greville. Thallus erect, 8 to 15 centimetres high, terete, repeatedly bifurcated, cartilaginous, of reddish black in colour, with haptera.

Admixture of *P. rotundus* occurs in harvests of *Furcellaria fastigiata*[1]). The *Polyides* can be distinguished from the *Furcellaria* by colour[2]). Anatomical investigations of both algae[3]) and typification of *Polyides* and *Furcellaria* have been reported[4]). *P. rotundus* occurs on the Irish Atlantic coast. BRENNAN wrote in 1950 "This species can be used in the manufacture of an agaroid substance and gives a much firmer gel than *Furcellaria*"[5]). *Polyides* could, perhaps, be utilized as raw material, if it should be found to occur in sufficient quantity.

A red alga from Chile has been submitted for investigation for industrial utilization; apparently this was a *Polyides* species[1]).

P. caprinus. Occurs on the coast of Finland[7]).

Literature

1) HOPPE, H. A., Unveröffentlichte Arbeiten und persönliche Mitteilungen
2) LEVRING, T., Diss. Lund, 1940
3) DENYS, G., Mitt. Bot. Staatsinstitut Hamburg 27: 1–31, 1909
4) DREW, K. M., J. Linnean Soc. London, Botany 55 (363): 744–752, 1958
5) BRENNAN, A. T., Inst. f. Ind. Res. and Stand., Dublin, 1950
6) DANGEARD, P., Coll. Int. du Centre Nat. de la Recherche Scient., Dinard, 1960
7) MATHIESEN, Memor. Soc. Fauna et Flora Fenn 34: 78–84, 1959

General Literature
FRITSCH, F. E., 1952; HOPPE, H. A., 1958; KUCKUCK, P., 1953; KYLIN, H., 1956; see Index of Literature.

Polysiphonia

Approximately 150 species, ubiquitous. Numerous species have not yet been described sufficiently in research literature.

P. violacea (Roth.) Grev. European coasts of the Atlantic Ocean, Baltic, and Mediterranean coasts. The type species. Thallus mostly erect or lying flat at first, and later erect, 2 to 20 centimetres high, sometimes reaching a length of more than one metre[1]), terete branched. Shoots mostly very long, slender, flexible.

An alga of this species collected in the Kieler Förde contained an antimicrobial substance[2]).

A number of *P.* species have been studied in detail. Apart from *P. violacea* inter alia *P. urceolata* and *P. nigrescens* have been described. SEGI published a monograph on *Polysiphonia* in 1951 and reported in 1959 on further *Polysiphonia* studies, describing the following Japanese species: *P. codiicola*, *P. ferulacea*, *P. kampsaxii*, *P. obsoleta*, *P. pulvinata*, *P. saratieri*, *P. scopulorum*, *P. subtilissima*, *P. tokidae*, *P. tongatensis*[4]). In 1960 he published the second part of these studies, with descriptions of a further 20 species[5]).

P. japonica, *P. morrowii*, and *P. urceolata* occur on the coast of Sakhalin[6]).

Cytological studies on *Polysiphonia* have been published in 1959[7]. The commercial significance of the *P.* species is, doubtless, not very great as yet, nevertheless numerous investigations deserving mention have been published.

P. fastigiata and *P. fructiculosa* contain sodium L-glycerate α D-mannoside[8]. The uptake of radioactive cobalt and vitamin B_{12} by *Polysiphonia*[9], the concentration of ribose nucleic acid, protein, anionic polysaccharide[10] and of fluorescing substances[11] have been investigated.

P. elongata. Atlantic and Baltic coasts. This alga contains antimicrobial substances[2] and considerable quantities of Na-Mannoglycerate[12].

P. fastigiata. Atlantic coasts, particularly of Great Britain and Ireland[30] [31]. This species has been investigated chemically[13] [14] [15]. It contains D-glucose[16], D-galactose [16] [17], glyceritol-glycoside[18], mytilitol[19], and antibiotic substances[20]. The seasonal rhythm of the antibiotic activity has also been studied[21]. Extraction gives a weak gel[30].

P. ferulacea. Japan and elsewhere, for example, on Pakistani coast[22]. An alga collected near Puerto Rico proved to have considerable antibiotic activity[23].

P. lanosa. Extracts have been investigated and the presence of antibiotic substances proved[24] [25] [26].

P. morrowii. A bromophenolic compound has been found in this alga[27].

P. nigrescens. Atlantic coasts from the Arctic Sea to France, Baltic coasts. The alga has been found to contain antimicrobial substances[2], and a growth factor was identified as Vitamin B_{12}[9].

P. platycarpa and *P. variegata.* Coasts of Pakistan[22].

P. urceolata. European and North American coasts. The alga has a dark red colour and is similar to the *P. atrorubescens* found on the North-Frisian mud banks[28].

P. gopnathensis. A new species collected from the south-east coast of Gujarat, Gulf of Cambay, India[29].

P. crassa, P. forcipata, P. senticulosa. Occurring on the coasts of Japan, have been mentioned by Segawa.

Physiological and biochemical investigations of *P. fastigiata, P. harveyi, P. lanosa, P. nigrescens, P. pulvinata, P. urceolata,* and *P. violacea* have been reported in Lewin (1962).

Literature

1) Fott, B., Algenkunde. Gustav Fischer Verlag, Jena, 1959
2) Roos, H., Kieler Meeresforschungen XIII (1): 1957
3) Levring, T., Studien über die Algenvegetation von Blekinge, Südschweden. Diss. Lund, 1940
4) Segi, T., Rep. Fac. Fish., Univ. of Mie, 3 (2): 257–266. 1959
5) Segi, T.; ibid. 3 (3): 608–626, 1960
6) Tokida, J., The Marine Algae of Southern Saghalien. Fac. of Fish., Hokkaido Univ., 2 (1), 1954
7) Yabu, H. and Kawamura, Fac. of Fish., Hokkaido Univ. 7 (1/2): 61–71, 1959
8) Black, W. A. P., Ann. Rep. Chem. Soc. (London), 1953. Vol. 50
9) Ericson, L. E., Chem. and Ind., 829–830, 1952
10) Werz, Naturwissenschaften 48 (7): 221–222. 1961
11) h'Eocha, C. O. et al., Third Int. Seaweed Symp., 1958
12) Augier, J., Rev. gen. bot. 60: 257–283. 1953. CA 49, 438b
13) Ross, A. G., J. Sci. Food Agric. (7): 333–335. 1953
14) Black, W. A. P., The Algae ex Altschul, "Processed Plant Proteins Foodstuffs". Academic Press Inc., New York, 1958
15) Wickberg, B., Acta Chem. Scand. 11 (3): 506–511, 1957
16) Smith, F. and Montgomery, R., The Chemi-

stry of Plant Gums and Mucilages. Reinhold Publishing Corporation, New York, 1959
17) BLACK, W. A. P. and CORNHILL, W. J., Chem. and Ind. 514–516, 1954
18) WICKBERG, B., Acta Chem. Scand. 12 (6): 1183–1186, 1958
19) VINCENT, D. L., Chem. and Ind. 1109–1110, 1960
20) CHESTERS, C. G. C. and STOTT, J. A., Second Int. Seaweed Symp., 1955
21) Institute of Seaweed Research. Ann. Rep. 1945–1955
22) ANAND, P. L., Marine Algae from Karachi. The University of the Panjab, Lahore, 1943
23) BURKHOLDER, P. R. et al., Bot. Mar. 2 (1/2): 149. 1960
24) Institute of Seaweed Research. Ann. Rep. 1956
25) ibid., 1957
26) ibid., 1958
27) SAITO, T. et al., Nippon Kagaku Zasshi 76. 478. 1955. CA 51. 22. 178101, 1957
28) KUCKUCK, P., Der Strandwanderer. J. F. Lehmanns Verlag, München, 1953
29) THIVY, F. and SREENIVASA Rao, P., Bot. Mar.
30) MARSHALL, S. M. and ORR, A. P., Natural Plant Hydrocolloids, 1954, 101
31) BRENNAN, A. T., Notes on some common Irish Seaweeds. Inst. for Ind. Res. and Stand., Dublin, 1950

General Literature

FRITSCH, F. E., 1952; KYLIN, H., 1956; LEWIN, R. A., 1962; PAX, F., 1962; SEGAWA, S., 1957; see Index of Literature.

Porphyra

Approximately 25 species, nearly ubiquitous, often in great abundance. It is very difficult to distinguish the individual species.

P. species mainly inhabit rocks between tide levels. Some have a very significant resistance to dry out and form entire zones in the upper littoral region.

KORNMANN[139]) refers to the confusion in the nomenclature of *Porphyra* species, the need for clarification in the taxonomy of the genus and for investigation into the life history of the individual species. He describes five *P.* species occuring on the coast of Heligoland:

P. leucosticta Thur., *P. linearis* Crev., *P. umbilicalis* (L.) Kütz., *P. purpurea* (Roth.) C. A. Ag., and a *P.* species (*amethystea* Kütz?). The type species is *P. umbilicalis* (L.) Kütz.

A significant aspect of *Porphyra* is the *Conchocelis* stage. In the course of recent years several papers have been published on that subject. Basic reports have followed the investigations of DREW in 1954. The following is a quotation of the summary of that work by KYLIN:

"One type of spore has been germinated in culture. On glass and in liquid culture-solution such spore give rise to creeping septate filaments – but showing no reproductive cells. Experiments show that when such spores or the filaments are brought in contact with the shell the spores penetrate the shell and give rise to the growth identical with the 'species'."

P. species have been cultivated in Japan and other coastal areas, for many centuries. Highly developed cultures and modern processing industries have arisen in the course of time[1–3]).

Conchocelis rosea Batters. is therefore a phase in the life-history of *P. umbilicalis* var. *laciniata* and not an autonomous species.

An article of KORNMANN "Von *Conchocelis* zu *Porphyra*" fills up a gap[141]). Japanese scientists report about artificial multiplication of *Conchocelis* thallus of *Porphyra*[142]) and about pigments of *Conchocelis*[143]).

Investigations have been carried out on the production and possible utilization of the polysaccharide extract known as porphyran.

A number of papers on *Porphyra* cultivation have been published mainly by Japanese authors, particularly respecting the influence of outside seawater, industrial waste waters, light intensity etc.[4–19]).

Observations on the culture in artificial seawater and effects of amino acids and purines on the growth and the metabolic disease "akagusare" have been reported[140]).

Further references will be mentioned in the paragraphs discussing the individual species.

A large number of investigations have been undertaken of the constituents of *Porphyra* species.

Investigations include the varietal and seasonal variation in the form of nitrogen[20]), because *P.* species have a relatively higher nitrogen content than other seaweeds (LEWIN, 1962), amino acids[21]), cation binding, and exchange in killed *Porphyra*[22]), vitamin[23]) and antibiotic content[24]).

P. amplissima. Japanese name: "Beni-tasa". It is recorded from the coasts of the Arctic and Subarctic seas. Kurile Islands, Sakhalin, Kamchatka, Pacific, and Atlantic coasts of North America. Grows on rocks, sometimes also freely drifting[25]). This species has a double layered thallus. FUKUHARA has reported on the holdfasts and their structure[136]).

P. angusta. Coasts of Japan.

P. atropurpurea. Coasts of Indonesia, the Philippines, and Hawaiian Islands, and in smaller quantities on the Coast of New England[53]). The alga is difficult to be collected, because it grows on rocks exposed to heavy seas. It is salted in preparation and regarded as a delicacy in Hawaii. Additional imports from Japan satisfy the local demand. *P. atropurpurea* is used medically for poultices and cataplasms[26]). This species has also been found in the Baltic[27]).

P. bulbopes. Coasts of Japan[136]).

P. capensis. Coasts of South Africa[1]) [28–30]), particularly on the west coast. Its polysaccharide has been studied[31–36]).

This alga contains 25 to 30 per cent of mucilage which can be regarded as a gelan[135]). The constituents have been reported in LEWIN (1962).

P. columbina. Antipodes, Campbell Islands, Auckland, New Zealand, South and eastern States of Australia, Chile[1]) [37]) [132]). Several forms are recorded from Chile. These, however, can for the most part be referred to *P. columbina*[37]). In New Zealand the fermented alga is used by the Maoris as an aperient[26]) [38]).

P. crassa. Coasts of Japan, Bay of Oshoro, Hokkaido. Sometimes epiphytic[39]).

P. crispata. East-Asian coasts. The constituents of the alga have been investigated[40–41]). LEWIN (1962).

P. dentata. Coasts of the Japanese islands[1]). A free-living species[36]). Japanese name "Oni-amanori". This species belongs to the "Iwa-nori" group. Studies on the artificial propagation in the Sea of Japan have been reported[137]).

P. denticulata. Coasts of Australia. This species has been found by LEVRING in 1953.

P. kunthiana. Coasts of Chile. This alga is eaten together with *Ulva latissima* and is known as "luche"[1]).

P. laciniata. Japanese islands, Great Britain, Ireland, France. *P. laciniata* belongs to the most important edible algae of Japan[41]). In Europe the alga is eaten either after being boiled or as a salad. In England it is described as "laver", in Scotland as "sloke". The use of this alga for human consumption has been known since the 15th century[1]). Its chemistry has been studied on several occasions[33]) [42]) [43]).

P. laciniata (Lightf.) Ag. according to KYLIN, is identical to *P. umbilicalis* (L.) Kütz. According to other authors, however, this alga is a variety of *P. umbilicalis*[36]).

P. lanceolata. Coasts of the U.S.A., from Washington to California, known as "red jabot laver"[1]).

P. leucosticta. Coasts of the Indian and Pacific ocean. Known in Hawaii as a delicacy under the name of "limu luan"[44]). This species is used in the laver-bread industry in Great Britain (South Wales)[45]).

KORNMANN writes on the development of the alga in culture experiments[138]).

P. linearis. Occurrence and culture experiments have been reported[46]).

P. lucasii. Coasts of Australia. This species has been found by LEVRING in 1953.

P. marginata. Coasts of East Asia. This species has been described only a few years ago[47]).

P. miniata. Arctic and Subarctic regions. This species, like the *P. amplissima* has a double layered thallus.

P. miniata f. *cuneiformia.* Northern Pacific coasts of America, South Chile[37]).

P. moriensis. Coast of Mori; grows epiphytically on other algae, often in depths of 3 to 8 metres. The alga is harvested in winter and spring and exported[48]).

P. naiadum. Coasts of U.S.A., from Washington to California, known as "red fringe"[1]).

The pigments of the alga have been investigated, and B-phycoerythrin has been isolated[49]).

The properties of phycobilin from *P. naiadum* have been investigated: *P. naiadum* (Cf. *Smithora naiadum)* var. *australis* contains biliproteins. Fresh extracts contain B-phycoerythrin, allophycocyanin and a purple complex. *P. naiadum* var. *naiadum* contains B-phycoerythrin, allo-phycocyanin, and C-phycocyanin[50]) [51]) [52]).

P. nereocystis. Pacific ocean. The alga is eaten in China[44]). It contains 50 per cent more vitamin C than oranges, and considerable quantities of B vitamins[1]) [53]). This species grows to 2 metres in length[36]).

P. ochotensis. Kurile Islands, Sakhalin, Japanese name: "Ana-amanori"[25]).

P. okamurai. Japan, Bay of Oshoro, Hokkaido, sometimes epiphytic[1]) [39]), sometimes free-living[36]), Japanese name: "Kuro-nori".

P. onoi. Japan, Bay of Oshoro, Hokkaido. Sometimes epiphytic[1]) [39]), sometimes free-living[36]). Japanese name: "Ono-nori".

P. perforata. Pacific coast of the U.S.A., from Alaska to Southern California. This alga grows on rocks in sheltered places, in depths of 60 centimetres to one metre. It develops thalli nearly 1 metre in length. *P. perforata* supplies "California laver" or "purple laver". Collection is easy: the plants are picked from the rocks and dried in the sun. Quantities harvested amount to several 100,000 pounds per year. The product is eaten by the Chinese population of the U.S.A., and is also shipped to China[1]) [36]) [44]) [53]). Cultures of the alga have been studied[54]), and other investigations carried out[55–59]) [144]).

P. pseudo-linearis. Japanese coasts, particularly of Honshû, Bay of Oshoro, Hokkaido, Kurile Islands, Sakhalin, and Korea. Japanese name: "Uruppui-nori". This species belongs to the "Iwa-nori" group. The alga grows in exposed areas, sometimes epiphytically. It is also cultivated[1]) [25]) [39]) [36]). A number of chemical investigations have been made on it[60–63], and studies on its artificial propagation on the coast of the Sea of Japan have been reported on[13].

P. surborbiculata. Pacific ocean. The alga is eaten in various preparations in China. It is known as "tsu choj"[44].

P. tasa. Japan[136].

P. tenera. This alga, known in Japan as "Amanori" or "Nori", is the most important and well-known Japanese *Porphyra.* It occurs not only on Japanese coasts, but also in many other East Asian areas, especially on the China coast. It is cultivated in numerous forms. The quantity of the dried final product is reported to be approximately 5.000 tons annually, which corresponds to a value of 85 to 100 million dollars[1] [36] [44] [53] [64] [65].

In the course of the last years Japanese scientists have studied the culture of *Porphyra tenera,* its ecology, spore formation, and growth, the environmental factors at the cultural grounds and the correlations between water temperature and geographical distribution, and the influence of industrial waters[66–86] [151–153]. The alga has been cultivated in the laboratory[154], and its physiology[87], metabolism[88], and biochemistry have been studied[89] [90] [155–156] (LEWIN, 1962). Chemical studies have been made on the air dried alga[91] and on the alga during preparation[92]. The polysaccharides have been studied; mannose and D-L-galactose were found[41] [93] [94]; other chemical investigations on the species include studies on the reducing sugar in "Asakusa-nori"[95], on the volatile constituents of the alga responsible for the characteristic odour[96] (the alga contains 0.037 to 0.085 per cent of volatile oils[135]), and on substances (amino acids) responsible for the flavour[97–99], on nucleic acids[100], and on pigments, chlorophylls, carotinoids, and chromoproteins[52] [101–109] [145–148].

The *Porphyra tenera* products of this species are described as "Hoshi-nori". The surface conditions of splint-mat and its improvement by silicone have been investigated[149]. The *Porphyra tenera* "round type" yields the product "Natsu-nori".

P. umbilicalis (L.) Kütz. To be regarded as the type *P.* species. It is described as "common porphyra", or in Japanese as "Chishima-kuronori". In England the alga is known as "laver", in Scotland as "slack", and in Ireland as "sloke". On the Atlantic coasts of Europe and North America it occurs in the upper littoral zone on rocks or on other algae.

In addition to f. *laciniata* various forms of the alga, especially f. *linearis* and f. *vulgaris*[25] occur on the East Asian coasts, particularly Hokkaido, Honshû, Kurile Islands, Sakhalin, Kamchatka, the Sea of Okhotsk, and on the Pacific coast of North America, coasts of South America, Hawaiian Islands, and on the coast of South Africa[1].

Thallus erect, leafy, flat, thin, one or two layered, lobed or split, margin serrate undulated or entire. The colour of the fresh alga varies from a greenish to a reddish brown or violet, often variagated. In a dried state it is very thin and violet in colour.

Particularly abundant populations of *P. umbilicalis* are found in early summer at Cape of Nichimotoro (Japan). The fishermen harvest the alga at low tide and use it to make "Asakusa-nori". Also on the coasts of the British Isle and Ireland the alga is harvested and eaten in various preparations[45] [110] [111] [133]; it serves as the raw-material of the "laver-bread" industry. On the Hawaiian Islands the alga is known as "limu", and is a popular delicacy. Reports have been published on the *Porphyra* populations on the Scottish coasts and on cultural experiments with *P. umbilicalis* f. *umbilicalis*[46] [112].

Studies on the *Conchocelis* phase of *P. umbilicalis*[113] [114] and cytological investigations have been published[25] [115]. The alga has been investigated chemically[116] with reference to protein and amino acid[117] [118] and polysaccharide content (mannan[33] [111] [119] [120] and galactose[41] being found) and to other carbohydrates[121–124] [134] [150] including sugar sulphates from mucilage[127] and the isolation of isofloridoside[127]. The constituent chemistry of this species is dealt with in LEWIN 1962.

P. variegata. Northern parts of Honshû, Hokkaido, Kurile Islands, Sakhalin, Bering Islands, Pacific coasts of North America, sometimes epiphytic. Japanese name "Tomoe-nori". The alga has been described by FUKUHARA[136].

P. vulgaris. Pacific and Atlantic Oceans. The alga is eaten in California and in Japan[44]. It contains vitamin C[23]. A number of authors regard *Porphyra vulgaris* to be a variety of *P. umbilicalis.*

P. yezoensis. Coasts of Japan, Bay of Oshoro, Hokkaido, sometimes epiphytic[1] [19] [39] [128], sometimes free-living[36]. Japanese name: "Susabi-nori".

Further *P.* species occur in various areas, for example on the coasts of the Philippines[129].

Several reports have been published on the development of a possibly new *P.* species found in "nori"-fields in Tokushima[130].

Two further new *P.* species and their subgeneric relationship have also been described[131].

Physiological and biochemical investigations of *P. capensis, P. laciniata, P. leucosticta, P. naiadum, P. nereocystis, P. perforata, P. tenera,* and *P. umbilicalis* have been reported by LEWIN (1962).

Literature

1) SCHACHAT, R. E. and GLICKSMAN, M., Some Lesser Known Seaweed Extracts ex Whistler, R. L. and BeMiller, J. N., Industrial Gums. Academic Press Inc., New York, 1959
2) YAMADA, Y., Coll. Int. du Centre Nat. de la Recherche Scient., Dinard, 1957
3) SCHMID, O. J. and HOPPE, H. A., Bot. Mar. III, Supplement, 1962
4) SUTO, S. and UMEBAYASHI, O., Bull. Jap. Soc. Sci. Fish. 19 (12): 1176–1178. 1954
5) SUTO, S. and UMEBAYASHI, O.; ibid. 19 (12): 1179–1182. 1954
6) KUROGI, M. and HIRANO, K., Bull. Tôhoku Reg. Fish. Res. Lab. 8. 27–44. 1956
7) ARASAKI, S., Plant Protection 10. 234–246. 1956
8) NOZAWA, K., Seibutsu Kagaku (Biol. Sc.) 8 (4): 154–156. 1956
9) FUKUHARA, E., Monthly Rep. Hokkaido Fish. Exper. Stat. 14 (12): 17–22. 1957
10) NOZAWA, K. and NOZAWA, Y., Nippon Suisangaku Kaihi 23. 423–426. 1957. CA. 50. 9526i
11) FUKAI, R. and SHIOKAWA, F., Bull. Tokai Reg. Fish. Res. Lab. 18. 21–30. 1957
12) FUJIYA, M. and CHIKINI, S., Bull. Naikai Reg. Fish. Res. Lab. 11. 53–67. 1958
13) TSENG, C. K. et al., Kexue Tongbao (Scientia) (5): 169–171. 1959
14) TSENG, C. K. et al.; ibid. (5): 171. 1959
15) CHANG et al., Shengwuxue Tongbao 2. 55–59. 1959
16) TAKESUE and MIYAZAWA, J., Shimonoseki Coll. Fish. 8 (2): 199–207. 1959
17) IWAMOTO, Bull. Jap. Soc. Phyc. 7 (1): 4–11. 1959
18) KUROGI, M., Bull. Tôhoku Reg. Fish. Res. Lab. 15. 33–42. 1959
19) KUROGI, M.; ibid. 15. 43–56. 1959
20) TAKAGI, M., Bull. Fac. Fish. Hokkaido Imp. Univ. 2 (1): 31–42. 1951
21) TAKAGI, M. and KURIYAMA; ibid. 10 (1): 72–76. 1959
22) EPPLEY, R. W., Exper. Cell. Res. 13 (1): (1): 173–174. 1957
23) AKAO, A. et al., Nihon Univ. Med. J. 16 (4): 1103–1107. 1957
24) CHESTERS, C. G. C. and STOTT, J. A., Second Int. Seaweed Symp., 1955
25) TOKIDA, J., Fac. of Fish., Hokkaido Univ., 2 (1). 1954
26) ZANEVELD, J. S., Ec. Bot. 13 (2): 89. 1959
27) HOFFMANN, C., Kieler Wiss. Meeresunters. N. F. 21. 1–9. 1928
28) GRAVES, J. M., Nature 175 (4452): 393–394. 1955
29) GRAVES, J. M., Phycol. Bull. 4. 12–13. 1956
30) ISAAC, Wm. E., Proc. Linnean Soc. London 168 (Parts 1/2): 61–65. 1957
31) NUNN, J. R. and v. HOLDT, M. M., Chem. and Ind. 467. 1956. CA 51. 1. 236a, 1957

32) Nunn, J. R. and v. Holdt, M. M., J. Chem. Soc. 1094–1097. 1957. CA 51. 14. 10385h. 1957
33) Smith, F. and Montgomery, R., The Chemimistry of Plant Gums and Mucilages. Reinhold Publishing Corporation, New York, 1959
34) Araki, Ch., Seaweed Polysaccharides ex Wolfrom, M. L., Carbohydrate Chemistry of Substances of Biological Interest. Pergamon Press Ltd., London, 1959
35) Janecke, H. and Kehr, W., Pharmaz. Ztg. 105 (25): 684. 1960
36) Chapman, V. J., Seaweeds and their Uses. Methuen & Co., Ltd., London, 1950
37) Levring, T., Contributions to the marine algal Flora of Chile. Lunds Univ. Årsskrift N. F. Avd. 2. 56 (10). 1960
38) Brooker, S. G. and Cooper, R. C., Ec. Bot. 15 (1): 1. 1961
39) Tokida, J. and Masaki, T., Fac. of Fish., Hokkaido Univ., 10 (3): 173–195. 1959
40) Su, J. C., J. Chinese Chem. Soc. (Taiwan) Ser. II, 3. 65–71. 1956. CA 512g. 1959
41) Black, W. A. P. and Cornhill, W. J., Chem. and Ind. 514–516. 1954
42) Matsui, H., J. Coll. Agr., Tokyo Imp. Univ. 5. 391–393. 1916
43) Thomas and Marman, Biochem. et Biophys. Acta 35 (2): 316–323. 1959
44) Uphof, J. C. Th., Dictionary of Economic Plants. H. R. Engelmann (J. Cramer), Weinheim, 1959
45) Hampson, M. A., Gr. Brit. Min. Agr. Fish and Food Invest. Ser. 2, 21 (7): 1–8. 1957
46) Conway, E., Fourth Int. Seaweed Symp., 1961
47) Tseng, C. K. and Chang, T. J., Acta Bot. Sinica 7 (1): 15–26. 1958
48) Ohmi, H., Fac. of Fish. Hokkaido Univ. 5 (3): 231–239 . . .
49) Airth, R. L. and Blinks, L. R., Biol. Bull. 111. 321–327. 1956
50) Airth, R. L. and Blinks, L. R., J. Gen. Physiol. 41 (1): 77–90. 1957
51) H'Eocha, C. O. and Haxo, F. T., Biochem. et Biophys. Acta 41. 516–520. 1960
52) H'Eocha, C. O. Arch. of Biochem. and Biophys. 73 (1): 207. 1958
53) Humm, H. J., The Red Algae of Economic Importance ex Tressler, D. K. et Lemon, J. Mc. W., Marine Products of Commerce. Reinhold Publishing Corporation, New York, 1951
54) Hollenberg, G. J., Am. J. Botany 45 (9): 653–656. 1958
55) Eppley, R. W., Ann. Arbor (Diss.) Publ. 23194. CA 49. 22. 16025e, 1955
56) Eppley, R. W. and Blinks, L. R., Plant Physiol. 32 (1): 63–64. 1957
57) Eppley, R. W., J. Gen. Physiol. 41 (5): 901–911. 1958. CA 52. 22. 20435b. 1958
58) Eppley, R. W.; ibid. 42 (2): 281–288. 1958. CA 53. 9. 8319a. 1959
59) Eppley, R. W.; ibid. 43 (1): 29–38. 1959
60) Takagi, M., Bull. Fac. Fish. Hokkaido Imp. Univ. 2. 297–301. 1952. CA 47. 21. 11356g. 1953
61) Takagi, M., Bull. Jap. Soc. Sci. Fish. 18. 483–487. 1953. CA 5602i. 1955
62) Takagi, M.; ibid. 19 (8): 889–892. 1953
63) Takagi, M., Bull. Fac. Fish. Hokkaido Imp. Univ. 3 (4): 253–258. 1953
64) Segi, T., Rep. Fac. of Fish. Univ. of Mie 3 (2): 251–256. 1959
65) Arasaki, S., Amemiya Inb. Publ. Fish. Sc. 805–818. 1957
66) Iwasaki, H. and Matsudaira, C., Bull. Jap. Soc. Sci. Fish. 20 (2): 112–119. 1954
67) Iwasaki, H. and Matsudaira, C.; ibid. 20 (5): 380–385. 1954
68) Yamasaki, H., Bull. Jap. Soc. Sci. Fish. 20 (6): 442–446. 1954
69) Yamasaki, H.; ibid. 20 (6): 447–450. 1954
70) Takeuchi, T. et al.; ibid. 20 (6): 487–489. 1954
71) Sudo, S. et al.; ibid. 20 (6): 490–493. 1954
72) Sudo, S.; ibid. 20 (6): 494–496. 1954
73) Tseng, C. K. and Chang, T. J., Scientia Sinica 4 (3): 375–398. 1955
74) Saito, Y., Bull. Jap. Soc. Sci. Fish 21 (4): 226–228. 1955
75) Arasaki, S. et al.; ibid. 22 (3): 167–171. 1956
76) Kurogi, M. and Hirano, K., Bull. Tôhoku Reg. Fish. Res. Lab. 8. 45–61. 1956
77) Yatuyanagi, K. et al., Rep. Yamaguchi Pref. Fish. Exper. Stat. 8. 1–20. 1956
78) Sano, T., Tôhoku-kaiku Suisan-Kenkyajo Kenkyû-hôkóku. 8. 62–69. 1956; 10. 89–94. 1957; CA 53. 2. 1479g. 1959; CA 53. 2. 1479h. 1959
79) Nozawa, K. and Nozawa, Y., Bull. Jap. Soc. Sci. Fish. 22 (11): 427–429. 1957
80) Yamasaki, H. et al.; ibid. 23 (4): 195–198. 1957
81) Tanaka, K.; ibid. 23 (4): 202–205. 1957
82) Tanaka, K.; ibid. 23 (10): 604–607. 1958
83) Migita, S., Bull. Fac. Fish. Nagasaki Univ. 7. 81–86. 1958
84) Iwasaki, H. and Matsudaira, C., Bull. Jap. Soc. Sci. Fish. 24 (6/7): 398–401. 1958
85) Matsumota, J. Fac. Fish. & Anim. Husb. Hiroshima Univ. 2 (2): 249–333. 1959
86) Sano, T. and Tanil, Bull. Tôhoku Reg. Fish. Res. Lab. 15. 57–66. 1959
87) Iwasaki, H. and Matsudara, C., Tôhoku J. Agr. Res. 8 (1): 47–54. 1957
88) Kanzawa, A. et al., Mem. Fac. Fish. Kagoshima Univ. 187–191. 1959
89) Yamasaki, H., Bull. Jap. Soc. Sci. Fish. 20. 1006–1009. 1955. CA. 50. 12. 881oh. 1956
90) Sato et al.; ibid. 25 (10): 798–802. 1959
91) Yamakawa, T.; ibid. 18. 478–482. 1953. CA 48. 4. 2184b. 1954

92) Takagi, M., Bull. Fac. Fish. Hokkaido Imp. Univ. 3 (4): 250. 1953
93) Black, W. A. P., The Algae ex Altschul, Processed Plant Proteins Foodstuffs. Academic Press Inc., New York, 1958
94) Black, W. A. P., Ann. Rep. Chem. Soc. 50. 1953
95) Yamazaki, H., Bull. Jap. Soc. Sci. Fish. 24 (12): 961–965. 1959
96) Katayama, T.; ibid. 22 (4): 244–247. 1956
97) Murata, K. et al., Bull. Fac. Fish. Hokkaido Univ. 7 (2): 130–133. 1956
98) Tsuchiya, Y. and Suzuki, Y., Bull. Jap. Soc. Sci. Fish. 20. 1092–1098. 1955. CA 50. 12. 8935h, 1956
99) Tsuchiya, Y. and Sasaki, T.; ibid. 23 (4): 230–233. 1957. CA 53. 6. 5534e. 1957
100) Katayama, T. and Fujiyama, T.; ibid. 23 (5): 249–254. 1957. CA 53. 6. 5420i. 1959
101) Fujiwara, T., J. Biochem. (Tokyo) 42. 411–417. 1955. CA 49. 22. 16025e. 1955
102) Fujiwara, T.; ibid. 43 (2): 195–203. 1956. CA 50. 17. 12189e. 1956
103) Fujiwara, T., J. Biochem. (Tokyo) 44. 723–733. 1957. CA 52. 5. 3879f. 1958
104) Fujiwara, T., Tokyo Univ. Amemya Jul. Publ. Fish. Sci. 829–840. 1957
105) Fujiwara, T., Third Int. Seaweed Symp., 1958
106) Fujiwara, T., Fourth Int. Seaweed Symp., 1961
107) Hashimoto and Masuhiro, Bull. Jap. Soc. Sci. Fish. 25 (7/9): 561–564. 1959
108) H'Eocha, C. O., Chemical Studies of Phycoerythrins and Phycocyanins ex Comparative Biochemistry of Photoreactive Systems. Academic Press Inc., New York, 1960
109) Sano, T., Tôhoku-kaiku Suisan-Kenkyajo Kenkyû-hôkoku 8. 62–69. 1956; 10. 89–94. 1957; CA 53. 2. 1479g. 1959; CA 53. 2. 1479h. 1959
110) Food Manufacture, 34 (11): 443–444. 1959
111) Whistler, R. L. and Smart, Ch. L., Polysaccharide Chemistry. Academic Press Inc., New York, 1953
112) Conway, E., Third Int. Seaweed Symp., 1958
113) Drew, K. M., Nature 164 (29): 748. 1949
114) Baker and Drew, K. M., Proc. Int. Bot. Congr., Stockholm, 1950. 7. 834–835. 1953
115) Krishnamurthy, V., J. Ag. Ann. Bot. (London) n. s. 23 (89): 147–176. 1959
116) Ross, A. G., J. Sci. Food Agric. (7): 333–335. 1953
117) Coulson, C. B., Chem. and Ind. 997–998. 1953
118) Coulson, C. B., J. Sci. Food Agric. 6. 674–682. 1955
119) Foster and Stacy, Polysaccharides from Algae and Lichens ex W. Ruhland. Handbuch der Pflanzenphysiologie, Springer Verlag, 1958
120) Young, E. G., Carbohydrate accumulation by lower plants (Rhodophyta); ibid.
121) Lindberg
122) Wickberg, B., Acta Chem. Scand. 12 (6): 1187–1201. 1958
123) Turvey, J. R. and Rees, D. A., Third Int. Seaweed Symp., 1958
124) Turvey, J. R., Coll. Int. du Centre Nat. de la Recherche Scient., Dinard, 1960
125) Stoll and Jucker, Heteroside ex W. Ruhland, Handbuch der Pflanzenphysiologie, Springer Verlag, 1958
126) Wickberg, B., Third Int. Seaweed Symp., 1958
127) Turvey, J. R. and Williams, Fourth Int. Seaweed Symp. 1961
128) Fukuhara, E., Bull. Jap. Soc. Phyc. 6 (1): 8–11. 1958
129) Velasquez, G. T., Philippine Seaweed Report, 1955
130) Katoh, T., Bull. Jap. Soc. Sci. Fish. 20 (6): 483–486. 1954
131) Mikami, H., Bot. Mag. (Tokyo) 69 (819): 340–345. 1956
132) Bergquist, P. L., Bot. Mar. I (3/4): 86. 1960
133) Brennan, A. T., Inst. f. Ind. Res. and Stand., Dublin, 1950
134) Peat, S. et al., J. Chem. Soc. 1590, 1961
135) Hegnauer, R., Chemotaxonomie der Pflanzen, Birkhäuser Verlag, Stuttgart, 1962
136) Fukuhara, E., Month. J. Hokkaido Fish. Exper. Stat. 17 (3): 38–40. 1960
137) Yoshinaga, H., Aquiculture 8 (2): 107–109. 1960;
Yoshinaga, H. and Yatsuyanaga, K.; ibid. 7 (3): 12–17. 1960
138) Kornmann, P., Helg. Wiss. Meeresunters. 8 (1): 167. 1961
139) Kornmann, P., Helg. Wiss. Meeresunters. 8 (1): 176. 1961
140) Teramoto, K. and Kinoshita, S., Bull. Jap. Soc. Phyc. 8 (2): 66–71. 1960; ibid. 8 (3): 90–95. 1960
141) Kornmann, P., Helg. Wiss. Meeresunters. 7 (4): 189–193. 1960
142) Akiyama, K., Bull. Tôhoku Reg. Fish. Lab. 17. 45–48. 1960
143) Sano, T.; ibid. 17. 38–44. 1960
144) Su, J. C. and Hassid, W. Z., J. Biol. Chem. 235 (9): 36–37. 1960
145) Yakushiji, E., Nature 185 (4706): 105–106. 1960
146) Katoh, S., Plant & Coll. Physiol. 1 (2): 91–98. 1960
147) Katoh, S., Nature 186 (4719): 138–139. 1960
148) Yakushiji, E. et al., Nature 185 (4706): 105–106. 1960
149) Umemoto, S. and Tanita, S., Bull. Tôhoku Reg. Fish. Res. Lab. 16. 135–139. 1960
150) Turvey, J. R. and Rees, D. A., Third Int. Seaweed Symp., 1958

151) SANO, T. and SATÔ, S., Bull. Tôhoku Reg. Fish. Res. Lab. 16. 127–134. 1960
152) ENDO, T. and MATSUDAIRA, Y., Bull. Jap. Soc. Sc. Fish. 26 (9): 871–876. 1960
153) FUJIYA, M. and CHIKIMI, S., Aquiculture 7 (3): 48–52. 1960
154) SUTO, S.; ibid. 7 (3): 7–11. 1960
155) TSURUGA, H. et al., Bull. Nakai Reg. Fish. Res. Lab. 13. 19–22. 1960; 13. 23–32. 1960; 13. 33–38. 1960
156) ITO, K. et al., Bull. Jap. Soc. Sc. Fish 26. (9): 938–943. 1960

General Literature

CHAPMAN, V. J., 1950; FOTT, B., 1959; FRITSCH, F. E., 1952; HOFFMANN, C., 1938; KUCKUCK, P., 1953; KYLIN, H., 1956; LEWIN, R. A., 1962; PAX, F., 1962; SCHMID, O. J. and HOPPE, H. A., 1962; SEGAWA, S., 1957; see Index of Literature.

Porphyridium

4 species, unicellular and microscopic, widespread.

P. cruentum Näg. Marine or more usually terrestrial in moist shady places by the sea. The type species; grows individually or in larger or smaller colonies. The globular or oval cells are surrounded by mucilage. The chromatophores of the type species are red. The cells contain starch. The alga often has been cultivated in the laboratory and has been investigated chemically[1–4] [7]. A polysaccharide has been found[5]. The hydrolysates contain glucose, galactose, xylose, and uronic acid. The pigments have been investigated and found to contain phycoerythrin and 2 phycocyanines[1] [6]. It seems to be possible that phycocolloid from *P.* will one day gain economic importance[1].

Physiological and biochemical investigations of *P. aerugineum* and *P. cruentum* have been reported in LEWIN (1962). The extracellular polysaccharides produced in cultures have been investigated.

Literature

1) HAXO, F. T. and H'EOCHA, C. O., Second Int. Seaweed Symp., 1955
2) H'EOCHA, C. O., Archives of Biochem. and Biophys. 73 (1): 207–219, 1958
3) H'EOCHA, C. O. and HAXO, F. T., Biochim. et Biophys. Acta 41: 516–520, 1960
4) DOUGHERTY, E. C. et al., Exptl. Cell Research 13: 171–173, 1957
5) JANECKE, H. and KEHR, W., Pharmaz. Ztg. 105 (25): 684, 1960
6) LATIMER, P. H. et al., ex Gaffron, H. et al., Research in Photosynthesis 107–112, 1957
7) THOMAS, J. B. and GOVINDJEE, Biophys. J. 1 (1): 63–75, 1960

General Literature

FRITSCH, F. E., 1952; KYLIN, H., 1956; LEWIN, R. A., 1962; see Index of Literature.

Pterocladia

Approximately 10 species are recorded from the Pacific and in a number of other oceanic regions.

P. lucida (R. Br.) J. Ag. Coasts of Australia, Tasmania, New Zealand (North Auckland, Bay of Plenty), the Chathams, Lord How and Norfolk Islands. The type species. Thallus erect, 10 to 30 centimetres high, double bladed and flattened, or compressed and lanceolate in form, pinnately branched, variable in form (Pl. XII: 29).

An agar raw material; utilized commercially[27]. Mostly found as "driftweed" and collected by the Maoris[1–10]. In the collected material *P. lucida* dominates the smaller *P. capillacea*[11]. This species is eaten in New Zealand by the Maoris[12].

P. capillacea (= *P. pinnata*). Coasts of Australia, New Zealand (mostly North Iceland, South Iceland, Bay of Plenty, Hokianga, East Cape districts), Japan, China, Brazil, Cameroon[13], coasts of Great Britain, Ireland, France, Algeria, Egypt. The taxonomy of this species has been worked out [26]. An agar raw material, particularly the New Zealand plant which is utilized economically; utilized in Japan in mixture with other agarophytes[11] [2] [3] [15] [5] [6] [16] [7] [8] [9] [14] [27].

P. capillacea from Egypt is getting increasing significance as raw material. The chemical composition of this material[17] and the seasonal variation in the composition of drift *P. capillacea* from Alexandria have been investigated[18].

P. complanata. U.S.A., coast of California; an agar raw material[3] [19].

P. densa. Coasts of Japan. An agar raw material, utilized economically in mixture with other agarophytes[7]. This alga supplies a hard type of agar[8].

P. nana. Coasts of Japan. An agar raw material, utilized economically in mixture with other agarophytes[7]. This alga supplies a hard type of agar[8].

P. pyramidale. U.S.A., coast of California. An agar raw material[27]. This alga has been investigated chemically and agarose proved to be present[3] [19–22].

P. tenuis. Coasts of Japan. An agar raw material, utilized commercially in mixture with other agarophytes[7] [23]. The alga contains polysaccharides and yields a hard type of agar[25] [8].

Physiological and biochemical investigations have been reported in Lewin (1962).

P. species. U.S.A., Pacific coasts, Hawaiian Islands. Two new species have been described which appear to be suitable as an agar raw material[24].

Literature

1) Tseng, C.K., Science 101 (2633): 597–602. 1945
2) Humm, H. J. and Wolf, F. A., Duke Univ. Press (3) 1946
3) Stoloff, L. and Lee, Ch. F., Fishery Leaflet 173, 1946. US Dep. of the Interior, Fish and Wildlife Service, Washington
4) Morel, Th., Medeling v. h. Centraal Inst. v. Techn. Onderzoekingen, Djakarta, 1949
5) Humm, H. J., The Red Algae of Economic Importance ex Tressler, D. K. and Lemon, J. Mc. W., Marine Products of Commerce. Reinhold Publishing Corporation, New York, 1951
6) Stoloff, L. and Silva, P., Ec. Bot. 11 (4): 327–330. 1957
7) Selby, H. H. and Selby, T. A., Agar ex Industrial Gums. Whistler, R. L. and BeMiller, J. N., Academic Press Inc., New York, 1959
8) Smith, F. and Montgomery, R., The Chemistry of Plant Gums and Mucilages. Reinhold Publishing Corporation, New York, 1959
9) Bergquist, P. L., Bot. Mar. I (3/4): 86, 1960
10) Dangeard, P., Coll. Int. du Centre Nat. de la Recherche Scient., Dinard, 1960
11) Moore, L. B., New Zealand J. of Science and Technol. 25 (5) Sec. B. 183–209. 1944
12) Uphof, J. C. Th., Dictionary of Economic Plants. H. R. Engelmann (J. Cramer), Weinheim. 1959
13) Lawson, G. W., J. West Afric. Science Ass. 1 (2): 1955
14) Fourment, R. et al., Bull. Soc. Hist. Nat. Afrique du Nord 32. 176–180. 1941
15) Brennan, A. T., Inst. f. Ind. Res. and Stand., Dublin, 1950
16) Dixon, P. S., Ann. Bot. (London) n. s. 22 (87): 353–368. 1958
17) El Shazly, K., Alexandria J. Agr. Research 3 (1): 41–50. 1955
18) El Shazly, K.; ibid. 4. 15–26. 1956. CA 52, 20, 17413h, 1958
19) Selby, H. H., Fishery Leaflet 118, 1948. US Dep. of the Interior, Fish and Wildlife Service, Washington

20) Ross, A. G., J. Sci. Food Agric. (7): 333–335. 1953
21) Yaphe, W., Can. J. Microbiol. 3. 987–993. 1957
22) Yaphe, W., Can. J. Bot. 37. 751–757. 1959
23) Araki, Ch., Seaweed Polysaccharides ex Wolfrom, M. L., Carbohydrate Chemistry of Substances of Biological Interest. Pergamon Press Ltd., London, 1959
24) Loomis, N. H., All. Hancock Found. Occ. Pap. (24): 1–10. 1960. Rev. de Biol. Mar., Valparaiso, 1960
25) Hegnauer, R., Chemotaxonomie der Pflanzen, Birkhäuser Verlag, Suttgart, 1962
26) Dixon, P. S., Bot. Not. 113 (3): 295–319. 1960
27) Stoloff, L., Ec. Bot. 16 (2): 86–94. 1962

General Literature
Chapman, V. J., 1950; Fott, B., 1959; Fritsch, F. E., 1952; Hoffmann, C., 1938; Hoppe, H. A., 1958; Hoppe, H. A. and Schmid, O. J., 1962; Kylin, H., 1956; Lewin, R. A., 1962; Segawa, S., 1957; see Index of Literature.

Pterosiphonia

Approximately 20 species are recorded from temperate and warmer oceans.

P. cloiophylla (Ag.) Falkenb. Coast of Cape Province. The type species. Creeping rhizome, thallus erect, 2 to 10 centimetres high, flattened, branched.

P. complanata. This alga has been investigated chemically, when considerable quantities of Na-mannoglycerates were found[1]).

P. thuyoides. Fluorescing substances have been isolated from this species.

P. bipinnata, *P. fibrillosa*, *P. pennata* are recorded from Japanese coasts.

Literature

1) Augier, J., Rev. gen. bot. 60: 257–283, 1953

General Literature: Fritsch, F. E., 1952; Kylin, H., 1956; Segawa, S., 1957; see Index of Literature.

Ptilota

4 species are recorded from the northern parts of the Atlantic and Pacific oceans.

P. plumosa (L.) Ag. Northern Atlantic. The type species. Thallus erect, 5 to 25 centimetres high, flattened, richly branched, often as an epiphyte on *Laminaria*. This alga has been chemically investigated[1]) [2]). Extraction gives a viscous extract[6]).

P. pectinata. Coasts of Honshû, Hokkaido, Kurile Islands, Sakhalin, Siberia on the Sea of Japan, northern Pacific and Atlantic coasts of the U. S. A. The alga has been described and investigated for several times[3]) [4]). It has been found to contain mannoside glycerate[5]).

Other East Asian species have also been studied[4]). *P. asplenoides* and *P. dentata* occur on Japanese coasts.

Results of physiological and biochemical investigations of *P. elegans* and *P. plumosa* have been reported by Lewin 1962).

Literature

1) Ross, G., J. Sci. Food Agric. (7): 333–335, 1953
2) Black, W. A. P., The Algae ex Altschul. Processed Plant Proteins Foodstuffs. Academic Press Inc., New York, 1958
3) Tazawa, N., Bull. Jap. Soc. Phyc. 6 (2): 68–74, 1958
4) Tokida, J., Fac. of Fish. Hokkaido Univ. 2 (1), 1954
5) Lindberg, B., Second Int. Seaweed Symp., 1955
6) Marshall, S. M. and Orr, A. P., Natural Plant Hydrocolloids, 1954, 101

General Literature
Fritsch, F. E., 1952; Kylin, H., 1956; Lewin, R. A., 1962; Segawa, S., 1957; see Index of Literature.

Rhodochorton

Approximately 20 species, in the temperate and colder oceans of the northern hemisphere.

R. Rothii (Turton.) Näg. Atlantic coasts of Europe and North America. The type species. Thallus tender, filamentous, irregularly branched, mostly erect, usually 5 rarely to 25 millimetres high, of a dark purple red colour. The phycoerythrin of the alga has been investigated[1]).

R. floridulum. North Sea coasts. Constituents of this alga have been investigated[2]). It contains xylose[3]) [4]) and B-phycoerythrin or a similar pigment[5]).

Results of physiological and biochemical investigations of *R. Rothii* and *R. floridulum* have been reported by Lewin (1962).

Literature

1) H'Eocha, C. O., Archives of Biochem. and Biophys. 73 (1): 207–219, 1958
2) Ross, A. G., J. Sci. Food Agric. (7): 333–335, 1963
3) Black, W. A. P. and Cornhill, W. J., Chem. and Ind. 1954, 514–516
4) Black, W. A. P., Ann. Rep. Chem. Soc. 1953. Vol. 50
5) H'Eocha, C. O. and Haxo, F. T., Biochim. et Biophys. Acta 41: 516–520, 1960

General Literature
Fritsch, F. E., 1952; Kuckuck, P., 1953; Kylin, H., 1956; Levring, T., 1940; Lewin, R. A., 1962; Segawa, S., 1957; see Index of Literature

Rhodoglossum

Approximately 10 species, on the coasts of Australia and the west coasts of North and South America.

R. lanceolatum (Harv.) J. Agardh. Coasts of Tasmania. The type species. Thallus erect, 3 to 50 centimetres high, flat to blade-shaped, much bifurcated or lobed. *R.* species contain carrageenan[1]).

R. affine. Coasts of U.S.A. (California). This alga has been investigated[2]). It contains carrageenan, with a share of kappa-carrageenan corresponding to that of *Chondrus crispus*[3]). Stoloff (1962) describes the polysacolloid also as a carrageenan[8]).

R. pulchrum. Coasts of Northern Honshû, Hokkaido, Kurile Islands, and Sakhalin. This alga belongs to raw materials used for "Ginnansô" in Japan[4]). Chemical investigations have revealed cholesterol[5]) [6]) and L-citrulline[7]).

Physiological and biochemical investigations have been reported by LEWIN (1962).

Literature

1) YOUNG, E. G. ex Ruhland, W., Handbuch der Pflanzenphysiologie, Springer Verlag, 1958
2) ROSS, A. G., J. Sci. Food Agric. (7): 333–335, 1953
3) YAPHE, W., Can. J. Bot. 37: 751–757, 1959
4) TOKIDA, J., Fac. of Fish. Hokkaido University 2 (1), 1954
5) TSUDA, K. et al., Science 126: 927–928, 1957
6) TSUDA, K. et al., Chem. et Pharm. Bull. 6 (1): 101–104, 1958
7) KURIYAMA, M. et al., Bull. Fac. Fish. Hokkaido Univ. 10 (3): 277–283, 1959
8) STOLOFF, L., Ec. Bot. 16 (2): 86–94. 1962

General Literature

FRITSCH, F. E., 1952; KYLIN, H., 1956; LEWIN, R. A., 1962; SEGAWA, S., 1957; see Index of Literature.

Rhodomela

Some 5 species are recorded from the colder oceans of the northern hemisphere.

R. subfusca (Woodw.) Ag. Atlantic coasts of Europe and North America, Baltic coasts. Coasts of Japan. The type species. Thallus erect, 10 to 20 centimetres high, terete, branched of tough consistency, bushy. The alga contains antimicrobial substances[1]), and floridorubin has been found in it[2]).

R. larix. Pacific coast of North America, Northern Honshû, Hokkaido, Kurile Islands, and Sakhalin. Coasts of Ochotsk and the Sea of Japan. This species has been found to have antibiotic activity[1]) [3]). It occurs particularly abundantly on the Japanese coast, and serves as a raw material for the production of bromine. The bromine content of the dried alga is about 3 per cent[4]).

R. confervoides. Fluorescing substances have been investigated in this species[5]).

For results of physiological and biochemical investigations of the three species see LEWIN (1962).

Literature

1) ROOS, H., Kieler Meeresforsch. XIII (1): 1957
2) AUGIER, J., Rev. gen. bot. 60: 257–283, 1953
3) MAUTNER, H. G. et al., J. Am. Pharm. Ass. Sci. Ed. 42: 294–296, 1953
4) TOKIDA, J., Fac. of Fish. Hokkaido Univ. 2 (1), 1954
5) H'EOCHA, C. O. et al., Third Int. Seaweed Symp., 1958

General Literature

FRITSCH, F. E., 1952; KUCKUCK, P., 1953; KYLIN, H., 1956; LEVRING, T., 1940; LEWIN, R. A., 1962; SEGAWA, S., 1957; see Index of Literature.

Rhodosurus

This alga has been found in a seawater sample from Las Palmas.

R. marinus. The type species. The plant forms an irregular, multicellular gelatinous stroma. The globular cells have a thin membrane. The colour is a dull, somewhat blueish carmine. The photosynthetic pigments have been investigated[1]). The pigment is a B-phycoerythrin or similar pigment[2]).

The alga has been cultivated in a synthetic medium[3]).

Literature

1) MOYSE, A., Coll. Int. du Centre Nat. de la Recherche Scient., Dinard, 1960

2) H'EOCHA, C. O. and HAXO, F. T., Biochim. et Biophys. Acta 41: 516–520, 1960

3) GIRAUD, G., C. R. Acad. Sc. (Paris) 246 (25): 3501–3504, 1958

General Literature

KYLIN, H., 1956; see Index of Literature.

Rhodymenia

Forty species are recorded, widespread in warmer oceans.

Rh. palmata (L.) Grev. Coasts of France (Brittany), Great Britain[1]), and Ireland[2]), Norway, Iceland, North Eastern coast of Canada (very abundant off Grand Manan Island, Bay of Fundy, the centre of the Canadian dulse industry[3]), Atlantic coasts of the United States[4]) [5]), Pacific coast from California to Alaska[6]), often growing epiphytically on *Nereocystis luetkeana.*

Coast of the Gulf of Guinea[7]), coast of Japan (Bay of Oshoro) in several varieties[8]). The occurence of the alga in tropical South and East Asia has not been confirmed, but it is reported to occur in Indonesia (cf. *Rh. indica)*[6]) [9]).

The type species, very variable in form, often growing on the stipes of *Laminaria hyperborea* Thallus erect, 2 to 40 centimetres high, flat, bifurcated, with split or irregular lobes, sometimes branched, usually stalked; of a dark red colour.

The alga is known in maritime countries under various popular names: "Dulse" (Scotland), "Dillisk", "Duilleasg" (Ireland), "Söl" (Iceland), "Sou-söll", "Søl", "Blôm" (Norway), "Goemon à vache" (France), "Darusu" (Japan), "Dulse" (Canada), "Red kall", "Neptun's Girdle" (North America) (Pl. XIII: 31; Pl. XVI: 43).

In 1930 a study has been published on "la constitution du principe sucre" in *Rh. palmata*[10]). The alga has since been studied in detail by several authors. It has been found to contain xylose, xylan, floridoside[11–20]). The hydrolysis of the soluble pentosans of *Rh. palmata* and wheat flour by ruminal micro-organisms have been investigated[21]). Protein and amino acid have been investigated[12]) [22–26]). The biological value of the protein has been found to be 42 per cent[12]). The content of vitamin C has been estimated by HUMM in 1951 to be 49 mg per 100 grammes of fresh weight[5]).

Modern analytical studies include carotene and carotenoids[27–30]) [16]), niacin[31]), vitamin B_{12} [32]) [33]), vitamin E and the folic acid group of vitamins[32]), lipids and enzymes [34]), antibiotics[35]), peptides[36–38]), and inorganic constituents[39]). Sodium and potassium transport[40]), and the structure of the cell wall[41]), including electron micrographic investigations of the floridian pit[42]) have also been studied.

Rh. palmata is eaten or used for animal fodder in many countries. The alga is harvested by hand, then washed and dried for human consumption. It is also eaten in a raw state. In Ireland the alga is sold on the weekly markets. In some countries it is used as a kind of chewing tobacco. In the northern countries *Rhodymenia* is eaten with butter together with potatoes or fish. In the Mediterranean *Rh.* is used as a garnish in ragout and other dishes, as red colourant and as a thickener. "Dulse" is a popular item of food on the Atlantic coast of North America, particularly in Canada[5] [43]).

Strangely enough, in Japan, the country where edible algae are of great importance, *Rhodymenia palmata* is not used at all[44]). In Kamchatka and Alaska it is fermented to produce an alcoholic beverage[6]); in Indonesia it is used as a vermifuge.

Rhodymenia is used in many countries for feeding domestic animals. The alga is particularly appreciated by sheep and goats, but also by cattle[4]). In Iceland the alga is also used for winter indoor feeding, particularly for cattle and sheep[5]) [45]). Horses also feed willingly on this alga[4]).

A number of other species of *Rhodymenia* are known to occur; they would probably repay investigations.

Rh. coacta. Coasts of Japan.

Rh. corallicola. West coast of Africa (Mauritania). This alga is approximately 5 centimetres long and approximately 4 millimetres wide[46]).

Rh. cuneifolia. Coasts of Chile (P. Arenas, Puerto Buenol, Juan Fernandez), southern coast of Argentina, Falkland Islands[47]).

Rh. Howeana. Coasts of Peru and Chile[47]).

Rh. indica. Coastal areas of South and East Asia, often growing together with *Rh. palmata.* This species is used as a vermifuge[6]) [9]).

Rh. indica var. *marginifera* is recorded from Java[9]).

Rh. intricata. Coasts of Japan (Bay of Oshoro)[8]).

Rh. ligulata. Mediterranean coasts.

Rh. palmatiformis. Chile, Fuegia, Patagonia, Falkland Islands, S. Georgia, Kerguelen, Campbell Islands[47]).

Rh. pertusa. Pacific coasts of the northern subarctic region, Bering Sea, Kamchatka, Sea of Okhotsk, Kurile Islands, Sakhalin, Korea, Japan (Honshû), Hokkaido, Bay of Oshoro – Japanese name "Anadarusu"[8]) [44]); Pacific coast of North America from Alaska to Washington, California, Puget Sound; indigenous name "Red eyelet silk".

Other *Rh.* species are recorded from the Californian coasts. Some of them grow in deep water on the South-American coasts[6]).

Several other *Rh.* species are also recorded from Indonesia.

Work has been carried out during the last years on the cultivation of *Rhodymenia*[48]).

For physiological and biochemical investigations of *Rh. palmata* and *Rh. pertusa* see Lewin (1962).

Literature

1) Marshall, S. M. and Orr, A. P., Natural Plant Hydrocolloids 1954, 101
2) Brennan, A. T., Inst. f. Ind. Res. and Stand., Dublin, 1950
3) Young, E. G. et al., J. Fish. Res. Bd. Canada, 16 (1). 1959
4) Hoffmann, C., Die praktische Bedeutung der Meeresalgen, Kieler Meeresforsch. III (1). 1938
5) Humm, H. J., The Red Algae of Economic Importance ex Tressler, D. K. and Lemon, J. Mc. W., Marine Products of Commerce. Reinhold Publishing Corporation, New York, 1951
6) Schachat, R. E. and Glicksman, M., Some Lesser Known Seaweed Extracts ex Whistler, R. L. and BeMiller, J. N., Industrial Gums. Academic Press Inc., New York, 1959

7) Lawson, G. W., J. Ecol. 44 (1): 1954
8) Tokida, J. and Masaki, T., Fac. Fish. Hokkaido Univ. 10 (3): 173–195, 1959
9) Zaneveld, J. S., Ec. Bot. 13 (2): 89, 1959
10) Colin, H. and Geugen, E., C. R. Acad. Sc. 191. 163–164. 1930
11) Ross, A. G., J. Sci. Food Agric. (7): 333–335. 1953
12) Black, W. A. P., The Algae ex Altschul. Processed Plant Proteins Foodstuffs. Academic Press Inc., New York, 1958
13) Black, W. A. P., Ann. Rep. Chem. Soc. 50. 1953
14) Black, W. A. P. and Cornhill, W. J., Chem. and Ind. 514–516. 1954
15) Smith, F. and Montgomery, R., The Chemistry of Plant Gums and Mucilages. Reinhold Publishing Corporation, New York, 1959
16) Institute of Seaweed Research, Ann. Rep. 1945–1955
17) Chanda, S. K. and Percival, E. G., Nature 166 (4227): 787–788. 1950
18) Young, E. G., Carbohydrate accumulation by lower plants (Rhodophyta) ex W. Ruhland, Handbuch der Pflanzenphysiologie, Springer Verlag, 1958
19) Lindberg, B., Second Int. Seaweed Symp., 1955
20) Stoll and Jucker, Heteroside ex W. Ruhland, Handbuch der Pflanzenphysiologie, Springer Verlag, 1958
21) Howard, B. H., Biochem. J. 67. 643–651. 1957. CA 52, 9, 74151, 1958
22) Coulson, C. B., Chem. and Ind. 1953, 971–972
23) Coulson, C. B., Chem. and Ind. 1953, 997–998
24) Coulson, C. B., J. Sci. Food Agric. 6. 674–682. 1955
25) Smith, D. B. and Young, E. G., Publ. 3731 of the Nat. Res. Council, Halifax, Canada
26) Institute of Seaweed Research, Ann. Rep. 1957
27) Owen, E. C., J. Sci. Food Agric. 5 (9): 449–453. 1954
28) Haug, A. and Larsen, B., Second Int. Seaweed Symp., 1955
29) Haug, A. and Larsen, B., Acta Chem. Scand. 10. 472–473. 1956
30) Haug, A. and Larsen, B., Norsk inst. f. tang- og tareforskning, Rep. 16, 1957
31) Larsen, B.; ibid. Rep. 19, 1958
32) Ericson, L. E., Almquist & Wiksella, Uppsala, 1953, 60–63
33) Scott, R. and Ericson, L. E., J. exp. Bot. 6 (18): 348. 1955
34) Duncan, W. M. et al., Biochem. J. 63. 44–51. 1956
35) Chesters, C. G. C. and Stott, J. A., Second Int. Seaweed Symp., 1955
36) Institute of Seaweed Research, Ann. Rep. 1956
37) ibid., 1960
38) ibid., 1961
39) Young, E. G. and Langille, W. M., Publ. Nr. 4636 of the Nat. Res. Council, Halifax, Canada, 1958
40) Dainty, J. and Mac Robbie, E. A. C., Proc. Roy. Phys. Soc. Edinburgh 24 (2): 38–43. 1955
41) Myers, A. and Preston, R. D., Proc. Roy. Soc. London 150B (941): 447–455. 1959
42) Myers, A. et al., Ann. Bot. n. s. 23 (90): 257–260. 1959
43) MacFarlane, C. I., Nova Scotia Res. Found., Halifax, 1956
44) Tokida, J., Fac. Fish. Hokkaido Univ. 2 (1), 1954
45) Chapman, V. J., Seaweeds and their Uses. Methuen & Co., Ltd., London, 1950
46) Seoane Camba
47) Levring, T., Contributions to the marine algal Flora of Chile. Lunds Univ. Årsskrift N. F. Avd. 2, 56 (10), 1960
48) Sparling, Am. J. Bot. 48 (6) Part I: 493–499. 1961

General Literature

Chapman, V. J., 1950; Fott, B., 1959; Fritsch, F. E., 1952; Hegnauer, R., 1962; Hoffmann, C., 1938; Kylin, H., 1956; Lewin, R. A., 1962; Pax, F., 1962; Schmid, O. J. and Hoppe, H. A., 1962; Segawa, S., 1957; Uphof, J. C. Th., 1959; see Index of Literature.

Rissoella

A single species is found out for the Mediterranean region.

R. verruculosa (Bertol.) J. Ag. Thallus flat, leaf-like, irregularly bifurcated or lobed. The alga is easy to harvest from spring to early summer on the Mediterranean coast of France and can be utilized as a raw material for the production of carrageenan[1]).

Literature

1) DANGEARD, P., Coll. Int. du Centre Nat. de la Recherche Scient., Dinard, 1960

General Literature
FRITSCH, F. E., 1952; KYLIN, H., 1956; see Index of Literature.

Rytiphlaea

A single species is recorded from the Mediterranean and adjoining areas of the Atlantic Ocean.

R. tinctoria (Clem.) Ag. Thallus erect, 7–18 centimetres high, narrow, bandshaped, flattened, branched, cartilaginous. The alga contains the water-soluble red pigment floridorubin, which turns into an orange-yellow on addition of acid[1]). Physiological and biochemical investigations have been reported by LEWIN (1962).

Literature

1) FOTT, B., Algenkunde, Gustav Fischer Verlag, Jena, 1959

General Literature
FRITSCH, F. E., 1952; KYLIN, H., 1956; LEWIN, R. A., 1962; see Index of Literature.

Sarcodia

Some 5 species are recorded from the Indian Ocean, the coasts of Australia, New Zealand, and Japan.

S. montagneana (Hook and Harv.) J. Agardh. Coasts of India and Indonesia, Pacific Ocean, particularly near New Zealand. The type species. Thallus erect, 1–20 centimetres high, leafy, flat, dichotomously branched or irregularly lobed. The alga is eaten in Indonesia, especially on the Moluccas[1]) [2]) [3]).

Literature

1) EISSES, J., Indonesian J. of Nat. Sc., Nr. 1, 2, 3, 1953
2) ZANEVELD, J. S., Chron. Nat. 105 (1): 1949
3) ZANEVELD, J. S., Ec. Bot. 13 (2): 89, 1959

General Literature
CHAPMAN, V. J., 1950; KYLIN, H., 1956; see Index of Literature.

Sarconema

Some 5 species are recorded from the Indian Ocean and Australia.

S. furcellatum Zanard. Coasts of the Red Sea and Pakistan. The type species. Thallus erect, 7–15 centimetres high, terete, dichotomously branched.

S. filiforme. Coasts of Pakistan[1]) and India (east coast of State of Madras). This alga contains an agaroid and may be utilized industrially as a raw material[2]). The content amounts to approximately 20%[3]).

S. furcatum and *S. scinaioides* are two additional species recorded from the coasts of Pakistan[1]).

Literature

1) ANAND, P. L., The University of the Panjab, Lahore, 1943
2) SELBY, H. H. and SELBY, T. A., Agar ex Industrial Gums. Academic Press Inc., New York, 1959
3) HOPPE, H. A., Unveröffentlichte Arbeiten und persönliche Mitteilungen

General Literature

KYLIN, H., 1956; see Index of Literature.

Scinaia

Some 10 species are recorded from the warmer seas.

S. furcellata (Turn.) Bivona. Mediterranean and coasts of the North Atlantic. The type species. Thallus terete or a little flattened, dichotomously branched.

S. turgida. The polysaccharide of the alga has been investigated[1]).

A number of *S.* species, for example, *S. complanata, S. Hatei, S. indica* are recorded from Pakistan near Manora and Karachi[2]).

S. cottonii, S. japonica, and *S. moniliformis* occur on the Japanese coasts.

Literature

1) JANECKE, H. and KEHR, W., Pharmaz. Ztg. 105 (25): 684, 1960
2) ANAND, P. L., The University of the Panjab, Lahore, 1943

General Literature

FRITSCH, F. E., 1952; KYLIN, H., 1956; SEGAWA, S., 1957; see Index of Literature.

Smithora

S. naiadum (= *Porphyra naiadum*) cf. p. 270. This alga has been placed in a new genus *Smithora* because of its unusual method of reproduction[1]). The constituents have been investigated[2]). Results of physiological and biochemical investigations have been reported by LEWIN (1962).

Literature

1) HOLLENBERG, G. J., Pac. Nat. 1 (1), 1959
2) H'EOCHA, C. O. and HAXO, F. T., Biochim. et Biophys. Acta 41: 516–520, 1960

General Literature

LEWIN, R. A., 1962; see Index of Literature.

Soliera

Some 5 species are recorded for the west coasts of Europe, the Indian Ocean, and Australia.

S. chordalis (Ag.) J. Agardh. West coast of Europe, coasts of Pakistan[1]). The type species.

Thallus erect, 10–15 centimetres high, terete, irregularly branched. On European coasts the alga is used together with other algae as a fertilizer.

S. mollis and *S. robusta* grow on Japanese coasts.

Other *S.* species are recorded from the eastern Mediterranean, the Red Sea, and East-African coast[2]).

Literature

1) Anand, P. L., The University of the Panjab, Lahore, 1943

2) Aleem, A. A., Meddel. från Göteborgs Botaniska Trädgard XVIII, 1950

General Literature

Chapman, V. J., 1950; Fritsch, F. E., 1952; Hoffmann, C., 1938; Kylin, H., 1956; Segawa, S., 1957; see Index of Literature.

Sphaerococcus

Several species, European coasts of the Atlantic and Mediterranean.

S. coronopifolius (Good. and Woodw.) Ag. The type species. Thallus erect, 8–15 centimetres high, flattened, branched.

S. cartilagineus. Coasts of China, Pacific Ocean. The alga is used as a medicament in China.

S. gelatinosus and *S. serra*. Coasts of East Asia. The alga is used in Malaya to make an agar.

Literature

Fritsch, F. E., 1952; Kylin, H., 1956; Uphof, J. C. Th., 1959; see Index of Literature.

Spyridia

Approximately 15 species are recorded from warmer seas.

S. filamentosa (Wulfen) Harv. Warmer parts of the Atlantic Ocean. Mediterranean coasts. Coasts of Japan. The type species. Thallus erect, 5–15 centimetres high, richly branched, terete. The Italian alga can be regarded as an agaroid raw material[1]).

Indian *S.* species likewise yield an agaroid[2]).

S. elongata. Recorded from Japan.

Literature

1) Selby, H. H. and Selby, T. A., Agar ex Industrial Gums. Academic Press Inc., New York, 1959

2) Hoppe, H. A., Unveröffentlichte Arbeiten und persönliche Mitteilungen

General Literature

Fritsch, F. E., 1952; Kylin, H., 1956; Segawa, S., 1957; see Index of Literature.

Suhria

2 species. The type species occurs on the South African coast, the other on the coasts of the Japanese islands.

S. vittata (L.) J. Ag. Thallus erect, 15 to 20 centimetres high, flat, lancetshaped, with a marked mid-rib, branches arising from the margins of the thallus. Both the margins and the mid-rib often carry small blades.

The alga is known on the South African coast as "red ribbons" and is eaten. Jelly products are made from it[1]). *S. vittata* has been proposed as a raw material for the agar industry[1–4]), but according to other statements it contains an agaroid[5–6]), or a gelan[7]), or a polysacolloid that is very similar to carrageenan[8–9]). STOLOFF (1962) describes it as an agar[12]). The alga when investigated chemically has been found to contain agarose[11]).

Literature

1) CHAPMAN, V. J., Seaweeds and their Uses. Methuen & Co., Ltd., London, 1950
2) HUMM, H. J. and WOLF, F. A., Duke University Press, Nr. 3/1946
3) FOTT, B., Algenkunde, Gustav Fischer Verlag, Jena, 1959
4) YAPHE, W., Can. J. Microbiol. 3: 987–993, 1957
5) HUMM, H. J., The Red Algae of Economic Importance ex Tressler, D. K. and Lemon, J. Mc. W., Reinhold Publishing Corporation, New York, 1951
6) SELBY, H. H. and SELBY, T. A., Agar ex Industrial Gums. Academic Press Inc., New York, 1959
7) STOLOFF, L. and SILVA, P., Ec. Bot. 11 (4): 327–330, 1957
8) SMITH, F. and MONTGOMERY, R., The Chemistry of Plant Gums and Mucilages. Reinhold Publishing Corporation, New York, 1959
9) STOLOFF, L. and LEE, CH. F., Fishery Leaflet 173, 1946 US Dep. of the Interior, Fish and Wildlife Service
10) ROSS, A. G., J. Sci. Food Agric. (7): 333–335, 1953
11) YAPHE, W., Can. J. Bot. 37: 751–757, 1959
12) STOLOFF, L., Ec. Bot. 16 (2): 86–94. 1962

General Literature
HOPPE, H. A., 1958; KYLIN, H., 1956; see Index of Literature.

Tuomeya

1 species, Atlantic Coast of North America and in running fresh water.

T. fluviatilis (Harv.). Thallus erect, up to 5 centimetres high, richly branched. The polysaccharide of the alga has been studied[1]). Physiological and biochemical studies have been reported in LEWIN (1962).

Literature

1) JANECKE, H. and KEHR, W., Pharmaz. Ztg. 105 (25): 684, 1960

General Literature
FRITSCH, F. E., 1952; KYLIN, H., 1956; LEWIN, R. A., 1962; see Index of Literature.

Turnerella

Some 5 species, northern part of the Pacific and Atlantic Oceans, the northern Arctic Ocean.

T. mertensiana (Post. and Rupr.) Schmitz. Northern part of Honshû, Hokkaido, Kurile Islands, Sakhalin, Coast of Camchatka. The type species. Thallus erect, 10 to 60 centimetres high, leafy, undivided or irregularly lobed.

This plant has been utilized in Japan (Nemuro) for the production of a phycocolloid since 1925. The alga grows in deeper waters, and the fishermen only harvest drifting specimens[1]).

Literature

1) TOKIDA, J., Fac. of Fish. Hokkaido University 2 (1): 1954

General Literature
FRITSCH, F. E., 1952; KYLIN, H., 1956; SEGAWA, S., 1957; see Index of Literature.

Wrangelia

Approximately 25 species, mainly from Australia, but several species from the Atlantic Ocean, from Malaya, and the Japanese Islands.

W. penicillata Ag. Mediterranean and warmer areas of the Atlantic Ocean. The type species. Thallus erect, 5 to 15 centimetres high, terete, branched.

W. penicillata, W. argus, and *W. bicuspidata* from Puerto Rico proved to have antibiotic activity[1]).

W. argus and *W. tagoi* are also recorded for Japan.

Literature

1) BURKHOLDER, P. R. et al., Bot. Mar. II (1/2): 149, 1960

General Literature
FRITSCH, F. E., 1952; KYLIN, H., 1956; SEGAWA, S., 1957; see Index of Literature.

Yatabella

A single species, recorded from Japan.

Y. hirsuta Okam. Thallus terete, initially erect, later lying flat, 15–20 centimetres long, irregularly branched. The alga contains a carrageenan[1]) of higher kappa-carrageenan content than *Chondrus crispus*[2]). STOLOFF (1962) describes the polysacolloid as a carrageenan[3]).

Literature

1) YOUNG, E. G. ex W. Ruhland, Handbuch der Pflanzenphysiologie, Springer Verlag, 1958
2) YAPHE, W., Can. J. Bot. 37: 751–757, 1959
3) STOLOFF, L., Ec. Bot. 16 (2): 86–94. 1962

General Literature
KYLIN, H., 1956; SEGAWA, S., 1957; see Index of Literature.

Commercial Products

By Heinz A. Hoppe and Otto J. Schmid

We have dealt in the chapter "Marine Algae as Raw Material" with the historical development from very old methods to modern industries. The following chapters describe the products obtained from marine algae and used as human and animal food, Kelp-products etc.

To-day the most important industrial products from marine algae are phycocolloids from Phaeophyceae and Rhodophyceae.

They are, above all, necessary auxiliaries in numerous industries as emulsifying agents, gelling agents, stabilizers, suspension-agents, and thickeners.

The term phycocolloid has come in use. 1945 it has been defined by TSENG as the polysaccharide-complexes from Rhodophyceae and Phaeophyceae, which form colloidal systems when dispersed in water. During the last years this collective name became more and more defined.

More terms are used, however, for the polysaccharide-complexes from Rhodophyceae, e.g. TSENG refers to the extracts from Rhodophyceae, namely agar, carrageenan etc. which form more or less firm jellies in aqueous solutions, as gelose.

STOLOFF and SILVA (1957) refer to the water extractable polysaccharides from Rhodophyceae as polysaccharide hydrocolloids and divide them into 3 groups, agars, carrageenans, and gelans.

Gelans are polysaccharides similar to carrageenans regarding viscosity and gelating properties, but without the ability to agglomerate casein and precipitate methylene blue. Furcellaran and hypnean are grouped with the gelans.

STOLOFF 1962 also used the collective name polysacolloids for the three groups agars, carrageenans, and gelans.

Agar, formerly for the most part called agar-agar, is a gel-forming extract from certain Rhodophyceae, the agarophytes, mainly *Acanthopeltis japonica*, Japan, *Ahnfeltia plicata*, USSR, *Gelidium* species, Japan, U.S.A. (California), South Africa, Spain, Portugal, Morocco, *Gracilaria* species, Japan, U.S.A. (North Carolina), and *Pterocladia* species, New Zealand.

Agaroids are the extracts from Rhodophyceae, the agaroidophytes, forming, in contrast to agar, softer jellies; phyllophoran and eucheuman belong to this group.

Carrageenan and its derivates are obtained from carrageenophytes, also Rhodophyceae. The most important raw material is carrageen, composed of *Chondrus crispus* and *Gigartina stellata*.

Further phycocolloids from Rhodophyceae are:

Gracilaria Gum from *Gracilaria* species

Agaran from *Gracilaria blodgettii*, *Gracilaria cornea*, *Gracilaria floridiana*

Funoran from *Gloiopeltis* species

Iridophycan from *Iridea* species
and some others of lesser importance.

Among the phycocolloids of Phaeophyceae, alginic acid and some of its numerous derivates, especially sodium alginate, are of great economic importance. These products are primarily industrial auxiliaries. Numerous Phaeophyceae, the alginophytes, are the raw material for these.

Further products from Phaeophyceae are fucoidan, fucosan, fucosterol, laminaran, and mannitol.

The single products are described. After the most important raw material and countries of production have been mentioned, there follows a short description of the manufacture, a characterization of the structure, features, and notes of applications.

Agar

Raw Materials and Occurence

Only the most important of the numerous Rhodophyceae, used in the various production areas as a source of agar, are to be mentioned here.

They are the following: *Acanthopeltis japonica*, East-Asian coastal areas. The alga supplies "Toriashi-agar". It is usually exploited in mixture with other Rhodophyceae.

Ahnfeltia plicata occuring mainly on the Russian coast of the White Sea, in Sakhalin, Korea, and the Japanese islands, but also in other areas. *A. plicata* is the most important raw material for Soviet Russian agar production.

In Japan *A. plicata* is used for the production of "Itani-agar". It often is exploited in mixture with other Rhodophyceae.

Ceramium species are utilized on Japanese coasts.

Gelidium. A number of *G.* species play a significant rôle as agar raw materials, above all *G. corneum* var. *sesquipedale*, occurring particularly on the Atlantic coasts of Spain, Portugal, and Morocco.

G. amansii mainly existing on the coasts of Japan, Korea, and China. This alga supplies "Tengusa-agar".

Numerous further Japanese *G.* species are utilized, for agar production mostly in mixture with other raw materials, e.g.

G. crinale, *G. divaricatum*, *G. elegans*, *G. japonicum*, *G. liatulum*, *G. pacificum*, *G. subcostatum*, *G. subfastigiatum*, *G. tenue*, *G. vagum*.

G. cartilagineum, coasts of South Africa, U.S.A. (California), Mexico. An important raw material of the American agar industry.

G. coulteri and *G. nudifrons* are likewise utilized in the U.S.A.

G. lingulatum, coast of Chile, the raw material for Chilenean agar. It can be taken for granted that other *G.* species are also utilized in Chile.

G. pristoides, South Africa, is used locally and also exported.

G. spinulosum, in addition to *G. corneum*, plays a rôle as a raw material in Morocco.

Gracilaria. The two most important of the hundred odd species are *G. confervoides*, above all on the coasts of South Africa, the Indian Ocean, East Asia, and the American continent, and *G. lichenoides* on the coasts of the Indian Ocean. The algae supply an agar or an agaroid, depending on provenance and quality.

The phycocolloid from *Gracilaria* is also described as "Gracilaria Gum".

G. confervoides is harvested and utilized in the U.S.A., on the Atlantic and Californian coasts, and likewise in Chile, Argentina, South Africa, and East Asia. The alga, like *G. lichenoides*, is not only made into agar, but also into other algal products, such as "Ceylon Moss" etc.

Pterocladia lucida, Australia, Tasmania, New Zealand, Japan, is used as agar raw material in local industries. The smaller *P. capillacea*, which also occurs on the Egyptian coasts is similarly utilized.

Definition

Agar, in former times mostly called agar-agar, is a gel forming extract of agarophytes. The article before deals with the most important of them. The name agar-agar is of Malayan origin and originally has been the term of the *Eucheuma* species.

The name agar should be reserved for the phycocolloids of the Rhodophyceae which according to the USP are insoluble in cold water, but soluble in hot water. So an agar-solution of 1.5 per cent is clear and after getting cold at a temperature of 32–39° C it forms a solid and elastic gel which does not dissolve below a temperature of 85° C. Similar phycocolloids which do not completely comply with this definition are called agaroids. Rhodophyceae as raw material for agaroids are called agaroidophytes.

Therefore eucheuman (cf. p. 329), a phycocolloid nowadays won out of the *Eucheuma* species, is an agaroid.

Countries of Production

Agar is the oldest and perhaps the best known of all algal products. It is therefore only natural, that numerous maritime countries with raw material resources are interested in the production of agar, particularly since agar in many areas can be produced as a cottage industry. However, with growing interest in agar, the production methods have been changed, for the market today demands a product consistently uniform in quality and exact specification.

Noteworthy surveys of agar production in the world have been published by Stoloff and Lee, 1946[1]), Humm, 1947 and 1951[2]) [3]), Chapman, 1950[4]), Selby, 1954[5]), Selby and Selby, 1959[6]). Hoppe and Schmid[7]) reported on the subject in Botanica Marina, 1962.

Europe

France. Until very recently France did not have any own agar industry. Reports in 1962 showed that a plant was to be established in Hendaye[8]).

Italy. There is a small agar industry in Italy. Reports have been made on the production of the Italian agar from *Gelidium* species[9]).

Portugal. The most important agar raw material of the Portuguese coast is *Gelidium corneum*. For a number of years there has been an agar industry which certainly can be further expanded since it has a very good raw material basis. Already in 1951 a report has been published of "L'industrie portugaise de l'agar-agar et les applications de ce produit dans les industries agricoles"[10]).

Spain. Spain is the greatest European agar producer, operating a number of factories, whose capacity has repeatedly been increased in recent years. The Spanish agar quality is good. The most important raw material is *Gelidium corneum* var. *sesquipedale*.

Cabrero Gomez has reported in 1951 on "Estudio de las algas marinas españoles desde el punto de vista de su aprovechamiento industrial"[11]), and Aravio-Torre in 1952 at the First International Seaweed Symposium has reported on the development of the Spanish agar industry[12]), which has since then become a significant economic factor. The same author and his coworkers have reported in 1959 on the treatment and quality of *Gelidium*-agar[13]).

Africa

Egypt. Apparently it is intended to establish an agar industry. *Pterocladia capillacea* can be used as a raw material.

Morocco. A number of valuable agar raw materials occurs in considerable quantity on the Moroccan coast. Utilization should above all refer to *Gelidium corneum* var *sesquipedale* and *Gelidium spinulosum*.

Moroccan agar is characterized by an excellent quality. Work has been published on Moroccan raw materials, inter alia, by Dangeard in 1957[14]) and Md. Gayral ("Algues de la côte atlantique marocaine").

South Africa. A number of red algae occur on the South African coast and constitute a valuable agar raw material, e.g. *Gelidium cartilagineum*, *G. pristoides*, *Gracilaria confervoides*, and *Suhria vittata*.

The raw materials are often exported. But there is also a domestic agar industry.

By the work of Isaac[15]) and other authors we are well informed of South African raw materials. (Cf. Hoppe, 1964[16])).

Asia

Ceylon. Raw materials for an agar production on Ceylon are *Gracilaria confervoides* and *G. lichenoides*. The algae occur particularly near Kalpitiya, Mannar, and Trincomalee. They supplied "Ceylon Moss" in former times.

Durairatnam and Medcof has reported on "Ceylon's Red Seaweed Resources" in 1954[17]).

China. No statistics are available on agar production in the People's Republic of China. National China, however, has a considerable agar production[18]).

India. A number of Rhodophyceae occurring on the Indian coast are potential agar producers, above all *Gracilaria* species. The algae are used in the country itself, and investigations have been made regarding the feasibility of establishing a modern agar industry.

Indonesia. Agar raw materials occur in South-East Asian island regions, various species. They are processed into agar for the domestic market or are exported, mainly to Japan. A large number of publications has informed us of Indonesian algae, e.g. "The Agar-content of Indonesian Seaweeds" (GROOT, 1947[19]), "De Bereiding von Agar-Agar" (MOREL, 1949[20]), "The Mucilages of some Indonesian Seaweeds" (EISSES, 1952[21]), "Seaweeds in the Indonesian Trade" (EISSES, 1953[22]), "The Utilization of Marine Algae in Tropical South and East Asia" (ZANEVELD, 1959[23]).

The most important raw materials are *Gelidium amansii*, *G. latifolium*, *Gracilaria confervoides*, *G. lichenoides* and others. *Eucheuma* and *Hypnea* spp. likewise occur there.

Japan. Agar production for many centuries was a world monopoly of the Japanese, and until 1939 Japan was the main producer. Production was done according to old traditional methods, which are still applied today for "home-made" agar. But a modern industry has also developed. Thus, for example, the impairment of products, which formerly occurred in mild frost-free weather, can be eliminated by today's modern technology. Numerous raw materials can be utilized. The most important ones are the following:

Gelidium amansii and *G. elegans* – "Tengusa", *Gelidium japonicum* – "Onikusa", *Acanthopeltis japonica* – "Dorakusa", "Toriashi", *Gelidium subcostatum* – "Hirakusa", *Gracilaria confervoides* – "Ogo", *Ceramium hypnaeoides* – "Ego", *Ceramium boydenii* – "Igisu", *Ahnfeltia plicata* – "Itani".

The best grades "Tengusa" and "Onikusa" contain 30 to 70 per cent of agar.

Numerous scientific publications appeared in the past while knowledge of raw materials, investigation of production methods, chemistry etc. still will be expanded.

The most important raw material areas are the prefectures of Shizuoka, Tokyo and Chiba. The main areas of agar production are Nagano (more than 30 per cent), Gifu and Osaka. In 1950 there were about 460 agar producers, mostly as a subsidiary manufacturing. They have formed associations[24]. The production of "Kanten" and "Tokoroten" for home consumption has been reported by TOKIDA, 1954[25].

Agar, as a pre-war monopoly of the Japanese, and the steadily increasing importance of this product have been the subject of a comprehensive work published by HUMM, 1947[26].

Korea. *Gelidium amansii* and other Rhodophyceae on the East-Asian coast are the raw materials of the Korean agar industry. This commodity is often offered in strips or as powder in various grades to European trade.

Harvesting, treatment of the raw material, and the production of the agar have been described in detail by SCHULZE (1958)[27].

Russia. The raw material of Russian agar produced on the coasts of the White Sea and Sakhalin, is *Ahnfeltia plicata*.

A number of Russian authors published works on agar production from *Ahnfeltia*[28–31].

Australia. The Australian agar industry has been established many years ago. The raw materials are chiefly *Pterocladia capillacea* and *P. lucida*.

New Zealand. New Zealand has its own agar raw material basis, which – as on the Australian coast – mainly consists of *Pterocladia lucida* and *Pt. capillacea* and supplies a very good agar.

MOORE (1944) has reported on "New Zealand Seaweed for Agar Manufacture"[32].

North and Central America

U. S. A. The most important raw material of Californian agar production is *Gelidium cartilagineum*. Other algae are also utilized on the Californian and on the Atlantic coasts.

Gelidium cartilagineum, on the Pacific coast, depending on the quality of the raw material, can contain 15 to 70 per cent of agar. The content is usually about 30 per cent higher in the summer months than in the winter months.

Production began in California in 1920. Further developments have been reported by SELBY, 1948[33]).

An agar-industry has also been established on the Atlantic coast of Florida and North Carolina. The most important raw material is *Gracilaria*[34]).

HUMM, 1944, has published a list of agar resources on the Atlantic ocean and the East Gulf Coasts[35]).

The same author has also dealt with the agar industry, in particular in California, as well as in North Carolina and Florida in his publication of 1951, "Marine Products of Commerce" by D. K. TRESSLER and MC. S. LEMON[36]).

Harvesting and treatment of raw materials, economic development, production methods, grade, and type specification, and fields of application have been described in detail by SELBY and SELBY in 1959 in WHISTLER-MILLER[6]).

Mexico. After the Second World War an agar industry was established in Mexico which, however, produced only small quantities and discontinued its operation[5]).

DAWSON et al. have reported in 1960[37]) on seaweeds beds along Southern California and North-Western Mexico.

The raw materials occur on the Pacific coast and can be used by the Californian agar industry[38]).

The algal resources in the Gulf of Mexico have been described by TAYLOR in 1953[39]).

Argentina. *Gracilaria confervoides* occurring on the Patagonian coast is a highly useful agar raw material being exported in large quantities. For some time Argentina has produced a small amount of this material.

Brazil. In view of the raw material resources it should be possible to establish an agar industry there. HUMM and WILLIAMS in 1948 have published a "Study of Agar from Brazilian Seaweeds"[40]).

NEUBERN DE TOLEDO in 1948–49 has published work on "Agar-agar do Brazil"[41]) and in 1953 on "Estudo experimental do agar-agar Brasileiro"[42]).

Chile. Agar is produced in Chile in a number of factories, but is also exported to Europe. The most important raw material is *Gelidium lingulatum.*

The main collection areas of the red algae for agar production are to be found on the coast of Coquimbo to Constitucion.

ETCHEVERRY[43]) [44]) and LEVRING[45]) have published some detailed works on the algae occurring off the Chilenian coast and the islands.

Uruguay. Considerable *Gelidium* resources are recorded for the coast of Uruguay (La Paloma), approximately 250 kilometres from Montevideo. The raw material is partly used locally for feeding purposes and partly exported. The intention is to create their own agar industry in Uruguay.

According to TSENG[46]), agar is the dried amorphous, gelatinous extract from *Gelidium* and other agarophytes, containing no nitrogen, which is insoluble in cold water, but soluble in hot water, a one-per-cent solution solidifying to form a solid gel at 30–50° C, and melting again at 80 to 100° C. In chemical respects TSENG regards agar as a sulphuric acid ester of a linear galactan.

Any references made relating to chemical composition of red algae can only be regarded as reliable when the type of alga from which the product investigated has originated is specifically indicated. Contradictions in the literature are primarily due to the fact that products of non-uniform composition were analysed. The products from *Eucheuma* and *Hypnea* must be distinguished from agar proper, as they clearly have properties entirely different from those of agar made from *Gelidium* or *Gracilaria.*

According to JONES & PEAT[47]) agar consists of a chain of 9-β-D-galactopyranose units linked in 1.4' bonds with a sulphated L-galactose (see Formula 1).

C. ARAKI[48]) has studied the structure of agar in detail. He confirmed of the fractionation of agar in two components after acetylation as reported by TAKAHASHI and SHIRAHAMA[49]). A portion of 70 per cent of the agar so treated could be dissolved in chloroform, while the residual 30 per cent remained undissolved. He described the soluble fraction as agarose acetate, the non-soluble fraction as agaropectin acetate. Likewise after methylation of agar two similar fractions arose. The sulphate fraction and ash of agaropectin is substantially higher than that of agarose. Also in physical respects the agaropectin differs in its lower gelating capability. As a sub-category applicable to both fractions ARAKI found agarobiose in which β-D-galactose and 3.6-anhydro-L-galactose are combined in a ratio of 1 : 4. This finding is based on investigations of an agar from *Gelidium amansii.* The link with the next agarobiose unit is via the 1.3' position. Also this intermediate link has been isolated by ARAKI and was given the designation neoagarobiose. The tetrasaccharide consisting of these two disaccharides was isolated from enzyme hydrolysates. The anhydrogalactoside bond is highly susceptible to acids.

An investigation into the presence of branching in the agarose molecule revealed none (C. ARAKI[50])). According to ARAKI the agarose acetate from *Gelidium amansii* contains 0.8 per cent of SO_4, while the acetate of agaropectin contains of 3.2 per cent (cf. ARAKI, op. cit.).

A new method of a production of agarose has been reported by HJERTEN[51]).

The sulphur content of agar was the object of investigations, although different results have been stated. PERCIVAL[52]) [53]), BARRY & DILLON[54]) have reported a sulphate content of 1.3 to 1.4 per cent. For agarose made from commercial agar ARAKI (op. cit.) he found 0.3 per cent, and for agaropectin isolated from the same material, he found 5.5 per cent of sulphate. H. H. SELBY[55]) denies the presence of significant quantities of sulphur in agar indicating values of about 0.1 per cent of sulphur. This author gives the following values for commercial agars of different origin: Spain 0.78 per cent sulphur, Morocco 1.0 per cent sulphur, Japan 0.57 per cent sulphur, Korea 1.14 per cent sulphur[56]).

A recent contribution to the knowledge of sulphur content of this material has been made by SPRINGER and MIDDENDORF[57]) 1961, who found 1.53 per cent of sulphur in an agar product from *Gelidium corneum* var. *sesquipedale.* Moreover, they have reported that

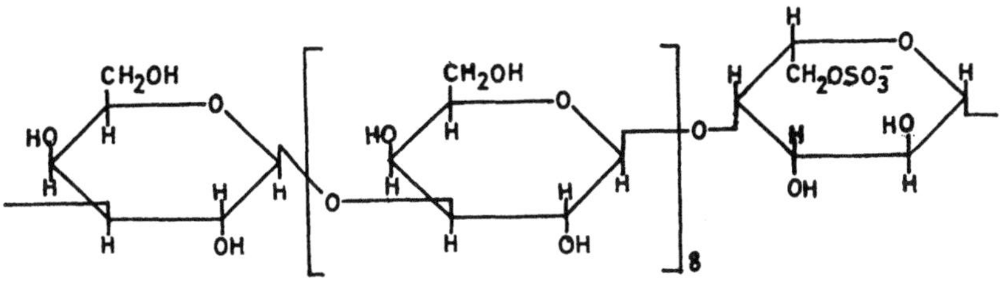

β-D-Galactopyranose

JONES & PEAT

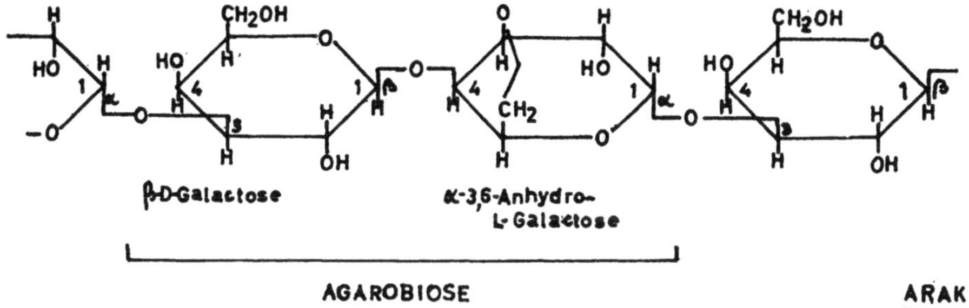

ARAKI

with these *Gelidium* agar materials there is one sulphuric acid radical per 7 monosaccharide units. The presence of sulphuric acid in agar now seems to be assured by the authors of this essay on the evidence of their own investigations.

The extent of esterification with sulphuric acid in the case of agar can be taken to be smaller than in the case of carrageenan. The differing reports made by individual workers are in all likelihood due to the fact that different algae have been analysed or that the habitats were different. The rigidity of the agar increases with the declining sulphate portion[58]) and a decline in the molar ratio galactose to 3.6-anhydrogalactose [81]).

According to Tseng[46]) agar is free from nitrogen. However, other analyses have been reported indicating up to 7 per cent of nitrogen in agar[59]) [60]) [61]). On the basis of the present knowledge it is likely that these are due to contamination, unless one assumes that amonium compounds of agar exist.

It should also be mentioned that Araki in 1937[62]) has found D-glucuronic acid. As a further constituent Hirase[63]) has found pyruvic acid, linked to galactose with its carboxyl group. The ratio is one unit of pyruvic acid for each 51 hexose units.

For the disintegration and decomposition of agar by enzymes reference should be made to page 378 of this book.

Agar has the quality of an ion exchanger, which has been studied in detail by T. Currie[64]).

A fractionation of agar is possible by extraction at various temperatures from the raw material. The size of the fractions fluctuates between 5 and 28 per cent. The fractions are chemically and physically different. With increasing extraction temperature the viscosity increases. Solvents appear to be unsuitable for fractionation (GLIKMAN[65]).

The ash content of agar according to USP can amount to 6.5 per cent of the dry weight.

Estimates of the molekular weight of agar range from 3,000 to 160,000 (SAMEC[66]), HOFFMAN[67]), MOROZOW[68]), H. H. SELBY & SELBY[69])). The differences are partly due to the different methods used for measuring and observation and partly to the different agar raw materials investigated. Also the amount of care taken to avoid hydrolysis during production may play a certain rôle.

Physical Properties

Agar is practically insoluble in cold water, although it swells noticably. It is colloidally soluble in boiling water. H. H. SELBY & TH. A. SELBY[70]) have given the data on the solubility in other solvents:

	Flocculated		Dry flake	
	25°	97°	25°	97°
Water	+++	++++	—	++++
Glycerol	+++	++++	—	+++
Ethanolamine	+++	++++	±	+++
Diethanolamine	+++	+++	—	+
Triethanolamine	+	++	—	—
Ethylene glycol	+	++	—	—
Diethylene glycol	—	—	—	+
Triethylene glycol	+	+++	—	—
Formamide	++++	++++	++	++++
Dimethylformamide	+	—	—	—
Hexamethylphosphoramide	—	+	—	—

The viscosity of agar depends on the raw material, the type of production, and electrolyte content. The shape of viscosity curves of American agars have been shown by H. H. SELBY and T. A. SELBY[70]) (cf. figure 76).

A method for the measurement of the synerese important for the utilization of agar has been given by STOLOFF[71]). He found a one to two per cent water discharge with various agar samples.

The phycocolloid content in agarophytes amounts to 15 to 30 per cent[72]).

It must be said, however, that agar varies very much in chemical and physical properties, depending on the kind of raw material from which it was made (cf. MURAKAMI[73])).

The gel-forming properties of agar are not influenced by the electrolytes[74]).

According to GLIKMAN & SHUBTSOVA[75]) agar made from *Ahnfeltia plicata* has a closer molecular weight scatter than agar of Japanese origin.

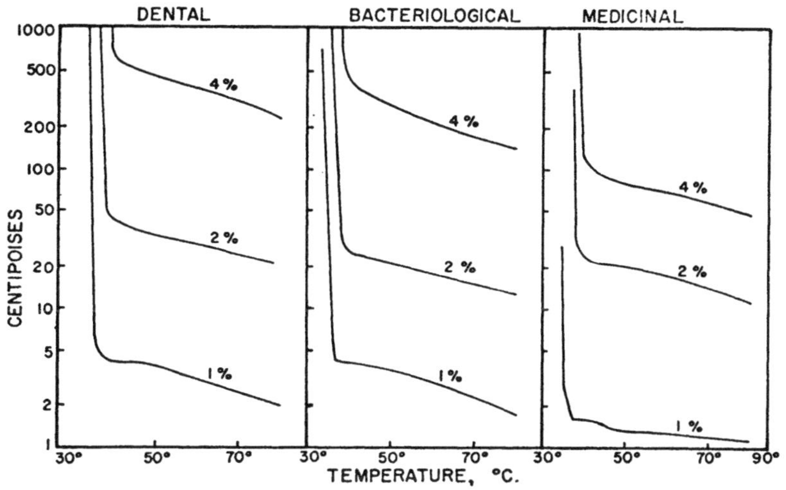

Figure 76: Viscosity of American agar sols; concentrations are given in percent solids. From H. H. Selby and Thelma A. Selby.

Production

The traditional production of agar in Japan comprises the following production phases: cleaning and purification of the sun-dried and bleached alga by several washing processes and subsequent renewed drying and bleaching in the sun, boiling of the alga for several hours, and slight acidification by the addition of acetic or sulphuric acid; repeated freezing and melting of the extract, by which it is purified from any ballast substances and concentrated. Thereafter it is dried in the form of cakes or filaments.

In modern processes (cf. Murolo[76]), Becher[77])) the comminuted algal material is boiled after thorough cleaning in autoclaves. The extract is de-colorized by means of active charcoal, and deodorized, filtered by means of filter presses, and evaporated under vacuum. The gel so obtained is crushed and frozen in freezing chambers. It is melted again by the addition of water, during which process the released contaminations are removed so that the purified agar flocculations finally can be dried in a drying tower. Japanese authors have also proposed an alkaline treatment, instead of the breaking down with acid[78]). The method is likely to have been modernized and improved, but no details have been made known.

Qualities

The trade makes a distinction between various agars according to the countries of production.

As mentioned, up to the Second World War, agar was practically the monopoly of the Japanese. Today, in addition to the Japanese agar, Spanish, Moroccan, and Korean agars are used particularly for European consumption. Agars of other origin play a less significant rôle, for example agar from Portugal, South Africa or Chile.

U.S. production is almost entirely consumed domestically.

A variety of different agar types is distinguished on provenance.

While formerly mostly Japanese agar was sold in strings (KOBE I–III, YOKOHAMA I–III) and in "bars" or "cuttings", the main contingent today is in a powderized form.

Of further distinction in quality are gel strength, the jelling and melting point, colour and the ash content, and, in the powdered grades, also mesh size. Very detailed reports on the subject have been published by SELBY and SELBY in 1959[6]). While, for example, Kobe agar has a greater gelation strength, Yokohama agar has a better solubility. Certificates of quality are based on analytical data.

In the Federal Republic of Germany the most imports of preferred qualities come from Morocco and Spain.

In the U.S.A. the American Agar & Chemical Co., San Diego[79]), produces four types of flake agar, which can also be obtained as powder. They are the bacteriological, medicinal, dental, and industrial types.

The industrial type is mainly used in the food industry, for the canning of jellied fowl, fish, and meat, for cheese and bakery products, confectionery, conserves etc.

Utilization

The most important property of agar for its industrial utilization is its considerable jelling strength, where even one per cent preparation gives very good gelation that can be cut.

In food technology agar is used for jelling and for thickening also in the confectionery industry, in bakery, and in the confectionery making, above all for such type of bakery and confectionery products that permit only a very low temperature. For the production of pudding, creams, and jellies preparations. For the preparation of cheese, as a stabilizer for mayonnaise and salad dressings. Also agar is used as a thickener for domestic uses. The greatest field of application is in the U.S.A. and in Australia for the fish and meat industries, in particular for canned products, as a protective against undesirable effects of the metal, and against shaking during transport etc. In the U.S.A. it is used for the thickening of jams, marmelades etc. Carrageenan has recently been preferred as a stabilizer of cocoa beverages, while for the production of ice cream sodium alginate and carrageenan are preferred.

It is also used as a clarifying agent for wines, beers, and liquors.

In pharmacy agar is used as a laxative for chronic constipation, often with the addition of anthraquinone drugs, as drug vehicle, and as the substratum for bacterial cultures agar also plays an important rôle. The type of agar used may influence the effects of phenols[80]).

In the cosmetic industry agar serves as a stabilizer for emulsions, as a constituent of skin creams, and ointments, lotions etc.

The photo-industry likewise makes use of agar. Technical qualities are used in the paper industry for glueing, in the textile industry as a finishing and sizing agent. The percentage share of agar consumption for technical purposes is very low compared with the foods and allied industries.

In prosthetic dentistry agar is also used. In other fields, for example in criminology, in museum preparations, for artists, and in other fields agar is used for the production of plastic materials.

In laboratory practice agar is also used as an auxiliary means for numerous investigations and analysis, for example microtomy, where agar is sometimes used as an embedding agent.

Literature

1) STOLOFF, L. and LEE, CH. F., Fishery Leaflet 173. 1946. US Dep. of the Interior, Washington DC
2) HUMM, H. J., Ec. Bot. 1 (3): 317–1947
3) HUMM, H. J., The Red Algae of Economic Importance ex Tressler, D. K. and Lemon, J. Mc. W. Marine Products of Commerce. Reinhold Publishing Corporation, New York, 1951
4) CHAPMAN, V. J., Seaweeds and their Uses. Methuen & Co. Ltd., London, 1950
5) SELBY, H. H., Natural Plant Hydrocolloids. Advances in Chemistry Series 11, 1954
6) SELBY, H. H. and SELBY, T. A., Agar ex Whistler, R. L. and BeMiller, J. N. Industrial Gums. Academic Press Inc., New York, 1959
7) HOPPE, H. A. and SCHMID, O. J., Bot. Mar. Suppl. 1962
8) Chemie Markt (17). 1962
9) BUZZONETTI, G. C., Chimica 12. 157–161. 1956. CA 17253e. 1956
10) SALEMA-VEIGINHA, A., Ind. agr. alim. 261–264. 1951
11) CABRERO-GOMAZ, F., Estudio de las algas marinas espanoles. C. S. I. C., Madrid, 1951
12) ARAVIO-TORRE, J., First Int. Seaweed Symp., 1952
13) ARAVIO-TORRE, J. and VILLEGAS, Rev. de Biol. Mar. Chile X. 212. 1960
14) DANGEARD, P., Coll. Int. du Centre Nat. de la Recherche Scient. Dinard, 1960
15) ISAAC, WM. E., Second Int. Seaweed Symp., 1955
16) HOPPE, H. A., Germany (38): 34–40. 1964
17) DURAIRATNAM, M. and MEDCOF, J. C., The Ceylon Trade J. 19. 1–6. 1954
18) SELBY, H. H. and SELBY, T. A., Agar ex Whistler, R. L. and BeMiller, J. N. Industrial Gums. Academic Press Inc., New York, 1959
19) GROOT, J. E. DE, Chronica Naturae 103. 1–12. 1947
20) MOREL, TH., Medeling van het Centraal Inst. v. Techn. Onderzoekingen, Djakarta, Oct. 1949
21) EISSES, J., J. f. Scientific Res. 2. 44–49. 1952
22) EISSES, J., Indonesian J. of Nat. Science (1, 2, 3), 1953
23) ZANEVELD, J. S., Ec. Bot. 13 (2): 89. 1959
24) Bericht der Hypothekenbank von Japan. Die Agar-Agar-Herstellung in Japan. Tokyo, Mai/Juni 1959 (17)
25) TOKIDA, J., Fac. of Fish. Hokkaido Univ. 2 (1). 1954
26) HUMM, H. J., Ec. Bot. 1 (3): 317. 1947
27) SCHULZE, W., D. Apoth. Ztg. 98 (15): 340. 1958
28) KIZEVETTER, I. V., Bull. Pacific Sci. Inst. Fish and Oceanogr. 13. 1–133. 1937
29) GRYUNER, V. and GORSHKOV, J., J. appl. Chem. USSR 10. 2054–2063. 1937
30) PLESKATSEVICH, U. P., Trudy Archangel Vos. Nauck Issl. Belogo Merya 234–254. 1938
31) KIZEVETTER, I. V., J. appl. Chem. USSR 14. 250–255. 1941
32) MOORE, L. B., New Zealand J. of Science and Technology 25 (5): Sec. B. 183–209. 1944
33) SELBY, H. H., Fishery Leaflet 118. 1948. US Dep. of the Interior, Washington, DC
34) WILLIAMS, R. H., Univ. of Miami, Marine Laboratory, 1949
35) HUMM, H. J., Science 100 (2593): 209–212. 1944
36) HUMM, H. J., The Red Algae of Economic Importance ex Tressler, D. K. and Lemon, J. Mc. W. Marine Products of Commerce. Reinhold Publishing Corporation, New York, 1951
37) DAWSON, E. Y. et al., Pacific Naturalist 1 (14): 1960. Rev. de Biol. Mar. X. 225. 1960
38) TAYLOR, Wm. R., Papers Mich. Acad. Sci. 39. 85–109. 1953
40) HUMM, H. J. and WILLIAMS, L. G., Amer. J. of Bot. 35 (5): 287–292. 1948
41) NEUBERN DE TOLEDO, T. A., Anais Fac. Farm. e Odontol. Univ. Sao Paulo 7. 91–103. 1948/1949
42) NEUBERN DE TOLEDO, T. A., Publs. Fac. Med. Univ. Sao Paulo, 1953
43) ETCHEVERRY, H., Algas marinas chilenas productoras de ficocoloides
44) ETCHEVERRY, H., Revista de Biologia Marina X (1, 2, 3). 1960
45) LEVRING, T., Contributions to the Marine Algal Flora of Chile. Lunds Univ. Arsskrift N. F. Avd. 2. 56 (10). 1960
46) TSENG, C. K., Science 101, 597, 1945
47) JONES, W. G. and PEAT, M. S., J. Chem. Soc. p. 225, 1942

48) ARAKI, C., a) J. Chem. Soc. Japan, 58, 1213 1937; b) IV. Intern. Congress Biochem., IV, 1958; c) Bull. Chem. Soc. Japan, 9, 543, 1956
49) TAKAHASHI, E. and SHIRAHAMO, K., J. Agr. Chem. Japan 8, 659, 1932
50) ARAKI, C. and HIRASE, S., Bull. Chem. Soc. Japan 33, 281–295, 1960
51) HJERTÉN, ST., Biochim. Biophys. Acta 62, Suppl. 1, 445, 1962
52) PERCIVAL, E. G. V., Nature 137, 997, 1936
53) PERCIVAL, E. G. V. and THOMPSON, T. G. H., J. Chem. Soc. p. 750, 1942
54) BARRY, V. C. and DILLON, T., Chem. & Ind. p. 167, 1944
55) SELBY, H. H. and SELBY, TH. A., Industrial Gums p. 32, 1959. Acad. Press, New York & London
56) SCHMID, O. J., Unpublished Data, 1962
57) SPRINGER, R. and MIDDENDORF, L., Arch. Pharmazie 294. 66, 693, 1961
58) SHUBTSOVA, J. G. and GLIKMAN, S. A., Nauch. Ezhegodnik za 1954 g Saratov Univ. Saratov 1955, 560–2, ref. Zhur. Khim. 1956, Abstr. 47165. CA 52, 20771d, 1958
59) HAGER, Handbuch Pharmaz. Praxis, p. 326 1938
60) GOMEZ, F. C., Estudio de las algas espanolas marinas, p. 64, 1951
61) SCHMID, O. J., Unpublished Data, 1959
62) ARAKI, C., J. Chem. Soc. Japan 58, 1213, 1937
63) HIRASE, S., Bull. Chem. Soc. Japan 30, 68, 70, 75, 1957
64) CURRIE, T., Chem & Ind. p. 116, 1955
65) GLIKMAN, S. A. and SHUBTSOVA, J. G., Colloid J. (russ.) 19, 281, 1957. Chem. Abstr. 51, 16048b, 1957
66) SAMEC, M. and ISAJEVIC, V., Kolloidchem. Beihefte 16, 285, 1922
67) HOFFMANN, W. F. and GORTNER, R. A., J. biol. Chem. 65, 371, 1925
68) MOROZOW, A. A., Colloid J. (russ.) 1, 37, 1935, Chem. Abstr. 30, 7968, 1936
69) SELBY, H. H. and SELBY, TH. A., loc. cit. p. 32
70) SELBY, H. H. and SELBY, TH. A., loc. cit. p. 37
71) STOLOFF, Fishery Leaflet, 307
72) KIRK-OTHMER, Encyclopedia of Chem. Technology 12, 118
73) MURAKAMI, Susumu, Sci. Repts. Saitama Univ., Ser. B, 2, 35, 1955. CA 51, 18161b, 1957
74) STOLOFF, L., Polysaccharides in Biology, IV Conference 1958, p. 288
75) GLIKMAN, S. A. and J. G. SHUBTSOVA, Uch. Zap. Saratorsk Gos. Univ. 75, 115, 1960. CA 60, 1915g, 1964
76) MUROLO, M. and MUROLO, V., French Patent 1.017. 156, 1952
77) BECKER, J., US P. 1712785, 1929
78) KOJIMA and FUNAKI, Bull. Jap. Soc. Sci. Fish. 16, 25, 1951. Kojima Y. et alii, Norinsho Suisan Koshusho Kenkya Hokoku 9, 43, 1960, CA 55, 9721f, 1962
79) American Agar and Chemical Co., San Diego. Types of American Agar. Characteristics and Uses
80) SANDS, J. G. et alii, ANTONIE VAN LEEUWENHOCK, J. Microbiol. Serol. 29, 386, 1963
81) TSUCHIYA, Y. and K. HONG., Abstr. V. Intern. Seaweed Symp. 33, 1965

General Literature

HEGNAUER, R., 1962; HOPPE, H. A., 1958; HOPPE, H. A., Drogisten-Fachblatt (6): 3. 1949; HOPPE, H. A., D. Drog. Ztg. (14): 362. 1950; HOPPE, H. A. and SCHMID, O. J., 1962; IDSON, B., Chem. Week July. 1956; Institute of Seaweed Research, 1956; LEWIN, R. A., 1962; PAX, F., 1962; SANFORD, F. B., Fishery Leaflet 469. 1958. US Dep. of the Interior, Washington DC; WHISTLER, R. L. and BE MILLER, J. N., 1959; see Index of Literature; see also *Acanthopeltis, Ahnfeltia, Ceramium, Gelidium, Gracilaria, Pterocladia*

Alginic Acid and Alginates

Raw Materials and Occurrence

Alginic acid is an acid membrane mucilage regarded as a typical constituent of brown algae.

Alginic acid is the base product of numerous economically interesting alginates and alginic acid derivates.

A high alginic acid content is not the only prerequisite for industrial utilization of brown algae but also sufficient size and density of crop as well as the presence of harvesting and production facilities in the raw material areas.

The brown algae of numerous coastal areas have been examined and new sources of raw material rendered accessible. In view of its growing significance this work is still being continued, and it has by now also covered coastal areas which had not been widely worked up in the past.

The continuation of these investigations is an urgent task, if only because of the variation in content due to season and other factors.

The most important raw material for alginate production is doubtless *Macrocystis pyrifera*, in view of the vastness of resources and size of the plants.

Laminaria hyperborea, *L. digitata*, *Ecklonia maxima*, *L. saccharina*, and *Nereocystis luetkeana* can also be considered to be important, also the so-called "rockweeds", especially *Ascophyllum nodosum*.

Further brown algae which may attain significance for alginate production under certain circumstances are the following:

Alaria species, *Durvillea* species, *Egregia* species, *Fucus* species, *Glossophora Kunthii*, and a number of other *Laminaria* species, mostly all on the East Asian coast:

Lessonia species, *Pelagophycus porra*, a number of *Sargassum* species, and *Turbinaria*.

Definitions

Alginic acid is a membrane mucilage and constituent of brown algae.

Alginate is the term used for a salt of alginic acid.

The term algin is used as a collective name for alginic acid and alginates, but also as a trade term for sodium alginate.

In the "Terminology of Seaweed Colloids" TSENG, 1945[1]), has given the following definition of the term 'algin': "It is logical to limit the use of 'algin' to the soda-extracted substance, i. e. sodium alginate, from kelps. The other alginous products may be designated as ammonium algin, calcium algin, etc.".

STEINER-McNEELY[2]) in 1954, wrote: "The term 'algin' generally designates the water soluble derivatives of alginic acid. Because the most common derivative is sodium alginate, the name algin is usually associated with that compound."

JEVTUSCHENKO (according to the translation by SCHULZEN, 1956[3])) makes a difference between alginic acid and algin precipitations of alginic acids.

McNEELY in 1959[4]) in "Industrial Gums", wrote: "The term 'algin' is the generic designation of the derivates of alginic acid. The most common algin compound is sodium alginate. Other commercially important compounds are the potassium, ammonium, and propylene glycol alginates".

Alginic acid is reported to exist in two forms, one soluble and the other insoluble in water. The former was obtained by dialysis of a sodium alginate solution acidified with acetic acid (JEVTUSCHENKO, 1954). The latter is the more stable form, precipitated from alkaline solutions on the addition of mineral acid. According to JEVTUSCHENKO, it is then in the form of an anhydride, for which he proposes that the term 'algin' should be used, retaining the name 'alginic acid' for the soluble acid.

In current practice, however, 'algin' usually refers to the extraction product 'sodium alginate' (LEWIN, KREGER, 1962[5])).

Alginic acid and alginates are produced in a number of countries, mostly from their own raw materials but in many cases considerable additional quantities of raw materials are also imported from other areas.

The world alginate production roughly doubled between 1954 and 1958, reaching approximately 7.500 to 8.000 tons in 1958. The largest producers were the U.S.A. with approximately 50 per cent, followed by Great Britain, France, and Norway. According to other statements, Norwegian production was approximately 9.000 tons only in 1956[6]).

The economic development of the seaweed industry has been reviewed in 1958 in a comprehensive work by HAKEN in CHEMISCHE INDUSTRIE, the development and level of algin production at the time being supported by relevant statistics[7]).

Precise production figures for individual countries, however, are available only in some few cases. A continuously rising trend has been recorded for the last years. This is due in part to the rising consumption of various known fields of application, but also to the opening up of new fields of application. In addition to the expansion of capacity in countries of previous production, the initiation of alginate production in other countries has been planned or is already on the way being established.

Australia. The establishment of an alginate industry was commenced in 1958 with raw material, mainly the large stocks of *Macrocystis pyrifera* off the coast of Tasmania.

Canada. The brown algae occuring on the coast of Canada have been analysed for their suitability as raw materials for alginates by Canadian experts. Considerable quantities of raw materials have been exported to the U.S.A. in recent times. For some time sodium, potassium, and ammonium alginates mainly have been of home production.

Federal Republic of Germany. The Federal Republic of Germany imported alginic acids and alginates mostly from France, Great Britain, and Norway. Smaller quantities came from the U.S.A., Japan, and Canada. The import in recent years has been about 100 tons per month on an average, with application in textile printing and the detergent industry accounting for the greater part of it.

Alginates of German provenance have been described by CZETSCH-LINDENWALD[8]).

France. French firms have an essential share in the world production of alginates (in 1961 it was approximately 16 per cent), and a rather rich range of products is offered by this industry. A considerable portion of the production is exported[9]).

Great Britain. The alginic acid industry has been expanded repeatedly in recent years. Indigenous raw materials occur mainly on the Scottish coast, but brown algae have also been imported, for example considerable quantities of *Laminaria* stipes from Ireland.

The development of the British alginate industry has been described in the ANNUAL REPORTS OF THE I.S.R.[10]). The British alginate industry has a great number of qualities to offer for pharmaceutical, cosmetic, food, and technological purposes. Considerable quantities have been exported.

Ireland. Production is insignificant in quantity, although there are considerable resources of raw material on the Irish coast. A technical quality is produced, which, however, hardly appears on the world market.

Japan has very rich resources of raw materials on its coasts. The various algae have been studied by Japanese scientists. The production of alginates plays a considerable rôle in the rather versatile algae industry of Japan. The largest portion of this production finds application in the domestic textile industry.

Morocco. In recent years Morocco has not only intensified scientific research of its enormous resources of raw material, but has also introduced to the market very good qualities of different phycocolloids[11]).

Netherland's production is confined to a number of grades for the food and allied industries, particularly for the production of ice cream.

Norway has considerable resources of raw material on its coasts for the production of alginates. The production programme of leading firms inter alia comprises 50 qualities of sodium, ammonium, and other alginates. Alginates used in the deep-freezing of fish, according to a Norwegian patent, have a very special place in this production (cf. MAASS[12])).

Spain, likewise, produces alginates. However, the alginate industry is of less importance than the production of Spanish agar from *Gelidium*.

South Africa has considerable resources of suitable raw materials for alginate production, especially *Ecklonia maxima.*

The establishment of a domestic alginate industry had been planned[11]).

U. S. A. The U.S.A. produce approximately 50 per cent of the world production of alginates. The range of grades for various applications is considerable. IDSON has reported in 1956 in detail on the commercial development of the phycocolloid industry in the U.S.A.[14]).

Well-known American scientists and experts provided the essential foundation for the development of the alginate industry in the U.S.A. and indeed in the whole world. See HUMM[15]), STEINER-MCNEELY[2]), MCNEELY[4]).

The most important raw material for alginate production in the U.S.A. is *Macrocystis pyrifera* on the Pacific coast of South California[4]) [16]).

U. S. S. R. Considerable resources of raw material for the production of alginic acid and alginates occur on some coasts of the U.S.S.R. for example on the coast of the White Sea and on the coasts of Eastern Asia. This products are manufactured in the U.S.S.R.[13]).

Structure

Alginic acid can be produced from a number of brown algae, a fact that has been known for some time. Already in 1881 STANFORD[17]) mentioned in his patent claims an alginic acid, which he had obtained from *Laminaria.* After initial erroneous assumptions on the constitution of this acid (nitrogen compound, glucuronic acid compound) NELSON & CRETCHER[18]) identified the uronic acid as a D-mannuronic acid. These findings with alginic acid obtained from *Macrocystis pyrifera* could be confirmed also by the investigation of alginic acid from other brown algae.

HIRST, JONES & JONES[19]) had confirmed LUNDE and co-workers[20]) earlier hypothesis of a β-(1–4) linkage of the mannuronic acid units. On the evidence of x-ray spectra, ASTBURY[21]) established, that two of the four mannuronic acids are orientated in opposition to the other two. The following Formula A of alginic acid results from these two findings:

FISCHER and DOERFEL[22]) found L-guluronic acid to be a further constituent. It is present in varying quantities depending on the type of brown alga, and the ratio of D-mannuronic acid to L-guluronic acid varies between 0.5 to 2.2 WHISTLER & KIRBY[23])

Formula A

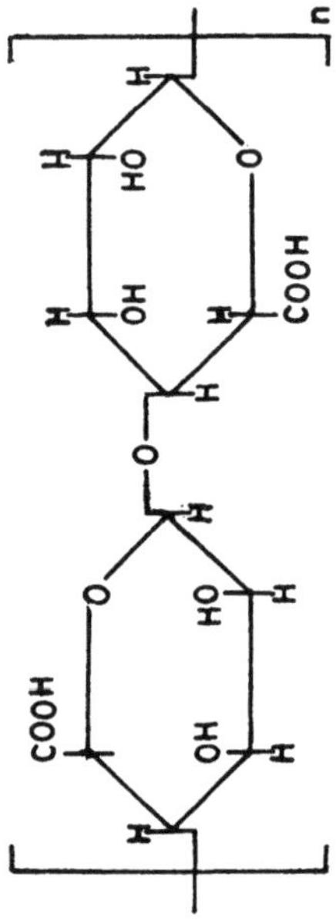

Formula B

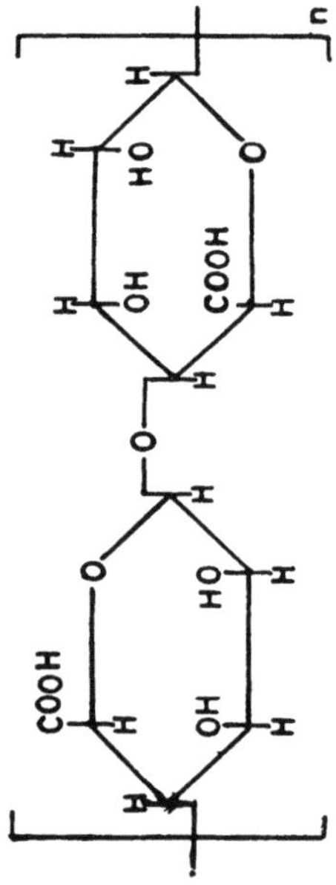

Formula C

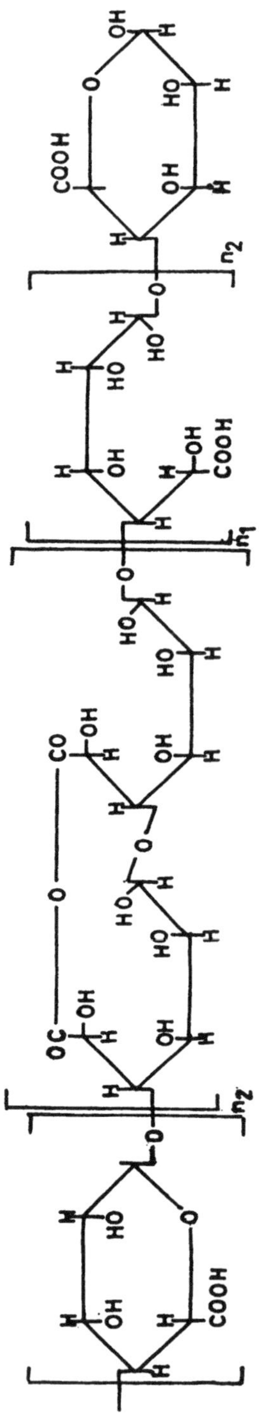

confirmed these findings also with alginic acids from *Macrocystis pyrifera*, so that the following formula B can be set up for alginic acid:

JEVTUSCHENKO[24]) supposes that in addition to the water soluble there is also a water insoluble alginic acid and attributes its appearance after precipitation of alginic acid from alginates by high concentrated acids to anhydride formation between two neighbouring carboxyl groups, and the absence of lactones. From this the structural Formula C would arise.

A very detailed description of the investigations of the constitution of this material has been given by SMITH & MONTGOMERY[25]) and MAASS[12]).

The composition of alginic acid was once more re-investigated by DRUMMOND, HIRST, and PERCIVAL[26]), and again they found L-guluronic acid and D-mannuronic acid. They infer a 1–4 bond from the presence of threaric acid after oxydation of alginic acid with per-iodic acid and bromine, however, they also think it possible, that also 1–3 bond exist. FREY and PRESTON[27]) proved with X-ray diagrams of alginic acids the presence of both guluronic and mannuronic acids; the diagram previously ascribed to mannuronic acid corresponds in fact to guluronic acid and the former acid is given a new diagram. It is worth noting that they state that the presence of poly-mannuronic acid is characteristic of intercellular material and the polyguluronic acid of the cell wall. The existence of alginic acids with varying proportions of guluronic and mannuronic acids has been corroborated by HAUG[28a]) who obtained two fractions by the treatment of alginic acid with potassium chlorine and manganese sulphate, differing in the proportions of the two residues. The fraction richer in guluronic acid also proved to have a strong affinity to calcium.

According to HAUG and LARSEN[28b]) alginic acid is not a single compound but a family of polysaccharides with a varying composition of guluronic and mannuronic acid. HAUG and LARSEN presented evidence for a substance accompanying alginic from *Ascophyllum nodosum*. Calling it ascophyllan they found fucose, xylose, uronic acid, sulphate ester, and a polypeptid as structural components[99]).

These new investigations shed fresh light on previously inexplicable findings and physico-chemical behaviour of alginic acid and its salts, and have also great bearing on industrial production and practical utilization of alginic acid.

Qualities

Alginic acid has a good reactivity and can be mixed with alkali, ammonium, amines, and organic bases to form water soluble and water insoluble salts with heavy metals and earth alkalis.

The water soluble salts, inter alia, are:

Ammonium alginates, iron alginates, lithium alginates, magnesium alginates, potassium alginates, sodium alginates

Magnesium and iron alginates are exceptional cases.

The water insoluble salts are:

Aluminium alginates, copper alginates, nickel alginates, silver alginates, zinc alginates.

In the presence of ammonia these salts, too, are water soluble.

The insoluble salts are:

Barium alginates, calcium alginates, lead alginates, mercury alginates, strontium alginates, wismut alginates.

Most alginates, like alginic acid itself, are insoluble in organic solvents. Further details have been reported by MAAS[12]) and SCHULZEN[29]).

The most important soluble alginates are the following:

Sodium alginates, potassium alginates, and ammonium alginates.

The best known insoluble alginate is calcium alginate.

A lot of alginic acid derivates are described in technical literature and in patents, such as

alginic acid ester (methyl-glycol-, propylen-glycol-, alcylen-glycol-, glycol-ester, ester-ether), amino acid compounds, amides, crosslink products, complex compounds, depolymerisates, mixed products with gums, gelatine etc.

Numerous patents have been published of the production of the numerous alginate compounds and derivates (cf. MAASS[12])).

SCHULZEN has reported on the properties of the salts of alginic acid, in particular of sodium alginate, metal alginate, ester alginate, and on methods of investigation for alginates and mixtures with other colloids[29]) [30]).

A distinction is made in degree of purity and milling according to whether the product is destined for food technology, pharmaceutical, cosmetic, and technical applications.

The viscosity is also adjusted by the producers according to the desired purpose of use and often ranges between 5 and 15,000 cp.

The largest internationally known manufacturers publish very detailed descriptions of the qualities they produce and their properties and applications. In many cases special products are used.

Cellulose-alginates are less purified commercial qualities, which often contain cellulose in addition to alginate and which are utilized for various industrial purposes.

Alginates in pharmaceutical industry:

Alginates of a very high degree of purity have been used as an excipient in pharmacy, and their application has been very wide and is increasing.

They are used as emulsifiers in watery emulsions with fats, oils, and waxes, as well as suspension agents or stabilizers.

Alginates serve as film-forming, as jelling, and as protective colloids.

They are used as fillers in the manufacture of tablets, pills, and lozenges, as vehicles and desintegrating agents.

Alginic acid is used as tablet swelling agent, as it has a very high swelling strength.

The base of many ointments is made with alginate (alkali- and amino-alginates). Alginates, being resorbable, also form the base of various dressing materials which can be left in the wounds. An alginate gauze is used as a blood-stopping plaster[3]).

The use of alginates in pharmaceutical industry has been reported by MAASS[12]) [32]).

VOGT published on the application of alginic acids and alginates in pharmaceutical technology[33]).

SCHWANDT has reported on alginates in ointment bases[34]).

BOLLIGER has published detailed results on the studies of properties of alginic acid and alginates and their use in pharmaceutical industry[35]).

Sodium alginate is also used in slimming agents. The alginate forms a jelly in the stomach, which produces the feeling of saturation.

Several reports have appeared in the pharmaceutical literature on the use of sodium alginate as a plasma substituting colloid[36–22]).

Calcium alginate is used for the production of capsules[43]).

Ammonium alginate wool is used as a filter for micro-organisms[44]).

Report has been published on the "Hemostatic, laxative, and toxic effects of alginic acid preparations"[45]).

Alginates in the cosmetic, soap, and detergent industries:

Alginates of different viscosity are an important auxiliary agent in cosmetic industry, in the soap and detergent sector[46]).

Alginates serve above all as thickening and dispersing agents in ointments, creams, jellies, liquid emulsions, lotions, tooth paste, compact powder, bathing preparations, as well as an additive in numerous hair cosmetics, such as hair dyes, hair fixing tonics, shampoos etc. Frequently used are sodium, calcium, propylen glycol, amino alginates, and alginic acid.

Alginates have a very good compatibility with other products used in the cosmetic industry, such as starches, dextrin, gums, pectin, glycerine, methyl cellulose etc.

Alginates are used as foam stabilizers in the soap and detergent industry. They influence the hardness of water favourably. Alginates increase the consistency of shaving creams, because they increase the milling properties of fine soaps.

An addition of alginates also has a favourable effect on liquid soaps and detergents.

It is made use of the ability of alginates to form films on the skin in skin protection ointments and creams.

Alcohol containing preparations make use of amino alginates, which, in contrast to the sodium alginates, are compatible with alcohol. Particularly the triethanolamine alginates. But also propylenglycol alginates are used in the cosmetic industry. The versatile application of alginates in these fields has been reported by MAASS[12]).

Of the earlier comprehensive surveys one should mention "alginates in drugs and cosmetics" by LESSER[47]), "Neue Grundstoffe in der Kosmetik" (New basic materials in cosmetics) by JANISTYN[48]), and "Alginate in der Kosmetik" (Alginates in cosmetics) by ROTHEMANN[49]).

Alginic acids and alginates in cosmetics also serve for the embedding and fixing of the aromatic and fragrant components.

Alginates in dental technology

Alginates are used in dental technology for making denture mouldings, as well as in denture fixatives[12]) [50]).

Alginates in food technology

Numerous special papers and references have been published on the application of alginates in the food technology. It seems to be obvious that only purified products of high quality can be used in this sector, since in many countries food law regulations must be observed.

Leading manufacturers in producer countries have developed various qualities most useful in practice, and they issue formulations and detailed instructions for use. Apart from the degree of purity such factors as viscosity, grain size, and colour are important for processing in practice.

Questions of food law aspects and of alginates as food ingredients have been studied by SCHADE in 1956[51]).

In 1958 DIEMAIR and WEICHEL[52]) have made a contribution to the knowledge and assessment of alginates. In this work after an investigation of more than 70 kinds of alginates criteria have been developed for the assessment with reference to external appearance, organoleptic assessment, viscosities, determination of molecular weight, and chemical investigation of alginates. Numerous applications in food technology have been referred to.

MAASS has reported in detail on the application of alginates and noted references to the international patent literature[12]) [32]).

GIDDEY has reported on the influence of constituents on the physical structure of homogeneous food products[53]).

In this survey only references to the various types of use can be given.

Bakery and confectionery products

Alginates improve the baking properties, and they are a constituent of baking emulsions. They are used to make sugar glazings, egg and fruit and other cream fillings etc. in confectionery manufacture, for making imitation fruit, and for fillings and meringues[12]).

Jelly products

Jelly products are made with water insoluble alginates (for example calcium alginate). In a number of countries alginates were suggested as a gelating agent in marmelades and jams.

Milk and dairy products

Alginates also have a very great range of application in the production of dairy products, whipped cream products, milkmixing products, milk-shake mixes, drinking cocoa and chocolates, puddings, cold-prepared pudding powders etc.

They are likewise used for making processed cheese, cheese cream, and soft cheese.

With milk products the essential function of alginates is the increase in viscosity and improvement in the whipping quality of cream, that they effect.

Alginates act as stabilizers in milk mixes, milk-shake mixes, and cocoa beverages. They often are used in a mixture with carrageenan.

The thickening and vitamin C stabilizing properties of alginate for dairy products have been studied by SCHADE[54]).

Also alginates are used as ingredients and additives in dessert and sauce powders, and also as gelling agents for jellied custards[55]).

Sweet industry

Alginates are used in the sweet-industry mainly as stabilizers and jelling agents.

Ice cream

For 20 years alginates have played a great rôle as stabilizers for ice creams of various types. In 1945 in the U.S.A. more than 50 per cent of all ice cream products were made with alginates. An essential advantage of the alginates is their ability to prevent the formation of ice crystals during the freezing process. Another advantage of alginate ice cream stabilizers is the uniform viscosity during the production process, good whipping ability, smooth melting, and full development of the aromatic components. An addition of 0.3 per cent is suggested. Alginates are also used for iced beverages.

Soups, sauces, and mayonnaise

Alginates are used as thickening agents and for the increase in viscosity of soups, sauces, mayonnaise, and mustard.

Fat and margarine industry

Alginates are used in margarine industry as emulsifiers and stabilizers.

Sodium alginates work as a protective colloid for fat-soluble vitamin A.

Beverage industry

Alginates are used as clarifying agents in wines and for other wine treatment processes. They are also used for the clarification of the raw liquor of sugar and similar products[56] [57] [58].

As stabilizers they prevent the settling of fruit beverages. Also in fruit juice powders and sherbets an alginate additive is of advantage.

Alginates act as foam stabilizers in beer and malt beer[4].

Dietetic foods

Alginates are used for products of low sugar content.

Meat and sausage industry

Meat and sausage products are given a longer shelf life with alginate films. Alginate string is used for tying up of rolled ham, sausages etc.

Artificial synthetic casings with an alginate basis have been developed particularly for small sausages. Comprehensive specialized literature has been published on this field of application (cf. Mass).

In the application of alginates as a binding agent for meat products, the respectivc law regulations have to be observed[12] [32].

Freezing processes

In Norway processes have been developed for the deep-freezing of fish in an alginate gel. Apart from fish, these can also be applied to meat, poultry, and fruit.

The Protane-freeze-gel method has been patented in many countries[59] [60] [61] [62].

Russian scientists report on the production of calcium alginate films and their application for the deep-freezing of meat[63].

Alginates in industry

The application of alginates and alginate compounds in industry as emulsifiers, jelling agents, stabilizers, suspension and thickening agents, as a protective colloid, as a film and filament former, have been studied in numerous comprehensive investigations published during the past 10 to 15 years. The following industries should be mentioned:

Adhesive industry, Briquette making, Building materials industry, Ceramics and China-ware-porcellain industry, Descaling, Electrode manufacture, Enamel industry, Explosive industry, Filter and absorption substances, Fire-protection agents, Floatation, Foam-rubber production, Foundry processes, Leather industry, Lubricant production, Mineral oil industry, Painting applications, Paper and cardboard industry, Pencil and colour crayon industry, Pesticides and insecticides, Photographic supplies, Polishing and floor treatment industry, Production of various fabrics, Rubber industry, Textile industry, Textile printing, Wood preservation.

Reference is to be made particularly to the following comprehensive studies (cf. List of Literature).

1951 H. J. HUMM: The Brown Algae-Algin from Kelps and Fucoids
1954 A. B. STEINER and W. H. MCNEELY: Algin in Review
1955 P. MERAT: The Alginates and their Industrial Uses
1956 Institute of Seaweed Research: Alginic Acid and the Alginates
1958 F. B. SANFORD: Seaweeds and their Uses
1958 H. A. HOPPE: Drogenkunde
1959 H. MAASS: Alginsäure und Alginate*)
1962 W. H. MCNEELY: Algin
1962 H. A. HOPPE and O. J. SCHMID: Meeresalgen für moderne Industrieprodukte
1962 H. SCHULZEN: Die chemisch-physikalischen Eigenschaften der Alginate
1962 H. MAASS: Ueber die Verwendungsmöglichkeiten von Alginaten.

Dyes and painting industry

Alginates, particularly ammonium alginates and sodium alginates, alginate ester of propylenglycol etc. are used for the production of paints on a water, emulsion and latex basis. They have the effect of providing for very good suspension and protection of pigments, stabilization of emulsion, flow and levelling characteristics, continuous film formation, and better adhesive properties. The alginates are compatible with starches, water soluble gums, carob or locust bean flour, caseine, gelatine etc. Lithopone, kaolin, titanium dioxide, and the most dyestuffs can be used in combination with alginates. However, the basic dyestuffs processed in acid media, are exceptional.

Brush-application and adhesive properties of paints and lacquers are improved by the addition of alginates[32]) [64]).

GRUBER has published a study on the use of alginates in paint and coating techniques[65]).

Propylenglycol-alginates are used in polyvinyl acetate emulsiones[66]), and vinyl resin paints are controlled to the desired viscosity by the use of alginates[67]).

* This standard work gives a total survey of the raw materials, methods of production, properties, methods of investigation, applications, and international literature.

The technical journal "Farbe und Lack" has reported on the properties and processability of alginates in paints and lacquer technology[68].

Sodium alginate was also used with good success by the steel industry for paint covers[69].

Building materials

Alginates are used as thickening agents for tar and asphalt emulsions for the construction of roads, runways, flooring materials etc. Application on moist roads thereby becomes possible, the additives improve adherence, and the formation of cracks is avoided to a great extent.

Sodium alginate additives improve the structure and plasticity of concrete, gipsum, and cement compounds. They work as delayagents in binding[32] [70] [71].

Alginates are used for the production of artificial wood, wall panelling, heat- and sound insulating materials, refractory compounds, security glass compounds, and joint sealing compounds.

Ammonium alginate is used for the production of ceramic compounds for the lining of furnaces etc. In the production of bricks alginates additives procure more thorough mixing, the reduction of the make-up water added, and faster drying.

Pencil and crayon industry

Alginates are used as a binder for pencil and crayon leads etc.

Briquette production

Technical alginates are used as binding agents for briquettes.

Electrode industry

If used in welding electrodes an addition of alginates prevents an excessively rapid hardening of the water glass, causes better suspension of the dry solids, and prevents the formation of knots and cracks[70].

Enamel industry

Enamel compounds are stabilized with alginate additives. Alginates are also used for surface coating and treatment.

Fire-protection compounds

Alginates are used for the improvement of fire extinguishing foams[32] [72].

Algin-thickened water has been used by the Forest Service for the fire-fighting from the ground and from aircraft. A combination of algin and di-ammonium phosphates has been successfully tested[73].

Textiles are made flame-resistant by means of alginates[32].

Filter and absorption substances

Sodium alginate cotton wool is used as a water soluble filter material for the determination of germs in the air[74].

Ammonium alginate cotton wool is used as a filter material for microorganisms[75].

Active carbon of a high absorption property was manufactured according to a British patent by coating of active carbon with alginic acid or metal alginates[76] [77].

Floatation

Alginates are used as a floatation substance in the treatment and preparation of ores[32].

Fabric production

Alginates are used for surface coatings and for impregnation in the manufacture of fabrics for sacks, bags, tarpaulins etc. and in the production of fish nets[32].

Alginate hessian fabrics are used for camouflage and military purposes in Great Britain[78] [79].

Alginate filaments, in particular calcium alginate rayon has been reported on by Ulrich at the Third International Seaweed Symposium, 1958 [80].

Foundry industry

Sodium alginate is used in iron foundries as an anti-adhesive substance, particularly in work with low temperatures. The mould parts are coated with an alginate solution, which has the effect of making the stripping of castings from the moulds easier after drying of the solution[32].

In moulds made of quartz sand an alginate addition has helped to attain constant moisture levels, and the binding characteristics of the core sand are increased[70].

Experiments on the use of sodium alginates in foundry practice were made in 1957–58 by the British Cast Steel Research Association[78] [81].

Wood protection

Technical sodium alginates can be used as a thickening agent for salt solutions used in the preservation of wood.

Rubber industry

Ammonium alginate has well proved its merits as a creaming up and stretching material for the concentration of rubber latex.

Alginates are also used as thickening agents and dispersing agents for fillers[32].

Ceramic, porcellain, and Chinaware industry

Alginate additions stabilize the pigment and glazing suspensions.

Prevention of boiler scale

Alginates have the effect of a protective colloid. Weakly alkaline algin solutions prevent corrosion and form a voluminous sludge that can be easily removed.

Adhesive industry

A paste is formed that can be used as an adhesive by infrared radiation of a 5-per-cent sodium alginate solution, previously treated with 50 per cent formic acid and controlled at pH 6–7 (Maass, 1962). Alginates are used as additives to vegetable and animal base glues, to make them applicable at low temperatures and to prevent any excessive penetration into the parts to be glued[19]). Gelatine-alginate glues have been produced[82]).

Alginates have also been used as adhesive for cellophane foils.

Experiments to increase the elasticity of water glass films have been made, and on this subject other authors have reported in the special journal [83]) [84]).

Leather industry

Alginates have manifold applications in the leather industry, for example leather treatment, in the grinding of split leather as an absorbent of dust, as a finishing and sizing agent, as an additive to glues, adhesives, and for the dying of leather.

The emulsions used in the leather industry are stabilized by alginate[70]).

After drying very smooth soft films arise on the surface. Also in rendering leather impenetrable by water alginates are used in combination with other chemicals required for such processes.

Alginates are also used as emulsifiers for leather preserving agents.

Mineral oil industry

By the use of alginates stable, aqueous emulsions are made out of mineral oils. Likewise they are used for producing clear mineral oils and tars for the pharmaceutical and cosmetic industries[32]) [85]).

Paper and cardboard industry

The use of alginates in paper glueing increases the gloss and softness of paper and facilitates waxing, printing, varnishing, and coating. Alginates are compatible with most other glues and adhesives, and produce a film-type adhesive action[32]) [70]).

Colloidal metal alginates serve as coating, impregnating agents and fillers for permeable material. Calcium alginate is used as a diluent and filler.

The grease proofness of parchment substitutes is increased by alginates. They are also used for the production of glossy papers.

In the wallpaper industry alginates are used as a thickening agent for printing inks.

In the international specialized press in recent years quite a number of publications appeared on the use of alginates in the paper industry, several of which are mentioned below.

"Sizing, Coating, Adhesives"[86]), "How to use Algin for Surface Sizing and Surface Control"[87]), "Effect of Sodium Alginate and C. M. C. and the Deflocculation of Fibre Suspension and Uniformity of Formation of the Resultant Paper"[88]), "The Application of Algin in Paper and Board Manufacture" (in this work special reference is made to the versatile application of alginates, and where the products can be used[89])), "Algin Sizing of Paperboard"[90]) [91]), "Algin Applications in Board and Liner-Manufacture"[92]).

Further articles and references have appeared in "World's Paper Trade", 1958, "Papeterie", 1958, "Das Papier, 1956–57, "Adhäsion", 1961–62.

The alginate manufacturers produced a number of special grades for the paper industry. Detailed investigations have been made and published on applications, for example in surface smoothing and levelling in grease-proof paper making and in the manufacture of corrogated cardboard etc. It should also be mentioned that experiments have been made on direct production of paper from marine algae by the vulcanization of algin[93]).

Photographic industry

In the photographic industry alginates are used for the coating of photographic material[12]).

Polishes and floor cleaning and treatment compounds

Alginates are used in polishes and floor polishes of various types. They are auxiliary substances for the polishing particularly of soft metals and alloys, as thickening agents for boot polish and as cleaning pastes etc.[32]).

Production of foam rubber

Alginates are used in the production process of voluminous foam rubber materials[70]).

Pesticides and insecticides

By the addition of alginates, particularly ammonium alginate, stability, dispersion, and toxic effect of pesticides and insecticides in suspensions are improved. The addition of sodium alginate serves as an adhesive in insecticides. Complex compounds are made with alginates, forming a weather-resistant stable coating when spread on plants. Alginate emulsions are also used for etching and caustic substances[32]).

Lubricants

Lubricants and coolants are often blended with alginates[32]).

Explosive industry

Alginates are also used by the explosive industry. Granulated explosive mixes are coated with sodium alginate[94]).

Textile industry

Alginates are suitable in the textile field for the refinement of fabrics, for finishing, sizing, impregnating, for the increase of mesh strength, for the fixation of base materials, for the achievement of water impermeability etc. MAASS and SCHULZEN[32]) [70]) have reported on the use of alginates in textile finishing and sizing.

Alginates can react and form compounds with quite a number of metals and salts. They are used as auxiliary substances and for the production of alginate filaments and strips (calcium-alginate yarn), films, foils, alginate rayon etc. Alkali resistant rayons made from chromium and beryllium alginate are highly flame-resistant. Calcium alginate rayon is, moreover, washable, and is used as a connection filament in hosiery production.

Alginates are used in textile printing, particulary as a thickening agent for printing paints. They dissolve in cold and warm water and do not coagulate. They can be mixed with other thickening pastes and substances, and enhance the clearness and colour of the prints.

The special advantages of alginates for textile printing were described by SCHULZEN and others:

Constant viscosity and consistency of the standardized types, quick dissolution, enhanced water binding properties, good processing capabilities, effective protective colloidal properties, hydrotropic effects, suspension, emulsifying and adhesive properties, evenness of prints, clearness of print, flexible film formation, high colour yield, easy washability etc. are some of the outstanding characteristics.

Alginic acids and their salts, their dye-behaviour vis-à-vis cold-based reaction dyestuffs and the dichloro-triazine type have been reported by ULRICH at the Colloque International at Dinard in 1960[95] [96].

Printing methods with alginates have been described in detail[70].

Alginates have been used in textile printing, such as wet printing, printing with pigment dyestuff, phtalogenic brilliant dyestuffs, azodyestuffs, and experience has been reported. The properties of the various products are of great significance in this regard[97].

In Japan alginic acid preparations are also used in textile printing, too[98].

Work has been published on the influence of thickening agents (alginates) for the padding of ultra-dispersed wet dyestuffs[99].

SCHULZEN in 1963 has reported on the chemistry and application of "Saicatex" methods in textile printing. This new method, according to experts, will have a great significance in the future. Also in the "Saicatex" method the use of alginates is of considerable importance[100].

Alginates also serve for preventing the smudging, marking of and run of colours in textile printing.

The alginate manufacturers produce an entire range of special grades for various applications and will advise potential customers on the basis of their rich experience and knowledge of material.

Miscellaneous uses

Alginates are used for making transparent films, for flux substances, for silver soldering, for the strengthening of liquid hydrocarbons, for the absorption of gases to solid vehicles or carriers, in oil drilling etc.

The very many possibilities for the application of algin compounds have scarcely been broached here. Numerous further industries are at present engaged in research and development. Suitable special products are in the process of being developed.

Literature

1) TSENG, C. K., Science 101 (2633): 597–602. 1945
2) STEINER, A. B. and MC. NEELY, W. H., Natural Plant Hydrocolloids, Advances in Chemistry, Series 11, 1954
3) SCHULZEN, H., Melliand Textilberichte 37 (9): 1087–1093. 1956
4) MCNEELY, W. H., Algin ex Whistler, R. L. and BeMiller, J. N., Industrial Gums, Academic Press Inc., New York, 1959
5) KREGER, ex Lewin, R. A., Physiology and Biochemistry of Algae, Academic Press Inc., New York, 1962
6) Chemische Industrie 12 (3): 111. 1960
7) HAKEN, W. v., Chemische Industrie (5). 1958
8) CZETSCH-LINDENWALD, H. v., Bot. Mar. IV (3/4): 255–260. 1962
9) Chemische Industrie 13 (10). 1961
10) Institute of Seaweed Research, Ann. Rep. 1945–1962
11) HOPPE, H. A., Germany (38): 34–40. 1964
12) MAAS, H., Alginsäure und Alginate, Strassenbau, Chemie und Technik Verlagsgesellschaft m.b.H., Heidelberg, 1959
13) Chemiker-Zeitung 81 (13): 430. 1957
14) IDSON, B., Chem. Week July, 1956
15) HUMM, H. J., The Brown Algae – Algin from Kelps and Fucoids ex Tressler, D. K. and

Lemon, J. Mc. W., Marine Products of Commerce, Reinhold Publishing Corporation, New York, 1951
16) Sanford, F. B., Fishery Leaflet 469. 1958, US Dep. of the Interior, Washington D. C.
17) Stanford, E. C. C., Patent Nr. 142, referred by H. Maass (10)
18) Nelson, W. L. and Cretcher, L. H., J. Am. Chem. Soc. 51, 1914, 1929
19) Hirst, E. L., Jones, J. K. W. and Jones, Winifred, J. Chem. Soc. 1880, 1939
20) Lunde, G., Heem, E. and Öy, E., Koll. Zschft. 83, 196, 1938. Kringstad, H. and Lunde, G., ibidem 83, 202, 1938
21) Astbury, W. T., Nature 155, 667, 1945
22) Fischer, F. G. and Dörfel, H., Hoppe Seyler's Zschft. physiol. Chem. 302, 186, 1955
23) Whistler, R. L. and Kirby, K. W., Hoppe Seyler's Zschft. physiol. Chem. 314, 46, 1959
24) Jewtuschenko, W. A., Kolloid. J. (russ.) 16, 255, 340, 1954, translated by Schulzen, Melliand Textilber. 37, 1087, 1956
25) Smith, F. and Montgomery, R., The Chemistry of Plant Gums and Mucilages, Reinhold Publ. Corp. New York, 1959
26) Drummond, W., Hirst, E. L. and Percival, Elisabeth, J. Chem. Soc. (London) p. 1806, 1962
Hirst, E. L., Percival, E. and Wold, J. K., ibidem 1493–9, 1964
Hirst, E. L. and Rees, D. A., ibidem 1182–7, 1965
27) Frey, Eva and Preston, R. D., Nature 196, 130, 1962
28a) Haug, A., Colloque. Intern. Centre Natl. Rech. Sci. Nr. 103, 163–172, 1960
28b) Haug, A. and Larsen, B., Proc. IV. Intern. Seaweed Symp., Pergamon Press 1961, p. 330
29) Schulzen, H., Bot. Mar. III, Suppl. 1962
30) Schulzen, H., Textil Praxis (1): 67–74. 1964
31) Seifen-Öle-Fette-Wachse, 87 (11): 306. 1961
32) Maass, H., Bot. Mar. III, Suppl. 1962
33) Vogt, Pharmaz. Zentralh. 93 (9): 323. 1954
34) Schwandt, D. Apoth. Ztg. 97 (32): 706. 1957
35) Bolliger, R., Schweiz. Apoth. Ztg. 96. 77–81 et 95–99. 1958
36) Ikeda, K. and Moroe, T., Tokyo med. News. 76209. 1959, CA. 53. 22 463a. 1959
37) Tomoda, M. et al., Jap. Pat. 4593. 1959, CA 53. 9584i. 1959
38) Robu, T., Hukuoka Acta Med. 42. 24–38. 1951 (cit. Maass, 1959)
39) Kiyoshi, J., Mem. Fac. Sci. Kynsyn Univ. Ser. C. 1. 109–178. 1952
40) Koch, K., Arneim. Forsch. 3. 520–525. 1953 (cit. Maass, 1959)
41) Hénaff, Produits pharmac. 9. 151–159. 1954 (cit, Maass, 1959)
42) Matsubayashi, H., Igaku Kenkyuu 23. 187–207. 1953, CA 479199. 1953
43) Seifen-Öle-Fette-Wachse (25): 774. 1959
44) Hammond, E. C., J. Gen. Microbiol. 19 (2): 267–270. 1958
45) Thienes, C. H. et al., Arch. intern. pharmacodynam. 111. 167–181. 1957
46) Hoppe, H. A., Seifen-Öle-Fette-Wachse 90 (21): 748–750. 1964
47) Lesser, M. A., Ec. Bot. 4 (4): 317–1950
48) Janistyn, H., Parfümerie und Kosmetik 36 (4): 150–154. 1955
49) Rothemann, Seifen-Öle-Fette-Wachse 85 (21): 638. 1959
50) Moncrieff, R. W., The Retail Chemist sept. 1951
51) Schade, Fette, Seifen, Anstrichmittel 58. 653. 1956
52) Diemair, W. and Weichel, H. H., Deutsche Lebensm. Rundschau 54 (3): 51–55. 1958; 54 (4): 76–79. 1958; 54 (7): 151–155. 1958
53) Giddey, Seifen-Öle-Fette-Wachse 88 (2): 35. 1962
54) Schade, H., Molkerei- und Käserei-Ztg. 5 (17): 2–4. 1954
55) Zabik, M. E. et al., Food Technol. 16 (12): 87–91. 1962, WFA 14 (3): 43. 1963
56) Kulzhinskaya, N. A., USSR Pat. 77. 159. 1949, CA 47 (19) 10172g. 1953
57) Kulzhinskaya, N. A., Vinodelie i Vinogradarstvo SSSR 11 (5): 30–31. 1951, CA 49 (2) 1275c. 1955
58) Deibner, L. and Bénard, P., Ann. inst. natl. recherche agron. Ser. E 7 (i): 103–109. 1958, CA 53. (4) 3591h. 1959
59) Olsen, A., Food. Manuf. 30. 267–270. 285. 1955, WFA 7 (1): 29. 1956
60) Plank, R., Refrig. Eng. 63. 57. 1955
61) Hutschenreuter, R., Jahresh. d. d. Fischwirtschaft 59. 1955
62) Berlin, A., Myasnaya Ind. SSSR. 28 (2): 44–47. 1957. CA 51 (21) 17007b. 1957
63) Slepchenko, I. R. et al., Sbornik Stud. Rabot. Moskov (4): 39–46. 1956
64) Chemische Industrie 7 (6): 321. 1955
65) Gruber, H., Fette, Seifen, Anstrichmittel 58 (12): 1083–1084. 1956
66) McDowel, R. H., Peintures, Pigments, Vernis (1): 35–38. 1956, Ref. Adhäsion (2): 66. 1957
67) Jubilee jr, B. D., Paint Oil, Color and Chemical Rev. (1): 6–7. 1958, Ref Adhäsion (6): 1958
68) Farbe und Lack 64 (5): 258. 1958
69) Institute of Seaweed Research, Ann. Rep. 1959
70) Schulzen, H., Textil-Industrie (22). 1957; ibid. (6, 8, 11, 14). 1958
71) Dalev, D. et al., C. r. Acad. bulg. Sci. 10 (2): 117–119. 1957, Ref. Kolloid Zeitschrift 163 (2). 1959
72) Mitchell, E. K. and McDowell, R. H., US Pat. 2. 902. 446. 1959, Ann. Rep. 1959
73) Chemical and Engineering News 40. 57–58, 61. 1962

74) RICHARDS, M., Nature 176. 559. 1955, Chemiker Ztg. 79. (20). 1955
75) HAMMOND, E. C., J. Gen. Microbiol. 19 (2): 267–270. 1958
76) Institute of Seaweed Research, Ann. Rep. 1961
77) ISHINO, F. and TOGAWA, S., B. P. 863. 768. 1961, Ann. Rep. 1961
78) Institute of Seaweed Research, Ann. Rep. 1957
79) ULRICH, H. M., Third Int. Seaweed Symp. 1958
80) Institute of Seaweed Research, Ann. Rep. 1958
81) Adhäsion (5). 1957
82) KREYLING, R. L., Adhesives Age (7): 35–37. 1959
83) SCHULZEN, H., Melliand Textilberichte 44 (1, 2, 3, 4, 5): 73–75, 179–184, 278–281, 385–387, 500–504, 1963
84) VALLANDIGHAM, V. V. et al., Pap. Ind. 33 (7): 788. 1951
85) VALLANDIGHAM, V. V. et al.; ibid. 33 (10): 1176–1177. 1952
86) SHOGENJI, T. et al., J. Jap. Tappi 8 (1): 568. 1954, 8 (2): 656–657. 1954
87) MAGNUSON, A. L., Tappi 39 (8): 142–143. 1956
88) BONANNO, A. D., Southern Pulp and Paper Mfr. (4): 44. 1962, Ref. Adhäsion (12). 1962
89) BONANNO, A. D., Paper Mill News (15): 8, 21. 1962, Ref. Adhäsion (12). 1962
90) BARTLETT, H. W. and MAGNUSON, A. L., TAPPI 39. 214–220. 1956, CA 50 (12) 9017c. 1956
91) Chemiker Ztg. 78. 166. 1954
92) Soc. francaise des explosifs, CA 10062s. 1957
93) ULRICH, H. M., Coll. Int. du Centre Nat. de la Recherche Scient., Dinard, 1960
94) ULRICH, H. M., Melliand Textilberichte 42 (1, 2): 81–85, 205–209. 1961
95) SCHULZEN, H.; ibid. 41 (7): 861–866. 1960
96) HATAMORI, T., Jap. Pat. 7395. 1955, CA 51. (22). 18637a. 1957
97) KERN, R., Textil-Praxis 13. 589–596. 1958
98) SCHULZEN, H., Melliand Textilberichte 44 (1, 2, 3, 4, 5): 73–75, 179–184, 278–281, 385–387, 500–504. 1963
99) HAUG, A. and LARSEN, B., a) Acta Chem. Scand. 17, 1653, 1963, b) Abstr. V. Intern. Seaweed Symp. 1965, p. 18.

General Literature

BRASCHWITZ, H., Adhäsion (5). 1958; FRIESER, E. P., Adhäsion (10). 1959; HOPPE, H. A., 1958; HOPPE, H. A. and SCHMID, O. J., 1962; HUMM, H. J., 1951; LEWIN, R. A., 1962; MAASS, H. 1959; MAASS, H., 1962; MAASS, H., Planta Medica 9 (1): 80–93. 1961; PAX, F., 1962; SCHMID, O. J., 1962; STEINER, A. B. and McNEELY, W. H., Natural Plant Hydrocolloids Advances in Chemistry, Series 11, 1954; WHISTLER, R. L. and BEMILLER, J. N., 1959; see Index of Literature.
see also *Alaria, Ascophyllum, Durvillea, Ecklonia, Egregia, Fucus, Glossophora, Laminaria, Lessonia, Macrocystis, Nereocystis, Pelagophycus, Sargassum, Turbinaria*

Algulose

Algulose is the term used for the insoluble cellulose-like constituents of algae, which are produced for example as a residue from the production of phycocolloids.

Literature

MAASS, H., Alginsäure und Alginate, Straßenbau, Chemie und Technik Verlagsgesellschaft m. b. H., Heidelberg, 1959

Aonori

Raw Materials and Occurrence

The raw materials of the "Aonori" products are the various *Monostroma* species from the Japanese coasts. These algae have been the subject of several reports. The most important regions of provenance are the bays of Tokyo and Oshoro. The algae cultivated mainly for this purpose, are *M. latissimum* and *M. nitidum* on Ise, Mikawa, and Tokyo Bays. Interesting modern methods of the culture of this alga have been reported by SEGI.

Definition and Countries of Production

"Aonori" is the Japanese name for the product.

Production, for the present time, is limited to Japan, although there are interesting possibilities of using *Monostroma* for edible algal products in other parts of the world.

Production

In addition to the various *Monostroma* species made into "Aonori" *Enteromorpha* species are often also used for this purpose. Production of this commodity has considerably increased in the last years.

Utilization

Fresh *Monostroma* species are eaten as a salad. The "Aonori" products are of rising significance in world nutrition, above all because of their vitamin content.

Literature

SCHMID, O. J. and HOPPE, H. A., Botanica Marina, Suppl., 1962; see also *Monostroma*

Awo-nori

Enteromorpha species occurs on the Japanese coasts. "Awo-nori", also called "Ohashi-nori", is an edible algal product in Japan.

Production

Products like those from *Monostroma* species are made from the partly cultivated Japanese *E.*-species. Used for human nutrition.

Literature: See *Enteromorpha*, *Monostroma*

British Agar

Raw Materials and Occurrence

Chondrus crispus and *Gigartina stellata:* Coasts of Great Britain and Ireland, and in other areas.

Definition

The term "British agar" has been used in Great Britain for a substance which today is known under the name of carrageenan or "Irish Moss Extract".

Literature

MARSHALL, S. M. and al., A Study of Certain British Seaweeds and their Utilization in the Preparation of Agar, His Majesty's Stationary Office, London, 1949; see also *Chondrus, Gigartina*

Carrageenan

Raw Materials and Occurrence

The most important raw materials for the production of carrageenan, the carrageenates, and their derivates are *Chondrus crispus* and various *Gigartina* species, particularly *G. stellata.*

Chondrus crispus occurs on the Atlantic coasts of Europe (Norway, Iceland, Scotland, Ireland, France, Northern Spain, Portugal) and North America (Canada, U.S.A.).

G. stellata, the most important type of this species, occurs in the same areas and often grows together with *Chondrus crispus.* Other *G.* species are also used for the utilization and production of carrageenan, particularly *G. acicularis, G. pistillata,* and *G. radula.*

Definition

Carrageenophytes are those Rhodophyceae which are used as raw material for the production of carrageenan and carrageenates. The trade name of the raw material, especially of *Chondrus crispus* and *Gigartina stellata* is carrageen, the extract is called Irish Moss Extract or carrageenan (= carrageenin).

The definition of carrageenan and the carrageenates is not always homogeneous. Though the investigations have not yet been finished, the following terms are used in the new technical literature:

Carrageenan is the unchanged extract of carrageenophytes. Carrageenates are the defined salts of carragenic acid after respective previous treatment.

Commercial carrageenates are not chemical pure salts of the carrageenic acid, but mixtures. So for example the commercial sodium carrageenate is a salt mixture of sodium, calcium, and potassium. For the various fractions of carrageenan (kappa- and lambda-carrageenan) see page 320, also for the cold soluble and hot soluble portions. This problem is partly very important for the application (cf. the article about textile industry, p. 324).

Countries of Production

Canada has significant raw material resources for the production of carrageenan. Whereas in former times the raw materials were exported to the U.S.A., Canada now has its own industry.

Denmark produces "Danish agar", a furcellaran. The production of carrageenan was taken up a few years ago. Qualities suitable for the food and allied industries are mainly produced, for example, for dairy products, for gelating materials in desserts, jams, fish jellies, as well as all grades which can be used as emulsifyers and stabiliziers in foods.

France is the largest carrageenan producer after the U.S.A. The number of defined types are offered for various applications. A considerable part of the production is exported. France is also the producer of carrageenates which are finding ever more important application in the textile industry.

Great Britain can be considered to be the third greatest carrageenan producer after the U.S.A. and France. A very great range of products is offered for a variety of applications by British traders. In Great Britain the carrageenan was originally described as "British Agar".

The development of the carrageenan industry in Great Britain has repeatedly been reported in the ANNUAL REPORTS OF THE INSTITUTE OF SEAWEED RESEARCH[1]) and by MARSHALL et. al.[2]).

Norway has its own raw materials of *Chondrus crispus* and *Gigartina*. Although alginate production keeps the first place among Norwegian algal industries, Norway also exports carrageenan.

The U.S.A. are the greatest carrageenan producers in the world. In 1958 v. HAKEN estimated production as approximately 1,200 tons, but the figure is much higher today[3]).

Numerous grades, types, and blends are produced for various applications.

Carrageenan production in the U.S.A. has been covered in a number of reports[4]) [5]) [6]) [7]) [8]).

Chemical Composition

"Irish Moss" *(Chondrus crispus* and *Gigartina stellata)* has been used in technology and medicine for a long time; it is the raw material for the manufacture of carrageenan and the carrageenates. Carrageenan is defined as the unchanged extract, while carrageenate is defined as a salt of carrageenic acid formed after treatment of the acid.

The carrageenan which can be extracted mainly from *Chondrus crispus* and *Gigartina* species differs from agar chiefly in its higher sulphate fraction and a higher ash content. Investigations have preferably been based on extracts from *Chondrus crispus*, and it is assumed that they are only slightly different from those of *Gigartina stellata*. Up to 30 to 40 per cent of D-galactose was found in hydrolysates of the polysaccharide. Moreover, small quantities of L-galactose, D-glucose, and 2-keto-D-gluconic acid were also found. The presence of the last mentioned compound is as controversial as that of raffinose and fructose.

It is widely accepted and corroborated today that carrageenan can indeed be separated into two fractions. SMITH and COOK[9]) isolated a branched chain of κ-fractions and a linear chain of λ-fractions. The κ-fraction is separated from the λ-fraction by precipitation with potassium chloride and amounts to 40 per cent of the carrageenan, the balance being the λ-fraction.

A fraction soluble in cold water and one soluble in hot water have also been found (P. HAAS and J. BUCHANAN e. a.[10]) [11])).

According to investigations of SPRINGER and MIDDENDORF[12]) their hot water extract stands for κ-fraction, and the λ-fraction corresponds to their cold water extract.

The objections raised originally by STOLOFF[16]) that the fractions in question were found in bleached commercial products may be regarded as superseded, because later investigations made with authentic basic material led to the same result.

Moreover, it should be noted that, in addition to the main fractions, smaller quantities of other polysaccharides must be present in carrageenan as the presence of monosaccharides in addition to galactose proves.

O'NEILL[13]) has suggested the following structural formula for κ-carrageenan:

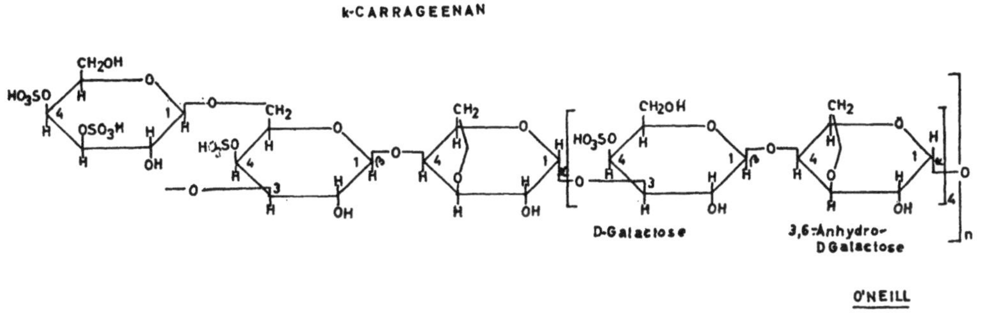

The sulphate content of κ-carrageenan is indicated to be 23–28 per cent and for λ-carrageenan 24–33 per cent. The main cations found were sodium, potassium, calcium, and magnesium. Calcium content is an important factor controlling the physical properties of carrageenan.

The results of their experiments cause K. MORGAN and A. N. O'NEILL to draw the following formula of λ-carrageenan:

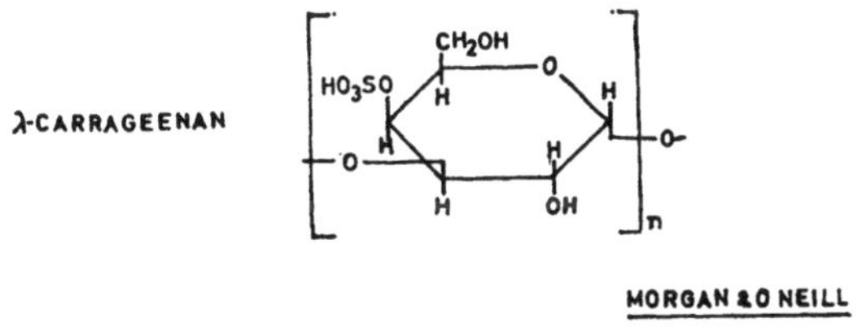

SPRINGER and MIDDENDORF[12]) have made solutions of carrageenic acid by means of cation exchange. The solutions are stable for only a few hours, as they disintegrate through spontaneous hydrolysis.

The molecular weight lies between 100– and 800,000. MASSON & GORING[17]) have reported a molecular weight of 2,500,000[11]) for an extract from *Chondrus crispus*. These authors mention two other possibilities of linkage between each of the 20 units.

Industrial Production

The industrial production methods differ from the methods for agar, because the extraction is much easier. The freeze-out-method, however, cannot be applied, because this polysaccharide immediately absorbs water upon melting. Carrageenan, according to Humm[6]), is produced in the following manner:

The delivered "Irish Moss" is checked for its content of carrageenan. After washing with fresh water to reduce the salt content, two to four parts of algae are boiled with one hundred parts of water. The separation of the carrageenan solution from the non-soluble residual portion of algae is made in a centrifuge. Active carbon and filter material is added under continuous agitation in a tank made of stainless steel. After the first run through the filter press more filtering agent is added and the whole filtered again. The concentration is then increased from 0.8 to 1 per cent to 2 per cent in a vacuum evaporator. Drying of the extract is effected on drum driers. The product, dried to a water content of only 5 per cent, is ground and by blending with other batches adjusted to a uniform final product.

According to the U.S. patents 2620334 and 2620335 the properties of carrageenan can be considerably changed by previous treatment with a salt solution. If the comminuted algae are treated in a bath of one per cent of sodium chloride for one hour at room temperature and then extracted in the usual manner, the viscosity of the final product rises, while its gelating power declines. The previous treatment of algae with a potassium chloride solution, on the other hand, increases the tendency to form gels.

The influence of the manner of extraction on the properties of carrageenan is emphasized by the results of D. A. J. Goring and E. G. Young[18]) who have found considerable differences in the properties of polysaccharide fractions from *Chondrus crispus* extracted at different temperatures.

Enzymatic hydrolysis will be dealt with in another part of this treatise.

The Properties of Carrageenan

Carrageenan forms colloidal solutions or dispersates, with water. The viscosity of commercial brands at 40° C and in a two-per-cent concentration has been reported by Stoloff[19a]) to be 3,000 centipoise for quality I. grade, 500 to 1,000 centipoise for medium qualities and 50 centipoise for products of lower viscosity. The solutions show thixotropy which becomes more strongly emergent with rising concentration of the gelation capabilities. The gels melt at substantially lower temperature, depending on the provenance and the concentration, at 27–41° C compared with 80–100° C in the case of agar. Gelation occurs only in the presence of certain ions, for example potassium, calcium, amonium, rubidium, and caesium. Stoloff has reported the following comparative figures for the gel strength of 1.5-per-cent solutions: Kobe-agar 350 g, high-gelating carrageenan 300 g (in 0.07 m KCl solutions[19b])). A characteristic feature of carrageenan is its capability of causing casein molecules to agglomerate. This is also utilized for practical purposes. It is connected with the ability of carrageenan to form coacervates with proteins (cf. Stoloff[20])).

A detailed study by Houck[21]) on anticoagulant and lipemia-clearing effects of algal polysaccharides shows that carrageenan has not such effect at a concentration of one milligramme per kilo of dog weight, while an extract from *Gigartina acicularis* has a very high anticoagulating effect in blood even exceeding that of heparin, but it also causes pathological changes in the organs.

Qualities

Gel strength and gelation temperature of carrageenan can be varied between wide limits to suit the purpose, for which it is to be used. Apart from its use in aqueous and milky liquids, it also finds application in media containing sugar and alcohol. This fact is of great significance in the food and allied industries.

The products are traded in various grades of viscosity and gelating and non-gelating quality. There are special grades for aqueous media and for milk and dairy products. Mixed products are very often found on the market in addition to pure carrageenan, for example carrageenan with addition of potassium salts, locust bean gum, guar gum or dextrose.

In 1954 GORING has published reports on the effects of hydrogen ion concentration on the viscosity and electrophoretic mobility of solutions of sodium carrageenate[22]).

SMITH et al. have reported some physical studies on carrageenan and carrageenan fractions in 1954[23]).

GORING and YOUNG have published a comparison of fractions in 1955[18]).

GORING has reported the effect on temperature and acidity on the dry storage of sodium carrageenate in 1956[24]).

SCHULZEN has reported on the chemical and physical properties of carrageenate in 1962[25]).

Utilization

Carrageenan in food and allied foods in technology

There are numerous possibilities of application for the carrageenan in this sector and a large number of studies have been published on them, for example by STOLOFF in WHISTLER-BEMILLER, 1959[8]), GLABE, 1956, 1957[26]) [27]), FITZGERALD, 1955[28]).

It finds application, for example, in bakery and condiment products, confectionaries, fillings for assorted chocolates, jelly and cream products, glazings and meringues, milk puddings, jelly, and custard powders, both cooked and instant puddings, cocoa and chocolate milk products, milk shakes.

The interaction between carrageenan and milk proteins has been studied by ANDERSEN[29]) [30]).

Knowledge of the influence of the various fractions on the viscosity of milk was of great significance in the practical application of carrageenan[31]).

Syrup products, whipped cream products, water gels for desserts, ice cream, and ice desserts are prepared with carrageenanes, as well as salads, sauces, soups etc.[32]).

It is used in the production of cheese, particularly in the soft and processed cheese industries and in the meat and fish industries. It is also used for clarification of beer, for fruit juices, and other beverages.

The influence of carrageenan on wheat flour products, for example spaghetti, is reported by GLABE et al.[33]). On the application in food manufacture reported O. CHRISTENSEN[50]).

The producers continuously publish formulations for numerous applications. Dietetic foods are likewise manufactured with the aid of carrageenanes.

Pharmaceutical industry

Carrageenanes are frequently used in the pharmaceutical industries as emulsifiers, for example in cod liver oil emulsions, as granulation and binding agents for example in tablets, elexiers, cough drops, syrups etc.

The use of carrageenan instead of agar in bacteriology has been studied[34]).

In the last years a number of interesting works about the application of carrageenans in the ulcer therapy have been published[25]) [35]) [42]). SCHNEIDER has suggested the importance of *Chondrus* respectively carrageenan for the therapy of diseases of the blood vessels[43]).

Carrageenanes in cosmetics

Carrageenanes are used in the cosmetic industry as a stabilizer, suspending and thickening agent[44]).

Reports have been published by DANO in 1952[45]) on their applications in cosmetics. Also some work appeared on the utilization of carrageen and carrageenan in the soap-producing industries[46]).

The products are often used in the manufacture of skin ointments, creams, lotions, toothpastes, shaving creams, hair tonics, foam stabilizers, sun protection oils etc.

Carrageenanes in various industries

The carrageen, also called "Malermoos", "Painter's Moss" has been used for some time in paint technology.

Carrageenanes are used by the paint manufacturers as stabilizers for pigments. They are also good film-forming agents.

Technical qualities of carrageenanes and carrageenates moreover are used in quite a number of other industries as stabilizers and thickening agents, for example in the industry producing polishing agents, leather industry, paper industry, textile printing.

Textile industry

The hot water extracts from *Chondrus crispus*, *Gigartina* species, and other Rhodophyceae according to SCHULZEN are not suitable for the textile industries.

The production of the cold water extracts for some time gave rise to a number of technical difficulties, principally concerning the clear separation of the cold and hot water fractions. This problem can now be regarded as solved.

At the beginning of manufacturing these phycocolloids difficulties arose in textile industry because of the coagulation of the hot water extractable fraction of the mixed product as initially supplied. These difficulties have been overcome in the modern cold-water extracts, used in textile industry, for printing, dying, finishing, and sizing. Detailed reports have been published[47] [48] on the properties, the behaviour, and the application technique.

The products are characterized by a good sizing property, pigmenting yields, and washability[49].

Finally carrageenanes are used for wax emulsions, graphite suspensions, in latex treatment, and for copying (hectographic) materials etc.

Literature

1) Institute of Seaweed Research, Ann. Rep. 1945–1962
2) Marshall, S. M. et al., A Study of Certain British Seaweeds and their Utilization in the Preparation of Agar, His Majesty's Stationary Office, London, 1949
3) Haken, W. v., Chem. Ind. (5). 1958
4) Idson, B., Chem. Week, July 1956
5) Godston, J., Food Ind. 51. 731. 1949
6) Humm, H. J., The Red Algae of Economic Importance ex Tressler, D. K. and Lemon, J. Mc. W., Marine Products of Commerce, Reinhold Publishing Corporation, New York, 1951
7) Stoloff, L., Natural Plant Hydrocolloids 92. 1954
8) Stoloff, L., Carrageenan ex Whistler, R. L. and BeMiller, J. N., Industrial Gums, Academic Press Inc., New York, 1959
9) Smith, D. B. and Cook, W. H., Arch. Biochem. & Biophys. 45, 232, 1953
10) Haas, P., Biochem. J. 15, 469, 1921
11) Buchanan, J. et al., J. Chem. Soc. (London) 1943, 51
12) Springer, R. and Middendorf, L., Arch. Pharmazie 294/66, 693, 1961
13) O'Neill, A. N., J. Amer. Chem. Soc. 77, 6324, 1955
14) Smith, D. B., O'Neill, A. N. and Perlin, A. S., Can. J. Chem. 33, 1352, 1955
15) Morgan, K. and O'Neill, A. N., Can. J. Chem. 37, 1201, 1959
16) Stoloff, L., Natural Plant Hydrocolloids pag. 93, Am. Chem. Soc. Adv. Chem. Series Nr. 11, Washington 1957
17) Musson, C. R. and Goring, D. A. J., Can. J. Chem. 33, 895, 1955
18) Goring, D. A. J. and Young, E. G., Can. J. Chem. 33, 480, 1955
19) a) Stoloff, L., in Industrial Gums, Academic Press, New York-London, p. 106, 1959; b) Stoloff, L., ibidem pag. 109
20) Stoloff, L., Polysaccharids in biology, IV. Conference, J. Macy jr. Found, New York 1959, pag. 301
21) Houck, J. C. et al., Proc. Soc. Exp. Biol. Med. 96, 528, 1957
22) Goring, D. A. J., J. Colloid Sci. 9 (2): 141–148. 1954
23) Smith, D. et al., Arch. of Biochem. and Biophys. 53 (1): 192–203. 1954
24) Goring, D. A. J., Canad. J. of Technol. 34. 39–41. 1956
25) Schulzen, H., Bot. Mar. Suppl. 1962
26) Glabe, E. F., Confectioners J. 82 (975): 14, 16, 46–47. 1956
27) Glabe, E. F. et al., Food Engineering 29 (1): 65. 1957
28) Fitzgerald, B. W. and Skauen, D. M., J. Am. Pharm. Ass. Sci. Ed. 44 (6): 358–360. 1955
29) Andersen, G., Fourth Int. Seaweed Symp., 1961
30) Andersen, G., Milchwissenschaft 17. 75–79. 1962
31) Smith, D. B., Can. J. Technol. 31 (10/12): 209. 1953
32) Stoloff, L. S., US Pat. 2.864.706. CA 8476b. 1959
33) Glabe, E. F. et al., Cereal Sci. Today 2. 159–162. 1957, CA 15026d, 1957
34) Walker, A. W. and Day, A. A., J. Bact. 45. 20. 1943
35) Institute of Seaweed Research, Ann. Rep. 1959
36) Anderson, W. and Watt, J., Inhibition of Peptic Activity. Protection against Histamine Ulceration in the Guinea-Pig and combination with gastric mucin by an algae Polyanion. J. Pharm. London 11 (318) 1959, Ann. Rep. JSR. 1959
37) Anderson, W. and Watt, J., The protective action of an algal Polyanion against experimentally produced Peptic Ulceration in the Guinea-Pig, J. Physiol. 147 (52): 1959, Ann. Rep. I. S. R., 1959
38) Ershoff, B. H. and Wells, A. F., Effects of Gum Guar, Locust Bean Gum and Carrageenan on Liver cholesterol of Cholesterol. Fed Rats. Proceedings. Soc. of Exp. Biology and Medicine, 110, 580, 582, 1962, WFA 15 (2) 45, 1964
39) Piper, D. W., Effect of a Sulphated Polysaccharide on Peptic Digestion, a study in vitro of Carrageenan. Gastroenterology 40. 638–643. 1961. WFA 15 (2) 45. 1964
40) Maita, Y. et al., Inhibiting Action of Carrageenan on crystalline Swine Pepsin. Bull. Fac.

of Fish, Hokkaido Univers. 13. 1–7. 1962. WFA 15 (1) 41, 1964
41) Vaughan, O. W. et al., Effects of Carrageenan on the Peptic Hydrolysis of various Proteins. J. Agric. Food. Chem. 10. 517–519. 1962. WFA. 15 (1) 41. 1964
42) Hawkins, W. W. and Leonard, V. C., Antithrombic Activity of Carrageenan in Human Blood. Can. J. Biochem. Physiol. 41 (5): 1325–1327. 1963. WFA 15 (1): 41. 1964
43) Schneider, G., Arzneipflanzen von aktueller medizinischer Bedeutung. D. Apoth. Ztg. 102 (47): 1529–1533. 1962
44) Becker, Ch., Am. Profess. Pharmacist 20. 939–944. 987. 1954, CA 1279a. 1955
45) Dano, J., Ind. Parfum. France, 7 (8): 227–228. 1952, Chem. Zentralbl. (40): 7749. 1952
46) Smith, P. J., Am. Perfumer Aromat. 67 (6): 67. 1956
47) Schulzen, H., Melliand 41 (5): 596–600. 1960; ibid. 41 (8): 992–995. 1960
48) Schulzen, H., Bot. Mar. Suppl. 1962
49) Schulzen, H., Melliand 41 (7): 861–866. 1960
50) Christensen, O., Food Manuf. 39, 44–7. 1964

General Literature

Black, W. A. P., Ann. Rep. Chem. Soc. Vol. 50. 1953; Hegnauer, R., 1962; Hoppe, H. A., 1958; Hoppe, H. A. and Schmid, O. J., 1962; Humm, H. J., 1951; Idson, B., Chem. Week, July 1956; Lewin, R. A., 1962; Pax, F., 1962; Sanford, F. B., Fishery Leaflet 469, 1958; Schmid, O. J., 1962; Schmitt, M., Coll. Int. du Centre Nat. de la Recherche Scient., Dinard, 1960; Stoloff, L., Carrageenan ex Whistler R. L. and BeMiller, J. N., 1959; see Index of Literature; see also *Chondrus*, *Gigartina*

Ceylon Moss

Raw Materials and Occurrence

Gracilaria lichenoides and *G. confervoides* (= *G. verrucosa*). Habitat: the coast of Ceylon.

Definition

"Ceylon Moss" is a term used for washed, bleached, and dried *G.* species. The product is also known as "Ceylon Agar". Produced in Ceylon, India.

Utilization

Gelating material for food and allied industries. As a pharmaceutical product used in the treatment of colds. The product was formerly exported to Europe and the U.S.A.

Literature

Durairatnam, M. and Medcof, J. C., The Ceylon Trade J. 19. 1–6. 1954
Durairatnam, M., Ceylon Dep. Fish., Progr. Rep. Biol. and Technol., (2): 54. 1956
Hoffmann, C., Die praktische Bedeutung der Meeresalgen, Kieler Meeresforschungen III (1), 1938
Hoppe, H. A., Drogenkunde, Cram, de Gruyter & Co., Hamburg, 1958
Hoppe, H. A. and Schmid, O. J., Bot. Mar. Supplement 1962
Schmid, O. J.; ibid.
Whistler, R. L. and BeMiller, J. N., Industrial Gums, Academic Press Inc., New York, 1959
see also *Gracilaria*, *Agar*

Chinese Moss

Raw Materials and Occurrence

Gracilaria confervoides: The alga occurs inter alia on numerous East-Asian coasts, particularly also on the coast of China.

Definition

"Chinese Moss" is the term used for a washed, bleached, and dried alga.
It is produced in China and will be used for gelation and as a thickening agent.

Literature: See *Gracilaria, Agar*

Cochayugo

Raw Materials, Occurence and Definition

Durvillea antartica. Habitat: the coasts of Chile. "Cochayugo" is a washed, dried, and afterwards roasted product made from the alga.

Utilization

Considerable quantities of this product are used for human food locally, either in the form of a vegetable, or as an admixture to soup, or as a "salted cabbage".

Literature: See *Durvillea*

Dulse

Raw Materials and Occurrence

"Dulse" is but one of the numerous popular names for various species of *Rhodymenia*, particularly *Rh. palmata. Rh. palmata* is widespread on European coasts and on the Atlantic and Pacific coasts of North America. Numerous *Rh.* species also occur in other coastal regions.

Detailed investigations of these species seem to be of great significance in view of their valuable constituents and their suitability for human and animal nutrition.

Definition

The term "Dulse" has come into use, above all, in Great Britain, Canada, and the U.S.A. for edible products of *Rhodymenia*.

Countries of Production

A "Dulse"-industry exists in Canada, where the alga occurs in considerable populations in the Bay of Fundy. *Rh. palmata* is used in many other countries for human consumption and as an addition to fodder.

Utilization

The algae are used in the fresh state as a salad or vegetable. However; they are also sold dried for human consumption after due preparation as a combination food with meat or fish dishes. In many areas, for example in Canada and in the U.S.A., preparations of *Rhodymenia* are regarded as a special delicacy. The interest in these dishes has risen in the last years. The dulsan obtained from *Rhodymenia* is mentioned on p. 328.

Literature

HOPPE, H. A. and SCHMID, O. J., Bot. Mar., Supplement, 1962
SCHACHAT, R. E. and GLICKSMAN, M., Some Lesser Known Seaweed Extracts ex Industrial Gums, Academic Press Inc., New York, 1959
SCHMID, O. J. and HOPPE, H. A., Bot. Mar., Supplement 1962
see also *Rhodymenia*

Dulsan

Raw Material, Occurrence and Definition

Rhodymenia palmata and other *Rh.* species are widespread on many coasts. Dulsan is the term used for polysaccharides isolated from *Rh. palmata.*

Utilization

During the Second World War in Great Britain experiments were made to isolate the polysaccharides from *Rh. palmata* in order to find a substitute for agar. But dulsan was found not to have any gel-forming properties, and therefore the product in the past has had no commercial significance. It appears more expedient to use the whole alga with all its constituents for food.

Literature

HOPPE, H. A. and SCHMID, O. J., Bot. Mar., Supplement 1962
SCHACHAT, R. E. and GLICKSMAN, M., Some Lesser Known Seaweed Extracts ex Industrial Gums, Academic Press Inc., New York, 1959
SCHMID, O. J., Bot. Mar., Supplement 1962
see also *Rhodymenia*

Eucheuman

Raw Materials and Occurrences

Eucheuma species occur above all in the South-East Asian coastal areas and around the islands, and at the East-African coast near Zanzibar. The most important species occurs in Indonesia, *E. muricatum*, which is referred in trade as "Singapore Weed".

E. serra and *E. cottonii* occur in East Africa and are sold as "Zanzibar Weed".

E. isiforme occurs on the coast of Florida and is likewise utilized commercially.

Definition

Eucheuman is the extract from the two *E.* groups, which are regarded as "Singapore Weed" and "Zanzibar Weed". These are agaroids. The old Malayan name for the *E.* species, it is true, was agar-agar, but according to modern scientific aspects the phycocolloid from the *E.* species does not longer come up to the requirements relating to an agar.

Countries of Production

Pure eucheuman is not yet produced in a large scale. The raw materials, often are used as an additional agar raw material (Japan) or utilized in other countries for the production of carrageenan.

Structure

Extracts from *E. muricatum* have been investigated by NAKAMURA[1]). After cold water extraction he was able to isolate 4 fractions, distinct in their total sulphur, total sugar, and aldose content. All four reacted negatively for ketose, negatively for pentose and positively for hexuronic acid. He established a ratio of glucuronic acid to anhydrogalactose to galactose of 1 : 3 : 12. He found properties of *Gloiopeltis* extracts to be very similar.

In *E. muricatum* the same author[2]) found the following constituents: galactose, 3.6-anhydrogalactose, xylose and galactose sulphate, with xylose, anhydrogalactose, and galactose in a ratio 1 : 3.5 : 12.5. He established that glucuronic acid amounted to 3.27 per cent of the total sugar[3]). Sulphur content in *E. cottonii*, *E. serra*, *E. muricatum* amounts to 6.1, 9.5 and 9.0 per cent[4]). The chemical composition of the polysaccharide is not yet fully clarified. SCHACHAT & GLICKSMAN have reported in detail on extracts from *Eucheuma* species[5]). KATSURA and coworkers described properties of the purified hotwater extract of *Eucheuma spinosum*[6]).

Qualities

While the eucheuman made from "Singapore Weed" if added at one-percent in water, forms a viscous solution, the eucheuman from the "Zanzibar Weed", forms soft elastic gels with thixotropic character.

Utilization

The range of utilization of eucheuman is between that of agar and carrageenan. The applications are mainly watery and milk-containing media in food technology, pharmacy, and cosmetic industry.

Literature

1) NAKAMURA, T., Bull. Japan. Soc. Sci. Fisheries 20, 501, 1954, Chem. Abstr. 49, 7878a, 1955 Bull. Japan. Soc. Sci. Fisheries 21, 1053, 1057, 1956, Chem. Abstr. 50, 12516, 1956
2) NAKAMURA, T., Bull. Japan. Soc. Sci. Fisheries 23, 647, 652, 1957–58, Chem. Abstr. 53, 7329e, 1959
3) NAKAMURA, T., Nippon Suisangaku Kaishi 24, 285, 1958–59, Chem. Abstr. 53, 11533b, 1959
4) SCHMID, O. J., Botanica Mar. 3, Suppl. 69, 1962
5) SCHACHAT, R. E. and M. GLICKSMAN, Industrial Gums (Whistler & BeMiller) p. 149, Academic Press 1959
6) KATSURA, K. et alii Kogyo Kagaku Zasshi 63, 332, 1960; CA 61, 12244e, 1964

General Literature
HOPPE, H. A. and SCHMID, O. J., 1962; SCHMID, O. J., 1962; WHISTLER, R. L. and BEMILLER, J. N., 1962; see Index of Literature; see also *Eucheuma*

Fucoidan

Raw Materials and Occurrences

Fucoidan probably occurs in nearly all brown algae.

The substance often occurs on the surface and on the cut faces of freshly harvested brown algae and forms mucous coatings or filaments of transparent mucous liquid.

Littoral brown algae, for example *Ascophyllum nodosum* or floating algae as *Macrocystis pyrifera* mostly contain more fucoidan than sub-littoral algae. The content declines very rapidly when the algae die. Fucoidan apparently is a substance used by the algae to protect them against the effects of drying out, when exposed. The fact that fucoidan is very high hygroscopic and that one can find it in higher concentrations in littoral algae, which often are not covered by water, than in marine algae occurring in greater depths corroborate this.

Fucus spiralis contains 9–11 per cent, *Ascophyllum nodosum* 6–8 per cent of fucoidan. *Pelvetia canaliculata* contains up to 20 per cent of fucoidan, calculated on the basis of the dry raw material[1]).

YOUNG has reported the maximum content of fucoidan in *Laminarias* and Fucaceae in the upper littoral to be 20 per cent, the lower limit to be 5 per cent in the sub-littoral. The content varies with the season. The upper limit is reached in the earlier winter months.

In British Columbia *Laminarias* have been found to have a fucoidan content of 4–7 per cent, with the highest content in March[2]).

According to O'COLLA *Laminarias*, *Fucus* species, and for example *Chordaria*, contain 5–20 per cent of fucoidan with a maximum content in autumn[3]).

The seasonal fluctuations in fucoidan content of *Ecklonia radiata* have also been investigated[4]).

Ascophyllum nodosum, according to O'COLLA contains a substance similar to fucoidan. It was isolated from the reproductive branches[3]).

Definition

Fucoidan (= Fucoidin) is an "intercellular-mucilage", a membrane-type of mucous and not a cell wall constituent (MCNEELY[5])). Fucoidan is a fucan and, besides fucose, also contains other sugars. O'COLLA calls it a polysaccharide sulphuric ester. The substance has only been studied in detail in recent years[3]).

Production

Fucoidan is not generally used as a commercial product, because it is very inconsistant in quality.

The previously known methods of production and purification treatments with acids, alkali and the drying process as well as the treatment of the raw material destroy most of the colloidal properties of the product obtained. Fucoidan, at any rate, would have to be produced from the fresh raw material.

In Great Britain fucoidan was made from *Fucus vesiculosus* and *Pelvetia canaliculata* by means of a chloroform extraction.

Fucoidan was isolated for the first time by KYLIN[6]) from *Laminarias* and *Fucus*. It is contained in brown algae at a rate of 2 to 9 per cent. Its constitution has been studied by various scientists. CONCHIE & PERCIVAL[7]) report the following composition for a highly purified fucoidan: 38.3 per cent sulphate, 56.7 per cent L-fucose, traces of galactose, xylose and uronic acid. According to their investigations and the results obtained by O'NEILL[8]) fucoidan is a polymer of anhydro-L-fucopyranose with a sulphate residue of C_4 and mainly 1–2 bonds. The composition of fucoidan, apparently, varies greatly. PERCIVAL & ROSS[9]) found in the case of *Himanthalia lorea* 57 per cent fucose, 4 per cent galactose, 15 per cent xylose, and 3 per cent uronic acid. *M.* QUILLET[10]) has reported for fucoidan of *Pelvetia canaliculata* the following data: 35 per cent fucose, 3 per cent glucose, 22 per cent galactose, 6 per cent arabinose, and small quantities of xylose, mannose, and uronic acid. The latter substances have been investigated least. A very thorough analytic treatment of various monosaccharides obtained after an acid hydrolysis of fucoidanes has also been made by *M.* QUILLET[10]). The share of fucose varies with eleven different brown algae between 31 and 72 per cent, the galactose between 5 and 31 per cent, of mannose between 4 and 20 per cent, of xylose between 3 and 29 per cent, of arabinose between 0 and 25 per cent. The nonesterified monosaccharides of the fucoidans of these 11 algae have accounted for between 3.3 and 7.1 per cent, only in *Pelvetia canaliculata* is the figure 15.4 per cent, outside these limits. From the results of the investigations QUILLET[10]) has drawn the following conclusions: there is not only one type of fucoidan, but whole families of fucoidans. L-fucose is a characteristic constituent of the polysaccharides of brown algae, just as L-rhamnose is of green and blue-green algae, and 3.6-anhydrogalactose of red algae. SPRINGER & coworkers[11b]) regularly found in their fucoidan fractions two components with the reaction of ω-desoxy sugars. O'COLLA[12]) on the basis of work of various authors, pointed out that in addition to the 1 : 2' bonds there must also be 1 : 3' bonds and

1 :4' bonds of the fucose units, and that the sulphate group can also be attached to the C_2 and C_3.

Fucoidan is characterized by a tendency to produce a filamentous, stringy type of mucilage[13]). The viscosity of fucoidan solutions is highly susceptible to ageing, diluted acids and bases. Cane sugar and glycerin have a conserving and preserving influence upon its mucilaginous properties[8]). By the exchange of acid residues with alginate fucoidan maintains the inner milieu of the algae. In contrast to BLACKS[14]) view of fucoidan as a substance specifically effective against the drying out of algae, QUILLET[16]) believes the swelling properties of fucoidan to be of secondary significance physiologically.

Fucoidan can be split up by bacterial enzymes[15]). GALLI & GIESE[16]) have found an enzyme of the same properties in a snail.

Particular attention has been directed to fucoidan because of its antilipaemic and coagulation inhibiting properties[11a,c]). It is interesting that these properties are not connected with the maintenance of unchanged viscosity.

The assumption that fucoidan constitutes the fertilization inhibitor in the case of sea urchins *(echinoidea)* was found not to be true according to the investigations of ULLA ESPING[17]).

SCHWEIGER[18]) has found L-fucose and D-galactose in a ratio of 18 : 1 in fucoidan obtained from *Macrocystis pyrifera*, and regards the small quantities of xylose and mannose found as contaminations. He considers it unlikely that fucoidan is a pure fucan sulphate.

Fucobiose is a disaccharide of fucoidan[21]).

Utilization

If there should be success in producing fucoidan in a stable form, it could obtain great economic significance due to its very high viscosity at very low concentration.

Fucoidan is the base material for fucose production.

In Great Britain experiments have been made to use fucoidan in the pharmaceutical industry. In therapy it can be used as a blood anticoagulant. It probably has a stronger antithrombotic effect than heparin[19]) [20]).

The properties of a purified fucoidan have been reported and compared with heparin[20]).

Literature

1) BLACK, W. A. P., Ann. Rep. Chem. Soc. Vol. 50. 1953
2) YOUNG, E. G., Carbohydrate Accumulation by lower Plants (Rhodophyta) ex W. Ruhland, Handbuch der Pflanzenanalyse, Springer Verlag, 1958
3) O'COLLA and LEWIN, R. A., Physiology and Biochemistry of Algae, Academic Press Inc., New York, 1962
4) STEWART, C. M. et al., Nature (London) 192. 1208. 1961
5) MCNEELY, W. H., Fucoidan ex Whistler, R. L. and BeMiller, J. N., Industrial Gums, Academic Press Inc., New York, 1959
6) KYLIN, H., Zschft. Physiol. Chemie 83, 171, 1913
7) CONCHIE, J. and PERCIVAL, E. G. V., J. Chem Soc. p. 827, 1950
8) O'NEILL, A. N., J. Am. Chem. Soc. 76, 5074, 1954
9) PERCIVAL, E. G. V. and ROSS, A. G., J. Chem. Soc. p. 717, 1949
10) QUILLET, M., Coll. Intern. Centre Nat. Rech. Scient. Nr. 103, p. 145, 1960/61
11) SPRINGER, G. F. et al., a) Proc. Soc. Exper. Biol. Med. 94, 404, 1957; b) Polysaccharids in Biology, 4. Conf. p. 284, J. Macy Jr. Found. New York, 1959; c) Naturwissenschaften 44, 265, 1957
12) O'COLLA, P. S., Physiology and Biochemistry of Algae, p. 350, Academic Press 1962

13) McNeely, W. H., Industrial Gums, p. 120, Academic Press 1959
14) Black, W. A. P., J. Sci. Food Agr. 5, 445, 1954
15) Japhe, W. and Morgan, M., Nature 183, 761, 1959
16) Galli, D. R. and Giese, A. C., J. Exptl. Zool. 140, 415, 1959
17) Esping, Ulla, Arkiv Kemi 11, 107, 1957
18) Schweiger, R. G., J. Org. Chem. 27, 4267, 4270, 1962
19) Springer, G. F. et al., Proc. Soc. exp. Biol. Med. N. Y. 94. 404. 1957
20) Bernardi, G. and Springer, G. F., J. biol. Chem. 237. 75. 1962
21) Côté, R. H., J. chem. Soc. 2248. 1959

General Literature

Institute of Seaweed Research, Ann. Rep. 1955, 1957–1959, 1961, 1962; Hegnauer, R., 1962; Hoppe, H. A. and Schmid, O. J., 1962; Lewin, R. A., 1962; Mautner, H. G., Ec. Bot. 8 (2): 174. 1954; Pax, F., 1962; Schmid, O. J., 1962; Tseng, C. K., Science 101 (2633): 597–602. 1945; Whistler, R. L. and BeMiller, J. N., 1962; see Index of Literature; see also *Ascophyllum, Chordaria, Ecklonia, Fucus, Laminaria, Macrocystis, Pelvetia*

Fucosan

Raw Materials and Occurrences

Fucosan is contained in the physodes or fucosan bladders of the Phaeophyceae, for example *Ascophyllum nodosum* and *Sargassum* species. These brown algae have a positive tannin reaction. The content in fucosan, for example, in the case of *Ascophyllum nodosum* is approximately 9 per cent of the dried raw material. Fucosan is a tanning substance.

Structure

Haug and Larsen have studied this substance in detail[1]) and have found that the reducting qualities of extracts from *Ascophyllum nodosum* ascribed to it, amount to 0.2–2.3 m. equ/g dry matter. It is likely to be contained in the physodes. It has a phenolic character and can be removed by Merck's skin powder. The fucosan has the property of a tanning substance. The authors, therefore, describe it as a "tannin". They estimate its content ration to be 9 g per 100 g of alga. Moreover, they ascribe the brown discolouring of algal extracts, algae, and algal meals to this substance.

Likewise Ogino and Taki[2]) describe a substance from *Sargassum ringgoldianum* as tannin which, according to their investigation, belongs to the catechol series in the classification of Freudenberg.

Utilization

Aqueous extracts, for example from *Sargassum ringgoldianum* are used for tanning in areas where the alga occurs in sufficient quantities. It appears quite possible that other algae, an equivalent content of tanning substances, could also be utilized for the very same purpose.

Literature

1) Haug, A. and Larsen, B., III. Intern. Seaweed Sympos., Galway, p. 70 (1958), Acta chemica Scand. 12, 650, 1958, Nature 181, 1224, 1958
2) Ogino, Ch. and Taki, Y., J. Tokyo Univ. Fisheries **43**, 1–5, 1957, Chem. Abstr. 52, 18668a, 1958

General Literature
HEGNAUER, R., 1962; Institute of Seaweed Research, Ann. Rep. 1958; LEWIN, R. A., 1962; MAUTNER, H. G., Ec. Bot. 8 (2): 174. 1954; see Index of Literature; see also *Ascophyllum, Sargassum*

Fucosterol

Raw Material and Occurrences

Fucosterol is a characteristic sterol of a number of Phaeophyceae. It also occurs in Chlorophyceae and Rhodophycae, for example in *Cladophora sauteri,* and in *Rhodymenia palmata.*

Definition and Structure

Fucosterol in a pure form was first isolated in 1934. It was named after *Fucus* from which it was made. Fucosterol has the numerical formula $C_{29}H_{48}O$ Δ^5, is unsaturated in the side chain and has a mp of 124° C and an optical rotation of $[a]_D$ –38°. See also page 380.

Production

Fucosterol is extracted with organic solvents from Phaeophyceae. It is a white, non-odorous, crystalline substance, insoluble in water, but soluble in organic solvents.

Utilization

Fucosterol may grow in significance as a base material for the manufacture of sexual hormones and for steroid synthesis.

Literature

HEGNAUER, R., Chemotaxonomie der Pflanzen, Birkhäuser Verlag, Stuttgart, 1962
MAUTNER, H. G., Ec. Bot. 8 (2): 174. 1954
See also *Cladophora, Fucus, Rhodymenia*

Funoran

Raw Materials and Occurrences

The most important funoran raw materials are *Gloiopeltis* species that occur above all, on the coasts of Japan and China, in particular *G. tenax*, *G. furcata*, and *G. complanata.*

Ahnfeltia concinna, Dumontia incrassata, Grateloupia species, *Gymnogongrus pinnulata,* and *Iridea* species have also been used for funoran production in East Asia.

Definition

In contrast to other phycocolloids funoran is not an extract, but rather it is the total alga which is made soluble by special treatment, the algae having previously been more or less bleached.

The product has been used in Japan and China for many centuries.

Countries of Production

The production of funoran is mainly confined to East Asia, the most important areas being Japan, near Osaka, where the product is referred to as "funori", and China (Chekiang) where it is referred to as "hailo". China, however, imports additional funoran from Japan.

Qualities

Trade makes a distinction between three qualities: "very light", "light-coloured" and a "darker" grade for technical purposes.

Structure

In 1920 TAKAHASHI[1]) found galactose in funoran and a sulphate fraction. This finding has since been confirmed for several times. ARAKI[2]) reported sulphate content of 18.5 per cent, and identified agarobiose as a major residue of the polysaccharide. The purified polysaccharide contains 3.6 anhydrogalactose, D-galactose, and L-galactose in a relationship of 8 : 12 : 1[3]). Aqueous solutions of the polysaccharides have an $[\alpha]^{18}_{D}$ of –20.6°. Funoran differs from other red algal extracts in not forming gelations[4]). The extract can be purified by precipitation with alcohol[5]). A comprehensive description of the properties of funoran has been given by SCACHAT & GLICKSMAN[6]).

Utilization

Funoran is used mainly because of its excellent adhesive and sizing properties, for which it is in widespread use. In cosmetic industries it is used in hair curling and hair dyeing preparations.

In the pharmaceutical industry funoran has been proposed as a blood anticoagulant.

In East Asia funoran is used for artistic purposes, for example for painting on very thin paper.

It is an important paper and textile sizing agent, on account of its rapid and good solubility. In the textile industry it is used as glazing agent and stiffer for silks, particularly of the widely popular black silk.

In the paint and dye industry it is used for the production of water colours and for porcelain and China painting.

Funoran is often used instead of starch, in the countries where it is produced.

It is also used to coat clay and cement.

Literature

1) Takahashi, E., J. Coll. Agr. Hokkaido Imp. Univ. 8, 183, 1920
2) Araki, C., Hirase, S. and Ito, T., Abstr. III. Intern. Seaweed Sympos. p. 81, 1958, Hirase, S., et al., Bull. Chem. Soc. Jap. 31, 428, 1958, Chem. Abstr. 53, 6361 a, 1959
3) Hirase, S. et al., Bull. Chem. Soc. Jap. 29, 985, 1956, Chem. Abstr. 52, 8972 g, 1958
4) Araki, C., Proc. IV. Intern. Congr. Bioch. 1, 28, 1959
5) Mita, K., Nippon Suisan Gakkaishi 22, 558, 1957, Chem. Abstr. 52, 7444, 1958
6) Schachat, R. E. and Glicksman, M., Industrial Gums by Whistler, R. L. and BeMiller, J. N., p. 170, 1959, Academic Press

General Literature
Chapman, V. J., 1950; Chiu, B. T., Second Int. Seaweed Symp., 1955; Hoppe, H. A. and Schmid, O. J., 1962; Humm, H. J., 1951; Sanford, F. B., 1958; Smith, F. and Montgomery, R., 1959; Tseng, C. K., Science 101 (2633): 597–602. 1945; Whistler, R. L. and BeMiller, J. N., 1959; see Index of Literature; see also *Ahnfeltia, Dumontia, Gloiopeltis, Grateloupia, Gymnogongrus, Iridea*

Furcellaran

Raw Materials and Occurrence

The raw material basis of furcellaran, also known as "Danish agar" is *Furcellaria fastigiata,* which is harvested in large quantities, above all, in the Kattegat.

The main area is located in the central part of the Kattegat off the northern coast of the peninsula of Djursland[1]).

F. fastigiata grows in Danish waters, mostly in depths of between 20 to 30 fathoms (about 35 to 55 metres).

It is dragged by trawlers by means of drag nets from the bottom of the sea, where it often accumulates in large quantity due to the water current. The 100 to 200 ton trawlers take up to 100 tons of fresh algae per day. Harvesting can be carried out the year round, but the colder months are preferable, because the algae are exposed to the dangers of decomposition less than during the warm season.

The alga also occurs on the Baltic coast of Schleswig-Holstein where increasing populations could be found in several areas after the Second World War (Hoffmann, Hoppe, Schmid). It occurs also on the Swedish coast (Levring).

Czapke (1960)[2]) has reported on "Investigations and the Possibility of Obtaining Furcellaran from Seaweeds in the Polish Coastal Waters", and he has delivered a paper on the product, which he still described as "agar-agar", at the Fourth International Seaweed Symposium, 1961[3]).

In 1950 Förster has reported on the product from *F. fastigiata* from the Baltic[4]).

Painter[5]) has reported the occurrence of *Furcellaria fastigiata* in the Baltic Sea[5]) and also its occurrence in commercial quantities in the Gulf of St. Lawrence, Canada. He has investigated the polysaccharides of the alga.

F. fastigiata is often harvested together with carrageenan on the Irish coast. The alga was formerly sorted out as "black carrageen". Today it is exported to Denmark.

40000 tons of *Furcellaria fastigiata* are available pro anno on the Russian Baltic Sea coast for the manufacture of furcellaran[15]).

Definition

Furcellaran = "Danish agar" is a galactoglycane sulphate isolated from *Furcellaria fastigiata*. The product is also described as "furgin"[6]).

Countries of Production

The first attempts at the manufacture of agar products were made in Denmark during the First World War, however, without any practical results. The investigations and experiments were taken up again in 1940.

Since then a number of processes have been developed and several large factories established.

LUND and BJERRE-PETERSEN have reported in 1952 at the FIRST INTERNATIONAL SEAWEED SYMPOSIUM in Edinburgh on developments between 1946 and 1952[7]). They reported again at Biarritz in 1961 showing the prospects of this algal industry to be very promising.

Structure and Properties

Furcelleran belongs to the group of the sulphate-containing galactanes. It consists mainly of D-galactose units, arranged in 1:3 links. Moreover 3.6 anhydrogalactose, glucose, xylose, and uronic acid have been found on hydrolysis of the polysaccharide[8]) [9]). The sulphate fraction according to Ross amounts to 10.8 per cent[10]). 5.1 per cent of sulphur was found in the alga itself and 4 per cent of sulphur in the polysaccharide[11]). According to PAINTER[5]) there is a very high disaccharide fraction consisting of galactose and 3.6 anhydrogalactose. This configuration is also found in agar. GRYUNER and SUINE[12]) have compared the colloid chemical properties of furcellaran with those of phyllophoran and *Ahnfeltia*-agar. They are better than the properties of the former and worse than those of the latter.

A detailed description of the physical and chemical properties has been given by SCHACHAT and GLICKSMAN[13]).

G. YOUNG[14]) refers to the similarity in chemical composition of furcellaran and a polysaccharide from *Hypnea spicifera.*

Qualities

Furcellaran is marketed in a number of grades for various uses and mostly known under proprietory names. A distinction is made between bleached and unbleached quality for aqueous and milky liquids. The greatest portion (over 90 per cent) is accounted for by bleached qualities.

Utilization

The main application of furcellaran is the production of milk pudding powders, it is used in condiments, confectionary products, and ice specialities, often together with other gums such as guar or locust bean gum and in dietetic preparations, in particular also in baby and diabetic food.

Furcellaran can also be used in the food canning industry, as it withstands heating in autoclaves, for example fish and meat cans, marmelade, jams, fruit, and other jellies can be prepared with furcellaran. Manufacturers are giving details and a number of formulations for the purposes.

In the cosmetic industry it is used for example in the manufacture of tooth paste.

Literature

1) Lund, S. and Bjerre-Petersen, E., Fourth Int. Seaweed Symp., 1961
2) Czapke, K., Prace Morsk. Inst. Ryback. Gdyni 11/B: 95–111. 1960
3) Czapke, K., Fourth Int. Seaweed Symp., 1961
4) Förster, Pharmaz. Zentralh. 89 (12): 409. 1950
5) Painter, T. J., Canad., J. of Chem. 38. 112. 1960
6) Christiansen, E., Danish Agar ex Whistler, R. L. and BeMiller, J. N., Industrial Gums, Academic Press Inc., New York, 1959
7) Lund, S. and Bjerre-Petersen, E., First Int. Seaweed Symp., 1952
8) Glancy, M. J. et al., III. Intern. Seaweed Symp. 1958, Abstracts pag. 78
9) Schachat, R. E. and Glicksman, M., Econ. Botany 13, 368, 1959
10) Ross, A. G., J. Sci. Food Agric. 4, 333, 1953
11) Schmid, O. J., Bot. Mar. 3, Suppl. 69, 1962
12) Gryuner, V. S. and Suine, O. E., CA. 57, 11650d, 1962
13) Schachat, R. E. and Glicksman, M., Industrial Gums by Whistler and BeMiller, p. 155 ff., Academic Press 1959
14) Young, G., Colloqu. intern. Centr. Nat. Rech. Sci. p. 176, 1961
15) Kirejew, M. S., Priroda, p. 100, 1965

General Literature

Hoppe, H. A. and Schmid, O. J., 1962; see Index of Literature; see also *Furcellaria*

Gelar

Raw Materials and Occurrence

Various brown and red algae of the Baltic, such as *Fucus* species, *Ceramium*, *Polysiphonia* etc.

Definition

The product should be used as agar substitute. It often has been also called "Baltic agar".

Utilization

Gelar should be used in bacteriology and in industry, according to reports published in the journal "Pharmazie" between 1948 and 1955. It does not, however, appear to have got any commercial significance.

Literature

FÖRSTER, K., Pharmaz. Zentralhalle 89 (12): 409. 1950
FÖRSTER, K., Pharmazie 10 (10): 608. 1955
KIRSCHNINCK, H., Pharmazie 3 (4): 171. 1948; 5(7):353.1950; 5(12):598.1950; 10(6):386.1955
PAX, F., Meeresprodukte, Gebrüder Bornträger, Berlin, 1962
See also *Ceramium, Fucus, Polysiphonia*

Ginnansô

Raw Materials, Occurrence, and Definition

The raw materials of ginnansô are above all *Iridea cornucopiae* and *Rhodoglossum pulchrum* both of which occur on the coast of Japan.

Neodilsea yendoana on Hokkaido, known as "Akaba" or "Akahata" and *Turnerella mertensiana* on Hokkaido, known as "Obaso", are also used for ginnansô production.

Gigartina ochotensis and *G. pacifica* are also utilized. Ginnansô belongs to the group of "algal-slimes", similar to funoran.

Countries of Production

Japan. From 1930 to 1936 annual production was between 10 and 45 tons (TOKIDA).

Utilization

Ginnansô is used in East Asia like funoran, on account of its adhesive properties. Before the First World War the price for ginnansô was one third of that of funoran.

Literature

TOKIDA, J., Fac. of Fish., Hokkaido Univ. 2 (1). 1954
See also *Gigartina, Iridea, Neodilsea, Rhodoglossum, Turnerella, Funoran*

Gracilaria Gum

Raw Materials and Occurrence

Gracilaria species (cf. p. 244).

The most important species are

G. confervoides (= *G. verrucosa*) from South Africa, Asia, Australia, North, Central and South America.

G. armata, particularly west coast of the U.S.A.

G. blodgettii, particularly Japan, U.S.A. (Florida, North Carolina).

G. cornea, U.S.A. (Florida), West Indies, North Brazil.
G. floridiana, U.S.A. (Florida).
G. lichenoides, Asia and Australia.
G. multipartita, particularly eastern North America, South India, and others.

The products made from the various *G.* species differ in their jelling temperature and gel strength.

The following table is taken from the work of R. E. Schachat and M. Glicksman.

Gelling Properties of Common *Gracilaria* species

	Gelling temperature	Concentration in per cent	Breaking load g*)
Commercial strip agar	37°	1	80
	37°	1.5	270
Japanese agar	38°	1	70
	38°	1.5	80
Gracilaria confervoides	43°	0.5	113
(North Carolina)	43°	1	407
	59°	1.5	650
Gracilaria confervoides	47°	1	120
(California)	47°	1.5	190
Gracilaria cornea	46°	1	0
(Florida)	46°	2	10
	46°	3	15
Gracilaria floridiana	48°	1	20
(Florida)	48°	2	110
	48°	3	210
Gracilaria confervoides	37°	1	187
(South Africa)	37°	1.5	337
Gracilaria confervoides	46°	1	95
(Australia)	46°	1.5	167

*) grams per cm of plunger circumference.
The raw material from North Carolina contains 26–28 per cent *G. confervoides* (Stoloff and Lee).

Definition

Some of *Gracilaria* species and varieties supply an agar, others supply an agaroid.

Knowledge of the chemical and physical properties of the phycocolloids from these marine algae has improved in recent years. It now appears possible to define various products more precisely, so that special fields of application can be found.

The products made from *Gracilaria* species are frequently described as "Gracilaria Gum" in the more recent literature. Further *Gracilaria* products are the following:

Ceylon Moss, Chinese Moss, Gulaman Dagat, Ogo-nori.

Countries of Production

Gracilaria products are mainly manufactured in the U.S.A., Brazil, Argentina, South Africa, Ceylon, India, Japan, Australia.

Utilization

These softer gelling materials are used particularly wherever harder gelling products, such as agar from *Gelidium corneum* are unsuitable, for example in various fields of food technology.

Literature

Hoppe, H. A. and Schmid, O. J., Bot. Mar., Supplement, 1962

Schachat, R. E. and Glicksman, M., Some Lesser Known Seaweed Extracts ex Industrial Gums, Academic Press Inc., New York, 1959

Stoloff, L. and Lee, Ch. F., Fishery Leaflet 173, March 1946. US Dep. of the Interior, Fish and Wildlife Service

See *Gracilaria, Agar*

Gulaman

Agardhiella tenera comes from the Philippines. Gulaman is the Philippine name for *A. tenera* and incidentally also for *Gracilaria lichenoides*. The material is used for making jelly products.

Literature

Whistler, R. L. and BeMiller, J. N., Industrial Gums, Academic Press Inc., New York, 1959

Zaneveld, J. S., Ec. Bot. 13 (2): 89. 1959

See also *Agardhiella, Gracilaria, Agar*

Gulaman-Dagat

Gracilaria confervoides comes from Indonesia. This material is used for making jelly products.

Literature

Whistler, R. L. and BeMiller, J. N., Industrial Gums, Academic Press Inc., New York, 1959

Zaneveld, J. S., Ec. Bot. 13 (2): 89. 1959

See also *Gracilaria, Agar*

Hijiki

Hizikia fusiforme on the coasts of Japan and Korea. "Hijiki" is the Japanese name for the alga. "Hijiki" is sold either fresh or dried and is used as a vegetable.

Literature: SCHMID, O. J. and HOPPE, H. A., Bot. Mar. Suppl. 1962. See also *Hizikia*

Hondawara

Sargassum enerve and perhaps also other Japanese *S.*-species. "Hondawara" is the term used to describe products from *S. enerve* used in Japan as an additive in soya sauces for human nutrition.

Literature: See *Sargassum*

Hypnean

Raw Materials and Occurrence

A number of *Hypnea* species on subtropic and tropical coasts can be regarded as a hypnean raw material. The most important species in the present time seems to be *H. musciformis*, which occurs on the Atlantic coast of the U.S.A. and on the coast of northern Brazil.

Definition

Hypnean is an extract from species of *Hypnea*. It forms gels of a very great firmness. Hypnean has often been considered to be a special kind of agar.

Countries of Production

The production of hypnean is in the present time limited to the U.S.A. and even there, as compared with other phycocolloids, it is not produced on a large scale, although the raw material is available in sufficient quantities. It can be anticipated that this phycocolloid will have great commercial importance.

Structure and Properties

According to the investigations of SMITH and COOK[1]) hypnean is similar in its structure to carrageenan in containing a κ- and a λ-fraction. HAMILTON and CARROLL[2]) have compared the polysaccharides of *Hypnea musciformis* and *Chondrus crispus*. They did not find any principal chemical differences. Their results corroborated those of SMITH and COOK. Hypnean has a stronger gel-forming property than other phycocolloids[3]). The stronger gelation is ascribed to the higher κ-fraction and the difference in the kation concentration and the hydrogen bond[2]).

Hypnean is soluble in cold distilled water, but not in cold water containing electrolytes. Production happens by hot water extraction and purification by repeated freezeout, as in the case of agar. Acid and potassium chloride facilitate extraction. Gel-formation is promoted by electrolytes and a little less by non-electrolytes such as sugar, glycerine, ethylene glycol etc. A detailed description of these interesting physical properties has been given by SCHACHAT and GLICKSMAN[4]).

Utilization

The pattern of utilization very much resembles to that of agar. The gel-strength far exceeds that of other known phycocolloids. The melting point can be influenced by concentration changes and by variation of its salt and electrolyte content.

Hypnean can also be used to produce particularly elastic gels in combination with other components[5]) and can also be used for fish freezing[6]).

Literature

1) SMITH, D. B. and COOK, W. H., Arch. Biochem. Biophys. 45, 232, 1953
2) HAMILTON, R. D. and CARROLL, J. J., Nature 196, 1200, 1962
3) HUMM, H. J. and WILLIAMS, L. G., Am. J. Bot. 35, 287, 1948, HUMM, H. J., U.S. Patent 2446091 ov 27th July, 1948
4) SCHACHAT, R. E. and GLICKSMAN, M., Industrial Gums by Whistler and BeMiller, p. 140, Academic Press 1959
5) BAKER, G. L., U.S. Patent 2.669519 of 16th February, 1954
6) STOLOFF, L. S., U.S. Patent 2567085 of 4th September, 1951

General Literature

HOPPE, H. A. and SCHMID, O. J., 1962; HUMM, H. J., 1951; SANFORD, F. B., 1958; SCHACHAT, R. E. and GLICKSMAN, M., Ec. Bot. 13 (4): 369. 1959; SCHMID, O. J., 1962; WHISTLER, R. L. and BEMILLER, J. N., 1959; YOUNG, E. G., Coll. Int. du Centre Nat. de la Recherche Scient. Dinard, 1960; see also *Hypnea*

Iridophycan

Raw Materials, Occurrence and Definition

The most important raw material for the production of iridophycan is *Iridea flaccida* on the Pacific coast. The terminology used for the individual *Iridea* (= *Iridophycus*) species is not uniform (cf. p.255). A number of other *Iridea* species in addition to *I. flaccida* are utilized as raw materials, for example *I. capensis*, South Africa, *I. cordata*, west coast of North America, *I. pulchra*, Pacific coast, *I. splendens*, coast of California and Oregon. Numerous further *Iridea* species can be regarded as possible raw materials. Iridophycan (= Iridophycin) is the extract from *Iridea.*

Countries of Production

The production of this phycocolloid began in 1945 in the U.S.A.
In Japan *Iridea* species are used for producing "funori" and "ginnansô".

Structure

Iridophycan is produced from *Iridea flaccida* (= *I. laminaroides* = *Iridophycus flaccidum*), and T. MORI[1]) classes it with the carrageenans to which it is doubtless related. HASSID[2]) investigated the polysaccharide isolated by T. MORI, and by hydrolysis he found D-galactopyranose units sulphated on C_6. T. MORI[3]) also found 1–3 bonds as well as the 1–4 bonds of the galactose units found by HASSID. At present it is not yet finally decided whether only one of these two types of bonds occur exclusively or whether the two arise side by side, as in the case of other galactanes. A further known detail is the α-position of the bonds[4]). According to YAPHE[5]) 36 per cent of the *Iridophycus* mucilage are of the type of κ-carrageenan. The sulphur content is recorded as 10.06 per cent for a purified barium salt of iridophycan[3]) and 8.7 per cent for an unpurified extract[6]). For the production of iridophycan the alga is extracted with hot water and purified by precipitation in alcohol[2]). A purification of iridophycan by the cold water extraction method was successful with barium salt[7]).

Iridophycan can be compared with carrageenan in its properties. The viscosity is shown in the following table[8]):

I. flaccida

Rotovisco Number of gears	centi poise 15°	20°
27	755	—
18	582	472
9	377	315
6	288	252
3	183	165
2	140	124
1	91	80

With rising shear the viscosity declines.

Utilization

Iridophycan recently increased in commercial significance, particularly in its use as cold-mix stabilizer.

It has found a widespread use in the U.S.A. as a stabilizer for chocolate and beverage syrups. Technologically, it is also used as a paper and cloth size, for cementing materials and for walls and partition walls, for wall colours, paints, and as a stabilizer for calcium paints.

In the pharmaceutical industry it could gain significance as an aqueous extract to prevent coagulation of blood.

Literature

1) Mori, T., Adv. Carbohydr. Chem. 8, 330, 1953
2) Hassid, W. Z., J. Amer. Chem. Soc. 55, 4163, 1933; 57, 2046, 1935; Plant Physiol. II., 461, 1936
3) Mori, T., J. Agr. Chem. Soc. Japan 19, 297, 1943; Chem. Abstracts 44, 7783, 1950
4) Mori, T. and Fumoto, T., J. Agr. Chem. Soc. Japan 23, 81, 1949
5) Yaphe, W., Can. J. Botany 37, 751, 1959
6) Schmid, O. J., Botan. Marina III. Suppl. 70, 1962
7) Mori, T. and Tsuchiya, Y., J. Agr. Chem. Soc. Japan 14, 609; Chem. Abstracts 32, 9175, 1938
8) Schmid, O. J., Unpublished Result 1961

General Literature

Black, W. A. P., Ann. Rep. Chem. Soc. Vol. 50. 1953; Hoppe, H. A. and Schmid, O. J., 1962; Lewin, R. A., 1962; Pax, F., 1962; Whistler, R. L. and BeMiller, J. N., 1959; Tseng, C. K., Science 101 (2633): 597–602. 1945; see also *Iridea*

Isingglass

Raw Materials and Occurrence

Gelidium species, East Asian coasts, particularly Japan and China.

Definition

"Isingglass" is the term used in Japan and China for "homemade" agar made from various *Gelidium* species, mostly for local and domestic consumption, less for exports.

Literature

Hoppe, H. A., Drogenkunde, Cram, de Gruyter & Co., Hamburg, 1958

Tokida, J., Fac. of Fish., Hokkaido Univ. 2 (1). 1954

Tseng, C. K., Science 101 (2633): 597–602. 1945

Zaneveld, J. S., Ec. Bot. 13 (2): 89. 1959

See also *Gelidium*

Kanten

"Kanten" is the Japanese name for agar. "Karafuto-Kanten" is the term used for an agar produced from *Ahnfeltia plicata*, particularly in Sakhalin.

Literature

Selby, H. H. and Selby, T. A., Agar ex Whistler, R. L. and BeMiller, J. N., Industrial Gums, Academic Press Inc., New York, 1959

Tokida, J., Fac of Fish., Hokkaido Univ., 2 (1). 1954

See also *Ahnfeltia*, Agar

Kausam

Alaria fistulosa and other *A.*-species. Coasts of Japan. "Kausam" or "Kauan" are the terms used in Japan for products of *Alaria* species as a foodstuff, similar to "Kombu" and "Wakame" products.

Literature: See *Alaria*

Kelp Products

The Historical Development and Definition

The meaning of the term "Kelp" has changed in the course of the centuries. Since the 17th century it has been used on the coasts of England to describe a compound obtained from the incineration of brown algae, used for the production of a mixture of potassium and sodium salts, and also glass making. This was called "Kelp". The word "Kelp" may have a Scottish origin. These factories were called "Kelp kilns".

The "Kelp" industry also had great significance in France in the time of Louis XIV and was regulated by special laws. In France the residues from the incineration processes of marine algae were called "soude de varech".

Towards the end of the 18th century Scotland produced approximately 20.000 tons of "Kelp" annually, for which approximately 400.000 tons of fresh marine algae were required.

After the invention of the Leblanc method for the production of soda in 1800 the "Kelp" industry declined but revived again after the discovery of iodine by Courtois in 1811. Marine algae were the most important source of iodine.

Even during the Second World War, in Japan the production of iodine and sodium salts from the ash of marine algae was referred to the "Kelp" industry.

In the U.S.A., on the other hand, "Kelp" was the term used for the large brown algae in the fresh state, e.g. "Horsetailkelp" *(Laminaria digitata)*, "Broadleaf kelp" *(Laminaria saccharina)*, and "Giant kelp" *(Macrocystis pyrifera)* (Tseng, 1945, Humm, 1951[1] [2])).

Sanford (1958[3])) made a difference between "Kelp", the large brown algae and "Rockweeds", the marine algae growing on the rocks between the low and higher water marks (for example Fucaceae).

In recent literature the term "Kelp" is used for the entirety of fresh brown algae, Maass (1959)[4]).

The term "Kelp-meal" is also used for "Seaweed-meal", i.e. algal meal.

The basic material for the production of iodine and sodium salts etc. is also referred to as "Kelp ash".

Countries of Production

"Kelp" factories were established about threehundred years ago on the coasts of Scotland and France. Dried brown algae were roughly comminuted, put into trenches up to ten metres long and about sixty centimetres deep and broad. The algae were then set on fire and ashed at a very high temperature, until a grey to black flux compound arose, from which the iodine and the salts were subsequently obtained. The freshly cut algae were preferred to drift weed, for it was well known even then that the content of valuable substances is higher in freshly harvested algae than in drifting algae thrown up on the beaches.

Five to ten tons of dried marine algae supplied one ton of "Kelp". This "Kelp" contains approximately 30 per cent of potassium and approximately 28 per cent of sodium salts and up to 1 per cent of iodine.

During the First World War there was a widespread "Kelp"-industry on the American Pacific coast for the production of acetone, ethylacetate, ethylpropionate, ethylbutyrate, iodine, and iodine salts, potassium chloride, potassium acetate, sodium chloride, and "Kelpchar", a kind of charcoal used as a de-colouring agent[3]).

The most important raw materials on the Pacific coasts were *Macrocystis pyrifera* and *Nereocystis luetkeana* as well as other great large brown algae, for example *Pelagophycus porra*.

Special ships were built for harvesting with underwater mowing machines to harvest up to several hundred tons of marine algae a day.

The algae were comminuted on the harvesting vessels and carried to fermentation vats, where they were subjected to a fermentation process of several weeks. The above mentioned products were then obtained from the fermentation liquid.

According to American estimates, reported by SANFORD, the raw material on the Pacific coast of the U. S. A. is sufficient to supply approximately six million tons of sodium chloride and approximately 20.000 tons of iodine salts per year.

Apart from the fermentation process, which is used by the HERCULES POWDER Co. for the production of acetone and other solvents and for iodine and potassium salts, there is also what is known as the "char"-process. This is a kind of distillation of dried algae in a closed distillation chamber, in which the combustible gases and the heavy oils are separated, and a residue of iodine, bromine, and potash is obtained.

Although the manufacture of marine algal products is no longer considered to be very profitable in the U. S. A. this does not mean that it is not profitable in other parts of the world, or will not become so again, where sufficient raw material, good harvesting possibilities and cheap labour are available. Ultimately the whole question is one for modern chemical engineering and technology to decide. Also American authors have pointed out that it is generally possible in the present situation that the "Kelp" industry will change one day. In 1951 new experiments were made in the U. S. S. R. for the production of alcohol, acetone, lactic acid, and other organic acids from marine algae (FOTT, 1959[5])).

Iodine

The iodine production from brown algae began in the first decades of the last century in France, following Courtois' discovery of the element in the ash of brown algae in 1811. At first drift algae were used, described as "goémon épave", until it was found that algae freshly harvested supplied "goémon poussant en mer" with much greater yield.

Iodine was produced from such material also in England and Scotland. The raw material used on European coasts is, above all, *Laminaria*.

These marine algae can store up to 30.000 times of the iodine content of ocean water. They contain up to 1 per cent of their dry substance in iodine, while the ocean water contains iodine in a concentration of only 0,000.003–0,000.007 per cent. It occurs in the marine algae mostly in the form of iodide, particularly as NaI and KI, but a small portion is organically bound. A detailed survey on the occurence of iodine in numerous brown algae, red algae, and green algae from the various coastal areas has been reported by VINOGRADOW in 1953[6]).

The occurrence of iodides and organically bound iodine has also been reported, inter alia, by KYLIN, 1929[7]), DANGEARD, 1957[8]), BLACK, 1958[9]).

The iodine content of *Laminaria digitata* has also been reported by SHAW in 1957[10]), at the COLLOQUES INTERNATIONAUX in Dinard.

The distribution of iodine compounds within the plant has been investigated in a number of algae (for example *Laminaria japonica*, *Ascophyllum nodosum*, *Laminaria hyperborea*[11]) [12]) [13])). Seasonal fluctuations and variations in iodine content were found above all in the marine algae of the sub-littoral. The time of the maximum content differs with the individual species. Young plants in general have a higher iodine content than older plants, and plants at exposed coasts are richer in iodine than those in bays. The lamina usually contains more iodine than the stipe[14]).

The iodine is generally absorbed very rapidly by the marine algae. Numerous investigations have been made on the subject.

With various marine fungi the iodine concentration is even higher than that of marine algae. Sometimes it reaches 14 per cent of the dry weight.

The method used for the production of iodine originally was rather primitive. Losses in iodine and other valuable components amounted sometimes to 50 per cent of the total. HOFFMANN has described in detail the processes practised in 1938 and made many proposals for improvements.

Today numerous patents are valid for the production of iodine from marine algae. Often, however, these are portmanteau processes whereby various products including alginic acid and alginates can be won. According to other patents iodine is isolated electrolytically or by extraction methods from marine algae (MAASS).

In Japan the extraction of iodine from marine algae has great significance even today. But it did not reach an industrial scale until about 1900. Up to the First World War approximetely 100 tons of iodine salts were produced per year, during the war the production rose by a factor of 2½ and then declined again. 1929 115 tons were produced.

Iodine is also produced in Sakhalin. The centre of the "Kelp" industry is Shiramishi. In the cold water region of the Nishinotoro peninsula there occur large quantities of marine algae with a high content of iodine and potassium salts e.g. *Alaria fistulosa*, *Arthrothamnus kurilensis*, *Laminaria diabolica* (TOKIDA[15])).

The algae of the White Sea have been investigated with regard to their iodine content, and methods on iodine extraction have been reported[16]).

In general, brown algae are utilized for the production of iodine. Iodine is less widespread in red algae. An exception to this are the *Phyllophora* species which contain 0.27 to 0.45 per cent and which are utilized in the U.S.S.R. for iodine production. Also *Ptilota* species contain iodine, and so do some green algae, for example *Codium intricatum* and *Enteromorpha compressa*[14]).

According to reports of KALLE (1962 in PAX[17]), today approximately 75 per cent of the world's requirements of iodine are produced from the mother liquor of Chile salpeter, which contains approximately 0.1 per cent iodine. The residual 25 per cent are produced from marine algae, particularly in Japan and Russia.

Bromine

The production of bromine salts from marine algae has not reached the significance of iodine production from these raw materials.

Ocean water contains 0,007 per cent of bromine on an average. It is strongly enriched in some marine algae. Bromine generally occurs in quantities of 0,03–0,2 per cent of the dry weight in red algae.

1941 it has been discovered in Kobe that the red algae *Rhodomela larix* has an unexpectedly high content of bromine. It contains approximately 3 per cent of the dry weight.

Investigations made in Hokkaido have proved that also *Rhodomela macracanthua*, *Odonthalia corymbifera* and other *Odonthalia* species, *Polysiphonia morrowii* and others are interesting raw materials for the production of bromine. Thus the content of *Odonthalia corymbifera* amounts to almost 6 per cent of the dry weight (TOKIDA[15]).

Brown algae accumulate bromine in a much smaller extent than do the various red algae.

Fungi and types of corals often contain a substantial quantity of bromine.

Potassium Salts

Numerous marine algae contain potassium.

In Japan there had been formerly a considerable production of potassium from marine algae, which, however, was carried out by a not very profitable ashing process.

Unfortunately figures are available only for 1911 to 1917, when production rose from approximately 500 tons to 8.500 tons annually.

On Sakhalin likewise a number of brown algae, inter alia, *Laminaria saccharina*, *Costaria costata*, *Sargassum* species were utilized for potassium salt production (TOKIDA[15]).

Reference has already been made to the production of potassium salts within the scope of the "Kelp" industry in the U.S.A.

Literature

1) TSENG, C. K., Science 101 (2633): 597–602. 1945
2) HUMM, H. J., The Brown Algae ex Tressler, D. K. and Lemon, J. Mc. W., Marine Products of Commerce, Reinhold Publishing Corporation, New York, 1951
3) SANFORD, F. B., Fishery Leaflet 469. 1958, US Dep. of the Interior, Washington, D. C.
4) MAASS, H., Alginsäure und Alginate, Straßenbau, Chemie und Technik Verlagsgesellschaft m.b.H. Heidelberg, 1959
5) FOTT, B., Algenkunde, Gustav Fischer Verlag, Jena, 1959
6) VINOGRADOV, A. P., The Elementary Chemical Composition of Marine Organisms, Sears Foundation for Marine Research, Yale University, New Haven, 1953
7) KYLIN, H., Ztschr. physiol. Chem. Bd. 186. 50–84. 1929
8) DANGEARD, P., Botaniste 41 (1/6): 187–207. 1957

9) BLACK, W. A. P., The Algae ex Altschul, Processed Plant Proteins Foodstuffs Academic Press Inc., New York, 1958
10) SHAW, T. I., Coll. Int. du Centre Nat. de la Recherche Scient., Dinard, 1960
11) MORITA, M. and SATO, H., J. Chem. Soc. Japan. Pure Chem. Sect. 75 (4): 416–418. 1954
12) BAILEY, N. and KELLY, S., Biol. Bull. 109 (1): 13–21. 1955
13) LARSEN, B. and HAUG, A., Bot. Mar. II (3/4): 250. 1961
14) SHAW, T. I. in LEWIN, R. A., Physiology and Biochemistry of Algae, Academic Press Inc., New York, 1962
15) TOKIDA, J., Fac. of Fish. Hokkaido Univ. 2 (1). 1954
16) SKROSNIKOWA, N. S., Arb. Archangelsk. wiss. Forsch. 157–174. 1938
17) PAX, F., Meeresprodukte, Gebrüder Bornträger, Berlin, 1962

General Literature

CHAPMAN, V. J., 1950; HEGNAUER, R., 1962; HOFFMANN, C., 1938; HOPPE, H. A. and SCHMID, O. J., 1962; see Index of Literature; see also *Alaria, Arthrothamnus, Ascophyllum, Codium, Costaria, Enteromorpha, Laminaria, Macrocystis, Nereocystis, Odontalia, Pelagophycus, Phyllophora, Polysisphonia, Ptilota, Rhodomela, Sargassum*

Kombu

Raw Materials and Occurrence

The raw materials of "Kombu" products are East-Asian *Laminaria* and some other brown algae. The most important *Laminaria* are the following:

L. angustata. For the production of "Kizami-kombu", the "Cutted-kombu", which is exported from Japan to China in considerable quantities.

L. diabolica. For the production of "Hanaori" and "Motozoroi-kombu", particularly *f. genuina* and *f. longipes.*

L. japonica. This species is one of the most valuable raw materials, which is being cultured in Japan and in China. It supplies "Shinori-kombu". The alga has a very high sugar content. It is also used for the production of sweets.

L. longipedalis supplies "Enage-kombu", which, however, is of a low value and of which only relatively small quantities are marketed.

L. longissima. This species is above all exported to China.

L. ochotensis. This is a very important raw material, as it supplies what is known as "Rishiri"- or "Hosome-kombu". The alga, particularly the Teshio quality, is characterized by a high sugar content. The "Nagahama-kombu" of Sakhalin is also a very valuable product.

L. religiosa. This species also supplies "Kombu" of good quality.

L. saccharina. "Sugar Wrack", as it is sometimes called, occurs on numerous European and North-American coastal areas. The East-Asian material occurs in numerous forms and is a very valuable "Kombu" raw material in these regions.

L. yezoensis. This species is also utilized, but appears to be of lesser commercial significance.

Russia utilizes the East-Asian *Laminaria* for food. China imports considerable quantities from the northern regions.

Apart from *Laminarias* a number of other brown algae are processed into "Kombu", for example:

Alaria. East-Asian *Alaria* species supply "Kausam" or "Kauan", but they are also used with *Laminarias* for "Kombu" production.

Artrothamnus. A. kurilensis occurs in major quantities in the East-Asian cold-water regions. The alga has a high sugar content; as a raw material it serves for the production of "Chishima-nekoashi-kombu". Also *A. bifidus* is used for "Nekoashi-kombu".

Kjellmaniella. K. gyrata and *K. crassifolia* grow in the cold-water coastal regions of East Asia and are edible. *K. crassifolia* supplies "Nagakiri-kombu".

Definition

"Kombu" is a collective term of the various edible products from *Laminarias* and other brown algae produced in Japan.

Sometimes in the literature use is made of the term "Kobu" instead of "Kombu".

Countries of Production

The most important production country is Japan (Hokkaido, Honshû). The most "Kombu" factories are located in Osaka, but Korea and Sakhalin are also areas of "Kombu" production. In 1958 Russians began to produce vegetable preserves on the East Asian coasts. These preserves were made with *Laminarias.* China imports large quantities of "Kombu", but today it also manufactures its own.

Constituent Substances

All import constituent substances required for human nutrition are contained in the brown algae used for "Kombu". In addition to their content of vitamins, trace elements, and salts the value of these products lies in their high sugar content.

Production, Qualities and Utilization

Algal harvest for "Kombu" production is made in the autumn months up to October. The fishermen go out in their boats and tear the algae from the rocks with long poles (but in many cases they are also twisted off rather than torn off). The algal harvest is then dried on the beaches. As soon as it is dried, the stipes and adhesive discs are cut off and the fronds sorted and bundled. "Kombu" factories produce various qualities of these products.

"Kizami-kombu" or "Ao-kombu" is a cut product, sometimes referred to as "Green Kombu". Dried algae are boiled and stirred, from 15 to 30 minutes, then dried again, stacked, pressed and bundled. Such bundles weigh over 100 kilos each. The bundles are cut afterwards, and once again pressed, some moisture being added. After some time the blocks are sliced and the cutting dried just sufficiently for them to remain pliable. The commodity is sold either in packages or boxes.

"Kuro-tororo" or "Black-Kombu" is made from thicker *Laminarias.* The dried algae are soaked in diluted vinegar. The black surface is scraped off with a knife. These pieces are referred to as "Black Kombu", which usually is less valuable. "Shiro-tororo" or

"White Kombu" is obtained by continuing the scraping process, which at length yields a whitish product.

The third quality arising from this process is the stringy type, "Oboro-kombu".

The residue is bundled, pressed, and planed. It is known as "Shirago-kombu".

L. japonica is cut up and roasted over a fire. This yields the product known as "Koiro-kombu", which is candied with sugar and is popular as a cake called "Kuvashi-kombu".

"Saimatsu-kombu" is a powder product, which is pressed to form cookies or biscuits and is sometimes candied.

"Cha-kombu" is made in a similar manner to "Green Kombu", and it looks and is drunk like tea.

"Green Kombu" is used for the preparation of fish and meat dishes, for soups and as a vegetable.

Powderized "Kombu" is used for bakeries, soups, sauces, and as an additive with rice.

The numerous varied recipes render these algal products interesting to Europeans.

The quantities of harvested *Laminarias* processed into "Kombu" before the war varied between 300.000 and 500.000 tons per year, supplying approximately 50 to 60.000 tons of "Kombu", of a value of about 5 to 7 million Yen.

These statistics, truly, are several years old, and now the production is much higher. Numerous scientific papers have been published on this subject.

Literature

HOFFMANN, C., Kieler Meeresforsch. III (1). 1938

HOPPE, H. A. and SCHMID, O. J., Bot. Mar. Suppl. 1962

SCHMID, O. J. and HOPPE, H. A., Bot. Mar. Suppl. 1962

TOKIDA, J., Fac. of Fish., Hokkaido Univ. 2 (1). 1954

See also *Alaria, Arthrothamnus, Kjellmaniella, Laminaria*

Laminaran

Raw Materials and Occurrence

Laminaran (= Laminarin) is found in the fronds of *Laminaria* and in smaller quantities in *Ascophyllum, Fucus* species, *Desmarestia* and others (cf. Raw Materials).

It is not present or only in small quantities in the stipes and in the stems of these algae[1]).

Only *Laminaria* can be regarded as of significance for industrial utilization of laminaran.

Investigations particularly of seasonal fluctuations in laminaran content have to be expressly mentioned here, as also some referring to other factors causing changes in percentage. The maximum content of laminaran was found to be present in autumn and in early winter. On an average it reaches 10 to 20 per cent of the dry weight of the alga. It can reach 36 per cent in the fronds of *L. cloustoni*. It is absent from the blades in late winter and in spring. The maximum content of laminaran in raw materials dried after harvesting immediately under controlled conditions amounted to 23–32 per cent for *L. hyperborea*, 13–25 per cent for *L. digitata* and 18–33 per cent for *L. saccharina*.

Marine algae dried in the open air show considerable losses in soluble carbohydrate content.

Black investigated the seasonal fluctuations in laminaran content of Scottish *Laminaria*.

Laminaria hyperborea. Entire algae have been investigated. There was no laminaran in June, but in September it had reached 18 per cent. The content in laminaran declines with increasing water depth [2]) [3]).

Laminaria digitata from the open sea contains no laminaran in May, 9 per cent in August. Algae in coastal offshore areas contain 2 per cent in April and 21 per cent in July[4]) [5]).

Laminaria saccharina from the open sea contains 2 per cent laminaran in March and 23 per cent in July. Algae near the coast contain 1 per cent in March and 24 per cent in August.

No laminaran was found in the stipes of *Laminaria* in these investigations[6]).

The content in the fronds reaches its maximum in late autumn (cf. Mannitol).

The laminaran content of *Fucus vesiculosus* is very low, only 2.6 per cent were found[7]).

Of the South-African marine algae only *Ecklonia maxima* contains any substantial amount of laminaran[8]).

E. radiata has also been investigated for seasonal variations[9]).

Eisenia bicyclis. The laminaran of this alga, in contrast to that of *Laminaria* does not contain mannitol[10]).

Powell has investigated a number of brown algae from Puget Sound for their laminaran content and established that *Laminaria* species and *Alaria* have the highest content. The content was most constant in the sporophylls of *Alaria*. His analyses of Phaeophyta of Vancouver Islands showed a content of 2–33,9% laminaran/dry matter[56]).

Definition

The glucane laminaran is a reserve carbohydrate of certain brown marine algae which is absent in other species. Laminaran plays a similar rôle for marine algae as starch for land plants. A difference is made between "soluble" laminaran which mostly occurs in *L. digitata* and "insoluble" laminaran from *L.cloustoni* and other species[12]) [13]).

Structure

Laminaran was discovered in 1885 by Schmiedeberg[21]).

Laminaran is a reserve carbohydrate of brown algae. It exists in two forms, one soluble and one insoluble in cold water[16]). T. Mori[22]) concluded from the results of several authors that there is no chemical distinction between the soluble and insoluble laminaran, and that both are exclusively composed of glucose. However, Hirst and coworkers[23]) have found that there are differences in the length of chain and in the mannitol proportion.

Solubility of laminaran is referred to molecular size, and aggregation of the molecules in the aqueous solution is held to be responsible for the differences in solubility[24]).

On the basis of results of physical methods of investigation Friedlaender and coworkers[25]) indicate that the molecular weight of the soluble laminaran obtained from *Laminaria digitata* is 5300, while for the insoluble or not easily soluble laminaran made from *Laminaria cloustoni* they indicate a molecular weight of 3,500. However, they also accentuated that different molecular weights are not the determinant factors for the different solubilities.

The molecular weights determined by various authors, according to a table by WOODWARD[26]) lay between 1,900 and 6,000. Statements on length of chain vary between 9 and 24 and more units. BROATCH and GREENWOOD[27]) have found that in the fractionation of methylated laminaran the molecular weight rises from 1,900 to 12,000.

The original assumption of a relatively simple structure, especially, of a 1 : 3 glucan, according to WOODWARD[26]), referring to recent results, can no longer be retained. 1 : 6 bonds were found in addition to 1 : 3 bonds, and an ever present mannitol fraction was proved by PEAT and coworkers[28]).

Other reports in literature are noted by WOODWARD[26]). These results have been confirmed by HIRST's group[23a-e]). Branching occurs in soluble and insoluble laminaran[57]).

D-mannose was found in laminaran in addition to mannitol[29]). Laminaran sulphate has the characteristic hydrolysis rate of other sulphate-containing polysaccharides of marine algae[30]).

Laminarose and laminaritol are components of laminarin. Laminaribiose was produced from laminarin by hydrolysis. LESTER has reported on the laminarose made by enzymatic decomposition of laminarin.

In 1961 DILLON has contributed a paper to the FOURTH INTERNATIONAL SEAWEED SYMPOSIUM on laminaric acid[20]).

The derivatives of laminaran have been the subject of some work by BLACK and DEWAR[19]).

The two components of laminaran, i.e. laminarose and laminaritol, reported by SMITH and coworkers[31]) [32]), which have been more closely investigated by GOLDSTEIN[33]) as MEEUSE[34]) states with every justification, have even yet not been sufficiently clarified in regard of their constitution. It can be accepted, however, that there is mannitol in the endposition[35]). As for branching the statements made by HANDA and NISIZAWA[36]) should be mentioned which say that only a linear chain can be assumed in laminaran made from *Eisenia bicyclis* due to the absence of 3.6-diglucosyl-glucose. Laminaran of this alga, however, appcars to have a special position, because no mannitol could be proved in it[36]).

The previous knowledge of the structure of laminaran could be confirmed by the synthesis of suitable oligosaccharides[37]).

The production of an amide of the laminaran sulphuric acid has been described in the British patent 911 484 of 28th November, 1962[38]). This has a ratio of sulphate-monosaccharide of 1.46 : 1, an anticoagulant activity of 1.4 units/mg and an antilipaemic activity of 50 units/mg.

Production

Work on the production of laminaran, its various forms, and derivatives, has been carried out after the Second World War by the INSTITUTE OF SEAWEEDS RESEARCH.

L. cloustoni (fronds) harvested in the months of August to September, were used as a raw material. The extraction was made with watery solvents from both fresh and dried raw materials. Special attention was devoted to sodium laminaran sulphate, which is interesting for various therapies[14] [15].

The isolation of laminaran from brown algae has been described by BLACK and al[16].

REID has reported on this subject to the SECOND INTERNATIONAL SEAWEED SYMPOSIUM[17].

The isolation of D-glucose from laminaran has been dealt with by BLACK and al[18].

Laminaran can be produced as a by-product for the production of sodium alginate. A number of international patents have been issued, likewise on the production of glucose.

Qualities

Two forms of laminaran are distinguished, the soluble and the insoluble form. The insoluble form is practically insoluble in cold water, but soluble in hot water. The soluble form is readily soluble in cold water and gives neutral solutions. Aqueous solutions of laminaran are non viscous, do not form gels and do not possess adhesive or film-forming properties. Both forms are virtually insoluble in all organic solvents (WOODWARD[43]).

Utilization

The application of sodium laminaran sulphate as a blood anticoagulant and antilipaemic agent has been studied for several years. DEWAR has reported on the subject at the SECOND INTERNATIONAL SEAWEED SYMPOSIUM in 1955[44] [45] [46].

Animal experiments in this field were carried out by Dr. BURT. This author has reported on her work at the THIRD INTERNATIONAL SEAWEED SYMPOSIUM in 1958[47].

Some further work on the subject are quoted here:

"The Anticoagulant Action in Blood of Sulphated Derivatives of Laminarin" (HAWKINS and O'NEILL, 1955[48]).

"The Anticoagulant Activity and Toxicity of Laminarin Sulphate" (ADAMS and THORPE, 1957[49]).

"The Physiological Activity of Laminarin Sulphate" (HAWKINS and LEONARD, 1958[50]).

"The Antilipaemic Activity of Laminarin Sulphate" (MOOKERJEA and HAWKINS, 1958[51]).

Experiments have shown that the sulphated polysaccharides from seaweeds possess antilipaemic activity, but no anticlotting properties[52].

A sulphated laminaran with high antilipaemic activity and a feeble anticlotting action was made according to a British patent[53].

ADAMS and others have reported on the effects of laminaran sulphate on bone growth[54].

It is also significant that compared with laminaran sulphate, decomposed laminaran sulphate from *Laminaria cloustoni* has the higher anticoagulant activity[39].

In animals the cholesterol level in the blood could be reduced by laminaran sulphate[40]. In animal experiments an inhibitive effect was found of laminaran sulphate on sarkoma 180[41].

The production of laminaran sulphate has been described by O'NEILL[42].

BLAINE worked on "Laminarin as a Talcum Powder Substitute" in "Surgery"[55].

A glucose is made from laminaran by hydrolysis and is used as a fodder.

2-deoxy-D-ribose was obtained from laminaran (KENNER).

There is evidence to show that laminaran could be used as a fuel for coloured smoke mixtures, as a filler in adhesive pastes and for custard powders.

Enzymatic hydrolysis of laminaran is treated in the chapter on "Enzymes" in this book.

Literature

1) Young, E. G., Carbohydrate Accumulation by lower Plants (Rhodophyta) ex W. Ruhland, Handbuch der Pflanzenphysiologie, Springer Verlag, 1958
2) Black, W. A. P., J. Soc. Chem. Ind. 67. 165–168. 1948
3) Black, W. A. P.; ibid., 69. 161–165. 1950
4) Black, W. A. P.; ibid., 67. 169–172. 1948
5) Black, W. A. P.; ibid., 67. 172–176. 1948
6) Black, W. A. P., Nature 161. 174. 1948
7) Black, W. A. P. et al., J. Sci. Food Agric. 754–764. 1955
8) Holdt, M. M. v. et al.; ibid., 193–196. 1955
9) Stewart, C. M. et al., Nature 192. 1208. 1961
10) Handa, N. and Nisizawa, K.; ibid., 192. 1708. 1961
11) Powell, J. H., Amer. J. Bot. 49. 673. 1962
12) Young, E. G., Carbohydrate Accumulation by lower Plants (Rhodophyta) ex W. Ruhland, Handbuch der Pflanzenphysiologie, Springer Verlag, 1958
13) Woodward, F. N., J. Sci. Food Agric. 477–487. 1951
14) Institute of Seaweed Research, Ann. Rep. 1955
15) ibid. 1956
16) Black, W. A. P. et al., J. appl. Chem. 1. 505. 1951
17) Reid, K. C., Second Int. Seaweed Symp., 1955
18) Black, W. A. P. et al., J. Sci. Food. Agric. 4. 58. 1953
19) Black, W. A. P. et Dewar, E. T.; ibid., 5. 176–183. 1954
20) Dillon, Th., Fourth Int. Seaweed Symp., 1961
21) Schmiedeberg, J. E. O., Tageblatt d. Naturforscherversammlung, p., 1885
22) Mori, T., Adv. Carbohydrat Chemist. 8, 344, 1953
23a) Anderson, F. B., Hirst, E. L. and Manners, D. J., Chem. & Ind. (London) p. 1178, 1957
23b) Annan, W. D., Hirst, E. L. and Manners, D. J., J. Chem. Soc. p. 984, 1962
23c) Connell, J. J., Hirst, E. L. and Percival, E. G. V., J. Chem. Soc. (London) 3494, 1950. I. Mitt.
23d) Percival, E. G. V. and Ross, A. G., J. Chem. Soc. (London) 720, 1951. II. Mitt.
23e) Anderson, F. B., Hirst, E. L., Manners, D. J. and Ross, A. G., J. Chem. Soc. 3233, 1958
24) Whistler, R. L. and Smart, C. L., Polysaccharide Chemistry, Academic Press Inc. New York 1953, p. 350
25) Friedlaender, M. H. G. et alii, Biochimica Biophysica Acta 14, 136, 1954
26) Woodward, F. N. in Industrial Gums, Academic Press 1959, p. 131
27) Broatch, W. N. and Greenwood, C. T., Chem. & Ind. 1015, 1956
28) Peat, St., Whelan, W. J. and Lawley, H. G., J. Chem. Soc. (London) pp. 724 a. 729, 1958
29) Smith, F. and Unrau, A. M., Chem & Ind. 636 (1959)
30) Rees, D. A., Bioch. J. 88, 343, 1963
31) Briggs, D. R., Garner, E. F. and Smith, F., Nature 178, 154, 1956
32) Lewis, B. and Smith, F., J. Am. Chem. Soc. 79, 3929, 1957
33) Goldstein, I. J., Smith, F. and Unrau, A. M., Chem. & Ind. (London) 124, 1959
34) Meeuse, B. J. D. in Physiology & Biochemistry of Algae, Academic Press, New York a. London 1962, p. 195
35) Whelan, W. J., Lawley, H. G. and Evans, J. M., Biochem. J. 61, X, 1955
36) Handa, N. and Nisizawa, K., Nature 192, 1078, 1961
37) Peat, St., Whelan, W. J. and Evans, J. M., J. Chem. Soc. (London) 175, 1960
38) Ronald, A. G. et alii, Brit. Patent 911484 for Boots Pure Drug Co. Ltd. vom 28. 11. 1962
39) Adams, S. S. et alii, J. Atherosclerosis 2, 314, 1962
40) DiLuzio, N. R., US Patent 3081226 vom 12. 3. 1963
41) Jolles, B. et alii, Brit. J. Cancer 17, 169, 1963
42) O'Neill, A. W., Can. J. Chem. 33, 1097, 1955
43) Woodward, F. N., Laminaran ex Whistler, R. L. and BeMiller, J. N., Industrial Gums, Academic Press Inc., New York, 1959
44) Dewar, E. T., Second Int. Seaweed Symp., 1955
45) Institute of Seaweed Research, Ann. Rep. 1958
46) Besterman, E. M. M. and Evans, J., Brit. med. J. I. 310. 1957
47) Burt, C. C., Third Int. Seaweed Symp., 1958
48) Hawkins, W. W. and O'Neill, A. N., Can. J. Biochem. Physiol. 33. 545–552. 1955
49) Adams, S. S. and Thorpe, H. M., J. Pharm. and Pharmacol. 9. 459–463. 1957
50) Hawkins, W. W. and Leonard, V. G., Can. J. Biochem. Physiol. 36. 161–170. 1958

51) Mookerjea, S. and Hawkins, W. W.; ibid., 36. 261–268. 1958
52) Murata, K., J. Geront. 13. 30. 1962
53) Gollin, R. A. et al., B. Pat. 911. 484
54) Adams, S. S. et al., Lancet, 618. 1958
55) Blaine, G., The Medical Press CCXXVI (5876). 1951
56) Powell, J. H. and Meeuse, B. J. D., Econ. Bot. 18, 164, 1964
57) Fleming, M. et alii Abstr. V. Intern. Seaweed Symp., 1965, p. 12

General Literature
Fott, B., 1959; Hegnauer, R., 1962; Hoppe, H. A. and Schmid, O. J., 1962; Lewin, R. A., 1962; Maass, H., 1959; Mautner, H. G., Ec. Bot. 8 (2): 174. 1954; Pax, F., 1962; Sanford, F. B., Fishery Leaflet 469. 1958; Schmid, O. J., 1962; Tseng, C. K., Science 101 (2633): 597–602. 1945; Whistler, R. L. and BeMiller, J. N., 1959; see Index of Literature; see also *Alaria*, *Ascophyllum*, *Desmarestia*, *Ecklonia*, *Eisenia*, *Fucus*, *Laminaria*

Laverbread

Raw Materials and Occurrence

A number of *Porphyra* species form the basic raw material of the "Laverbread" industry in Europe on the coast of Wales and other countries. These industries, however, have not risen to the significance of the Japanese "Nori" industry, although they are of some importance.

P. atropurpurea. Coasts of Indonesia and of the Pacific islands. This alga is regarded as a special delicacy in Hawaii.

P. kunthiana. Coasts of Chile. This alga is eaten together with *Ulva latissima*.

P. lanceolata. Pacific coast of the U.S.A. This alga is a raw material of the "Laverbread" industry and is known by the name of "red jabot laver".

P. leucosticta. Coasts of the North Sea and of the Indian and Pacific Oceans. This alga is used as a raw material of the "Laverbread" industry in Great Britain. In Hawaii it is regarded as a special delicacy.

P. naiadum. Pacific coast of the U.S.A., and known by the name of "red fringe".

P. perforata. Pacific coast of the U.S.A., known as "California-" or "purple-laver". The alga, which develops large thalli, is harvested and processed in a considerable quantity. It is regarded highly as a food by East-Asians living in the U.S.A.

P. umbilicalis. This alga is considered to be the most characteristic species of *Porphyra*. It is a very important raw material in the "Nori" industry of East Asia; it is also the basis of the "Laverbread" industry in Great Britain.

Definition

"Laverbread" is the term used above all on the coasts of Great Britain (Wales and Scotland) for the preparations of *P. umbilicalis* made locally. It will be produced in Great Britain and in the U.S.A.

Utilization

The use of algae for human consumption has been known in Scotland and South Wales ever since the 15th century. Today the potential of the products is interesting wider groups, and manufacturers import additional raw materials from other areas. "Laverbread" is either boiled or otherwise cooked and eaten with lemon juice, spices, and butter.

Today the name "Laverbread" is very often also used to name other edible algal products.

Literature

HOFFMANN, C., Kieler Meeresforsch. III (1). 1938
SCHMID, O. J. and HOPPE, H. A., Bot. Mar. Suppl. 1962
WHISTLER, R. L. and BEMILLER, J. N., Industrial Gums, Academic Press Inc., New York, 1959
See also *Porphyra*, *Nori*

Lechuga de samba

Raw Materials and Occurrence

Ulva lactuca and other *U.* species. Pacific coast of Latin America and other coastal areas of the world.

Definition and Utilization

"Lechuga de samba" is the Chilean term for the alga. It is often used as salad, as a vegetable or for soups. *Ulva* species are also found in other areas of the world and are used for human consumption.

Literature: SCHMID, O. J. and HOPPE, H. A., Bot. Mar. Suppl. 1962. See also *Ulva*

Makassar Agar

Eucheuma muricatum. Indonesia. "Makassar" or "Java Agar" or "East-Indian Carrageen" is the gelatinous substance produced from algae by boiling. It is used like agar.

Literature: See *Eucheuma*, Agar, Eucheuman

Mannitol

Raw Materials and Occurrence

Mannitol is found in brown algae and is a constituent of the cell sap[1]) [2]).

Mannitol is a very important sugar alcohol originally isolated by STENHOUSE in 1844.

In 1953 and subsequent years LINDBERG and coworkers isolated several mannitol compounds from brown algae[3]).

Mannitol also occurs in a number of other plants, for example in *Fraxinus ornus*, *Olea glandulifera*[4]).

The content of mannitol is subjected to a very strong seasonal fluctuation[1]). The function of mannitol in the plant itself is as yet undefined. Perhaps it assists in the synthesis of alginic acid and laminaran[5]).

Fucus vesiculosus contains about 9.0 per cent of mannitol[6]).

Laminaria hyperborea contains about 5 per cent in May and up to 18 per cent in July.

L. digitata from the open sea contains 6 per cent in March and up to 26 per cent in August. The same species from the coast contains 5 per cent in January and up to 21 per cent in July[7]) [8]) [9]).

L. digitata is used for the production of mannitol in France[10]).

L. saccharina from the open sea contains 5 per cent in January and up to 22 per cent in May. *L. saccharina* from the coast contains 9 per cent in February and up to 23 per cent in July[11]).

The stipes of *Laminaria* have only a very small mannitol content. The fronds have a maximum in summer.

Fronds of *Laminaria hyperborea* from a depth of approximately 7.5 metres gave a maximum mannitol content of 36.7 per cent dry weight.

Thus, not only seasonal fluctuations occur in mannitol content, but the content also depends on the depths in which the algae are growing. In general, the content declines with increasing depth, reaching a maximum between six and ten metres, and then declining again. Detailed investigations on fluctuations in composition of algae of the Scottish coast have been made by BLACK[12]).

Investigations have been made on the variations in mannitol content of a number of South-African marine algae, e.g. *L. pallida*, *Ecklonia maxima*, *Bifurcaria brassiciformis*[13]). The mannitol content of *Ecklonia radiata* at various seasons has also been investigated[14]).

Definition and Structure

Mannitol (= Mannit) is a sugar alcohol of the hexite series also occurring as Mannitan (anhydride). Mannitol was found in marine algae not only in a pure state, but also in various lowmolecular compounds[1]) [2]) [3]). The formation of isotopically labelled mannitol has been investigated too[4]).

Production

Work on the production of mannitol from marine algae has been carried out after the Second World War by the INSTITUTE OF SEAWEED RESEARCH.

L. hyperborea fronds have been used as raw material[20]).

The isolation of mannitol from brown algae has been described by BLACK and others[21]). REID has reported on the subject at the SECOND INTERNATIONAL SEAWEED SYMPOSIUM. The fronds of *Laminaria hyperborea*, harvested in the months of July to October, are a suitable raw material for the commercial production of mannitol[22]).

Mannitol is produced from marine algae in the U.S.A., Great Britain, France, and Japan. A number of patents have been issued on this subject.

Several patents should be mentioned here for the production of mannitol from algae: British Patent Nr. 757463 of 19th September, 1956; British Patent Nr. 729677 of 11th May, 1955; Japanese Patent Nr. 2975 of 29th May, 1954; British Patent Nr. 727013 of 30th March, 1955.

Large quantities of mannitol, however, are produced from other raw materials.

Utilization

The application for mannitol is highly diversified. In pharmacy, in the production of tablets, for diabetic food etc. For making chewing gum, as a dusting powder etc.

In the paint and varnish industry, leather and paper production, pyrotechnics and explosives industry.

It can be used for plasticizers. In organic synthesis and in the plastics industry for the production of resins.

In the production of yeast from marine algae the content of mannitol and sugar is of advantage.

Literature

1) Black, W. A. P., Ann. Rep. Chem. Soc. Vol. 50. 1953
2) Meeuse and Lewin, R. A., Physiology and Biochemistry of Algae, Academic Press Inc., New York, 1962
3) Hegnauer, R., Chemotaxonomie der Pflanzen, Birkhäuser Verlag, Stuttgart, 1962
4) Hoppe, H. A., Drogenkunde, Cram, de Gruyter & Co., Hamburg, 1958
5) Young, E. G., Carbohydrate Accumulation by lower Plants (Rhodophyta) ex W. Ruhland, Handbuch der Pflanzenphysiologie, Springer Verlag, 1958
6) Black, W. A. P. et al., J. Sci. Food Agric. 754–764. 1955
7) Black, W. A. P., J. Soc. Chem. Ind. 67. 165–168. 1948
8) Black, W. A. P.; ibid., 67. 169–172. 1948
9) Black, W. A. P.; ibid., 67. 172–176. 1948
10) Perrin, M., Chimie et Industrie 74 (2). 1955
11) Black, W. A. P., Nature 161. 174. 1948
12) Black, W. A. P., J. Soc. Chem. Ind. 69. 161–165. 1950
13) Holdt, M. M. von et al., J. Sci. Food Agric. 193–196. 1955
14) Stewart, C. M. et al., Nature (London) 192. 1208. 1961
15) Lindberg, B. et al., Acta chem. Scand. 7. 1119. 1953
16) Lindberg, B. et al.; ibid., 8 (817): 1547. 1954
17) Lindberg, B. et al.; ibid., 9 (168): 1093. 1955
18) Martinez Nadal, N. et al., J. Pharm. Sci. 52. 498. 1963
19) Bidwell, R. G. S. et al., Science 128. 776. 1958
20) Institute of Seaweed Research, Ann. Rep. 1955. 1956
21) Black, W. A. P., J. appl. Chem. 1. 414. 1951
22) Reid, K. C., Second Int. Seaweed Symp., 1955

General Literature

Fott, B., 1959; Hoppe, H. A. and Schmid, O. J., 1962; Maass, H., 1959; Mautner, H. G., Ec. Bot. 8 (2): 174. 1954; Pax, F., 1962; Schmid, O. J., 1962; see Index of Literature; see also *Bifurcaria, Ecklonia, Fucus, Laminaria*

Miru

Codium species. Coasts of Japan. "Miru" products are dried or salted preparations of *C. divaricatum* "Kiru-miru", *C. fragile* "Miru", *C. intricatum* "Motsure-miru" etc. They are used as foodstuff.

Literature: See *Codium*

Nori

Raw Materials and Occurrence

Japan, China, Korea, and a large number of islands in the Pacific Ocean, along the coasts of Australia, New Zealand, and South America; in all these regions the use of marine algae and marine algal products for human consumption has been known for many centuries, and has often been a substantial economic factor as, for example, in Japan.

Fish and marine algae were important for the nutrition of the population in the earliest times in East Asia. Both, for example, in Japan already two thousand years ago belonged to the religious sacrifices given to the gods in various religious rites. Fishing, also algal fishing, in those times as today was a substantial source of income. The "be", as the fishermen's guilds are known in Japan, were mentioned for the first time in literature in 274 A.D. The Japanese and Chinese reports from that time refer repeatedly to the significance of the products of the ocean – fish and algae.

Nori is made from species of *Porphyra*, a tender genus of red algae. It is used for food, particularly as a vegetable or salad, and enjoys increasing popularity in many parts of the world, not only in East Asia.

Several of the numerous *Porphyra* species on record as occuring on the East Asian coasts are important raw materials for the production of "Nori" (cf. p. 268).

P. tenera. "Nori" or "Amanori" is the economically most important East Asian *P.* species, being the raw material for "Hoshi-nori" production. The alga is widely cultivated, and numerous reports have been published about its constituent substances. "Natsu-nori" is the term used for preparations made of *P. tenera* of the "round type".

P. umbilicalis. "Chishima-kuronori" occurs in various forms in East-Asian coastal areas, but it is present also on the Atlantic coasts of Europe and North America, where it forms the basis of the "laverbread" industry (cf. p.357). The alga grows in great quantities on the Cape of Nichimotoro (Japan), where it is used for making "Asakusa-nori".

The following other Japanese *P.* species are utilized for "Nori"-production:

P. laciniata. Occurs on the coasts of the Japanese islands. This alga, however, is also an inhabitant of European coasts. In Japan it is one of the most important edible algae.

P. nereocystis. Coasts of the Pacific ocean. This alga contains fifty per cent more vitamin C than oranges and it also has a high content of B vitamins.

P. pseudo-linearis. Habitat on the coasts of Japan. This alga is cultivated for the production of Nori.

P. suborbiculata. Occurs on the coasts of the Pacific Ocean, and is known in China as "tsu choj".

P. variegata (= "Tomoe-nori") and *P. vulgaris*. These species are cultivated in Japan. Their vitamin C-content is high.

Of lesser importance for Nori production are the following: *P. amplissima* "Beni-tasa". This alga grows on the coasts of the Arctic and sub-Arctic Oceans.

P. bulbopes, *P. crassa*, *P. crispata*, *P. dentata*, *P. ochotensis* (= "Ana-amanori"). *P. okamurai* (= "Kuro-nori"). *P. onoi* (= "Ono-nori") *P. Tasa*, *P. yezoensis* (= "Susabi-nori").

The group of "Iwa-nori" is held to include *P. dentata* and *P. pseudo-linearis*.

Studies have been made on these algal cultures on the Japanese coasts.

Definition

"Nori" is a collective term used for the various edible *Porphyra* products, made in East Asia, especially in Japan.

Countries of Production

Nori-like products are also made in East-Asian countries besides of Japan:

On the coast of Southern Korea there are widespread *Porphyra* cultures. The most important trade products, characterized by very good quality, are Dong-laver, Choo-laver, and Pong-laver. The Japanese have laid out cultures of *P.* on the coasts of Chittagong, East Pakistan, while *Porphyra* products are also manufactured for human nutrition in Europe, North America, and other countries.

Constituents

Reference is made on the pages 268 and 365 on the constituent substances in *Porphyra* important in human nutrition. The value of Nori is due to the high content of protein, vitamins, and minerals.

Production, Quality and Utilization

The Japanese fishermen lower bundles of bamboo or oak bush, known as "Hibi", into the sea in areas of high salinity, preferably at low water in the autumn months. It has long been established that the spore formation in *Porphyra* is much greater in water of high salinity than in fresher water. As the algae, however, grow better in the latter, the bundles are removed after development of the germlings to a favourable place for growth. Bundles of bushes are used because the microscopic spores will best adhere to rough substrata and develop most favourably there.

When the algae are mature, they are harvested from their substrata at low tide. In many cases the fishermen harvest from boats.

For quite a number of years, instead of bamboo branches or bundles, nets have been used on which other algae, e.g. *Enteromorpha* spp. have been first made to grow. Because of the growing importance of the *Porphyra* industry for the Japanese economy, great attention is devoted to the problem of rational methods of the breeding of algal cultures.

Numerous investigations have been published on the various species and varieties of *Porphyra*, on its growth and the effect of the varying composition of sea water on growth, on culture methods, on diseases, their prevention and cure, on harvesting and on chemical composition etc.

For processing the algae are first thoroughly washed in fresh water, and then cut to small pieces. Then the comminuted algae are spread on thin bamboo mats and airdried. The dried algal paste is introduced to the trade in the form of thin, transparant red-violet leaves, where the individual comminuted particles are still easily visible. These "sheets" have a size of 0.472 square feet (= approximately 25 x 25 centimetre) and will usually be bundled, mostly one hundred sheets to a bunch to form a unit known as a "sok". Also roasted products are marketed, often in very elegant packaging.

Before being used Nori will usually be roasted, during which process the red-violet colour changes to green. It is a highly popular ingredient in various preparations, in sauces, soups, and many other dishes, for instance. Quite a number of these preparations are being offered; "Sushi" is particularly popular. Even greater is the number of very palatable recipes for rice dishes and fish dishes, and for bakery products etc.

The polysaccharides, the porphyranes, extracted from *Porphyra* species are dealt with on page 365.

Literature

Chapman, V. J., Seaweeds and their Uses, Methuen & Co., Ltd., London, 1950

Hoffmann, C., Kieler Meeresforsch. III (1). 1938

Schmid, O. J. and Hoppe, H. A., Bot. Mar. Suppl. 1962

Tokida, J., Fac. of Fish., Hokkaido Univ. 2 (1): 1954

Whistler, R. L. and BeMiller, J. N., Industrial Gums, Academic Press Inc., New York, 1959

See also *Enteromorpha*, *Porphyra*

Ogo-nori

Gracilaria confervoides. East-Asian coasts. "Ogo-nori" and "Oginori" are the East-Asian names of the product made in Japan and Malaya. The algae are washed and treated with hot water and dried. The product is used as a thickening agent, particularly for domestic use.

Literature

Hoppe, H. A. and Schmid, O. J., Bot. Mar. Suppl. 1962

Schmid, O. J. and Hoppe, H. A., Bot. Mar. Suppl. 1962

Tokida, J., Fac of Fish., Hokkaido Univ. 2 (1). 1954

Whistler, R. L. and BeMiller, J. N., Industrial Gums, Academic Press Inc., New York, 1959

See also *Gracilaria*

Phyllophoran

Raw Materials, Occurrence and Definition

The raw materials of phyllophoran are *Phyllophora rubens* and *Ph. nervosa*. A difference has been established in both quality and quantity of phyllophoran in the raw material between spring and autumn[1]). Phyllophoran is an agaroid.

Production Countries

Phyllophoran has been produced since 1930 in large quantities in the U.S.S.R. on the Black Sea coast, near Odessa.

Structure

A comprehensive survey on phyllophoran has been given by V. GRUNER[2]). The purified agaroid contains 13 to 15 per cent of mineral constituents and can be broken down into a fraction (40 per cent) soluble in cold water and an insoluble fraction (60 per cent), the latter being the actual agaroid. The insoluble fraction has an $[\alpha]_D^{20}$ + 52.5° and the soluble fraction an $[\alpha]_D^{20}$ + 20°. The insoluble fraction mainly consists of galactose with 22.16 per cent sulphate, while the soluble fraction mainly consists of glucose and laevulose with 15.5 per cent sulphate.

The following structural formula was proposed by V. GRUNER[2]):

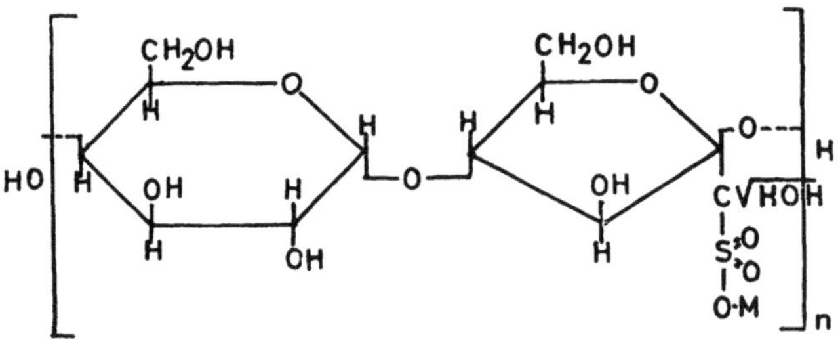

GRUNER

With an interchange of cations from Li, via Na, K, Ca to Ba, viscosity declines, while swelling increases[3]) [4]).

Properties

Phyllophoran is completely soluble in hot water. The solution is without smell and taste. Gelation temperature and the melting temperature of the gel increase with rising concentration. The addition of electrolytes and of alcohol, glycerin and sugar increases both temperatures and the gel strength. Starch, dextrin etc. reduce the gel strength. The molecular weight is about 5.000.

Utilization

In general the range of application of phyllophoran in Russia is identical with that of agar and agaroids in other countries. The phyllophoran is used in food technology, for example for the confectionary condiments, fruit jellies, marmelades, and jams etc., and as a thickener for sauces and soups, and in the ice cream and dairy industries.

Literature

1) LEBEDEV, S. I., Third Int. Seaweed Symp., 1958

2) GRUNER, V., Coll. Int. du Centre Nat. de la Recherche Scient. Dinard, (103): 183. 1960

3) Stavrov, S. N. and Morozov, H. A., Ukr. Khim. Zhur. 23. 721. 1957, CA. 20. 13370b. 1958
4) Morozov, H. A. and Stavrov, S. N., Kolloid Zhur. 20. 194. 1958, CA 52. 16840a. 1958

General Literature
Chapman, V. J., 1950; Fott, B., 1959; Hoppe, H. A. and Schmid, O. J., 1962; Humm, H. J., 1951; Tseng, C. K., Science 101 (2633): 597–602. 1945; see Index of Literature; see also *Phyllophora*

Pigments from Marine Algae

Ceramium-Pigment

Ceramium rubrum. Coasts of northern Atlantic and coasts of Japan (cf. p. 213). In recent times the pigment has been used in cosmetics and as a dyestuff for textiles[1].

Floridorubin

Rytiphlaea tinctoria. Mediterranean and adjoining oceans (cf. p. 283). The alga contains, instead of phycoerythrin the watersoluble dyestuff floridorubin which is converted into orange-yellow when acids are added. The compound is obtained by crystallization from an ethanol solution. The product is used as a dyestuff[2] [3].

Plocamium-Pigment

Plocamium coccineum. Coasts of the Atlantic (cf. p. 265). The alga contains phycoerythrin. The alga was used by the Romans in classical times because of its red pigment, for cosmetics. Today the pigment is still used in the cosmetic industry.

Literature

1) Stoloff, L., Ec. Bot. 16 (2): 86–94. 1962
2) Fott, B., Algenkunde, Gustav Fischer Verlag, Jena, 1959
3) Pax, F., Meeresprodukte, Gebrüder Bornträger, Berlin, 1962
See also *Ceramium*, *Plocamium*, *Rytiphlaea*

Porphyran

Raw Materials, Occurrence and Definition

Porphyra species occur in various coastal regions of the world (cf. p. 268).

Content and composition differ in the individual species. Approximately 30 per cent of porphyran has been obtained from *P. capensis*. Porphyran is the substance extracted from species of *Porphyra* by means of hot water.

Utilization

Porphyran has not yet been utilized commercially. SCHACHAT and GLICKSMAN, however, consider that it will be of importance because of its interesting structure.

The significance of *Porphyra* in human nutrition as a highquality protein and vitamin carrier has been reported elsewhere (cf. p. 268 and 361).

Structure

Apparently there are porphyrans of different chemical composition. J. K. N. JONES has isolated a polysaccharide which yielded fucose and galactose on hydrolysis[1]). He found a water insoluble mannan as a copper complex, a β-1-4 mannopyranose chain branched at every 12 unit. In the porphyran from *Porphyra capensis*, NUNN and von HOLDT[2]) have established a molecular relationship of a D-galactose, L-galactose, 6-0-methyl-D-galactose, and 3.6 anhydrogalactose of 1 : 1 : 2 : 1. DL-galactose was found in *Porphyra tenera*[3]) [4]). KINTARO[5]) has established DL-galactose and mannose as constituents of porphyran from *Porphyra laciniata*.

J. R. TURVEY[6]) first found L-galactose-6-sulphate in *Porphyra umbilicalis*. Moreover, he established a content of 10 per cent of sulphate esters. From his analysis he concluded that the galactan from this alga is either a complex polysaccharide or a mixture of polysaccharides. Five units (D-galactose, L-galactose and L-galactose-6-sulphate) are connected to a core of a few 3–6 anhydrogalactose units.

The chain ends are often 3.6 anhydrogalactoside-6-0-methylgalactose or D-galactose. In water-insoluble polysaccharide from *Porphyra*, REES and CONWAY[7]) have found galactose, 6-0-methylgalactose, 3.6 anhydrogalactose, and in addition to ester sulphates they have found a precursor ester sulphate 1.2 or 1.4 linked with L-galactose-6-sulphate, which is converted enzymatically into 3.6 anhydrogalactose. In other reports[8]), based on an analysis of forty samples of *Porphyra* species, the same authors report the following quantitative different results:

3.6 anhydro-L-galactose	5–19 per cent
ester sulphate	3–18 per cent
6-0-methyl D-galactose	3–28 per cent
D-galactose	24–45 per cent

Anhydro sugar and galactose sulphate account for one half of the sugar units, D-galactose and methyl galactose the other half. 3.6 anhydrogalactose givens the solutions a higher viscosity and gelating power, and it occurs to a greater extent among the species of *Porphyra* that are subjected to heavy mechanical stress in their natural environment.

An insoluble polysaccharide, investigated in detail by SU and HASSID[9]), proved to be a compound of galactose, 6-0-methyl-D-galactose, 3.6 anhydrogalactose, and sulphate in a ratio of 1 : 1 : 2 : 1, where the D and L isomeres of the galactose have a relationship of 1.3 : 1.

Literature

1) JONES, J. K. N., J. Chem. Soc. 3292, 1950
2) NUNN, J. R. and VON HOLDT, M. M., J. Chem. Soc. 1094, 1957
3) MIWA, T. Jap. J. Botany II, 41 (1940), Chem. Abstr. 35, 1941
4) HAYASHI, K., Hôgaku (Sci. of Agr.) 2, 226, 1948. Chem. Abstr. 44, 2086, 1950
5) KINTARO, O. u. B. Tellens, Berichte 34, 1422, 1901
6) TURVEY, J. R., Colloqu. Intern. Centre Nat. Rechn. Sci. 103, 29, 1060
7) REES, D. A. and CONWAY, ELSIE, Nature 195, 398, 1962
8) REES, D. A. and CONWAY, E., Biochem. J. 84, 411, 1962
9) SU, J.-CH. and HASSIDI, W. Z., Biochemistry 1, 468, 1962

General Literature

SCHACHAT, R. E. and GLICKSMAN, M., Some Lesser Known Seaweed Extracts, ex Whistler, R. L. and BeMiller, J. N., Industrial Gums, Academic Press Inc., New York, 1959; SCHMID, O. J. and HOPPE, H. A., 1962; see Index of Literature; see also *Porphyra*

Sarumen

Raw Materials and Occurrence

Japanese *Alaria* species are processed into "Sarumen". "Sarumen" is an edible algal product from Japan.

Production and Utilization

The algae are dried and prepared in a similar manner to "Kombu" products. Different *Alaria* species are made into "Kombu" by mixing with *Laminaria* species and are utilized as a food product.

Literature

SCHMID, O. J. and HOPPE, H. A., Bot. Mar. Suppl. 1962

TOKIDA, J., Fac. of Fish., Hokkaido Univ. 2 (1) 1954
See also *Alaria*

Seatron

Raw Materials, Occurrence and Definition

Nereocystis luetkeana Pacific coast of North America. "Seatron" is the name used for de-salted, aromatized, and candied pieces of stipe and floating bladders, marketed locally. They are used as a food and a confectionary product.

Literature: SCHMID, O. J. and HOPPE, H. A., Bot. Mar. Suppl. 1962. See also *Nereocystis*

Tokoroten

"Tokoroten" is the name used in Japan for an agar produced for local consumption.

Literature: TOKIDA, J., Fac. of Fish., Hokkaido Univ. 2 (1). 1954. See also Agar

Tsunomata

Raw Materials and Occurrence

Tsunomata is produced from East-Asian *Chondrus* species, such as *Ch. armatus*, *Ch. elatus*, *Ch. ocellatus*, *Ch. pinnulatus* etc. (cf. p. 218).

Definition and Countries of Production

Tsunomata is an "algal slime", similar to funori and produced in Japan. Compared with funoran and ginnansô, production is very low.

Utilization

Tsunomata is used in East Asia, like funori, because of its adhesive properties, but also as a gelating agent for domestic purposes.

Literature

TOKIDA, J., Fac. of Fish., Hokkaido Univ., 2 (1). 1954

See also *Chondrus*, Funoran, Ginnansô

Wakame

Raw Materials, Occurrence and Definition

Undaria species. East Asian coasts. "Wakame" are the algae treated according to special processes. The best known grades are "Kizami-wakame" and "Ito-wakame".

Production

Methods of harvesting of the algae and production of "Wakame" are described on p. 203. Utilized as a foodstuff.

Literature: SCHMID, O. J. and HOPPE, H. A., Bot. Mar. Suppl. 1962. See also *Undaria*

Various Substances

By Otto J. Schmid

Antibiotics

The occurrence of an antibiotic substance from marine algae apparently has been reported for the first time by SHIRAHAMA[1]), who found a substance in Sargaceae *Cystophyllum hakodatense* that had an effect upon *Lactobacillus bulgaricus* and *L. helveticus*. McCutcheon and coworkers[2]) made extracts from *Cymathaerae triplicata* inhibiting *Bacillus subtilis*. The extract obtained by PRATT and coworkers[3]) from various marine algae was found to be soluble in ether, and it inhibited *Staphylococcus aureus*, *Escherichia coli*, and *Pseudomonas aeruginosa*. The ethereal extract obtained by Mautner and coworkers[4]) from *Rhodomela larix* contained a bromated phenolic substance as an active substance. The bromine content of the extract amounted to 38 per cent.

CHESTER and SCOTT[5]) extracted inhibitor substances from red algae by means of ether. SAITO and SAMESHIMA[6]) applied methyl alcohol for extraction and isolated three antibiotically active substances, inter alia, a phenolic substance containing 32.2 per cent of bromine. KAMIMOTO[7]) extracted various marine algae and checked their efficiency against acid-resistant bacteria. His results have been compiled in the following table for easier reference:

Cold water extract	*Mycobact. phlei*	*Mycobact. avium*	*Mycobact. tuberc.*	*Timothy Bac.*
Ulva pertusa	S	S		
Chrysymenia wrightii		—	—	
Codium fragile		+		S
Sargassum thunbergii			m	
Ceramium boyderii		+		m

Hot water extract	*Mycobact. phlei*	*Mycobact. avium*	*Mycobact. tuberc.*	*Timothy* Bac.
Ulva pertusa	—	—		
Chrysymenia wrightii		m	m	

Hot water extract	*Mycobact. phlei*	*Mycobact. avium*	*Mycobact. tuberc.*	*Timothy* Bac.
Plocamium telfairiae			s	
Dermospongia			(+)	
Sargassum Thunbergii				+

Ether extract	*Mycobact. phlei*	*Mycobact. avium*	*Mycobact. tuberc.*
Ulva pertusa	c	c	
Chrysymenia wrightii		m	m
Gelidium amansii			s
Zostera marina			c
Dermospongia			m
Codium fragile		s	

Acetone extract	*Mycobact. phlei*	*Mycobact. avium*	*Mycobact. tuberc.*
Ulva pertusa	—	—	
Chrysymenia wrightii		m	m
Grateloupia turutu			m
Gelidium capillaris		c	c
Sargassum thunbergii	s?	s?	s*

*) bactericidal

Alcohol extract	*Mycobact. phlei*	*Mycobact. avium*	*Mycobact. tuberc.*
Ulva pertusa	c	c	
Chrysymenia wrightii		—	—
Gelidium amansii			s
Undaria pinnatifida			c
Codium fragile		s	

Legend: s = slightly inhibiting; m = inhibiting, without statement of details; c = considerably inhibiting; + = promoting growth

H. Roos[8]) has found in watery raw extracts of 22 marine algae antimicrobial effects, which, with a very few exceptions, were susceptible to heat. From *Ascophyllum nodosum* an inhibiting substance has been extracted[9]) that was active as a bactericide against a number of germs. An inhibitor from *Sargassum natans* has been the subject of a first report by Nadal and coworkers[10]). Further work revealed that this was an acid, volatile, and oily substance, active against *St. aureus*, *B. subtilis*, *E. coli*, *Pr. vulgaris*, *Ps. aeruginosa*, *M. smegmatis*, and *C. albicans*. Its name is "Sarganine"[11a]). Moreover, it seems[11b]) that this is a complexity of substances of which component A has fungicidal, antibacterial, and cytotoxic properties, while component B has only cytotoxic effects. A similar substance, known as chonalgine, has been extracted by the same authors from *Chondria littoralis*. A research team headed by P. R. Burkholder likewise has investigated marine algae in Puerto Rico and found antibiotic activity in 66 of 150 species, but also growth-promoting effects, for example in *Goniaulax*[12]). Seasonal fluctuations of antibiotic activity have been investigated by Almodovar[13]), and he found significant differences during the year in only 2 of 12 investigated species. The antibiotic factors may be of different nature for the same alga, as was proved by Olesen and coworkers[14]). The authors have particularly accentuated the possible riches of the oceans in antibiotics-producing plants.

In the 35 marine algae investigated, extracts from 11 of these have been found to be active against fungi[15]; 4 species of these are to be subjected to a further research series.

Antibacterial substances have been reviewed by SIEBURTH[16].

The rôle of antibacterial activity for survival of "gulf Weeds" is discussed by CONOVER and SIEBURTH [17].

Literature

1) SHIRAHAMA, K. L., J. Fac. Agr. Hokkaido Imp. Univ. 49, 1–93, 1942
2) MC CUTCHEON, R. et al., J. Amer. Pharmac. Assoc. Sci. Ed. 38, 196, 1949
3) PRATT, R. et al., J. Amer. Pharmac. Assoc. Sci. Ed. 40, 575, 1951
4) MAUTNER, H. G., J. Amer. Pharmac. Assoc. Sci. Ed. 42, 294, 1953
5) CHESTER, C. G. C. and SCOTT, J. A., Ann. Repts. Inst. Seaweed Res. 1953, 1954, 1955, II. Intern. Seaweed Symp. p. 49–53, 1956
6) SAITO, K. and SAMESHIMA, M., Nippon Nôogei-Kagaku Kaishi 29, 427–30 (1955) Chem. Abstr. 52, 9299c, 1958
7) KAMIMOTO KASUHIKO, Nippon Saikingaku Zasshi 11, 307–13, 1956, Chem. Abstr. 51, 18138i, 1957
8) ROOS, HANNELORE, Kieler Meeresforschungen 13, 41, 1957
9) VACCA, D. D. u. WALSH, R. A., J. Amer. Pharmac. Assoc. Sci. Ed. 43, 24, 1954
10) MARTINEZ NADAL, N. G., Inf. Bull. Res. Dept. Univ. Puerto Rico Coll. Agric. Mech. Arts p. 39 (1958/59), J. Amer. Pharmac. Assoc. Sci. Ed. 50, 356, 1961
11a) MARTINEZ NADAL, N. G. et al., Antimicrobial Agents and Chemotherapy p. 68–72, 1963
11b) MARTINEZ NADAL, N. G. et al., Abstr. IV. Conf. Antimicrob. Ag. & Chemotherapy p. 82, 1964, Bot. Marina 9, 21 and 62, 1966
12) BURKHOLDER, P. R. et al., Bot. Marina 2, 149, 1960
13) ALMODOVAR, L. R., Bot. Marina 6, 143, 1964
14) OLESEN, P. E. et al., Bot. Marina 6, 224, 1964
15) WELCH, ANNE M., J. Bact. 83, 97, 1962
16) SIEBURTH, J. MCN., Develop. ind. Microbiol. 5, 124–34, 1963
17) CONOVER, J. T. and SIEBURTH, J. M., Bot. Marina 6, 147, 1964

Amino Acids, Peptides, and Proteins

COULSON[1]) has found that marine algae contain more peptides than free amino acids and that therefore they are different from the freshwater plants and land plants. In *Chondrus crispus* YOUNG and SMITH[2]) have found the following free amino acids: Alanine, arginine, asparaginic acid, citrulline, glutamic acid, glycine, isoleucine, leucine, methionine, phenylalanine, serine, taurine, threonine. Those missing were histidine, lysine, proline, tyrosine and valine, which are found in a bound form in the whole plant. *Gigartina stellata* has a considerable proportion of free proline. *Ascophyllum nodosum*, *Pelvetia caniculata*, *Fucus vesiculosus*, and *Laminaria saccharina* contain less free amino acids, however, *Laminaria cloustoni* has more of them[1]). This finding is corroborated by the results of CHANNING and YOUNG[3]), who likewise found a few free amino acids in *Pelvetia caniculata*, viz. in *Laminaria cloustoni* and *Rhodymenia palmata*, mainly alanine, asparaginic acid, and glutamic acid, and in the latter alga also proline, leucine, and valine. In *Porphyra tenera* mainly the following free amino acids are present: alanine and glutamic acid[4]). LEE[5]) found 28 free amino acids in brown algae, of which 4 could not be identified. There are differences between the orders, families, genera and species, not, however, by season and habitat. Asparaginic acid, glutamic acid, serine, glycine, arginine, asparagine, threonine, alanine, glutamine, proline, valine, leucine, and isoleucine were found in all samples. Most frequently the following were found: ornithine, citrulline, histidine, tyrosine, tryptophane, phenylalanine, and cystine, less, however, lysine, and methionine. Also in various algae from the Bombay region almost all amino acids present in proteins in a free form and peptides have been found, while some amino acids were only found in a free condition[6]). The amino acids in the free form and the peptides vary seasonally in *Enteromorpha prolifera* and *Ulva lactuca*[7]).

Combined Amino Acid in Various Algae[1])

The results express nitrogen of the amino acids as percentages of total N in hydrolysate

Amino acid	*F. vesiculosus*	*A. nodosum*[2])	*Chondrus crispus*[3])	*R. palmata*[4])	*U. lactuca*[2])
Alanine	5,4	5,3	3,6	7,9	8,2
Arginine	8,2	8,0	28,0	10,8	10,1
Aspartic acid	9,0	6,9	3,7	7,2	7,8
Citrulline	0,0	0,0	3,3	0,0	0,0
Cystine	+	+	+	+	+
Glycine	5,4	5,0	3,1	7,5	7,2
Glutamic acid	11,0	10,0	3,4	6,2	6,6
Histidine	1,6	1,3	1,1	1,1	0,8
Isoleucine	3,0	2,8	1,7	3,1	2,8
Leucine	5,0	4,6	2,6	4,6	5,0
Lysine	6,0	4,9	3,3	6,4	5,8
Methionine	0,4	0,7	0,5	0,6	1,0
Ornithine	0,0	0,0	6,3	0,0	0,0
Phenylalanine	2,6	2,3	1,3	3,1	3,0
Proline	3,3	2,6	2,1	3,6	3,4
Serine	3,5	3,0	2,2	4,8	3,9
Threonine	3,3	2,8	2,0	3,9	4,2
Tryptophan	+	+	+	+	+
Tyrosine	1,2	0,9	1,0	1,6	1,6
Valine	3,9	3,7	2,6	5,1	4,9
Ammonia	16,0	14,6	6,9	9,9	7,9
Total	88,8	79,4	78,7	87,4	84,2

A recent investigation of protein obtained from *Porphyra* (Japanese commercial product) established the presence of the following amino acids after 48 hours of hydrolysis in 6 n -HCl. Histidine could not be evaluated here, because of the insufficient separation effect in the chromatogramme[8]).

22 amino acids have been identified in acid and alkaline hydrolysates of red algae, brown algae, and green algae by Ericson and Carlson[9]).

A tryptophane content has been reported by the above mentioned authors[9]) and by M. Takagi[10]).

---

[1]) Smith, D. G. and Young, E. G., J. biol. Chem. 217, 845 (1955)
[2]) Single determination
[3]) Mean of three determinations; for aspartic acid threonine, serine, and proline, mean of five
[4]) Mean of two determinations

Porphyra (red algae, treated for human consumption, Japan, 48 hours hydrolysis in 6 n-HCl, after E. KOFRANYI)

Per cent amino acids
N x 6,25 = 100

Lysine	4,81	Alanine	12.04
Arginine	4,55	Cystine	0,00
Ammonia	1,55	Valine	5,24
Asp. Acid	7,25	Isoleucine	3,73
Threonine	4,20	Leucine	6,47
Serine	3,97	Tyrosine	1,80
Glut. Acid	9,87	Phenylalanine	3,37
Proline	4,45	Methionine	1,75
Glycine	5,55		

In the two red algae *Polyides rotundus* and *Rhodymenia palmata* Russian authors[11]) have found more cystine and diaminocarbonic acid and less threonine, valine, and leucine in the first alga than in the second species.

Quantitative determinations of amino acids of *Fucus vesiculosus*[12]) and of *Sargassum*[13]), have also been made.

Valine has been found after oxidation in the form of a component of the fluorescing dyestuff from Rhodomelaceae, apart from 4-chlorophthalic acid[14]).

Of the rarer amino acids β-alanine and γ-amino-butyric acids have been found in the hydrolyses of marine algae[15]). As a ninhydrine-reactive substance 1.4 thiazane-3-carboxylic acid-S-oxide has been found in the red alga *Chondria crassicaulis*. An occurrence of this substance in green algae or brown algae has not been established yet[16]). D-cysteinolic acid has been discovered in green marine algae[21]).

For peptides R. SCOTT[17]) has found a widely ramified structure in *Rhodymenia palmata*, as short peptide chains are bonded to a central core with their carboxyl end.

Finally mention should be made of the investigation of protein from *Chondrus crispus* for its biological value, that is comparable to that of an egg albumin[18]).

As was remarked in an earlier survey on constituents[19]), modern knowledge relating to N-containing substances of marine algae is still insufficient. Modern investigational methods in the field of protein would certainly find a highly rewarding field of application here. The critical remarks made by FOWDEN[20]) on the not always sufficient analytical reliability of the investigations that have been made in the past, unfortunately, are accepted by the present author.

Literature

1) COULSON, C. B., Chem. u. Ind. p. 971, 997, 1953
2) YOUNG, E. G. and SMITH, D. G., J. biol Chem. 233, 406, 1958
3) CHANNING, D. M. and YOUNG, G. T., J. Chem. Soc. p. 2481–91, 1953
4) OGINO, CH., J. Tokyo Univ. Fish. 41, 107 (1955), Chem. Abstr. 53, 7436E, 1958
5) LEE, M.-J. et alii, Seoul Univ. J. Ser. D. 10, 1–7, 1961
6) LEWIS, E. J. and GONZALVES, ELLA A., Ann. Botany 26, 301, 1962
7) LEWIS, E. J. and GONZALVES, ELLA A., Ann. Botany 26, 317, 1962
8) KOFRÁNYI, E., Personal communication, 1964
9) ERICSON, L.-E. and CARLSON, BLENDA, Arkiv Kemi 6/49, 511, 1953
10) TAKAGI, M., Bull. Fac. Fish. Hokkaido Univ. 4, 86, 1953, CA. 49, 1151b, 1955

11) Sevenkov, G. P. and Zlochevskaya, J. V., Nanchn. Dokl. Vysshei Shkoly Biol. Nauki Nr. 4, 151 (1962), CA 58, 8236c, 1963
12) Smith, D. G. and Young, E. G., J. biol. Chem. 205, 849, 1953
13) Mazur, A. and Clarke, H. T., J. biol. Chem. 123, 729, 1938
14) O'hEocha, C., Colloqu. Intern. Centre Nat. Rech. Sci. 1960/61, p. 121
15) Smith, J. Intern. Botan. Congr., Repts. and Communs, Paris, VIII/17, 34, 1954
16) Kuriyama, M., Nature 192, 969, 1961
17) Scott, R., Ann. Rep. Inst. Seaweed Res. p. 13, 1956; p. 13, 1957
18) Larsen, B. H. and Hawkins, W. W., J. Sci. Food and Agric. 12, 523, 1961
19) Schmid, O. J., Botanica Marina I, 58, 1959
20) Fowden, L., Physiology and Biochemistry of Algae by R. A. Lewin, Academic Press 1962, p. 189
21) Keiji Iti, Nippon Suisan Gakkaishi 29, 771, 1963 C. A. 61, 3417f. (1964)

Other Carbohydrates, Sugar Alcohols, and Glycosides

β-1,3 linked xylan is contained in cell walls of *Porphyra umbilicalis*[14])

Lindberg and McPherson[1]) have found mannitol, glucose, fructose, l-mannitol-glucoside, 1.6-mannitol-di-β-glucoside, Laminaribiose, and Laminarotriose in *Laminaria cloustoni.* From *Fucus vesiculosus* Lindberg[2]) isolated mannitol, l-D-mannitolglucopyranoside, 1.6-D-mannitol-di-(β-D-glucopyranosid), trehalose, glucose, fructose, and sucrose. A compound extracted from red algae has been identified by Boveng and coworkers[3]) as 2-D-glycerine acid α-D-mannosid. Laminitol, a C-methylinositol and isomer with mytilitol forms *Laminaria cloust*[4]). Volemite (hexahydroheptane) was isolated from *Porphyra umbilicalis*[5]).

Of the sugar alcohols sorbitol has been made from *Bostrychia scorpoides*[6]), dulcitol likewise from *Bostrychia scorpoides*[6]) and *Iridea laminaroides*[7]). Scyllitol has been found in *Porphyra umbilicalis*[6]), laminitol likewise in *Porphyra umbilicalis* and *Gelidium cartilagineum*[5]) [8]). The mentioned mytilitol (C-methyl scyllosinositol) has also been found in *Polysiphonia fastigiata*[9]). Trehalose has been found for several times in red algae, for example in *Bostrychia scorpoides*[10]). The glycoside floridosid and isofloridosid have been isolated from various red algae[11]) [12]) [5]). These glycosides have been regarded as reserve carbohydrates[13]). About the biochemical rôle of the substances mentioned in this section it can be said that they are partly metabolic products, partly reserve carbohydrates. The differences between pectin, pectinlike substances versus carbohydrates of marine algae are discussed by Percival and McDowell[15]).

Literature

1) Lindberg, B. and McPherson, J., Acta Chem. Scand. 8, 1547–50, 1954
2) Lindberg, B., Acta Chem. Scand., 7, 1119–22, 1953
3) Bouveng et alii, Acta Chem. Scand. 9, 807, 1955
4) Lindberg, B. and McPherson, J., Acta Chem Scand. 8, 1875, 1954
5) Lindberg, B., Acta Chem. Scand. 9, 1097, 1955
6) Haas, P. and Hill, T. G., Bioch. J. 25, 1470 (1931); 26, 987, 1932
7) Hassid, W. Z., Plant Physiol. 8, 480 (1933); 11, 461, 1936
8) Lindberg, B., II. Intern. Seaweed Sympos. p. 35–57, Pergamon Press 1956
9) Wickberg, B., Abstr. III. Intern. Seaweed Symp., p. 67, O'Gorman Ltd., Galway 1958
10) Lindberg, B., Acta Chem. Scand. 9, 1093, 1955
11) Kylin, H., Z. physiol. Chem. 94, 337, 1915
12) Colin, H. and Guéguen, E., Compt. rend. 191, 163, 1930; 197, 1688, 1933
13) Turvey, J. R., Colloqu. Intern. Centre Nat. Rech. Scient. 103, 30, 1960, 1961
14) Frei, Eva and Preston, R. D.: Proc. Roy. Soc. London, Ser. B, 160, (980), 293–313, 1964
15) Percival, Elisabeth and R. McDowell, Chemistry and Enzymology of Marine Algae Polysaccharides, Academic Press, 1967

Chlorophylls, Carotinoides, Phycobilins, and other Pigments

A list of chlorophyll contents in green algae, brown algae, and red algae has been compiled by RABINOWITCH[1]). The level of knowledge until 1962 and development trends of research efforts on chlorophylls in algae are the subject of a detailed report by L. BOGORAD[2]). Reports have been published by KAPPANNA and coworkers[3]) pertaining to the chlorophyll content of marine algae on the Suarasthra Coast of India, according to which the ratio chlorophyll a : b fluctuated between 0.784 to 2.123 for 12 species. The properties of a purified chlorophyll C from *Sargassum flavicans* have been described by S. W. JEFFREY[4]).

$E \frac{0{,}1\%}{1\,\text{cm}}$ at 628–630 mμ = 15.8, the maxima are 628, 580, 442 mμ, and the magnesium content amounts to 2.31 ± 0.045%. The more intensive photosynthesis of red algae in the Black Sea is ascribed to the much higher specific activity of their chlorophylls[5]). The photo-synthesizing pigments of *Phyllophora nervosa* are subject to seasonal fluctuations[6]). The carotinoides in algae have been the subject of a comprehensive report by T.O.M. NAKAYAMA[7]). His work was supplemented by the reference made by O. C. OWEN[8]) that like land plants marine algae have a closer correlation of quantity with chlorophyll than with the main constituents of the dry substances of plants. With regard to their functions, that author discusses the possibility of the existence of a protective function in the photo-oxidation process, a rôle as energy transport agent, and only an indirect function in the cell metabolism relating to cell components not yet discernible. Reference should also be made to the work by A. JENSEN[9]) on fucoxanthin and fucoxanthol, and the colour reactions of carotinoids to mercuric chloride[10]).

The correlation of yellow pigment complexes from filtrates of cultures of marine algae and from sea water has been the subject of a report of JENTSCH & REICHERT[11]). They were not able to confirm the carotinoid nature indicated by FOGG & BOALCH[12]). The water-soluble pigment components strongly increase the oxygen uptake in raw sea water. The authors assumed a utilization of decomposition products of marine algae by marine bacteria.

The interesting chromoproteins of marine algae have been investigated particularly by O'HEOCHA[13]), who also gives a survey on modern knowledge on that subject[14]). The chromophorous group seems to be very close to bile pigments. The protein share indicates certain special properties. In hydrolysates of phycoerythrin made from *Porphyra tenera* FUJIWARA[15]) found xylose, arabinose, and galactose.

From the available results of investigations BRODY & BRODY[16]) have concluded that the accessory pigments of algae have a function as energy absorbers, while also transferring energy to chlorophyll a. The same view is held also by YARTSEVA[5]).

The occurence of these pigments adjusts the marine algae to different spectral compositions of radiant energy (cf. p. 14ff.).

Cytochromes have been found in green, brown and red seaweeds[17]) and investigated especially in *Porphyra tenera*[17]) [18]).

Literature

1) Rabinowitsch, E. I. "Photosynthesis and related processes, Interscience, New York 1945, I p. 410
2) Bogorad, L., in Physiology and Biochemistry of Algae, R. A. Lewin p. 385–408, 1962
3) Kappanna, A. N. et al., Current Sci (India) 31, 380–81 (1962) CA. 59, 5493d, 1963
4) Jeffrey, S. W., Biochem. J. 86, 313, 1963
5) Yartseva, I. A., Fiziol. Rast 10 (3), 288–94, 1963, CA. 59, 7857c, 1963
6) Lebedev, S. J., Problemy Fotosinteza, Doklady 2 – oi (Vtoroi) Konf. 1957, p. 409–13 (Publ. 1959) CA. 56, 720a, 1962
7) Nakayama, T. O. M., Physiology and Biochemistry of Algae, R. A. Lewin p. 409–420, Academic Press 1962
8) Owen, E. C., J. Sci. Food, Agricult. 5, 449–453, 1954
9) Jensen, A., Acta Chem. Scand. 15, 1604, 1961; 18, 840, 1964
10) Yamamoto, H. Y. et al., Anal. Chem. 33, Chem. 33, 1792, 1961
11) Yentsch, Ch. S. and Reichert, G. A., Bot. Mar. 3, 65, 1962
12) Fogg, G. E. and Boalch, G. T., Nature 181, 789, 1958
13) O'hEocha, C., Arch. Biochem. Biophys. 73, 207, 1958
14) O'hEocha, C., Physiology & Biochemistry of Algae by R. A. Lewin, Academic Press 1962, pp. 421–435; Proc. roy. Irish Acad. Sect. B. 63, 191–200, 1964
15) Fujiwara, T., Proc. IV. Internat. Seaweed Symp., Pergamon Press 1964, p. 312 ff.
16) Brody, Marcia and Brody, S. S., Physiology and Biochemistry of Algae, R. A. Lewin, Academic Press 1962, p. 9
17) Gibbs, M., in Physiology and Biochemistry of Alga, R. A. Lewin, Academic Press 1962 p. 77
18) Kato, S., Plant cell Physiol. (Tokyo) i, 29–38, 1959 CA 61, 12373f. 1964

Enzymes, Metabolic Problems

The knowledge of enzymes and their specific activity in marine algae is of general interest not only for the natural sciences, but also of considerable significance for the utilization of their constituents. After all the results of an extraction are influenced during harvesting, storage, and drying by decomposition and conversion processes. The quality also of the commercially utilizable products depends on such processes.

Research of enzymes in marine algae is rendered difficult by the very fact that extraction is difficult, and has to be selective depending on the type of enzyme investigated. In addition there is the presence of substances in brown algae and red algae which either directly or indirectly effect the results of the determination of enzyme activity or which disturb the optical part of the analysis work[1]). The present level of knowledge of enzymes in marine algae is still unsatisfactory.

The main cause for this fact are the major differences in respecting the chemistry of the more important carbohydrates, which also naturally require considerably diverse enzyme systems for their development and decomposition. Some works on carbohydrate and energy metabolism have been published by Watanabe[2]), Bean and Hassid[3]), Duncan and coworkers[4]), Bishop[5]), Richter[6]) and Jacobi[7]). The latter has provided a survey of enzymes of carbohydrate metabolism in marine green and red algae, that deserves reprinting here[1]).

Finally the work should be mentioned of Brody and Brody, Holm-Hansen, Spruit, Gibbs[8]), Jacobi[1]), and Ohmann[42]). A more recent study on photosynthesis of littoral algae has been published by Blinks[9]).

Reports have been made on nigerase, an enzyme for the hydrolysis of α (1–3)-bonds, in *Porphyra umbilicalis* by Peat and Rees[10]) and in *Cladophora rupestris* by Hutson and Manners[11]). An α-amylase and the presence of maltase in *Cladophora rupestris* has

Enzymes of Carbohydrate Metabolism in Marine Green and Red Algae

Alga	Hexo-kinase	Aldo-lase	Glyceraldehyde-phosphate dehydrogenase		Alcohol dehydro-genase	Lactic acid dehydro-genase	Malic acid dehydro-genase	Malic enzyme	Glucose-6-phosphate dehydro-genase
			DPN	TPN					
Chlorophyta									
Ulva lactuca	—	+	+	+	—	—	+	+	+
Chaetomorpha linum	—	+	+	+	—	—	+		+
Bryopsis plumosa		+	+	+	—	—	+		+
Rhodophyta									
Delesseria sanguinea	—	—	—	—	—	—	+	+	—
Phycodrys rubens		—	—	—	—	—	+	+	—
Polysiphonia violacea	—	—	—	—	—	—	+		—
Cystoclonium purpurascens		—	—	—	—	—	+	+	—

been reported by MANNERS[12]). The catalase activity of marine algae has been investigated by TAKAGI[13]); he found out that this was highest in *Monostroma angicaba* and *Porphyra pseudolinearis*.

The enzymes and the process of nitrogen metabolism have been the subject of several studies. Reports have been published by TAKAGI and MURATA[14]) on nitrate reductase. Aminoacid-dehydrogenases have been investigated by WATANABE[15]) and by JACOBI[16]). JACOBI[16]) also studied transamination. REUTER after JACOBI[1]) has found ureases in marine algae. A study has been published[16]) on proteinases from *Porphyra* sp. and *Eisenia* sp. Nitrogen assimilation has been described comprehensively by SYRETT[18]). As with the uptake of elementary nitrogen (nitrogen fixation[19])) these studies mainly related to microphytic algae of fresh and sea water.

VILLEBERT[20]) found allantoinase activity in 45 of the 50 species of the green, red, and brown algae. It varies between a 5–15 mg allantoin transformed/g total N.

Investigations by FOGARTY and REES[21]) have shown ways in which sulfate is transferred enzymatically to polysaccharides.

A citrate-susceptible sulfatase and another enzyme effecting a reduction in viscosity of porphyran[10]) have been isolated from *Porphyra umbilicalis*. *Rhodymenia palmata* seems to contain an enzyme decomposing carotene[22]).

At this point some enzymes should be mentioned which have not been isolated from algae, but which are of a bacterial or animal origin. From basidiomycetes Qu M 806 an enzyme has been derived which hydrolytically attacks the non-reducing end of the water-insoluble laminaran[23]). From Bacillus circulans HORIKOSHI and coworkers[24]) have been able to isolate an enzyme found to hydrolyze laminaran. Reports have been published by CHESTERS and BULL[25]) on laminarinases from fungi. Fucoidin is enzymatically split at a ratio of 24.2 per cent by the non-cellular medium of *Pseudomonas carrageenovora*[26]). κ-carrageen is hydrolyzed by an extracellular enzyme of a marine gram-negative rod. The enzyme is active also in extracts from *Furcellaria fastigiata*, *Hypnea musciformis*, *Gigartina acicularis*, and one species of *Iridophycus*[27]). An agarase from *Pseudomonas atlantica* hydrolyzes both commercial agar and extracts from agarophytes[28]). ISHIMATSU and coworkers[29]) have made a number of investigations on agar-liquefying bacteria and their hydrolytic enzymes, the most recent of which (XVIII) has been published in 1959. Further bacteria capable of decomposing agar have been isolated by HIDAKA and SAMESHIMA from the surface of the red alga *Porphyra tenera*[30]). As it can be assumed that bacteria able to decompose phycocolloids are present on the surface of other algae utilized industrially, there is the possibility of enzymes being entrained in the extraction process and of the final products being changed quantitatively by them during storage (cf. Ref. 10). Culture liquids from *Vibrio purpureus* were likewise found to split disaccharides, trisaccharides, and tetrasaccharides from agar[31]).

Bacteria hydrolyzing alginic acid and their enzymes have been investigated by WAKSMAN and coworkers[32]), by ANDO and INOUE[33]), and by WILLIAMS and EAGON[34]).

Mention should also be made of investigations made by FUJISAWA and SUKEGAWA on measures for the suppression of bacterial flora and the production of agar, on which some results have been published in 1956–1958[35]). As marine aerobic cellulose-decomposing bacteria also can use mannitol, alginate, and agar as nutrients[36]), they must also have the respective enzymes.

Different resistance of the chromoproteins of *Porphyra* sp. to *Pseudomonas*-proteinase has been found by MERKEL et alii[41]).

Animal enzyme systems capable of decomposing polysaccharides of marine algae have been investigated by

YSAMI and SAITO[37]) – Algin
GALLI and GIESE[38]) – Laminaran, alginic acid, fucoidan, iridophycan
HUANG and GIESE[39]) – Entire algae and their various constituents
MURAKAMI[40]) – Agar

Further research on the enzymes described will have great theoretical and practical value.

Literature

1) JACOBI, G., Physiology and Biochemistry of Algae, R. A. Lewin, Academic Press 1962, p. 125 ff.
2) WATANABE, A., Acta Phytochimica (Japan) 6, 315, 1932; 9, 235, 1937; 9, 255, 1937
3) BEAN, R. C. and HASSID, W. Z., J. biol. Chem. 212, 411, 1955; 218, 425, 1956; Science, 124, 171, 1956
4) DUNCAN, W. A. et alii, Biochem. J. 63, 44, 1956; 69, 343, 1958; 73, 295, 1959
5) BISHOP, N. J. et alii, Plant Physiol. 34, 551, 1959
6) RICHTER, G., Z. Naturforschung 12b, 662, 1957
7) JACOBI, G., Planta 49, 1, 561, 1957, Kieler Meeresforsch. 13, 212, 1957; 14, 247, 1958; 15, 161, 1959
8) Physiology and Biochemistry of Algae by R. A. Lewin pp. 1, 25, 47, 61, 91, Academic Press 1962
9) BLINKS, L. R., Protoplasma 57, 126–36, 1963
10) PEAT, S. und REES, D. A., Biochem. J. 79, 7, 1961
11) HUTSON, D. H. and MANNERS, D. J., Biochem. J. 89, 91 P, 1963
12) MANNERS, D. J., Compt. rend. IV. Congr. Intern. Alg. Mar. 1961 (publ. 1964) p. 348
13) TAKAGI, M., Bull. Japan. Soc. Sci. Fisheries 18, 483, 1953
14) TAKAGI, M. and MURATA, K., Bull. Fac. Fisheries Hokkaido Univ. 4, 296, 306, 310 (1954); 5, 173, 176, (1954); 6, 25, 29, 33, 1955
15) WATANABE, A., Acta Phytochimica (Japan) 15, 129, 1949
16) JACOBI, G., Planta 49, 561, 1957
17) TAZAWA, Y. and MIWA, A., Botan. Mag. (Tokyo) 66, 77, 1953
18) SYRETT, P. J., Physiology and Biochemistry, R. A. Lewin, Academic Press 1962, p. 171
19) FOGG, G. E., Physiol. and Biochem., R. A. Lewin, Acad. Press 1962, p. 161
20) VILLEBERT, S., Compt. rend. 246, 1452, 1948
21) FOGARTY, M. and REES, W. R., Compt. rend. IV. Congr. Intern. Alg. Mar. 1961 (publ. 1964) p. 306
22) HAUG, A. and LARSEN, N. B., II. Intern. Seaweed Sympos. Trondheim, p. 16 (publ. 1956)
23) NELSON, T. et alii, Can. J. Chem. 41, 1671, 1963
24) HORIKOSHI, K. et alii, Biochem. Biophys. Acta 73, 267, 1963
25) CHESTERS, C. G. C. and BULL, A. T., Biochem. J. 86, 28, 31, 38, 1963
26) YAPHE, W. and MORGAN, M., Nature 183, 761, 1959
27) YAPHE, W. and BAXTER, B., Appl. Microbiol. 3, 380, 1955
28) YAPHE, W., Can. J. Microbiol. 3, 987, 1957
29) ISHIMATSU, K. et alii, Kagaku to Kogyo (Osaka) 33, 409, 1959, CA. 55, 2787a, 1961
30) HIDAKA, T. and SAMESHIMA, M., Kagoshima Daigaku Suisan Gakuba Kiyô 3, 158–64, 1953, CA. 51, 14884i, 1957
31) MITSUYOSHI, Y. and WATANABE, K., Hyógo Nóka Daigaku Kenkyú Hókoku 3, 53 (1957) CA. 52, 19198d, 1958
32) WAKSMAN, S. A. et alii, J. Bacteriol. 28, 213
33) ANDO, Y. and JNOUE, K., Nippon Suisan Gakkaishi 27, 339, 342, 1961, CA. 57, 1390a, 1962
34) WILLIAMS, A. K. and EAGON, R. G., Can. J. Microbiol. 8, 649, 1962
35) FUJISAWA, H. and SUKEGAWA, T., I. Bull. Japan Soc. Sci. Fisheries 22, 30 (1956) CA. 52, 4105b, 1958; III. Bull. Japan Soc. Sci. Fisheries 23, 641 (1958) CA. 53, 6483d, 1959; IV. Bull. Japan Soc. Sci. Fisheries 24, 19 (1958) CA. 53, 7461a, 1959; V. Bull. Japan Soc. Sci. Fisheries 24, 342 (1958) CA. 53, 11715e, 1959
36) KADOTA, H., Mem. Coll. Agr. Kyoto Univ. 74, 1–128 (1956) CA. 50, 14043, 1956
37) YSAMI, T. and SAITO, T., Agr. Biol. Chem. (Tokyo) 26, 115, 1962, CA. 57, 6309, 1962
38) GALLI, D. R. and GIESE, A. C., J. Exptl. Zool. 140, 415, 1959
39) HUANG, HELEN and GIESE, A. C., Science 127, 475, 1958
40) MURAKAMI, S., Sci. Repts. Saitama Univ., Ser. B. 2, 191, 1956, CA. 51, 18161c, 1957
41) MERKEL, I. R. et alii, J. Bacteriol. 88, 974–80, 1964
42) OHMANN, E., Publ. Staz. Zool. Napoli 33, 331–40, 1963. CA. 61, 3418, 1964

Fats and Lipoids

According to Russel-Wells[1]) the fat content of the brown algae is between 0.16 and 6.3 per cent, while red algae, according to Ross[2]), have 0.4 to 3.2 per cent of fat. Considerable progress in the analysis of fats has been made by the introductions of gas-chromatographic methods, which have helped to enhance our knowledge of the exact composition. The fatty acids of three red algae, according to Laur-Henry[3]), can be separated as follows:

Percentage of total fatty acids:

	Rhodymenia palmata	*Furcellaria fastigiata*		*Polysiphonia elongata*
		DEGS	Apiezon	
C_8	0,5	1,7	1,7	0,7
C_{10}	—	—	—	—
C_{12}	0,1	2,3	2,3	0,7
C_{14} myristine				
C_{14} Myristoleine				
C_{14} total	10,7/13,8	8,4	5,0	4,0
C_{16} / C_{16} total	31,5	57,1		57,5
C_{16} unsaturated	—	1,1		
C_{18}	1,5/8,0	3,3	3,5	1,7
C_{18} oleic acid	5,6/9,0	11,5	16,9	5,3
C_{18} 2x unsaturated	1,4	0,8	0,7	2,9
C_{18} 3x unsaturated				
C_{20} polyunsaturated	48,7	15,7		28,4

That author compared in another paper[4]) the results of *Rhodymenia palmata* with the freshwater alga *Lemanea nodosa* Kütz and found C_{16} and C_{20} fatty acids in exactly the reversed relationship (30 per cent and 50 per cent as compared with 50 per cent and 30 per cent).

Further analyses of the fatty Acids of Green Algae, Brown Algae and Red Algae have been published by E. Klenk and coworkers[5]).

In green algae the unsaturated C_{18} fatty acids are more strongly represented, in which also tetraic acid was found. But also the unsaturated C_{16} fatty acids have a very high proportion.

In the red algae C_{16} and C_{18} are not very important, while C_{20} polyenic acids dominate. The fatty acids from the brown algae range somewhere in the middle between green algae and red algae. The synthesis of unsaturated fatty acids has been investigated by J. A. Erwin and D. Hulanicka[12]) in *Porphyridium cruentum*.

With other methods Wagner and Pohl[6]) have investigated the polyenic fatty acids of *Codium* Stackhouse, *Ulva lactuca*, *Lessonia fusca*, and *Porphyra columbina* Ag. and found that they contain considerable quantities of these acids, mainly C_{18} and C_{20} unsaturated fatty acids. The differences between the algae and the particular findings have been corroborated highly by the results of other authors.

Analyses of the fatty Acids of Green Algae, Brown Algae, and Red Algae[5]):

Types of algae	Algae dry matter g	Unsapo-nifiable matter g	Total fatty acids g	Saturated fatty acids g	Saturated fatty acids iodine number	Unsaturated fatty acids g	Unsaturated fatty acids iodine number
Green algae:							
Codium fragile	161	1,1	5,1	1,9	19	3,2	214
Enteromorpha compressa	653	4,1	22,2	5,1	43	14,0	256
Brown algae:							
Fucus serratus	1,900	5	31	13,2	—	25,9	—
Fucus platycarpus	2,000	5	32	13,1	—	26,1	—
Fucus vesiculosus	1,750	5	32,5	12,9	—	25,1	—
Red algae:							
Rhodomela subfusca	666	1,4	7,0	4,0	22	4,2	244
Ceramium rubrum	780	2,4	0,9	4,0	12	4,7	200
Plocamium coccineum a	150	1,3	1,9	0,6	18	1,3	174
Plocamium coccineum b	661	8,8	8,2	3,1	17	5,1	209

A comprehensive description of sterols present in the marine algae is given by J. D. A. MILLER[7]).

Content of Sterols in Marine Algae:

Sitosterol: *Enteromorpha compressa, Ulva lactuca, Cladophora sauteri, Gelidium corneum, Chondrus crispus, Gigartina stellata, Phyllophora membranifolia, Plocamium coccineum, Ceramium rubrum, Polysiphonia lanosa (s. fastigiata).*

Fucosterol: All previously investigated species, those of *Eisenia bicyclis* and *Sargassum ringgoldianum* contain moreover Sargasterol, moreover *Porphyra umbilicalis, Rhodymenia palmata, Polysiphonia nigrescens.*

Cholesterol: *Acanthopeltis jap., Gelidium amansii, G. japonicum, G. subcostatum, Pterocladia tenuis, Gloiopeltis furcata, Grateloupia elliptica, Tichocarpus crinitus, Chondrus giganteus, Ch. ocellatus, Iridophycus cornucopiae, Rhodoglossum pulchrum, Cocloseira pacifica, Rhodomela larix.*

Chalinasterol: *Acanthopeltis japonica, Gelidium amansii, Pterocladia tenuis.*

22-Dehydrocholesterol: *Hypnea japonica.*

Biosynthesis of the lateral ethylidene side chain of fucosterol has been described by VILLANUEVA et alii[13]).

MILLER refers to the significant contrast between the sterols in Japanese and British red algae, as the former contain cholesterol, the latter fucosterol or sitosterol.

Sterols of *Fucus gardneri* and from *Sargassum muticum* in animal experiments significantly reduced the plasma-cholesterine level[8]).

BENSON and SHIBUYA[9]) have tabulated the contents of galactosyl glycerids and other phosphatidyl compounds in 5 red algae and 2 green algae.

In *Ptilota pectinata* and *Prophyra umbilicalis* taurine has been found to occur, while in *Gelidium cartilagineum* taurine, n-methyltaurine and di-N-methyltaurine as well as choline sulphate have been established[10]). Taurine occurs furthermore in *Porphyra perforata*[11]).

Literature

1) RUSSEL-WELLS, B., Nature 129, 654, 1932
2) ROSS, A. G., J. Sci. Food Agric. 4, 333, 1953
3) LAUR, HENRY, M.-Hélène, Compt. rend. Acad. Sci. 253, 966, 1961
4) LAUR, HENRY, M.-Hélène, Compt. rend. IV. Congr. Intern. Alg. Mar. p. 345, Academ. Press 1961–64
5) KLENK, E. et alii, Hoppe Seyler' Zeitschrift f. physiol. Chemie 334, 44–59, 1963
6) WAGNER, H. and POHL, P., Naturwiss. 51, 163, 1964
7) MILLER, J. A. D., Physiol. and Biochem. Algae by R. A. Lewin, p. 362, Academic Press 1962
8) REINER, E. et alii, Can. J. Bioch. Physiol. 40, 1401, 1962
9) BENSON, A. A. and SHIBUYA, J., Physiol. and Biochem. Algae by R. A. Lewin, p. 379
10) LINDBERG, B. Acta Chem. Scand. 9, 1323, 1955
11) SU, J.-CH. and HASSID, W. Z., Biochemistry 1, 474, 1962
12) ERWIN, J. A. and HULANICKA, D., Comp. Biochem. Physiol. 12, 191–207, 1964
13) VILLANUEVA, V. R. et alii, Bull. Soc. Chim. Franc. 1964, 1423–4

Floridean Starch

Raw Materials and Occurrence

Floridean starch is a reserve polysaccharide found to be present in various Rhodophyceae, for example in *Dilsea edulis* (cf. p. 225) and *Furcellaria fastigiata* (cf. p. 230).

Structure

Floridean starch, a reserve carbohydrate of red algae has been described in detail by J. D. MEEUSE in LEWIN, R. A. This is an α-1-4 polyglucane, which apparently is closely associated with amylopectin, although α-1-3 bonds, too, have been proved to be existing.

Utilization

Although floridean starch today is not yet a commercial product, it appears advisable in this section of "algal products" to refer to this common constituent of red algae.

Literature

LEWIN, R. A., Physiology and Biochemistry of Algae, Academic Press Inc., New York 1962

WHISTLER, R. L. and BEMILLER, J. N., Industrial Gums, Academic Press Inc., New York 1962

See also *Dilsea*, *Furcellaria*

Fucose

Raw Materials and Occurrences

Fucose is a constituent of fucoidan (quod vide), which occurs in brown algae. *Fucus vesiculosus* contains 9.2 per cent and *Laminaria hyperborea* 3.14 per cent (dry weight) of fucose[1]). *Pelvetia canaliculata*, dried and pulverized, yielded 28% L-fucose[2]). Fucose occurs also in fungi (HEGNAUER, 1962).

Production

The isolation of fucose from *P. canaliculata* has been described by BLACK et al.[2]), the production from *Fucus serratus* has been made by TRABERT[3]).

Literature

1) BLACK, W. A. P. et al., J. Sci. Food Agric. 1955. 754–764
2) BLACK, W. A. P. et al., J. Sci. Food Agric. 4. 85–91. 1953
3) TRABERT, C. H., Archiv d. Pharmazie 293. 278, 1960

General Literature
Institute of Seaweed Research, Ann. Rep. 1955
ibid., Ann. Rep. 1960
HEGNAUER, R., 1962
MCNEELY, W. H., Fucoidan ex Industrial Gums, Academic Press. Inc., New York, 1959
See also *Fucus*, *Laminaria*

Inorganic Components

The most comprehensive survey on inorganic components of marine algae comes from VINOGRADOV[1]). The following data were taken from that study:

Water content

	in % of living matter
Chlorophyceae	
Enteromorpha sp.	70,83
Ulva sp.	78,0
Monostroma sp.	82,60
Phaeophyceae	
Nereocystis luetkana	91,58–92,80
Macrocystis pyrifera	86,08–87,00
Laminaria digitata	73,07–87,50
Laminaria saccharina	85,38–87,99
Fucus vesiculosus	68.17–76,55
Ascophyllum nodosum	70,52–80,47
Cystoseira fibrosa	75,0
Rhodophyceae	
Rhodymenia palmata	78,20–86,25
Chondrus crispus	75,37–80,84
Porphyra laciniata	86,0
Gigartina mamillosa	68,0
Furcellaria fastigiata	78,36
Phyllophora membranifera	66,18
Ahnfeltia plicata	59,04

Ash content

Phaeophyceae	Ash content in % of living matter	content Ash in % of dry matter
Nereocystis luetkana	4,76	50,57
Macrocystis pyrifera	4,96	35,62
Laminaria spp.	3,81–7,27	18,40–41,2
Fucus vesiculosus	4,89–6,38	16,18–22,20
Ascophyllum nodosum	6,14	23,6
Rhodophyceae		
Chondrus crispus	5,0–6,4	24,90–27,7
Rhodymenia palmata	3,9	26,9
Porphyra laciniata	2,25	15,7
Rhodomela pinastroides	9,7	53,40
Iridaea sp.	6,25	31,4

Composition of Ash

	Na_2O	K_2O	MgO	CaO	SO_3	P_2O_5	Cl	Fe_2O_3	SiO_2
Phaeophyceae									
Nereocystis luetkana	–14,12	36,90	2,42	2,76	3,62	1,35	38,50	0,16	—
Macrocystis pyrifera	–13,75	34,0	3,58	6,70	6,40	1,50	33,70	0,40	—
Laminaria	8,45	6,81	5,18	5,45	2,33	1,88	17,23	0,21	0,34
digitata	–26,55	–44,74	–12,91	–11,86	–22,01	–4,47	–38,67	–0,62	–3,10
Laminaria	4,79	8,14	2,15	1,35	4,15	0,65	0,53	0,062	0,21
saccharina	–24,39	–37,86	–14,49	–24,62	–32,67	–4,31	–34,63	–1,06	–1,14
Fucus	8,67	4,85	0,84	3,53	11,42	1,36	16,90	0,18	0,27
vesiculosus	–34,12	–21,41	–15,19	–28,45	–32,56	–5,86	–26,11	–8,58	–16,55
Ascophyllum	17,68	10,07	6,91	7,46	23,47	0,88	12,24	0,28	0,12
nodosum	–26,59	–21,53	–10,93	–12,80	–32,04	–1,83	–18,39	–10,86	–3,09
Rhodophyceae									
Chondrus crispus	18,73	17,32	11,35	7,16	41,24	0,41	3,79	—	—
Furcellaria	15,71	9,24	8,23	7,25	30,92	1,94	5,24	1,17	2,75
fastigiata	–23,47	–20,24	–10,46	–17,35	–36,62	–2,21	–8,39		
Gracilaria conferoides	26,37	3,94	2,00	6,25	7,61	4,61	23,53	14,96	14,60
Iridaea edulis	16,93	23,42	?	20,46	25,20	13,01	0,98	—	—
Delesseria	20,59	8,61	1,87	4,35	2,44	2,36	3,31	0,63	0,30
sanguinea	–23,17	–14,90	–6,46	–16,53	–44,18	–46,49	–4,58		

It should be noted that these values also include analyses obtained a very long time ago.

An investigation of the methods for the determination of water and ash content, and the sources of error pertaining thereto have been made by CLENDENNING[2]). He particularly refers to changes in ash content by the nitric or sulphuric acid often used as a catalyst. His observations on the water bound by polysaccharides should also be included with the analyses of marine algae.

The following data of the iodine content of marine algae have been published[1]):

Iodine content in per cent	Living matter	Dry matter	Ash
Phaeophyceae			
Laminaria saccharina	0,043–0,15	0,14–0,496	0,55–5,0
Laminaria digitata	0,045–0,15	0,10–1,7	0,758–5,5
Macrocystis pyrifera	—	0,13–0,27	—
Nereocystis luetkana	—	0,07–0,22	—
Fucus vesiculosus	0,0027–0,003	0,0113–0,026	0,028–1,4
Ascophyllum nodosum	0,004–0,011	0,0396–0,198	0,056–0,7934
Rhodophyceae			
Porphyra spp.	0,0008	0,0018–0,062	
Gelidium spp.	—	0,0075–0,177	
Gracilaria confervoides	—	0,0018	
Gigartina spp.		0,006–0,123	
Chondrus spp.	0,0018–0,0069	0,0023–0,0765	
Phyllophora spp.	—	0,27–0,58	
Iridaea spp.	0,0083	0,036–0,092	
Furcellaria fastigiata	0,0012	trace – 0,081	0,06
Eucheuma spp.	—	0.0009–0,0035	
Chlorophyceae			
Enteromorpha	—	0,0007–0,022	
Ulva spp.	—	0,0022–0,055	
Monostroma sp.	0,0008	0,0016–0,008	

Also more recent investigations, for example those by A. N. Kappanna[3]) on Indian algae, have confirmed earlier findings, according to which the iodine content of brown algae is higher than that of red algae and green algae.

Larsen and Haug[4]) once again investigated the distribution of iodine in *Laminaria* tissue and found up to 4.0 g of iodine per 100 gramme of dry matter in the peripheral tissue of the lower stipes of *L. hyperborea*.

Work on the production of iodine from algae is still going on[5]).

According to investigations made by Morita and Sato iodine is not contained in the algae as an alkyl iodine[6]). Tong and Chaikoff[7]) have found a complexing of iodine in organic form within four hours. Roche and Andre, however, have found that radio iodine was stored mostly in organic form[8]). Monoiodotyrosine and triiodothyronine have been found by Scott[9]), and diiodotyrosine by Roche and Lafon[10]). Further references to the presence of iodine and its metabolism have been published by T. J. Shaw[11]).

SUNDE's[12]) investigations on iodine containing algae as fodder are significant: the iodine content of eggs of hens fed with brown algae was 125-910 μg/egg compared with 6 μ g/egg for reference hen. 80–90 per cent of the iodine have been found in the yolk of the eggs.

The utilization of seaweed meal as source of iodine for dairy cows has been the subject of reports by E. BOOTH[13]).

The bromide content of marine algae are outlined in the following table[1]):

	Per cent bromide in living weight	Per cent bromide in dry matter
Phaeophyceae		
Laminaria saccharina	0,032	0,034
Laminaria digitata	0,05	0,138
Fucus vesiculosus	0,010	0,094
Cystoseira barbata		0,098
Ascophyllum nodosum	0,017	
Rhodophyceae		
Rhodymenia palmata	0,011	
Delesseria sinuosa	0,0256	
Odonthalia dentata	0,674	3,0
Chlorophyceae		
Monostroma fuscum	0,017	
Enteromorpha sp.	0,0031	

The accumulation of bromine from seawater by algae is 10^1 times, the accumulation of iodine from seawater, however, 10^5 times.

SAITO and ANDO[14]) have identified a bromine compound in *Polysiphonia morrowii* Harv. as 5-bromo-3.4-dihydroxy benzaldehyde. For a bromated phenol compound, cf. Section on Antibiotics.

Arsenic

Marine organisms contain 10 times as much arsenic as do land organisms[1]).

Apart from arsenic also boron has been found in agar (according to[1]).

Arsenic Content in mg/100 g of Dry Matter[1])

Laminaria digitata	0,05–5,55
Laminaria saccharina	0,01–9,5
Fucus vesiculosus	0,01–6,0
Cystoseira sp.	0,04
Chondrus crispus	0,07–1,20
Laurencia obtusa	0,50
Gigartina mamillosa	1,80
Ulva lactuca	0,015
Cladophora utriculosa	0,008–0,08
Enteromorpha compressa	1,2

Other elements shown to be present in marine algae (according to[1])) are the following: Aluminium, iron, manganese, copper, titanium, zinc, lead, molybdenum, cobalt, nickel, tin, mercury, silver, gold, vanadium, chromium, boron, bismuth, gallium, antimony, germanium, beryllium, tungsten, praseodymium, neodymium, samarium, cadmium, cerium, lanthanum, yttrium, lithium, rubidium, strontium, barium, thallium, fluorine, radium.

References in addition to those in VINOGRADOV are to be found in the "Chemical Studies of the Ocean" by Japanese authors, abstracts of which appeared in "Chemical Abstracts" 1959–1963.

The selective accumulation of particular elements from sea water gives many problems. Caesium^{134} is accumulated n x 100 times by *Fucus vesiculosus* and n x 100 times by *Rhodymenia palmata*[15]). Since light exerts a clearly promotive influence upon the uptake, it can be assumed that the process depends on metabolism. Yttrium^{91}-accumulation in *Ceramium rubrum*, *Bryopis plumosa*, *Enteromorpha compressa*, and *Ulva rigida* amounts to n x 100 – n x 1000 times of the respective content in sea water[16]). GUTKNECHT[17]) thinks that he has found a correlation between photosynthesis and the uptake of zinc in *Ulva lactuca*, but regards this influence as secondary to that of diffusion and pH. R. W. EPPLEY[18]) wrote on the uptake of radio-active wastes by fresh water and marine algae. In their investigations relating to the uptake of radio-isotopes from waste waters of a nuclear pile, HENDERSON and coworkers[19]) have found out that the uptake of P^{32} is most marked at some distance from the atomic pile.

As example for the investigations in retardation of the uptake of isotopes by seaweed-extracts may be mentioned that alginic acid in the food decreases the bone uptake of Sr^{89}.[20]).

Literature

1) VINOGRADOV, A. P., The Elementary Chemical Composition of Marine Organisms, Sears Foundation for Marine Research, New Haven 1953, 17ff.
2) CLENDENNING, K. A., Bot. Marina 4, 204, 1962
3) KAPPANNA, A. N. u. V. Sitakara Rao, J. Sci. Ind. Res. 21B, 559, 1962
4) LARSEN, B. and HAUG, A., Bot. Marina 2, 250, 1961
5) GLOESS, M. P., French Patent 978.099, 1951
6) MORITA, M. and SATO, H., Nippon Kagaku Zasshi 75, 416, 1954, CA. 5, 11491 c/1957
7) TONG, W. and CHAIKOFF, J. L., J. Biol. Chem. 215, 473, 1955
8) ROCHE, J. and SIMONNE ANDRE, Compt. rend. Soc. Biol. 156, 1968–71, 1962
9) SCOTT, R., Nature 173, 1098–99, 1954
10) ROCHE, J. and LAFON, M., Compt. rend. 229, 481, 1949
11) SHAW, T. J., Physiology and Biochemistry of Algae by R. A. Lewin, p. 247, Academic Press 1962
12) SUNDE, J., II. Intern. Seaweed Sympos. pp. 109–13, 1955/56
13) BOOTH, E., Compt. rend. IV. Congr. Intern. Alg. mar. 385ff., Pergamon Press 1964
14) SAITO, Ts. and ANDO, Y., Nippon Kagaku Zasshi 76, 478/1955. CA. 51, 178101i, 1957
15) SCOTT, R., Proc. 2nd Radioisotope Conf. Oxford, I. Med. & Physiol. Appl. p. 373–80, 1954
16) POLYKARPOV, G. G. and AKAMSIN, A. D., Trudy Sevastopol Biol. Stantsii im A. O. Kovalevskogo Akad. Nauk USSR 13, 299–301, 1960, CA. 57, 3829c, 1962
17) GUTKNECHT, J., Limnol. Oceanogr. 6, 426–31, 1961
18) EPPLEY, R. W., Physiology and Biochemistry of Algae, R. A. Lewin, Academic Press 1962, p. 839

19) Henderson, C. et al., US Atomic Energy Comm. Wash 275 p. 153–65, 1954

20) Paul, T. M. et alii, Can. Med. Assoc. J. 91 553, 1964

Vitamins

Many studies have been published on the vitamin content of marine algae. Vitamin A has been proved for brown algae[1]), and its presence is most likely in red algae. The presence of preliminary phases of the vitamin, carotenes, has been described by several authors. In addition to the work mentioned in the Chapter Chlorophylls . . . p. 375 the survey published by Larsen and Haug[2]) is reprinted here, for it has made use also of the results of other researchers.

	Carotene		Total Carotene in ppm
	α a)	β	
Ahnfeltia plicata	+(+)	+(+)	69
Ceramium rubrum	+(+)	+(+)	326
Chondrus crispus	+(+)	+(+)	84
Dellesseria sanguinea	+(—)	+(+)	270
Dumontia incrassata	+	+	222
Furcellaria fastig.	+(—)	+(+)	135
Gigartina stellata	(+?)	+(+)	86
Membranoptera alata	+	+	67
Ptilota plumosa	+	+	65
Phycodrys sinuosa	+	?	
Rhodymenia palmata	+(—)	+(+)	360
Callithamnion arbuscula	—	+	200
Corallina officinalis	—(—)	+(+)	/
Polysiphonia fastig.	—(—)	+(+)	245
Polysiphonia urceolata	—	+	/
Porphyra sp.	—(—)	+(+)	85

a) The symbols in parentheses refer to earlier results

Seybold and Egle[3]) have found 249–285 mg carotene/1000 g dryweight in *Rhodymenia palmata* and 93–167 mg/1000 g of dryweight in *Chondrus crispus*. Haug and Larsen[4]) have studied the influence of the drying processes upon the carotene content in the same alga and found that it declined rapidly with protracted drying processes, in contrast to brown algae. This is of practical significance for red alga utilization in, for example, feeding products. The vitamins of the B-complex have been studied on several occasions[1]) [5]).

The thiamine content of *Furcellaria fastigiata* has been found to amount to 1.62, of *Chondrus crispus* to 2.88, of *Rhodomela subfusca* to 3.04, and of *Porphyra laciniata* to 2.82 μg/g dryweight[6]).

KANAZAWA[7]) has found quantities of 0.27 to 4.60 μg thiamine/g dryweight in marine algae and an abundant quantity in *Gracilaria testorii*, *Gloiopeltis tenax*, and *Chondrus ocellatus*.

As much as 3.7 mg riboflavine/100 g dryweight is found in *Porphyra tenera*[8]).

Riboflavine has been found to be abundant in *Porphyra tenera*, *Gelidium amansii*, *Chondrus ocellatus*, and *Gloiopeltis tenax*; generally the quantities found ranged between 0.84 and 23.08 μg/g dryweight.

Niacin has been found by ERICSON and CARLSON[9]) in marine algae in quantities of 1–63 μg/g dryweight. This agrees with KANAZAWA's[7]) findings who found 3.51–68.33 μg of Niacin per gramme of dryweight.

Also for pantothenic acid good corroboration of the results indicated by the two above mentioned authors (0.25–12.21 μg[7]) and 0.18–12.5 μg[9])) has been established.

As regards vitamin B_6 no work has been found in respective scientific literature. However, it cannot be assumed that this vitamin is absent from marine algae.

ERICSON[10]) gives 0.004 to 2.8 μg/g of dryweight for vitamin B_{12}, a value higher than the highest mentioned in KANAZAWA's investigations[7]) (3.3–290.8 mμg/g dryweight).

The values for folic acid vary between 46.0–857.1 mμg/g dryweight[7]) and 0.5–2.0 μg/g dryweight[10]).

According to HASHIMOTO and MAEDA[11]) values for fresh marine algae vary between 0.1 and 20.0 μg per cent.

Red and green algae are a better vitamin B source than brown algae[10]) [12]).

Vitamin C values for green, red, and brown algae vary between 3–135 mg per 100 gramme of fresh tissue[13]). This confirms earlier findings by NORRIS[5]), viz. that *Fucus vesiculosus* can attain as high a value as 77 mg/100 g of wet weight.

Vitamin D has been proved to be present in brown algae[1]). JOHNSON and LEVRING have found 54 U. per g oil extracted from *Fucus vesiculosus*[1]).

The δ-tocopherol content in marine algae varies between 1–34 mg per 100 g of dryweight[14]).

See chapter Fats and Lipoids p. 380 for the fatty acids of the socalled vitamin F.

In view of the importance of vitamins in marine algae used for feeding purposes, MAUTNER[1a]) appears to be correct in advocating much more qualitative and quantitative research work in this field.

Literature

1) JOHNSON, N. G. and LEVRING, T., Svensk. Hydrogr. Biol. Kom. Skriftl. III, 1 : 3, 1947
1a) MAUTNER, H. G., Econ. Botany 8, 174, 1954
2) LARSEN, B. and HAUG, A., Acta Chem. Scand. 10, 470, 1956
3) SEYBOLD, A. and EGLE, K., Jahrb. wiss. Botan. 86, 50, 1938
4) HAUG, A. and LARSEN, B., II. Intern. Seaweed Symp. pag. 17, 1956
5) NORRIS, E. R. et al., J. Nutr. 13, 425–33, 1937
6) GERDES, G., Arch. Mikrobiol. 16, 53, 1951
7) KANAZAWA, A., Kagoshima Daigaku Suisan Gakubu Kiyô 10, 38–69 (1961) CA. 58, 5986, 1963
8) TSUJIMURA, M. et al., J. Agric. Chem. Soc. Japan 26, 11, 1952
9) ERICSON, L.-E. and CARLSON, BLENDA, Arkiv Kemi 6, 511, 1953
10) ERICSON, L.-E., Thesis Uppsala 1953
11) HASHIMOTO, Y. and MAEDA, Y., Bull. Japan. Soc. Sci. Fisheries 19, 141–4, 1953
12) TEERI, A. E. and BIEBER, R. E., Science 127, 1500, 1958
13) CREAC'H, P. V. and BARAUD, Y., C. r. soc. biol. 148, 105–7, 1954
14) BROWN, F., Chem. and Ind. 1953 p. 174

Miscellaneous

This section makes reference to works of various subjects that have not been mentioned in the other chapters.

Volatile constituents of marine algae have been established to be as follows by KATAYAMA, T.:

formic, acetic, propionic, butyric, valeric, caproic, caprylic, palmitic acid, p-cresol benzaldehyde, valeraldehyde, methylfurfural, furfural, α-pinene, D-limonene, linalool, geraniol [1a]).

Apart from the described substances: myristic acid, 4 fluoresceing substances [1b]), dimethylsulfide and arylic acid in some green and red algae, not, however, in the brown algae [1c]).

The same author discusses their phylogenetic and biochemical significance, for example as fragrant substances ([1c]), and gives a survey on the present level of knowledge in this field ([1d]).

Constituents of marine algae are of importance also in medical respect. Their use particularly as antilipaemic and anticoagulants[2]) [3]) [4]) [5]) has been investigated. A survey has been published by IRIS L. LIEBER[6]). These investigations in pharmaceutical fields, however, apparently have not yet gained any real significance. Animal experiments and investigations of the effects of decomposed carrageenan for acute gastric ulceration with positive effects have been described[7]). Sargaline, an extract substance from *Sargassum confusum* has been reported to reduce the blood sugar level[8]). The anthelminthic effect of extract substances of marine red algae has been described by several authors. H. A. HOPPE[9]) ascribes this effect to *Alsidium helminthochortum.* Domoic acid from *Chondria armata*[10]) and kainic acid from *Digenea simplex*[11]) is effective against Ascarides. According to GERBER and coworkers[12]) agar and carrageenan protect chicken embryos against infection by influenza B and mumps-virus.

Granuloma will arise[3]) after injection of agar. The formation of abscesses after injection of carrageenan is suitable for the testing of inflammation inhibiting substances[14]) [15]). Carrageenan favours the development of sarcomas after feeding of 2-acetamidofluorene[16]). On the other hand sulfated degraded laminaran has been found to inhibit the growth of sarcoma 180 when injected at the transplantation point or in the tumour[17]). An inhibitive effect upon tumours has also been reported for the seaweed extract Algasol T 331[18]).

The same preparation CLAUDIO and STENDARDO consider as a valuable auxiliary medicament in the treatment of tumours[19]).

Growth substances like auxins and gibberellins in algae are described by CONRAD and SALTMAN[20]) and recently by Joyce A. MOWAT[21]). It remains to further research to find out which substances are responsible for the interesting observation published by AITKEN, SENN, and MARTIN[22]) that algal extracts improve the frost resistance of plants.

Caulerpicin, responsible for the peppery taste of some *Caulerpa* sp., is a long chain saturated hydroxylamide ($C_{43}H_{87}O_2N$). 600 mg of the obvious toxic substance could be isolated from 1500 g dry *Caulerpa racemosa*[23]).

Literature

1) KATAYAMA, T. a) Nippon Suisan Gakkaishi 24, 346, 925, 1959; CA. 53, 11532c, 1959; CA. 57, 15512f, 1962; b) Nippon Suisan Gakkaishi 26, 818, 1960, CA. 55, 15630h, 1961; c) Nippon Suisan Gakkaishi 27, 75, 1961; d) Physiology and Biochemistry of Algae by R. A. Lewin, Acad. Press 1962 p. 467

2) CONSTATIMIDES, P. et alii, Arch. Intern. pharmacodyn. 99, 334, 1954

3) HAWKINS, W. W. and O'NEILL, A. N., Can. J. Biochem. Phys. 33, 545, 1955

4) HOUCK, J. C. et alii, Proc. Soc. Exptl. Biol. Med. 96, 528, 1957

5) MARIA D. G. PINEDA and DOMINICA MONTEQUI, Bol. inst. espan. oceanogr. Nr. 72/1955

5a) GÜVEN, K. C. and AKTIN, E., Bot. Marina 7, 1, 1964

6) LIEBER, IRIS I., Prensa Med. Arg. 40, 1815, 1962

7) ANDERSON, W. and SOMAN, P. D., Nature 199, 389, 1963

8) SAITO, K. and NAKAMURA, J. Chem. Soc. Japan. Pure Chem. Sect. 72, 992, 1950

9) HOPPE, H. A., Drogenkunde, Cram, de Gruyter, Hamburg 1958, p. 48

10) TAKEMOTO, Ts., Jap. J. Pharmacy & Chem. 3k, 461, 1959

11) MURAKAMI, S., TAKEMOTO, T. and SHIMIZU, Z., J. Pharm. Soc. Jap. 73, 1026, 1953

12) GERBER, P. et alii, Proc. Soc. Exptl. Biol. Med. 99, 590, 1958

13) KONNO, K. and TETSUKA, T., J. Biochem. (Japan) 53, 231, 1963

14) BENITZ, K. F. and HALL, L. M., Arch. Intern. Pharmacodyn. 144, 185, 1963

15) GOLDSTEIN, S. and SCHNALL, M., Arch. Intern. Pharmacodyn. 144, 269, 1963

16) WALPOLE, A. C., Morphol. Precursors of Cancer. Proc. Intern. Conf. Perugia 1961, p. 83–88 (publ. 1962)

17) JOLLES, B. et alii, Brit. J. Cancer, 109, 15, 1963

18) ROSSITO, G., Abstr. III. Intern. Seaweed Symp. Galway, p. 46, 1958

19) CLAUDIO, F. and STENDARDO, B., Abstr. V. Intern. Seaweed Symp., p. 7, 1965

20) CONRAD, H. M. and SALTMAN, P. in Physiology and Biochemistry of Algae by R. A. Lewin, p. 663, Acad. Press 1962

21) MOWAT, JOYCE A., Bot. Mar. 8, 149, 1965

22) AITKEN, J. B. et alii, Research Ser. No. 24 South Carolina Agric. Station

23) DOTY, M. S. and Gertrudes AGUILARSANTOS, Nature, 211, 990, 1966

General Literature

ARAKI, Ch., Seaweed Polysaccharides. Ex Wolfrom, M. L., Carbohydrate Chemistry of Substances of Biological Interest. Pergamon Press Ltd., London, 1959

BLACK, W. A. P., The Algae. Ex Altschul, Processed Plant Proteins Foodstuffs. Academic Press Inc., New York, 1958

BRENNAN, A. T., Notes on some common Irish Seaweeds, Inst. f. Ind. Res. and Stand., Dublin, 1950

CHAPMAN, V. J., Seaweeds and their Ulses. Methuen & Co. Ltd., London, 1950

COLLOQUES INTERNATIONAUX du Centre National de la Recherche Scientifique, Dinard, Écologie des algues marines, 1957. Ex Chimie et physico-chimie des principes immédiats tirés des algues, 1961

FOSTER ET STACEY, Polysaccharides from Algae and Lichens. Ex W. Ruhland (ed.), Handbuch der Pflanzenphysiologie, Springer-Verlag, 1958

FOTT, B., Algenkunde. Gustav Fischer Verlag, Jena 1959

FRITSCH, F. E., The Structure and Reproduction of the Algae. University Press, Cambridge, 1952 (II) and 1956 (I)

HEGNAUER, R., Chemotaxonomie der Pflanzen, Band I. Birkhäuser Verlag, Basel u. Stuttgart, 1962

HENGLEIN, Uron- und Polytuonsäuren. Ex W. Ruhland (ed.) Handbuch der Pflanzenphysiologie, Springer-Verlag 1958.

HIRST, E. L. et JONES, J. K. N., Gums and Mucilages of Plants. Ex W. Ruhland, Handbuch der Pflanzenphysiologie, Springer-Verlag, 1958.

HOFFMANN, C., Die praktische Bedeutung der Meeresalgen. Kieler Meeresforschungen. Band III (1). 1938

HOPPE, H. A., Drogenkunde. Handbuch der pflanzlichen und tierischen Rohstoffe. Cram, de Gruyter & Co. Hamburg, 1958

HOPPE, H. A., Geschichtliche Entwicklung der Auswertung von Meeresalgen. Botanica Marina, Vol. III Suppl. 1962

HOPPE, H. A. et SCHMID, O. J., Meeresalgen als moderne Industrieprodukte. Botanica Marina, Vol. III, Suppl. 1962

HOPPE, H. A., Beiträge zur Terminologie der Phykokolloide aus Rhodophyceen. Botanica Marina IX (1/2). 5. 1966

HUMM, H. J., The Brown Algae-Algin from Kelps and Fucoids. Ex Tressler, D. K. et Lemon, J. Mc. W., Marine Products of Commerce. Reinhold Publishing Corporation, New York, 1951

HUMM, H. J., The Red Algae of Economic Importance. Agar and Related Phycocolloids. Ex Tressler, D. K. et Lemon, Mc. W., Marine Products of Commerce. Reinhold Publishing Corporation, New York, 1951

INSTITUTE OF SEAWEED RESEARCH, Annual Reports 1945–1966

INTERNATIONAL SEAWEED SYMPOSIUM; Second. Trondheim. 1955 — Third. Galway. 1958 — Fourth. Biarritz. 1961 — Fifth. Halifax. 1965. Pergamon Press Ltd.

KUCKUCK, P., Der Strandwanderer, 6. Aufl. J. F. Lehmanns Verlag, München, 1953

KYLIN, H., Die Gattungen der Rhodophyceen. C. W. K. Gleerups Förlag, Lund, 1956

LEVRING, T., Studien über die Algenvegetation von Blekinge, Südschweden. Diss. Lund, 1940

LEVRING, T., Contributions to the Marine Algal Flora of Chile. Lunds Univers. Årsskrift N. F. Avd. 2. Bd. 56 (10), 1960

LEWIN, R. A., Physiology and Biochemistry of Algae. Academic Press Inc., New York, 1962

MAASS, H., Alginsäure und Alginate. Straßenbau, Chemie und Technik. Verlagsgesellschaft mbH. Heidelberg, 1959

MARSHALL, S. M. ET AL, A Study of Certain British Seaweeds and their Utilization in the Preparation of Agar. His Majesty's Stationary Office, London, 1949

MAUTNER, H. G., The Chemistry of Brown Algae, Ec. Bot. 8 (2), 1954
MIYABE, K., On the Laminariaceae of Hokkaido. J. of the Sapporo Agric. College, Vol. I, 1957
NOVA SCOTIA RESEARCH FOUNDATION, Utilization of Seaweed. Selected Bibliography on Algae, No. 2. 1953 — Nr. 3. 1955 — Nr. 4. 1958 — Nr. 5. 1960 — Nr. 6. 1965 — Nr. 7. 1966
OLTMANNS, F., Morphologie und Biologie der Algen, Jena, 1922
PAX, F., Meeresprodukte. Ein Handwörterbuch der marinen Rohstoffe. Gebr. Bornträger, Berlin, 1962
PERCIVAL, E. ET R. H. MCDOWELL, Chemistry and Enzymology of Marine Algal Polysaccharides, Academic Press, London and New York, 1967
SANFORD, F., Seaweeds and their Uses. Fishery Leaflet 469, 1958. U. S. Dep. of the Interior, Washington, D. C.
SEGAWA, S., The Seaweeds of Japan. Hoikusha, Osaka, Japan, 1957
SMITH, F. ET MONTGOMERY, R., The Chemistry of Plant Gums and Mucilages. Reinhold Publishing Corporation, New York, 1959
SCHMID, O. J., Die Inhaltsstoffe der Meeresalgen. Botanica Marina, Vol. III, Suppl. 1962
SCHOTT, GERHARD, Geographie des Atlantischen Ozeans. Hamburg, 1912
SCHULZEN, H., Die chemisch-physikalischen Eigenschaften der Alginate und Carragheenate. Botanica Marina, Vol. III Suppl., 1962
STOLL ET JUCKER, Heteroside. Ex W. Ruhland (ed.) Handbuch der Pflanzenphysiologie, Springer-Verlag, 1958
TOKIDA, J., The Marine Algae of Southern Saghalien. Fac. of Fish. Hokkaido University, Vol. 2 (1), 1954
UPHOF, J. C. Th., Dictionary of Economic Plants. H. R. Engelmann (J. Cramer), Weinheim, 1959
VINOGRADOV, A. P., Moscow, USSR, The Elementary Chemical Composition of Marine Organisms. Sears Foundation for Marine Research, Yale University, New Haven, 1953
WHISTLER, R. L. ET BEMILLER, J. N., Industrial Gums. Academic Press Inc., New York and London, 1959
WHISTLER, R. L. ET SMART, Ch. L., Polysaccharide Chemistry. Academic Press Inc., New York, 1953
YOUNG, E. G., Carbohydrate Accumulation by Lower Plants (Rhodophyta) Ex W. Ruhland (ed.), Handbuch der Pflanzenphysiologie, Springer Verlag, 1958
YEMM, E. W., The Plant Proteins and Peptides and their Localization in Cells and Fissues. Ex W. Ruhland (ed.), Handbuch der Pflanzenphysiologie, Springer-Verlag, 1958
ZANEVELD, J. S., The Utilization of Marine Algae in Tropical South and East Asia. Ec. Bot. 13 (2). 1959
Archives „Studiengesellschaft zur Erforschung von Meeresalgen e. V., Hamburg“ and BOTANICA MARINA, International Review for Seaweed Research and Utilization, 1958–1967.

Taxonomic Index

A figure in italics names the page with the main text on the alga in question.

Subject Index

A figure in italics names the page with the main text on the subject in question.

BOTANICA MARINA SUPPLEMENTS

VOL. II Suppl.: Marine Phytoplankton in New Zealand Waters

By VIVIENNE CASSIE. 8°, with 6 figures in the text and 119 pictures on 8 plates, 54 pages, 1961, DM 21,—.

In samples collected between the sea surface and twenty metres' depth off the New Zealand coastline in the period 1957 to 1959. 240 species (including 198 phytoplankters) have been identified. The 23 localities samples have been grouped according to a) current and b) environment.

VOL. III Suppl.: Meeresalgen — Industrielle Bedeutung und Verwendung

Gr. 8°, 122 Seiten mit 47 Abbildungen, 1962, kartoniert DM 35,—.

Inhalt: LEVRING, T., The Vegetation of the Sea — HOPPE, H. A., Geschichtliche Entwicklung der Auswertung von Meeresalgen — HOPPE, H. A., SCHMID, O. J., Die Meeresalgen als moderne Industrieprodukte — SCHMID, O. J., Die Inhaltsstoffe der Meeresalgen — SCHULZEN, H., Die chemisch-physikalischen Eigenschaften der Alginate und Carraghenate — MAASS, H., Über die Verwendungsmöglichkeiten von Alginaten — SCHMID, O. J., HOPPE, H. A., Meeresalgen als Dünge-, Futter- und Nahrungsmittel.

VOL. IX Suppl.: Meeresalgen für die menschliche Ernährung

Gr. 8°, 46 Seiten mit 6 Abbildungen und 1 Tafel, 1966, kartoniert DM 12,—.

Ausgehend von den Problemen der Welternährung behandeln die Autoren Günther Loose, Heinz A. Hoppe und Otto J. Schmid die Einsatzmöglichkeiten von Algen, ihren Inhaltsstoffen und von Algenprodukten für die menschliche Ernährung. Ein Abschnitt befaßt sich auch mit der Züchtung und Kultivierung mariner Algen.

VOL. XI Suppl.: Die Diatomeenassoziationen der Santa-Lucia-Lagune in Natal (Südafrika)

Von B. CHOLNOKY. Gr. 8°, 127 Seiten mit 165 Abbildungen und einer 2seitigen Karte, 1968, kartoniert DM 38,—.

54 Probenentnahmen im Gebiet der Santa-Lucia-Lagune und ihrer Zuflüsse zeigen, daß die heutige Flora der Lagune hauptsächlich aus den osmotisch widerstandsfähigen und für Salze permeablen Diatomeenarten des marinen Litorals besteht, während die Zuflüsse in ihrem Mündungsgebiet in nächster Nähe der Lagune Süß- und Brackwasserarten aufweisen. Diese Feststellung steht in Übereinstimmung mit der beobachteten Richtung der Wasserströmung und zeigt, daß auch die Sedimente überwiegend marinen Ursprungs sind. Außer diesem wichtigen Resultat für die Beurteilung der zunehmenden Versandung der Santa-Lucia-Lagune, das zugleich für ähnlich gelagerte Gebiete von Bedeutung ist, hat der Verfasser die floristischen und systematischen Ergebnisse seiner Untersuchungen an den gefundenen Diatomeenarten ausführlich abgehandelt.

CRAM, DE GRUYTER & CO., HAMBURG

Bei Fragen zur Produktsicherheit wenden Sie sich bitte an:
If you have any questions regarding product safety,
please contact:

Walter de Gruyter GmbH
Genthiner Straße 13
10785 Berlin
productsafety@degruyterbrill.com